Lecture Notes in Energy

Volume 79

Lecture Notes in Energy (LNE) is a series that reports on new developments in the study of energy: from science and engineering to the analysis of energy policy. The series' scope includes but is not limited to, renewable and green energy, nuclear, fossil fuels and carbon capture, energy systems, energy storage and harvesting, batteries and fuel cells, power systems, energy efficiency, energy in buildings, energy policy, as well as energy-related topics in economics, management and transportation. Books published in LNE are original and timely and bridge between advanced textbooks and the forefront of research. Readers of LNE include postgraduate students and non-specialist researchers wishing to gain an accessible introduction to a field of research as well as professionals and researchers with a need for an up-to-date reference book on a well-defined topic. The series publishes single- and multi-authored volumes as well as advanced textbooks.

Indexed in Scopus and EI Compendex The Springer Energy board welcomes your book proposal. Please get in touch with the series via Anthony Doyle, Executive Editor, Springer (anthony.doyle@springer.com)

More information about this series at http://www.springer.com/series/8874

Mohammad Reza Hesamzadeh ·
Juan Rosellón · Ingo Vogelsang
Editors

Transmission Network Investment in Liberalized Power Markets

Springer

Editors
Mohammad Reza Hesamzadeh
KTH Royal Institute of Technology
Stockholm, Sweden

Juan Rosellón
Centro de Investigación y Docencia Económicas (CIDE)
Mexico City, Mexico

Ingo Vogelsang
Boston University
Boston, MA, USA

ISSN 2195-1284 ISSN 2195-1292 (electronic)
Lecture Notes in Energy
ISBN 978-3-030-47928-2 ISBN 978-3-030-47929-9 (eBook)
https://doi.org/10.1007/978-3-030-47929-9

This Springer imprint is published by the registered company Springer Nature Switzerland AG
The registered company address is: Gewerbestrasse 11, 6330 Cham, Switzerland

Contents

An Introduction to Transmission Network Investment in the New Market Regime

M. R. Hesamzadeh, J. Rosellon, and I. Vogelsang

Over the course of the last 50 years, the electricity industry has gone through several fundamental changes. Many countries have liberalized their electricity industry. The historically vertically integrated electricity sector is now broken down into five parts: generation companies, transmission and distribution companies, retailers, and system operation. Competition has been introduced in the generation sector for efficient operation and investment decisions. This is while the operation and investment decisions of the transmission and distribution network have been placed under the control of regulators and system operators.

Achieving efficient investment in both generation and transmission assets requires close coordination between transmission and generation investment decisions. The determination of the optimal capacity, sequence and timing of transmission network investments is known as the transmission planning problem.

Transmission planning is a complex problem which involves analysis of a transmission investment plan under many future demand and supply scenarios. In principle, the transmission planning problem is well understood in the context of a vertically integrated electricity industry. In this context, additional transmission capacity allows for more efficient dispatch, it can substitute for and complement investment in the generation sector, and it reduces the need for synchronous reserve by allowing those reserves to be procured over a wider area.

M. R. Hesamzadeh (✉)
KTH, Stockholm, Sweden
e-mail: mrhesamzadeh@gmail.com

J. Rosellon
CIDE, Mexico City, Mexico

DIW, Berlin, Germany

I. Vogelsang
Boston University, Boston, USA

M. R. Hesamzadeh et al. (eds.), *Transmission Network Investment in Liberalized Power Markets*, Lecture Notes in Energy 79,
https://doi.org/10.1007/978-3-030-47929-9_1

In theory, if the liberalized electricity market is sufficiently competitive, the same tools and techniques developed for transmission investment in a vertically integrated electricity can be used for transmission investment in the liberalized electricity markets. However, due to imperfect markets and imperfect regulation, new issues have arisen. Besides, the increasing share of intermittent renewable generation capacities has added extra complexity to the transmission investment problem.

Both liberalization of the electricity industry and the government policies for increasing the share of intermittent renewable generation have changed almost all aspects of the transmission investment problem. Accordingly, in this new regime, the transmission investment problem needs to be revisited for adjusted definition and theory. New challenges and their proposed solutions, new mathematical models and solution algorithms, new regulatory designs and successful real-life stories deserve a comprehensive discussion. This is exactly the research gap the editors of this edited volume wish to fill. Coming themselves from economics and engineering backgrounds, the editors have sought prominent contributors from both those disciplines in order to provide a comprehensive picture of the transmission investment landscape. As a result, a number of the main issues are treated by several chapters from different perspectives. This should allow the readers to understand the differences in approach not only by discipline but also by more theoretical and by more applied angles.

Part 1 of the book provides seven chapters on the basic economics and engineering of transmission network investment.

In chapter "Definition and Theory of Transmission Network Planning" as the first of these chapters, Mohammad Majidi and Ross Baldick provide a definition of transmission investment under the new market regime. They explain the new concepts introduced in the engineering literature with regards to the transmission investment problem in the liberalized electricity markets. In their chapter, they first introduce different factors affecting the transmission investment problem. Subsequently, different models of transmission investment problem are reviewed. Then, they present different literatures on the transmission investment problem and different optimization models for transmission investment. The vertical, horizontal and hybrid decomposition techniques for solving large-scale transmission investment models are carefully explained. The authors at the end provide a general framework for transmission expansion planning, and they carefully study the computational challenges and their potential solutions. A stylized version of the Electricity Reliability Council of Texas (ERCOT) transmission network is introduced, and numerical analysis is provided to demonstrate the capabilities of the optimization-based transmission investment studies.

Chapter "Regulated Expansion of the Power Transmission Grid," by Thomas-Olivier Léautier, provides a thorough and clear presentation of the economics of electricity transmission, both verbally and formally. It goes beyond that by deriving new results on incentive regulation and setting those in perspective with the literature and, in particular, with the institutional context of transmission regulation in various countries. His main results complement and can be contrasted with those of

Biggar and Hesamzadeh (chapter "Merchant Transmission Investment Using Generalised Financial Transmission Rights") and Vogelsang (chapter "A Simple Merchant-Regulatory Incentive Mechanism Applied to Electricity Transmission Pricing and Investment: The Case of H-R-G-V") chapters.

The first result is that locational marginal pricing (LMP) of electricity transmission leads to suboptimal grid expansion, because a Transco as a monopolist in this case benefits from some, although not unlimited, increase in congestion. This has justified the practice of channeling back congestion rents to users (in hopefully non-distortive ways). However, as Vogelsang's chapter shows, LMP can be a productive part of a welfare-optimal transmission pricing regime. It only has to be accompanied by fixed or complementary charges that are inversely related to the congestion prices.

The second, Léautier's own important result is that optimal investment can be achieved if the Transco is made financially responsible for the congestion cost at the margin. As shown in Vogelsang's chapter, this is the negative of the H-R-G-V mechanism, where the Transco is rewarded by the reduction in congestion cost rather than by being penalized for the increase in congestion cost. Because of the potentially large financial amounts involved, Léautier's approach has to guard against the Transco making unacceptable losses, while the H-R-G-V approach has to guard against unacceptable profits. Interestingly, coming from a different, non-regulatory perspective, the Biggar and Hesamzadeh chapter also uses the change in congestion cost as an incentive device. In this case, for a monopoly trader trying to offer generalized FTRs that allow for maximal hedging by generators and loads in an uncertain environment.

Léautier's third result is that fully efficient transmission charges at the margin do not cover all transmission costs given his showing that transmission grids operate in the range of economies of scale. This justifies fixed or complementary charges, for the derivation of which Léautier suggests a menu approach, while the Vogelsang chapter discusses several other approaches starting out with that suggested by H-R-G-V.

Léautier provides two important institutional insights regarding the possibility to implement incentive approaches, such as his or the H-R-G-V mechanism. The first and encouraging one is that the mechanisms should not only work in a nodal pricing environment but also for zonal pricing and counter-trading and re-dispatch situations. The second, less encouraging one, is that such general incentive schemes may not work in environments where several actors are jointly responsible for transmission investments and operations, such as the USA with multiple ownership of grids and independent system operators (ISOs) in charge of operation or grids that cross international boundaries. This is something that is also emphasized by Joskow in his chapter. H-R-G-V acknowledges this point by assigning the ISO a very limited and observable role, but this remains a major issue.

Frank Wolak's chapter "Transmission Planning and Operation in the Wholesale Market Regime," further explains the differences between transmission planning and operation under the vertically integrated monopoly regime and the wholesale market regime. It introduces the concepts of engineering reliability, under the first regime, and economic reliability, under the second one. The engineering reliability concept has been well studied in the engineering literature, and it has been the primary factor for transmission planning in the vertically integrated monopoly regime. In contrast, the concept of economic reliability is to some extent new. An economically reliable transmission network has sufficient transmission capacity at each node such that the generation firms located at different nodes face significant competition from other generation firms to cause them to offer their energy at close to marginal cost in the majority of hours of the year. Accordingly, the economically efficient level of expanded transmission capacity in the wholesale market regime is found to be greater than the one in the vertically integrated monopoly regimes. Two simple models are provided to show that the optimal transmission capacity is higher under the wholesale market regime than under the vertically integrated monopoly regime. Subsequently, the chapter proposes a general forward-looking methodology for transmission planning in the wholesale market regime. In the forward-looking framework, a methodology for calculating the distribution of realized economic benefits from an upgrade in the wholesale market regime is proposed.

As it is clear from this chapter, the sophistication of the economic modeling required to assess the value of transmission upgrades in the wholesale market regime is much greater than that required for the vertically integrated regime. Different factors causing this sophistication are introduced and discussed. At the end, Wolak introduces the interesting concept of the insurance value of transmission expansions. Transmission upgrades can significantly reduce the likelihood of future extreme prices and act as an insurance against extreme prices. The author then concludes with some advice on improving the transmission planning and operation processes in wholesale market regimes.

Michel Rivier and Luis Olmos in chapter "Cost Allocation Issues in Transmission Network Investment" focus on cost allocation issues in transmission network investment. They argue that the current transmission network pricing techniques are overly simple in most power systems worldwide. In most cases, the transmission charge is a flat volumetric charge ($/kWh) which is added to the consumer's tariff. Often the generators are excluded from the transmission network charges. This chapter calls for a better network tariff to cope with significant transmission network investments resulting from the decarbonization of the power systems. They propose and support the idea of introducing a system of forward-looking and cost-reflective locational transmission charges to incentivize the efficient siting of the network users. The chapter then goes on with presenting the basic principles that should guide the design of complementary transmission charges. They emphasize the forward-looking, cost-reflective and beneficiary-pays principles as the main attributes of an efficient transmission charging system. They close their chapter with case examples of cost allocation approaches in regional markets. European schemes, US schemes, and the Central American schemes are reviewed and carefully discussed.

In chapter "Transmission Planning, Investment and Cost Allocation in U.S. ISO Markets" Richard O'Neill comprehensively reviews the transmission planning, investment and cost allocation mechanisms in the US ISO markets. He reviews the current approaches for transmission planning in competitive environments and presents a modeling approach for regional planning in US markets. The chapter starts with the definition of a reliable and economically efficient transmission planning process and discusses the value of price-responsive demand in the transmission planning process. Then, it treats market power and the interaction between market power and transmission planning processes. The transmission planning problem is subject to many uncertain factors which need to be properly considered in the planning process. Natural gas price uncertainty, weather uncertainty and technological innovation uncertainty are among the main uncertain parameters in the transmission planning process. The new emerging uncertainties add more complexity to the process and a less predictable evolution of power systems. Different models of the transmission planning problem are compared, and several tool boxes used by US ISOs for the transmission planning problem are cited. The merchant-regulatory approach for transmission investment and some recent developments (which are more extensively discussed in the Vogelsang's chapter) are touched upon. The theory of transmission investment cost allocation and some illustrative examples close this chapter.

In chapter "Transmission Planning and Co-optimization with Market-Based Generation and Storage Investment" Qingyu Xu and Benjamin Hobbs follow the co-optimization advice of O'Neill's chapter "Transmission Planning, Investment and Cost Allocation in U.S. ISO Markets" and propose a comprehensive mixed-integer linear program for co-optimizing transmission planning with market-based generation and storage investment. In the proposed optimization model, the transmission investment planning is performed while considering how the generation and storage investments will respond to the planned transmission investment. The proactive and forward-looking transmission planning advocated in Wolak's chapter "Transmission Planning and Operation in the Wholesale Market Regime" is here formulated as a mixed-integer linear program that co-optimizes transmission-generation-storage expansion. Then, they use a case study of the Western Electricity Coordinating Council (WECC) in the USA to show how the inclusion of storage co-optimization will change transmission investment planning. The economic benefit of such co-optimization is also quantified. It shows that accounting for storage expansion in transmission investment planning will help transmission planners to avoid over-building or underbuilding lines and will help efficient siting and sizing of generation and storage devices. In other words, their results show that storage can be both a complement and a substitute for transmission expansion.

Henrik Bylling, Trine Krogh Boomsa and Steven Gabriel in chapter "A Parametric Programming Approach to Bilevel Merchant Electricity Transmission Investment" discuss a mathematical complementary model for merchant transmission investment. They assume a bilevel program where in the upper level a merchant investor collects the congestion rents and in the lower level the ISO runs an economic dispatch model. They recast their bilevel problem as a mathematical program with equilibrium constraint (MPEC) and propose a parametric programming approach to solve their proposed bilinear bilevel program of merchant transmission investment. Their proposed technique takes advantage of decomposition with respect to both time periods and scenarios. They benchmark their approach against the MILP and nonlinear programming approaches to show the efficiency of their technique. Their proposed solution approach is applied to a case study of transmission expansion in the Nordic region.

Part 2 includes five chapters putting merchant investment in transmission networks into the perspective of other investment approaches.

Frank Felder's chapter "Market Versus Planning Approaches to Transmission and Distribution Investment" reflects the difficulties policymakers face in dealing with transmission expansion in an environment where the distribution and generation side have strong impacts on transmission and vice versa. He views this interaction in the context of both economic efficiency and diverse objectives of policymakers. Felder argues that due to vertical economies of scale and scope transmission and distribution grids cannot be viewed in isolation but have to be seen as a single integrated system. This also holds for individual generation or line investments that, because of network externalities and economies of scale in transmission investment, have an impact on the system as a whole. In this context, Felder brings out the conflicts between federal and state jurisdictions in the USA which appear to be most drastic if a low-cost state has an interest in blocking transmission projects for electricity exports to high-cost states. Felder shows that states can counteract federal transmission policies via the (state-regulated) distribution system.

Felder further points out that merchant transmission investment and other transmission investments are usually evaluated on the basis of congestion reduction, while there exists no market mechanism for reliability purposes. Neglecting the dual reliability and economic benefits would, in his view, provide an additional reason why merchant transmission would be associated with underinvestment. Trying to provide further revenues to merchant investors via backstop recovery may only work in favor of merchant investment if the backstop recovery concentrates on the beneficiaries of the project and does not socialize those costs.

Backed by a number of detailed empirical case studies Paul Joskow's chapter "Competition for Electric Transmission Projects in the U.S.: FERC Order 1000" provides a comprehensive assessment of the competitive transmission procurement model made possible in the USA through FERC Order 1000. In particular, Joskow lays out the empirical details of the various cases in which this model has been applied and it provides the conceptual background for this approach, which substantially differs but is nevertheless related to the merchant transmission model, on the one hand, and the franchise bidding literature, on the other. It takes from the merchant approach the possibility for non-incumbent transmission companies to play an active role in building transmission lines. On the other hand, it takes from the franchise bidding approach the competitive bidding. However, this is only for individual lines and upgrades rather than the whole grid, and ex post conventional regulatory tools are applied. Thus, it is characterized by ex ante competition and ex post rate-of-return regulation. As a result, its main feature is not to incentivize transmission companies to reduce costs but rather to provide information about the costs of various approaches potentially allowing for the selection of the least cost solution. Joskow emphasizes that incentive regulation of transmission companies could be quite difficult in the USA where multiple ownerships of each of the regional grids and management by non-owning ISOs prevent assigning cost responsibilities.

Joskow's empirical cases bring out how the competitive procurement approach has been applied in practice and what its shortcomings in the USA have been. These shortcomings are deemed responsible for the limited share that this approach has had since Order 1000. Joskow suggests improvements that may help it to gain more widespread acceptance both in the USA and elsewhere. A major insight from the case studies is the large discrepancy between costs of different proposals for the same project and the difference in approaches used by the different bidders. This is an important accomplishment, given that FERC does not usually research the reasonableness of the costs of proposed network investments. Thus, the main benefit of the competitive transmission procurement model for the USA has been to complement rather than substitute for regulation.

In chapter "Merchant Transmission Investment Using Generalised Financial Transmission Rights," Darryl Biggar and Mohammad Hesamzadeh explain a new mechanism for realizing the merchant transmission investment in AC power networks. This merchant transmission investment mechanism is developed around the new concept of generalized Financial Transmission Right (FTR) contracts. The authors show that the conventional FTR contracts suffer from two shortcomings: (a) The FTR payment is not congestion revenue exact. The literature on FTR contracts just proves that the FTR payments are revenue adequate. This means the congestion rent collected in the spot market is greater than the payment which the ISO needs to do under the signed FTR contracts in the FTR auction. However, for FTR contracts to act as a proper hedge mechanism against the inter-nodal pricing risk, the FTR payments should be exactly equal to the congestion rent collected by the ISO in the spot market. If the equality does not hold, then traders in the market offering hedge contracts to generators and loads face some spot market price risk. (b) Even if the equality of the FTR payments and congestion rent holds, the FTR contracts are fixed-volume contracts. These fixed-volume FTR contracts are of less value for the current electricity markets with a huge penetration of the intermittent renewable generation where generators and loads have varying generation level in the spot market. The fixed-volume FTR contract is useful for a base-load generator which has a good estimation of his generation level in the spot market when it signs the FTR contracts in the hedge market. However, it is very hard for an intermittent wind generator to know exactly how much it produces in the spot market to sign the FTR contract for. The same situation exists for a gas power plant.

Biggar and Hesamzadeh first analytically demonstrate these two drawbacks of conventional FTR contracts. In order to address these two issues, they propose the concept of Generalized FTR (G-FTR) contracts. The G-FTR contracts have a varying volume as compared to the fixed-volume FTR contracts. This property of the G-FTR contracts makes them suitable for intermittent generators to properly hedge their profit against the volatile spot market prices. The other very important property of G-FTR contract is that the G-FTR payment is always exactly equal to the congestion rent collected by the ISO in the spot market. In other words, the G-FTR contracts are congestion revenue exact (while the FTR contracts are congestion revenue adequate). The congestion revenue exactness of the G-FTR contracts puts the ISO in a secure financial situation in regards to the signed G-FTR contracts. The system changes between the G-FTR auction stage and the spot market stage do not put the ISO in a financial deficit situation to support its signed G-FTR payments. The sum of the G-FTR payments which needs to be paid by ISO is always exactly equal to the congestion rent collected by the ISO in the spot market stage. The congestion revenue exactness of the G-FTR contracts is a very important property which does not hold for standard FTR contracts.

Ingo Vogelsang's chapter "A Simple Merchant-Regulatory Incentive Mechanism Applied to Electricity Transmission Pricing and Investment: The Case of H-R-G-V" treats electricity transmission investment from an incentive regulation perspective. While incentive regulation has been widely applied to particular aspects of electricity regulation, with the exception of a limited period in the UK (mentioned in the Léautier chapter) no comprehensive incentive regulation of transmission investment is known to us. Vogelsang analyzes the potential practical applicability of such a comprehensive mechanism that was originally developed by the editors of this volume and S.A. Gabriel. It was called H-R-G-V after its four authors. In a fully competitive environment without environmental problems and without distributional issues, this mechanism incentivizes a monopoly Transco to invest efficiently. The trick is to provide the Transco with all the economic benefits from its investments. The chapter shows how H-R-G-V can or needs to be adapted to situations where the above assumptions do not hold. While the mechanism turns out to be quite robust, some changes would become necessary and second-best results have to be accepted.

Vogelsang's and Leautiér's chapters bring out the crucial role of providing the Transco with a reward closely linked to social surplus in order to induce welfare-optimal behavior. This role also comes out in the Biggar and Hesamzadeh chapter, where the trader responsible for perfect hedging also is given title over the change in total surplus. In a similar vein, the Revier and Olmos chapter suggests "allocating" not directly attributable costs of transmission investments to stakeholders according to the benefits they receive.

In chapter "Game-Theoretic Modeling of Merchant Transmission Investments," Dimitrios Papadaskalopoulos, Ying Fan, Antonio de Paola, Rodrigo Moreno, Goran Strbac and David Angeli address the topic of merchant transmission investment. They motivate merchant transmission investment as a promising alternative to the traditional centralized planning approach and consider it an important step toward the full liberalization of the electricity industry. Their research questions are, "Which entities are likely to undertake merchant transmission investments?" and "Will the merchant planning approach result in the social-welfare maximizing outcome?". To answer these, they propose a non-cooperative game-theoretic modeling framework to model the strategic interaction between multiple merchant investors. It consists of two non-cooperative game-theoretic models for finding the Nash equilibrium of transmission investment between multiple merchant investors. They apply their game-theoretic model to a simple 2-node system. This case study shows that the merchant network investments will be mostly undertaken by generation companies in areas with low nodal prices and by load serving entities in the areas with high nodal prices to collect more congestion revenue. Solving the Nash equilibrium for a set of merchant transmission investors turns out to be mathematically complex.

To address the computational complexity, they approximate the set of merchant investors as a continuum. This approximation allows the authors to derive mathematical conditions for the existence of a Nash equilibrium in an analytical fashion. They derive conditions under which the merchant transmission investment can result in a social welfare maximizing investment. These conditions are (1) fixed investment costs are neglected and (2) the network is radial without any loop. Since these conditions do not generally hold in reality, they find that even a fully competitive merchant transmission investment framework is not capable of maximizing social welfare. This finding is consistent with the Joskow chapter "Competition for Electric Transmission Projects in the U.S.: FERC Order 1000" as well as with the Biggar and Hesamzadeh chapter "Merchant Transmission Investment Using Generalised Financial Transmission Rights".

Part 3 provides four chapters on the interaction and coordination between electricity transmission networks and related investments in renewables, gas networks and electricity distribution networks.

In chapter "Transmission Investment and Renewable Integration" Hugh Rudnick and Constantin Velásquez argue that timely transmission capacity plays a central role for renewable energy integration. They discuss the main elements of transmission planning under transition to renewable generation technologies. Such a transition typically presents coordination problems between transmission and generation investments. The fluctuating nature of generation supply and demand under renewable generation integration may misguide the development of transmission expansion. But small geographically dispersed renewable projects may benefit from coordinated transmission expansions, such as the case of wind farms in Texas. However, other international experiences—such as those from Australia, Brazil and Chile—show that transmission investment may turn out to be quite difficult due to complexities in planning, coordination and allocation of risks and costs among market players. Solutions to optimal transmission expansion should then target both economic efficiency and practical feasibility. In this task, Rudnick and Velásquez argue that optimization models should support actual practical transmission planning, informing policy decision makers on qualitative and quantitative assessments for scenarios and expansion plans. Furthermore, optimization models could be improved by adding temporal and spatial correlation constraints of renewable resources. And practitioners should aim to determine the optimal expansion plan and not be intimidated by sophisticated models in real-life policy applications. Many planners rely on heuristic criteria of transmission investment instead of using stochastic programming, robust optimization and multi-objective optimization. Of course, Rudnick and Velásquez admit that transmission expansion processes should be ultimately determined by a mix of policy decision criteria, including regulation, technical, economic, market-fundamentals, social and environmental elements, as well as adequate governance and institutional design

In chapter "The Impact of Transmission Development on a 100% Renewable Electricity Supply—A Spatial Case Study on the German Power System" Jens Weibezahn, Mario Kendziorski, Hendrik Kramer and Christian von Hirschhausen analyze the German energy system assuming an extremely high share of distributed renewable sources (100%). They specifically analyze how such a system fares under different transmission (and storage) regimes, including: (i) status quo in 2022, (ii) an extended network in 2035 with and without HVDC lines and (iii) a full-fledged copperplate without any congestion. With such a purpose—and using real data from the German electricity system—they develop a model of optimal generation and storage investment and operation, where transmission expansion is exogenous. Their results suggest that the system may accommodate a high share of renewables by installing large amounts of short-term and long-term storage capacities. Wind would mainly be placed in the North and solar PV rather in the South. Likewise, a higher level of transmission expansion leads to lower requirements of distributed renewable capacities—mainly of rooftop solar and onshore wind—while only offshore wind benefits from increased transmission capacity. Finally, transmission congestion should not get in the way of renewable integration. While some network congestion could be observed in the 2022 and the 2035 w/o HVDC scenarios, it should be of minor importance. Thus, from the perspective of feasible transmission investment, Germany could be on its way to a 100% renewable generation portfolio.

In chapter "Coordination of Gas and Electricity Transmission Investment Decisions" Seabron Adamson, Drake Hernandez and Herb Rakebrand analyze policy issues arising from different investment models in New England for coordinated natural gas and electricity transmission. This US Northeast region has been characterized by a lack of investment in gas pipelines. This contrasts with the case of the US interstate pipeline industry which has been quite successful in attracting the signing of long-term contracts to develop large-scale pipeline networks. Such contracts have allowed FERC to signal the need for new capacity that pipeline firms are willing to build due to their expected and virtually assured stable revenues which recover large sunk costs. In contrast, given the lack of long-term natural gas transportation contracts in New England, pipelines there still need strong incentives to invest in capacity to serve electric generation markets. This has also translated into high gas and electricity prices. The lack of gas pipelines could in principle be ameliorated by new electricity transmission from Québec, which has substantial hydro generation. However, this has proven to be costly and slow and hence has not provided an alternative to new gas pipeline construction. A number of policy measures have been tried to solve this problem, such as capacity payments, winter reliability programs and market capacity rules. However, they have not been successful in promoting the new long-term capacity contracts necessary to support new pipeline projects into the region

Chapter "The Emergence of Smart and Flexible Distribution Systems" by Derek Bunn and Jesus Nieto-Martin provides a comprehensive discussion of the emergence of smart and flexible distribution systems. Electricity networks are now more complex to manage mainly due to rapid development of the distributed renewable generation and also the electrification of the transport system. Accordingly, the aging distribution networks are becoming less suitable for the emerging markets in the distribution network highlighting a need for stakeholders to upgrade and adapt to the new market requirements. Smart technologies at distribution level can provide more efficient operation of the distribution assets, and the smart markets can incentivize the aggregated small-scale participants to sell various flexibility services to the distribution network providers. In this chapter, the authors review the approaches that the distribution system operators can adopt to enable them to support a greater volume of demand, generation and storage to be connected to the distribution network in a smarter and more active setting. The active engagement of the distribution system operators (DSOs) in the wholesale power markets in terms of providing different flexibility services has a material impact on transmission investment decisions. Active distribution system operators by providing different flexibility services to the market may delay the need to investment in additional transmission capacities. This deferred transmission investment resulting from active distribution system operators will improve the social welfare generated by the market as a whole. The expensive new transmission towers and lines will be replaced by cheap smart distribution-level technologies. Through the smart market implementation at the distribution level, the local generation and consumption are traded and less import or export of energy with the main transmission network is needed.

Bunn and Nieto-Martin provide a comprehensive chapter on future smart and flexible distribution systems and different distribution network management approaches. The automated load transfer, meshed network and voltage control, energy storage and dynamic asset rating are explained as the smart grid techniques which are available in modern distribution networks. They comprehensively analyze these techniques and how they can help achieve a more efficient operation of the distribution networks. Then, they focus on the flexibility service which can be provided by the distribution system operators to transmission system operators. Distribution system operator market models are also discussed. They start with the conventional market model and then discuss different market model which can capture the complex interaction between the TSO, DSO and distribution-level participants. The chapter concludes by highlighting that access to smart technologies and smart markets are crucial for DSOs as they actively move from their traditional passive role to an active one.

Part 4 contains two chapters on practical experiences with transmission investment. While practical experiences are also provided in many other chapters, notably the one by Joskow, the following two chapters address the practical experience in New Zealand and the Nordic electricity market comprehensively.

Lewis Evans' chapter "Practical Experiences with Transmission Investment in the New Zealand Electricity Market" provides an assessment and analysis of the New Zealand transmission sector and its achievements and challenges for transmission investments. This chapter nicely complements that by Persson and Tangerås on the Nordic transmission system (chapter "Transmission Network Investment Across National Borders: The Liberalized Nordic Electricity Market"), both of which are characterized by dominant hydro generation in some geographic areas and fossil fuels and renewables in other areas requiring transmission links between those areas. In the case of New Zealand, the link between the hydro-based South Island and the fuel and renewable-based North Island is crucial for the country's electricity balance. The New Zealand transmission regime has gone through large institutional changes in the last decades, moving from a state-dominated to a market-dominated to the current regulatory regime with a state-owned regulated Transco ("Transpower") and market-based generation and loads. During the market-dominated regime transmission ownership by a "club" of generators and loads was discussed but not implemented.

The pivotal HVDC inter-island link allows for balancing generation in both islands, where the dominant flow is hydro-generated electricity from South to North but with some North-South transmission during dry years. While this capacity itself increases generation competition on both islands, this is further helped by long-term electricity swaps between the islands. The current electricity wholesale includes spot and hedge markets, where about 85% of trades through the spot market are hedged. This underlines the importance of inventing good hedge instruments as suggested in chapter "Merchant Transmission Investment Using Generalised Financial Transmission Rights" by Biggar and Hesamzadeh in this volume.

An important feature of regulatory price setting in New Zealand is the tool of using elaborate input methodologies as the basis for costing of networks and their usage. This tool both provides critical information for stakeholders about the way prices will be calculated and, importantly, it acts as a commitment device, because input methodologies can only be changed in long intervals and with very high transaction costs. The input methodologies in the case of electricity transmission specify a building-block approach to regulation that somewhat resembles US-style rate-of-return regulation but with some built-in efficiency and incentive features. It also includes service quality besides price as major features.

While Transpower at times took initiative for new transmission investments, most of the investment planning is driven by stakeholder input. For a time transmission investments were hindered by conflicts between two regulatory bodies that were in charge of different aspects of Transpower's activities, one for the price level and the other for the price structure. Thus, for a time Transpower could not increase its price level if investment required that. The resulting investment backlog was only resolved after reaching an agreement on price increases. While today the price level is based on the building-blocks approach, the price structure tries to follow the benefits created by new transmission investments.

The Lars Persson and Thomas Tangerås chapter "Transmission Network Investment Across National Borders: The Liberal-Ized Nordic Electricity Market" on the liberalized Nordic electricity market complements the Evans chapter by also treating a region with a strong hydro generation component. The role of the New Zealand inter-island link, however, is in the Nordic countries replaced by international links. This creates coordination problems that in principle can be handled in New Zealand by regulatory fiat. Persson and Tangerås treat the resulting coordination problems and solutions both by describing and analyzing the path-breaking and innovative Nordic solutions but also by providing the required theoretical background, both of which could be used by other countries and for coordination in the federal environment of countries like the USA.

In their description and analysis of the Nordic market, the authors show how the international agreement has grown from first only Norway and Sweden to all Scandinavian countries and to finally include Estonia, Latvia and Lithuania. The main institutional development here was the early creation of Nord Pool as the World's first international wholesale power exchange with its trading platform Elspot, over which in 2017 94% of the total electricity production of the Nord Pool member countries was traded. Trading in 15 price areas from time to time generates congestion rents that are distributed across transmission owners.

An important feature of the Nordic market is the complementarity between hydro power located mostly in Norway and intermittent renewables generation located in the other member countries of Nord Pool (Denmark in particular). This complementarity reduces conflicts of interest between those countries in reaching agreements about the necessity of building additional transmission lines. In a number of theoretical models, the authors characterize such potential conflicts and how they can be overcome in various cooperation models. While solutions are shown to be possible, they are difficult to achieve in multi-country contexts.

Basic Economics and Engineering of Transmission Network Investment

Definition and Theory of Transmission Network Planning

M. Majidi and R. Baldick

Sets and Indices

N_b	Set of buses; index *k, n*
N_g	Set of all generators; index *g*
N_{wg}	Set of all wind generators; index *g*
N_l	Set of all lines (existing and candidate); index *l, m*
N_o	Set of all existing lines; index *l, m*
N_n	Set of all candidate lines; index *l, m*
L_k	Set of lines connected to bus *k*
G_k	Set of all generators connected to bus *k*
N_s^{ω}	Set of system operation states under scenario ω; index *c* ($c = 1$ represents the normal operation condition)
υ	Superscript/index for iteration number
Ω	Set of scenarios; index ω
$\mathscr{I}$	Set of classes
$\mathscr{I}_i$	Set of scenarios in class *i*
$\mathscr{S}^i$	Set of clusters for class *i*
$\mathscr{S}_j^i$	Set of scenarios in cluster *j* for class *i*
$\mathscr{B}$	Set of bundles
$\mathscr{B}_i$	Set of scenarios in bundle *i*
$\vert\ \vert$	Size of a set

M. Majidi · R. Baldick (✉)
University of Texas at Austin, Austin, TX, USA
e-mail: baldick@ece.utexas.edu

M. R. Hesamzadeh et al. (eds.), *Transmission Network Investment in Liberalized Power Markets*, Lecture Notes in Energy 79,
https://doi.org/10.1007/978-3-030-47929-9_2

Parameters

q_i	Per MWh load shedding penalty at bus i
γ_g	Per MWh wind curtailment penalty for wind farm g
Co_g	Per MWh generation cost for generator g
ζ_l	Annual cost of line l construction
d_k	Demand at bus k
B	Diagonal matrix of line suseptance
$P_g^{\max}/P_g^{\min}$	Maximum/Minimum capacity of generator g
$f_l^{\max}/f_l^{\min}$	Maximum/Minimum capacity of line l
C^{ω}	Matrix of contingencies (operation states) that specifies the status of lines under different contingencies (1 for in service and 0 for out of service lines) for scenario ω; index c
ϑ	Variable freezing parameter
ρ_l	Penalty factor for line l in PH algorithm
κ	Size of each bundle
d	Size of a TEP optimization problem
SC	Number of structural constraints for a TEP problem
CV	Number of continues variables for a TEP problem
BV	Number of binary variables for a TEP problem

Random Variables

$\tilde{\xi}$	Random variables (load and wind)

Decision Variables

$r_{k,c}$	Load curtailment at bus k under operating state c
CW_g	Wind curtailment for wind farm g
p_g	Output power of generator g
$f_{l,c}$	Power flow in line l under operation state c
$\theta_{i,c}$	Voltage angle at bus i under operating state c $\Delta\theta_{l,c}$ is voltage angle difference across line l under operating state c. $\Delta\theta_{l,c}$= $\theta_{k,c}$-$\theta_{n,c}$ for line l from bus k to bus n
x_l	Binary decision variable for line l
$\boldsymbol{x}^{\omega}$	Binary decision variables vector for scenario ω
$\boldsymbol{x}^{\mathscr{B}_i}$	Binary decision variables vector for bundle $\mathscr{B}_i$
$\boldsymbol{W}_{\mathscr{B}_i}$	Multiplier vector for bundle $\mathscr{B}_i$ in PH algorithm
$\mathscr{Z}$	Binary variables matrix for clustering
$\mathscr{H}$	Binary variables matrix for bundling

1 Introduction

The transmission network is the backbone of the electric power system. Increasing penetration of renewable resources, energy storage devices, mobile and flexible demand, along with new public policies makes the future much more uncertain for transmission expansion planning (TEP). As the transmission network is a monopoly infrastructure, it is critical to expand and operate this network at minimum cost while keeping a high level of reliability. This is particularly the case in jurisdictions such as Electric Reliability Council of Texas (ERCOT) where investment and operation costs are distributed between all electricity users in the region.

Transmission expansion planning is the process of deciding which equipment should be selected, where it should be installed, and when is the best time to install it. Villasana et al. (1985) provide a hierarchy of three questions that should be answered in transmission planning:

(a) What new facilities should be installed so that future operation will not be limited by transmission capacity?
(b) What new transmission facilities can be economically justified versus the higher operation costs if new facilities were not installed?
(c) What new generation sites can be justified versus new transmission facilities or higher operation costs?

These three questions specify main components of the objective function in TEP. In question (a), the objective function is to invest in the transmission network as much as we need to supply all demand without the transmission network affecting generation dispatch or demand supply. It is sometimes called reliability planning, in which the main concern is satisfying network reliability criteria. Unit operation set points are mainly defined based on experience or least cost. In the case of using lower operating cost units as much as possible, we will have the least operation cost but we may need to invest highly in transmission expansion, posing the question of whether the investment is cost-effective.

In the next hierarchy level (question b), the impact of operation cost on decision making for TEP is considered, which means it might be economical to dispatch some expensive power plants to supply demand instead of building some new transmission lines to dispatch all cheap power plants. The second question provides a better modeling property compared to the first one as it economically adjusts transmission investment cost and power systems operation cost, but it is computationally more expensive.

In question (c), which has the highest rank in the hierarchy, not only the impact of operation cost but also the impact of investment in generation sector on TEP is evaluated. In other words, it might be economical to invest on the generation side (for example, building new power plants close to demand centers) instead of the transmission side to supply the demand. It provides a better expansion plan (from economical perspective); however, it is much more computationally expensive, and planners would need to have the authority to make decisions about the location/capacity of new power plants.

Since generation expansion decisions are usually made by individual private investors in vertically unbundled electricity industries, the consideration of generation investment may be beyond the control of transmission planners. In this chapter, our main focus will be on the second question, and we assume we know the location and capacity of future generation units (with uncertainties). In principle, generation expansion could be added to the formulation.

1.1 Factors Affecting Transmission Expansion Planning

TEP studies are performed for different timescales, including, for example, near-term (for five years or shorter) and long-term (for more than five years), and for each timescale different parameters with different level of detail are considered. The main issues that affect TEP can be categorized into four groups, namely environmental issues, policy and regulatory issues, uncertainties, and network modeling, and these are explained briefly in the following.

1.1.1 Environmental Issues

Environmental concerns/limitations may directly affect transmission planning especially for line routing in particular areas such as regions with wildlife and endangered species, wetlands, national parks, historic areas, and military areas.

Furthermore, there are some environmental concerns that indirectly affect transmission planning such as limitation on pollution production by power plants in different areas that will shift future generation mix toward more renewables, and access to water resources necessary for building and operating power plants. These factors will directly affect the generation expansion (both generation mix and location), and consequently, transmission expansion planning will be affected.

1.1.2 Policy and Regulatory Issues

Policy-makers can affect TEP in several different ways such as who should pay for transmission network upgrades, how the cost should be distributed among them, what the transmission usage tariffs should be, electricity market price caps, and penalties for pollutions. This is discussed in more detail in Sect. 1.2.

1.1.3 Uncertainties

There are several uncertainties that affect TEP and should be addressed during the planning stage. They mainly can be categorized as micro- and macro-uncertainties:

- Macro-uncertainties such as future changes in economic growth, market rules, carbon emission issues, fuel price, generation mix/location and capacity, technology revolutions, etc.
- Micro-uncertainties such as load and intermittent resource variations, availability of power plants and transmission lines in real time, market price, behavior of market participants, etc.

The micro-uncertainty may be well represented by probability distributions, and an expected cost framework may be sufficient to capture main issues. In contrast, the macro-uncertainties may not have well-defined probability distributions, and risk may be much more important in this context, motivating approaches such as robust optimization (Bertsimas et al. 2011; Ruiz and Conejo 2015).

1.1.4 Power System Modeling

The modeling of the power system can have a significant impact on TEP studies. It affects the accuracy of results and computational time required for solving the problem. Main modeling factors are briefly reviewed in the following:

- Steady-state power flow formulation: It can be divided into three main categories: transportation model in which only the first Kirchhoff's law is satisfied; the DC model that satisfies both first and second Kirchhoff's laws, while ignoring network losses and reactive power requirements; and the AC model, which is the most accurate model for power system steady-state modeling and considers network losses and reactive power requirements as well as the first and the second Kirchhoff's laws. There are also some hybrid models that are mainly driven from one of these three main models such as DC model with linear approximation of network losses or linearized AC model with loss and reactive power modeling.
- Transmission network model: Transmission network can be modeled as non-controllable or controllable. In the non-controllable model, the topology of the network is fixed, and in the controllable model, it is possible to use switching, phase shifters, FACTS devices, special protection schemes, and other available tools to control and manage flow on branches.
- Generation model: There are several parameters that affect a power plant's operation, i.e., its maximum and minimum capacity limits, ramp rate capability, minimum up and down time, and some limits that are driven by specific generation technologies like total energy limit for hydropower plants (based on their reservoir capacity).
- Demand model: There are two different ways to model load, i.e., elastic or inelastic. In the elastic model, demand can be controlled with different signals such as the

market price, but in the inelastic model, demand is modeled as a fixed quantity that should be supplied, if possible, and only curtailed in case of scarcity.

- Operation states: Normal and under contingency are two different types of operation states that can be evaluated in power system analysis (for both steady-state and transient analysis).
- Market model: There are several different aspects in market modeling like ideal versus real markets, day-ahead versus real time that may affect system operation costs and TEP.

1.2 Transmission Investment Financing and Coordination

Transmission system operation and expansion are heavily regulated because of their critical role in power system reliability and their natural monopoly. Although it might be owned/operated by different companies/organizations, the transmission network is an interconnected infrastructure in many countries and regions; therefore, coordination between owners/operators for efficient expansion and operation is critical to maintain power systems reliability and security while economically modeling future uncertainties. In this section, we briefly overview different regulatory schemes for coordination between transmission owners for capacity expansion, investment financing and cost recovery. For discussion regarding generation and load interconnection regulations and procedures, interested readers are referred to Regairz et al. (2017) for a more detailed review.

1.2.1 Transmission Organization Models

As discussed in (Regairz et al. 2017), transmission network ownership and operation model can be divided into three main organizational structures as follows:

- Vertically Integrated Utility (VIU) Model: In this model, which was a dominant model before electricity industry deregulation/restructuring, one company owns all generation, transmission, and distribution grid assets in a particular geographical area, and is a solely responsible for supplying its customers.
- Transmission System Operator (TSO) Model: In this model, which is common in the Europe, generation and customer supply are separated from transmission system to maintain the full independence of TSOs. In this model, a TSO is the owner and solely responsible for operation and expansion of the grid in its area.
- Independent System Operator (ISO) Model: In this model, which is common in the USA, not only is the transmission sector separated from generation and supply, but also its operation is separated from its ownership to enhance the independence of the system operator. In this model, ISO is responsible for systems and market operation, short-term and long-term resource adequacy and transmission expansion planning; however, the ISO does not own any transmission, genera-

tion or supply assets. Transmission owner companies own transmission assets and are responsible for maintenance and for most transmission operations, under the authority of the ISO.

1.2.2 Transmission Coordination: Planning and Investments

In this section, we briefly overview responsibility for transmission expansion plans development and investment financing (for each transmission organization model from Sect. 1.2.1) and how these activities are coordinated when expansion projects cross multiple transmission owners' territories.

- For vertically integrated utilities, the utility is responsible to perform transmission and generation expansion studies for its area and will select/approve the cost-effective expansion plans to be built. Their performance might be overseen by a local government or a regulatory agency. Depending on their interconnection with neighboring networks, they may be required to meet some external reliability and security requirements as well. For example, vertically integrated utilities in the USA, connected to the bulk power system network, should meet NERC reliability requirements for power system planning and operation. The North American Electric Reliability Corporation (NERC) is a not-for-profit international regulatory authority whose mission is to assure the effective and efficient reduction of risks to the reliability and security of the grid. It develops and enforces reliability standards that span the continental USA, Canada, and the northern portion of Baja California, Mexico (NERC 2019). Moreover, in this structure, the vertically integrated company itself is responsible for financing selected plans usually on the basis of a state-regulator approved rate of return on investment, based on cost of service to be discussed below.
- TSO performs planning studies for the network within its area and will send the results to a regulatory board for approval. TSO makes the investment to build approved expansion projects and will operate and maintain them. In Europe, the Ten Year Network Development Plan (TYNDP) presents a forward-looking non-binding proposal for electricity transmission infrastructure investments across 34 European countries (Regairz et al. 2017). For projects between countries, investment decisions are made based on specific agreements between parties who benefit from or are affected by the project.
- In the ISO model, ISOs are mostly responsible for performing transmission expansion planning studies. However, all stakeholders including generation owners, load serving entities, and transmission owners can participate in the planning process by submitting their proposals for transmission upgrades to the ISOs for their review/selection. The final expansion plans are sent to transmission owners for construction after they are approved by a board of directors or a regulatory agency. In the USA, CAISO, SPP, ERCOT, MISO, PJM, NYISO, ISO-NE are major ISOs. The Federal Energy Regulatory Commission (FERC) is a federal government agency that regulates the interstate transmission of natural gas, oil,

and electricity (FERC 2019). Except for ERCOT, all other ISOs in the USA are under FERC's jurisdiction.

1.2.3 Tariffs and Regulatory

Transmission and distribution sectors of electric power systems remained regulated in many countries even after thirty years of electricity industry reform. Traditionally, a regulated firm's budget constraint was formed based on the assumption that regulators have perfect information about technologies, costs, and consumer demands. However, in reality, regulators have imperfect information about costs and service quality opportunities, and in many cases, the regulated firm has more information than the regulator, which is a disadvantage for regulators. Four main approaches for compensating a regulated firm's costs, namely cost of service, price cap, incentive, and merchant-regulatory mechanisms, are briefly discussed in the following.

Cost of service is one of the widely used approaches for compensating regulated firms. In this method, it is effectively guaranteed that essentially all operation and investment costs that actually occurred will be compensated. Although it provides incentive to invest more on the grid maintenance and expansion, it does not provide any incentive for improving performance and reducing costs. *Price cap* regulatory mechanism is designed to provide incentives for managers to reduce costs and improve performance. Because of uncertainties in firm's actual realized costs, a low price cap may not cover all their costs. As regulators should consider financial viability, a high price cap should be selected to cover uncertainties, but this may decrease the efficiency of this approach (Joskow 2006). *Incentive* regulatory mechanism is designed to address this issue by providing a menu of options for different situations. A comprehensive review of incentive regulatory mechanism is provided in (Armstrong and Sappington 2005; Blackmon 1994; Sappington and Sibley 1988). *Merchant-regulatory* mechanism allows a combination of regulated and merchant investments. It provides more flexibility on planning and project approval stages but introduces cost recovery risks for merchant-based projects as there is no guarantee for their cost recovery. Hogan et al. (2010) and Rosellon and Weigt (2011) discussed this mechanism in detail.

For transmission network investment cost recovery, different mechanisms can be used. In the Europe with TSO model, regulated tariffs using incentive-based mechanism is used to recover transmission related costs (including investment and operation). In the USA with ISO models, a combination of regulated tariffs and merchant regulatory is used to recover investment and operation costs at transmission level. In regions with vertically integrated utility model, cost of service and price cap regulatory mechanisms are mainly used to guarantee cost compensations. For more details, interested readers are referred to references (Vogelsang and Finsinger 1979; Vogelsang 2001; Hesamzadeh et al. 2018).

Whatever the regulatory mechanism, there is an implicit assumption that the system is planned according to some criterion to achieve a particular objective. Historically, transmission planning has not, however, utilized systematic optimization approaches, but rather has involved expert knowledge and trial and error. The rest of this chapter focuses on systematic transmission expansion planning that is aimed at explicitly finding an optimal plan with less reliance on expert knowledge. It is organized as follows: in Sect. 2, a literature review on TEP studies with major focus on different TEP formulations, reliability, and uncertainty modeling are provided. In Sect. 3, stochastic and robust TEP optimization formulation along with different decomposition techniques are discussed. Then, we review a general framework for solving large-scale TEP studies and evaluate computational challenges from different perspectives in Sect. 4. In Sect. 5, numerical results on solving real-size networks are discussed. This chapter is based, in part, on (Majidi-Qadikolai 2017).

2 Literature Survey

As discussed in Sect. 1.1, there are significant factors affecting transmission expansion planning, which make TEP a multi-dimensional and very complex problem. A major question is how to model/formulate all those parameters, and more importantly how to solve TEP for large-scale networks. Making assumptions and simplifications are inevitable, and we seek to do so in a way that does not fundamentally invalidate the analysis. Environmental, legal, policy, and regulatory issues mostly can be considered in near-term TEP/line design stage and can be partially addressed in developing candidate lines for long-term TEP. Therefore, we can model their impacts outside of TEP optimization formulation and thereby significantly reduce TEP problem size. Uncertainties can be captured either by developing different possible scenarios or by developing uncertainty boundaries and using robust optimization techniques. Villasana et al. (1985) discussed different levels of complexity of the TEP optimization problem as follows:

Level I: Considering all quantities deterministic (future load, generation, and fuel price), static model (one planning horizon), single operation condition (normal operation), all variables as continuous (continuous line capacity for expansion);

Level II: Deterministic quantities, static model, single operation condition, mixed-integer problem (MIP) statement (binary decision variables for building transmission lines);

Level III: deterministic quantities, static model, multi-operation conditions (normal and under contingency operation states), MIP statement;

Level IV: Deterministic quantities, dynamic model (multi-planning horizons), multi-operation conditions, MIP statement;

Level V: Stochastic quantities (uncertainties in load, generation, and fuel price), dynamic model, multi-operation conditions, MIP statement.

By moving from level I to level V, the model will be more accurate and closer to reality, but much more complicated and challenging to solve. By using the DC model, stage I represents a continuous linear optimization problem. Adding integer variables makes it a mixed-integer programming (MIP) problem in level II. Level III adds contingency analysis into TEP that significantly increases the problem size and can easily make TEP optimization problem intractable. TEP moves from static to dynamic in level IV that increases the number of binary variables in the optimization formulation, and TEP is modeled as stochastic dynamic TEP in level V.

2.1 Solution Methods

Using some expert knowledge (EK) for solving large-scale TEP optimization problem is inevitable with current existing machines and software. But there are different points of view on how EK should be integrated into the transmission planning decision making process. Historically, decisions are mainly made by experts based on their expertise instead of using an optimization-based method. A second approach integrates EK into the TEP decision-making process by using EK to choose the worst case for planning, choose the list of possible contingencies, or reduce the list of candidate lines. A third approach converts EK into some criteria (where applicable) and tries to integrate them into a TEP optimization framework. Compared to the second approach, this method is systematic and tractable on the one hand, and more challenging from the modeling perspective on the other hand. The fourth approach tries to use EK as little as possible and solve the problem through pure mathematical formulation. These purely mathematically driven methods are usually computationally very expensive and are not practical for large-scale problems.

In heuristic models, approaches one and two, the TEP problem is solved through several steps of generating, evaluating, and selecting expansion plans, with or without the user's help (Latorre et al. 2003). One of the common heuristic methods is to use sensitivity analysis to select additional circuits (Latorre-Bayona and Perez-Arriaga 1994; Majidi-Qadikolai and Baldick 2015; Monticelli et al. 1982; Pereira and Pinto 1985). MISO Midcontinent ISO (2016), ERCOT ERCOT System Planning (2016), and CAISO Market & Infrastructure Development (2016) are three examples of independent system operators in the USA that use different heuristic methods for TEP.

In optimization-based methods, approaches three and four, a mathematical formulation for TEP is developed and the problem is solved using classical optimization programming techniques. Optimization-based methods are computationally very expensive and have historically been thought to be impractical for large-scale TEP problems (Latorre et al. 2003; Munoz et al. 2015). However, modern computing systems and optimization software, together with novel formulations, have begun to make optimization-based methods practical for large-scale planning. Several methods are proposed to formulate the TEP problem.

Using linear approximation of AC power flow equations is one of the most popular simplifications for modeling nonlinear power flow equations in high-level TEP studies. The accuracy of linear approximation of power flow equations (DC model) is evaluated in (Van Hertem et al. 2006; Baldick et al. 2005; Overbye et al. 2004). In Van Hertem et al. (2006), authors compared the results of AC and DC power flow results for the IEEE 300-bus system and showed the error between DC and AC results will be less than 5% when the assumptions of DC power flow are satisfied. Baldick et al. (2005) performed sensitivity analysis in power systems with DC and AC models and demonstrated that it provides a relatively reliable approximation of the behavior of the system. Overbye et al. (2004) showed that locational marginal prices (LMPs) that drive the economic analysis of power system operation will not be significantly affected when the AC model is approximated with the DC model so long as various assumptions are satisfied.

In Villasana et al. (1985) and Garver (1970), transmission planning is formulated as a simple linear programming (LP) problem with continuous decision variables. Villasana et al. (1985) proposed a LP method with continuous variables for optimal transmission planning by minimizing load curtailment. As transmission line capacity is lumpy, considering capacity to be a continuous variable is not accurate. Villanasa (1984) proposed a mixed-integer programming (MIP) formulation using binary decision variables for selecting new lines with DC power flow approximation. This method is more accurate in representing new line capacities, but the proposed formulation is not computationally efficient.

Kirchoff's second law is represented with two inequalities in a mixed-integer disjunctive model, each related to one possible flow direction in (Bahiense et al. 2001). This technique increases the number of constraints and provides better conditioning properties by tightening constraints. Bahiense et al. (2001) also used GRASP meta-heuristic method to provide an upper bound feasible solution. In Alguacil et al. (2003), power network losses are integrated into TEP optimization problem using piecewise linear loss function for each line. It provides more accurate power system model for planning purpose while preserving linearity and may affect the selected expansion plan for networks with relatively high losses such as systems with long transmission lines. However, the simulation time for this case is increased around five times compared to the case without losses.

Benders decomposition (BD) is used in several contexts as a powerful tool for decreasing simulation time for solving large-scale optimization problems. Mathematical formulation for implementing Benders decomposition for transmission and generation expansion planning was developed by EPRI in 1988 (Granville et al. 1988). Gomory cuts are added to Benders cuts in (Binato et al. 2001) to improve the performance of BD for large-scale MIP problems. To overcome the non-convexity of transmission planning problem Romero and Monticelli (1994); Rosellon and Weigt (2011) proposed a three-phase hierarchical decomposition method to find the global optimal answer. They used BD to solve each phase and transferred Benders cuts into the next phase to integrate different phases. Park and Baldick (2013) considered load and wind as dependent and uncertain variables and used a two-stage stochastic model and sequential approximation technique to solve TEP optimization problems with BD. A dynamic transmission expansion planning is formulated in (Munoz et al. 2014) and authors compared the performance of stochastic programming with deterministic and heuristic methods. Munoz et al. (2013) evaluated the impact of different approximations on TEP with renewable portfolio standards. Munoz et al. (2014) and Munoz and Watson (2015) proposed a new approach for multi-regional transmission and generation expansion planning with Benders decomposition technique, which is enhanced by developing new lower bounding constraints that increase convergence speed. They applied the model to large-scale networks with a relatively large number of scenarios to capture uncertainties and evaluated the impact of optimality gap on simulation time. A complex mathematical model for centralized transmission planning and decentralized generation expansion planning is developed in (Jin and Ryan 2014). To represent the interaction between generator and transmission planners during transmission and generation expansion planning, game theory-based approaches are used by (Tohidi and Hesamzadeh 2014; Tohidi et al. 2017a, b; Ruiz and Contreras 2007; Yen et al. 2000). To decrease computational efforts, all above-mentioned references ignored contingency analysis in their proposed methods for transmission planning. So, there is no guarantee that selected optimal plans by these papers satisfy reliability requirements.

2.2 Power System Adequacy and Reliability

The power system should be adequate and reliable. Based on North American Electric Reliability Corporation (NERC) definition "Adequacy is the ability of the electric system to supply the aggregate electric power and energy requirements of the electricity consumers at all times, taking into account scheduled and reasonably expected unscheduled outages of system components" and "Operating reliability is the ability of the electric system to withstand sudden disturbances such as electric short circuits or unanticipated loss of system components" (NERC 2007). In standard 51, NERC categorized system adequacy and security into four levels A-D (NERC 2005). Level A refers to system performance under normal conditions (no contingency), and in level B, system performance following the loss of a single bulk system element is evaluated. In Levels C and D, system performance under loss of two or more bulk system components and extreme events are evaluated, respectively. Categories A-C should be evaluated for near-term and long-term planning, and category D should be considered for near-term planning only.

The power system should be planned and be operated in a way to be able to supply all loads under normal conditions and in case of a single outage in system components (levels A and B). This is called the $N-1$ criterion (Electric Reliability Council of Texas 2014; NERC 2005). To satisfy this standard, system operators usually use security-constrained optimal power flow (SCOPF) or security-constrained unit commitment (SCUC) to dispatch/commit power plants. Post-contingency redispatch (Monticelli et al. 1987), congestion management (Majidi et al. 2008), transmission switching (Hedman et al. 2008; Majidi-Qadikolai and Baldick 2015; Ruiz et al. 2012a, b), or using FACTS devices (Majidi et al. 2008; Ziaee et al. 2017) are techniques used to add flexibility to transmission operation and subsequently reduce operation costs. In Monticelli et al. (1987), a new algorithm for security-constrained optimal power flow (SCOPF) is proposed that considers post-contingency corrective rescheduling to decrease dispatch costs. To integrate transmission switching in the system operation, Ruiz et al. (2012a) used the flow cancelation technique to model switching. They showed that this technique is faster than using binary variables to change the status of lines in topology control when the number of switching lines in limited.

Various researchers use either the $N-1$ criterion or probabilistic approaches such as loss of load probability (LOLP) or loss of load expectations (LOLE) for power system adequacy and security evaluation. Leite da Silva et al. (2010) explained drawbacks of each method and evaluated the impact of considering different reliability criteria on TEP. They performed numerical analysis for the Garver 6-bus system (Garver 1970) to compare the performance of these methods. The result shows that TEP with $N-1$ criterion requires more investment compared to TEP with probabilistic approaches as it should supply the demand under all single contingencies. Loss of load cost (LOLC) as a reliability index is calculated for the selected plan for both cases, and LOLC for TEP with $N-1$ criterion is much less than LOLC for TEP with the probabilistic approach, showing the impact of extra investment on improving system reliability. By considering $N-1$ criterion, the system quality and reliability indexes will be less sensitive to load variations and components' rate of outage compared to probabilistic approaches.

O'Neill et al proposed a comprehensive mathematical formulation for dynamic optimal power system planning and investment by integrating unit commitment, transmission switching, and $N-1$ contingency analysis into a power system operation cost formulation in (O'Neill et al. 2011). But as the authors mentioned in their paper, it is a very complex and computationally expensive model even for a very small case study, so it is not practical for large-scale networks at this time. More practical formulations for TEP optimization with $N-1$ contingency analysis are formulated in (Rudkevich 2012; Khodaei et al. 2010; Moreno et al. 2013; Zhang et al. 2012; Majidi-Qadikolai and Baldick 2016a,b, 2018). Rudkevich (2012) proposed a nodal capacity market framework for generation and transmission expansion planning. He used the flow cancelation technique to represent a fixed list of contingencies in a reliability dispatch formulation, in which all resources are dispatched at zero costs and load shedding will be penalized at value of lost load (VOLL) price. Khodaei et al. (2010) proposed a three-stage transmission and generation expansion planning optimization formulation with Benders decomposition technique and considered contingency analysis for all existing and candidate lines and integrated transmission switching to alleviate violations in line flows. In Carrion et al. (2007), transmission expansion and reinforcement are formulated as a stochastic optimization problem to reduce vulnerability of the system in case of deliberate attacks.

2.3 *Uncertainties*

Fast technology changes, new policies, increasing penetration of mobile/flexible demand along with intermittent nature of renewable resources make it hard to accurately predict future generation mix/location and demand as inputs for TEP studies;

therefore, these uncertainties should be explicitly modeled/evaluated in TEP process by system planners. It should be emphasized that developing a single expansion plan using methods that heavily depend on engineering judgment can result in a plan that is costly and inefficient when the implications of uncertainties are considered. Munoz et al. (2014), Munoz and Watson (2015) and Cedeño and Arora (2011) evaluated the impact of ignoring uncertainties on transmission planning by comparing the results of deterministic, heuristic, and stochastic TEP for different case studies. Their result shows that stochastic TEP may select some lines that will not be selected by either deterministic or heuristic methods.

The TEP optimization problem can be formulated as a two-stage stochastic resource allocation problem (a class of mixed-integer stochastic programming) to explicitly model uncertainties using a finite set of scenarios (Kall and Woodruff 1994). In this formulation, in the first stage, a decision about building a new transmission line is made, and the impact of this decision on power system operation under different scenarios is evaluated in the second stage. To capture all macro- and micro-uncertainties, usually a large number of scenarios are generated in the early stages of planning (there are different methods to generate scenarios to represent uncertainties such as Monte Carlo method (used by (Akbari et al. 2011)) and using historical data with statistical modeling (used by (Park and Baldick 2013)), and different clustering techniques are developed to reduce the number of scenarios (Munoz and Watson 2015; Park and Baldick 2013). There are also some commercial packages such as (SCENRED GAMS 2002) that can be used for this purpose. Akbari et al. (2011) integrated Available Transmission Capacity (ATC) constraints into a multi-stage stochastic TEP problem. They used GAMS/SCENRED as a tool to reduce a very large number of randomly generated scenarios and solved TEP with all contingencies for the IEEE-24 bus system. The impact of adding ATC constraints to TEP is evaluated; however, the performance of the model for large-scale systems is not discussed. Alvarez Lopez et al. (2007) integrated uncertainties and risks in load, availability of generation and transmission lines into a stochastic generation and transmission capacity expansion planning problem and formulated it as a non-linear mixed-integer optimization problem. A probabilistic method for capturing uncertainties in TEP is proposed in (Buygi et al. 2004). They developed probabilistic locational marginal pricing (LMP) index and suggested value-based criteria, i.e., decreasing congestion cost and reducing weighted deviation of mean of LMPs for selecting new transmission lines. In Zhang et al. (2015), Benders decomposition with aggregated multi-cuts is used to solve TEP under uncertainties. Pringles et al. (2015) used least-square Monte Carlo dynamic programming to solve stochastic TEP. They deployed sensitivity analysis to determine decision regions to execute, postpone, or reject transmission investment candidates.

Although formulating TEP as a two-stage stochastic optimization problem provides a strong modeling capability (Guo Chen et al. 2012; Majidi-Qadikolai and Baldick 2016a; Munoz and Watson 2015; Park and Baldick 2013), solving the extensive form (EF) of this problem is not tractable even for medium size problems especially when $N-1$ contingency analysis is added to the problem. Therefore, decomposition and heuristic techniques should be used for solving TEP for medium to large-scale systems.

Robust optimization is another method to integrate uncertainties into the TEP formulation. In robust optimization, uncertainties are represented using a range for each uncertain parameter or a budget of uncertainty for collections of uncertain parameters instead of developing scenarios (as used by stochastic optimization), and it finds a plan that is robust for the worst-case scenario. In this case, the final result is usually too conservative, which motivates an adaptive robust optimization (Bertsimas et al. 2011) formulation with budget limit constraints to mitigate the level of robustness (conservativeness of results). Ruiz and Conejo (2015), Garcia-Bertrand and Minguez (2016), Minguez and Garcia-Bertrand (2016) formulated the TEP problem as an adaptive robust optimization.

3 Transmission Expansion Planning Formulation and Decomposition Techniques

As stated in Sect. 2, the transmission expansion problem can be formulated as static (single-stage) or dynamic (multi-stage), deterministic or probabilistic, stochastic or robust. In this section, we investigate static TEP with stochastic/robust optimization techniques to address uncertainties. For mathematical formulations, variable/parameter definitions are provided in the beginning of this chapter.

3.1 Two-Stage Stochastic TEP Formulation

As discussed in Sect. 2.3, stochastic programming is one of the widely used methods to model uncertainties (by developing different scenarios) in the decision-making process for resource allocation problems. To capture uncertainties, different scenario generation/reduction methods might be used to finalize the input scenario set. The quality of scenarios is critical and can significantly affect the selected expansion plan. For example, in ERCOT, historical data along with workshops with stakeholders are used to develop scenarios for long-term TEP (ERCOT System Planning 2014). It should be mentioned that minimizing the expected value is a better criterion for micro-uncertainties in cases where probability distributions can be estimated from empirical data. The two-stage stochastic TEP is formulated as follows:

$$Z^* = \min_{x}\{\zeta^{\mathsf{T}}\boldsymbol{x} + \mathbb{E}\min_{y\in\Xi} Q(\boldsymbol{x}, \tilde{\xi}, y)\} \tag{1}$$

$$\text{st.} \quad \boldsymbol{x} \in \{0, 1\}^{|N_l|} \tag{2}$$

where $\boldsymbol{x}$ is the first stage binary decision variable, $\tilde{\xi}$ is a random variable vector for second stage uncertainties, y is the second stage continuous decision variables vector, and Ξ defines the feasible region for variable y. $\mathbb{E}\min_{y} Q(\boldsymbol{x}, \tilde{\xi}, y)$ represents the expected value of operation costs including load shedding and wind curtailment penalty and generation costs for TEP problem formulation with the expectation taken over the random variable $\tilde{\xi}$. This expected value is approximated with a weighted sum of a limited number of scenarios as follows (Ermoliev and Wets 1988):

$$\mathbb{E}\min_{y} Q(\boldsymbol{x}, \tilde{\xi}, y) \approx \sum_{\omega\in\Omega} P^{\omega} \min_{y^{\omega}} Q(\boldsymbol{x}, \xi^{\omega}, y^{\omega}) \tag{3}$$

where $\min_{y^{\omega}} Q(\boldsymbol{x}, \xi^{\omega}, y^{\omega})$ is the optimal value of power system operation over choices of second stage variables for a given scenario ω, and Ω is a discrete approximation to the distribution of $\tilde{\xi}$ (Majidi-Qadikolai and Baldick 2016a). The extensive form of the two-stage stochastic TEP can be written as follows:

$$Z^* = \min_{x, y^{\omega}} \left\{ \zeta^{\mathsf{T}}\boldsymbol{x} + \sum_{\Omega} P^{\omega} \left[\sum_{N_s}\Big(\sum_{N_b} q_k r^{\omega}_{k,c}\Big) + \sum_{N_{wg}} \gamma_g CW^{\omega}_g + \sum_{N_g} Co_g p^{\omega}_g \right] \right\} \tag{4}$$

$$\text{st.} \quad -\sum_{L_k} f^{\omega}_{l,c} + \sum_{G_k} p^{\omega}_g + r^{\omega}_{k,c} = d^{\omega}_k \tag{5}$$

$$-M_l(1 - C_{l,c}x_l) \le f^{\omega}_{l,c} - B_{l,l}\Delta\theta^{\omega}_{l,c} \tag{6}$$

$$M_l(1 - C_{l,c}x_l) \ge f^{\omega}_{l,c} - B_{l,l}\Delta\theta^{\omega}_{l,c} \tag{7}$$

$$CW^{\omega}_g \ge (P^{max,\omega}_g - p^{\omega}_g) \tag{8}$$

$$(C_{l,c}x_l) f^{min}_l \le f^{\omega}_{l,c} \le f^{max}_l (C_{l,c}x_l) \tag{9}$$

$$P^{min}_g \le p^{\omega}_g \le P^{max}_g \tag{10}$$

$$0 \le r^{\omega}_{k,c} \le d_k \tag{11}$$

$$-\frac{\pi}{2} \le \theta^{\omega}_{k,c} \le \frac{\pi}{2} \tag{12}$$

$$CW^{\omega}_g \ge 0 \tag{13}$$

$$x_l = 1, \quad \forall l \in N_o \tag{14}$$

$$x_l \in \{0, 1\}, \quad \forall l \in N_l \tag{15}$$

In (4), y^{ω} is the second stage decision variables vector that includes power generation (p_g^{ω}), load shedding ($r_{k,c}^{\omega}$), wind curtailments (CW_g^{ω}), branch flows ($f_{l,c}^{\omega}$), and voltage angles ($\theta_{k,c}^{\omega}$) for all scenarios and all operation states. The formulation minimizes the objective function over all first stage (x) and second stage (y^{ω}) decision variables and is constrained by (5)–(15). Equation (5) enforces power balance at each bus. Equations (6) and (7) represent power flow in transmission lines using the big-M technique. Equation (8) measures wind curtailment at each bus. Equation (9) shows flow ($f_{l,c}^{\omega}$) in branches should be between maximum and minimum capacity limits. Equations (10)–(12) enforce power plants' dispatch p_g^{ω}, load shedding $r_{k,c}^{\omega}$, and voltage angles $\theta_{k,c}^{\omega}$, respectively, to be between their minimum and maximum limits. Equation (13) enforces nonnegativity of wind curtailment. Equation (14) sets decision variables for existing lines to 1. Equation (15) enforces that x_l is a binary decision variable for transmission lines ($x_l = 1$ when line l is built and $x_l = 0$ when line l is not built).

Depending on the size of the network and the number of scenarios, solving the extensive form of problem (1) can be extremely computationally expensive. Therefore, decomposition techniques are used to find a near-optimal answer for large-scale problems.

3.2 *Robust Optimization TEP Formulation*

Robust optimization is a technique for modeling uncertainties and finding reliable solutions for the worst-case scenario. As discussed in Sect. 2.3, adaptive robust optimization can be used to adjust the level of robustness. Jabr (2013) and Ruiz and Conejo (2015) used this technique for TEP studies. Robust TEP can be formulated as three-level optimization problem as follows:

$$Z^* = \min_{x}\{\zeta^{\mathsf{T}}\boldsymbol{x} + \max_{\xi\in\mathscr{D}} [\min_{y\in\Xi} Q(\boldsymbol{x}, \xi, y)]\} \tag{16}$$

$$\text{st.} \quad \boldsymbol{x} \in \{0, 1\}^{|N_l|} \tag{17}$$

In objective function (16), in the first level, the best transmission expansion plan ($\boldsymbol{x}$) is selected by minimizing the total system cost. In the second level, a realization of uncertain variables $\boldsymbol{\xi}$ is selected from uncertainty set $\mathscr{D}$ that maximizes system operation costs ($Q(\boldsymbol{x}, \xi, y)$) to represent the worst-case scenario. In the third level, based on selected x and $\boldsymbol{\xi}$ from the first and the second levels, system operator tries to find the best values for third-level decision variables y (from its feasible set Ξ) to minimize system operation cost.

The result of robust optimization-based TEP is sensitive to uncertainty set definition; therefore, as stated in (Ruiz and Conejo 2015), having a careful definition of uncertainty set $\mathscr{D}$ is critical for an effective representation of uncertainties. A polyhedral uncertainty set is common to represent load and generation uncertainties. It can be described using the following constraints:

$$\xi_k \in [\xi_k^{min}, \xi_k^{max}] \tag{18}$$

$$\frac{\sum |\xi_k^{ref} - \xi_k|}{\sum |\xi_k^{max} - \xi_k^{min}|} \leq UB_a \tag{19}$$

Equation (18) shows each uncertain parameter (load or generation here) may change between a minimum and a maximum value. Equation (19) is added to the robust optimization formulation to control the level of robustness (adaptive robust optimization). It is usually defined at regional/area level to mitigate the worst-case scenario. For example, it is less likely that outputs of all wind farms located at the same region face 100% deviation from their reference value at the same time. In this equation, ξ_k^{ref} is a reference point to measure divisions ($\xi_k^{min} \leq \xi_k^{ref} \leq \xi_k^{max}$), and UB_a is uncertainty budget limit that can have a value between 0 and 1 ($0 \leq UB_a \leq 1$).

The extended form of robust optimization formulation formulation can be written as follows (Ruiz and Conejo 2015):

$$Z^* = \min_x \left\{ \zeta^\intercal \boldsymbol{x} + \max_\xi \left[\min_y \sum_{N_s} (\sum_{N_b} q_k r_{k,c}) + \sum_{N_{wg}} \gamma_g CW_g + \sum_{N_g} Co_g p_g \right] \right\} \tag{20}$$

$$\text{st.} \quad -\sum_{L_k} f_{l,c} + \sum_{G_k} p_g + r_{k,c} = d_k \tag{21}$$

$$-M_l(1 - C_{l,c}x_l) \leq f_{l,c} - B_{l,l}\Delta\theta_{l,c} \tag{22}$$

$$M_l(1 - C_{l,c}x_l) \geq f_{l,c} - B_{l,l}\Delta\theta_{l,c} \tag{23}$$

$$CW_g \geq (P_g^{max} - p_g) \tag{24}$$

$$(C_{l,c}x_l) f_l^{min} \leq f_{l,c} \leq f_l^{max}(C_{l,c}x_l) \tag{25}$$

$$P_g^{min} \leq p_g \leq P_g^{max} \tag{26}$$

$$0 \leq r_{k,c} \leq d_k \tag{27}$$

$$-\frac{\pi}{2} \leq \theta_{k,c} \leq \frac{\pi}{2} \tag{28}$$

$$CW_g \geq 0 \tag{29}$$

$$\text{st.} \quad 0 \leq P_g^{max} \leq P_g^{\hat{max}} \tag{30}$$

$$d_k^{min} \leq d_k \leq d_k^{max} \tag{31}$$

$$\frac{\sum (P_g^{\hat{max}} - P_g^{max})}{\sum P_g^{\hat{max}}} \leq UB_a^G \tag{32}$$

$$\frac{\sum (d_k - d_k^{min})}{\sum d_k^{max} - d_k^{min}} \leq UB_a^D \tag{33}$$

$$\text{st.} \quad x_l = 1, \quad \forall l \in N_o \tag{34}$$

$$x_l \in \{0, 1\}, \quad \forall l \in N_l \tag{35}$$

In the objective function (20), y is the third-level decision variables vector that includes power generation (p_g), load shedding ($r_{k,c}$), wind curtailments (CW_g), branch flows ($f_{l,c}$), and voltage angles ($\theta_{k,c}$) for all scenarios and all operation states. Constraints (21)–(29) form the feasible region Ξ. The second level decision variable ξ and includes the worst realization of demand (d_k^{max}) and generation ($\hat{P}_g^{max}$). Constraints (30)–(33) form the feasible region for uncertain variables ($\mathscr{D}$). Equations (30) and (31) limit minimum and maximum generation and load deviation at each bus, respectively. Equations (32) and (33) are uncertainty budget limits for generation and load at area a. Equations (34) and (35) limit values of the first level decision variable (x) to be 0 or 1 and set their value equal to 1 for all existing branches.

Decomposition-based formulations for the robust optimization TEP are developed in (Jabr 2013; Ruiz and Conejo 2015).

3.3 Constraint Filtering and Optimization Problem Size Reduction

Constraints define the feasible region of an optimization problem. In many cases, only a small subset of modeled constraints contribute in forming the final feasible region, and others can be removed from optimization problem without affecting the final optimal result. The key issue is finding which constraints can be removed. The following simple linear programming example with two variables is used for illustration purpose.

$$Z = \min_{y_1, y_2} \; 2y_1 + 5y_2 \tag{36}$$

$$\text{st.} \quad y_1 + 2y_2 \leq 6 \tag{37}$$

$$y_1 - y_2 \leq 0 \tag{38}$$

$$y_1 \leq 3 \tag{39}$$

$$y_2 \leq 5 \tag{40}$$

$$y_1 \geq 0 \tag{41}$$

$$y_2 \geq 0 \tag{42}$$

Constraints (37)–(42) limit the choice of y_1 and y_2 values by defining the feasible region for these two variables. These constraints and the formed feasible region are shown in Fig. 1. Lines $C1$ to $C6$ represent constraints (37)–(42), respectively, and the yellow triangle demonstrate the feasible region. The optimal solution is shown as the bullet. For this optimization problem, $C3$ and $C4$ do not contribute in forming the feasible region; therefore, removing them will reduce the problem size without affecting the optimal solution.

In power system operation, most of the constraints are not necessary for forming the feasible region. For example in ERCOT, there were only about 400 contingency

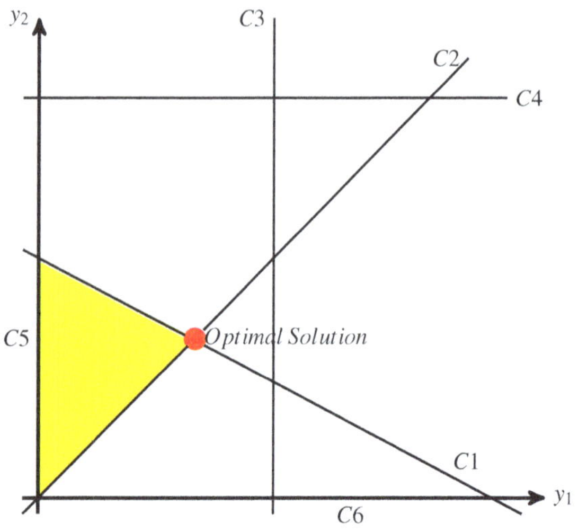

Fig. 1 Constraints and the feasible region

constraints (out of tens of millions of possible constraints) that were binding at some time during 2013 (Potomac Economics 2014). Usually during very low load/low wind periods, a single outage of any line will not cause overload on other lines in most power systems. In other words, constraints related to those contingencies will be dominated by other constraints in the optimization problem and will not affect the feasible region and the optimal answer. Therefore, for this particular case we can ignore contingencies and solve OPF instead of SCOPF. As constraints related to contingencies are dominated, results of OPF will be feasible for SCOPF as well. Although eliminating passive constraints can significantly reduce problem size, finding all active constraints forming the feasible region is challenging.

Ardakani and Bouffard (2013) developed a technique called umbrella constraint identification to find all necessary and sufficient constraints for DC-SCOPF formulation. Abiri-Jahromi and Bouffard (2017a, b) developed loadability set to find necessary constraints for minimal representation of the feasible region for SCOPF by projecting demand-generation-network spaces onto the demand space only. Madani et al. (2017) have found a minimal subset of security constraints for a general SCUC formulation that guarantees the satisfaction of all security constraints. This formulation does not depend on commitment decision for generators and can handle load and generation forecast errors. Majidi-Qadikolai and Baldick (2016a, b) developed heuristic algorithms for SCOPF contingency constraint reduction. This method does not guarantee to find the minimal subset, but it can significantly decrease the problem size and it is computationally very cheap, and it can be used for both deterministic and stochastic formulations.

3.4 Decomposition Techniques

Solving the extensive form of a two-stage stochastic TEP optimization problem for large-scale networks is not practically feasible; therefore, *Horizontal* or *Vertical* decomposition techniques or both can be used to decompose the original problem for large systems. These techniques are discussed in this section.

3.4.1 Vertical Decomposition

Benders decomposition (BD) is one of the widely used *vertical* decomposition technique for solving two-stage stochastic TEP (Benders 1962). It divides the original problem into two parts, i.e., master and subproblem and uses "cuts" from dual of the subproblem to model its constraints in the master problem (Granville et al. 1988). References Granville et al. (1988), Park and Baldick (2013), Guo Chen et al. (2012), Zhang et al. (2015), Akbari et al. (2011), Munoz et al. (2014), Khodaei et al. (2010) applied BD to solve TEP optimization problem.

Although in several papers it is claimed that BD is easily scalable (for TEP) and can be used for real-size problems, Munoz et al. (2014) showed that even for medium size networks when the number of scenarios is large (50 or more), an optimality gap between 3% to 6% would need to be accepted in the BD algorithm to get the result in a reasonable time. For large-scale problems, the subproblem itself will be hard to solve, and a large number of iterations between master and subproblem is required to meet optimality gap requirements. This drawback worsens when reliability constraints are added to the TEP problem, in which subproblems should be solved for normal and under contingency operation states for all scenarios.

The column-and-constraint generation method (also called cutting-plane method) is another *vertical* decomposition technique that can be used to decompose a two-stage problem. In this method, primal "cuts" are used to represent the subproblem constraints in the master problem instead of dual cuts used by BD. Convergence guarantees and other properties of this method are explained in (Jiang et al. 2013; Zeng and Zhao 2013). Jabr (2013) and Ruiz and Conejo (2015) used BD and cutting-plane decomposition techniques, respectively, for solving robust TEP.

The following generic two-stage stochastic linear program is used to explain mathematical formulation for BD algorithm.

$$SLP = \min_{x,y} \; cx + \sum_{\omega \in \Omega} p^{\omega} f^{\omega} y^{\omega} \tag{43}$$

$$\text{st.} \quad Ax = b \tag{44}$$

$$-B^{\omega} x + D^{\omega} y^{\omega} = d^{\omega}, \quad \forall \omega \in \Omega \tag{45}$$

$$x \in \mathscr{X}, \; y^{\omega} \geq 0, \quad \forall \omega \in \Omega \tag{46}$$

The BD algorithm decomposes the SLP into two problems, i.e., master problem and subproblem, and solves them iteratively. The master problem includes first stage decision variable/constraints and a relaxed version of the second stage constraints.

$$Master = \min_{x,\theta} \; cx + \phi \tag{47}$$

$$\text{st.} \;\; Ax \;\; = b \tag{48}$$

$$-G^i x + \phi \geq g^i, \quad i = 1, \ldots, l \tag{49}$$

$$x \in \mathscr{X} \tag{50}$$

In the subproblem, at iteration i the first stage decision variable (x) is fixed, and the problem is solved for the second stage decision variable (y^ω).

$$Subproblem = \min \sum_{\omega \in \Omega} p^\omega f^\omega y^\omega \tag{51}$$

$$\text{st.} \;\; D^\omega y^\omega = d^\omega + B^\omega x \; : \; \pi^\omega, \quad \forall \omega \in \Omega \tag{52}$$

$$y^\omega \geq 0, \quad \forall \omega \in \Omega \tag{53}$$

After solving the subproblem and assuming it is feasible, coefficients in (54) and (55) are calculated. These coefficients are used to form optimality cuts (equation (49)) that will be sent to the master problem for the next iteration. The standard BD algorithm is summarized in Fig. 2.

Inputs:Data to define the SLP problem (43)–(46), and ε as error tolerance
Output:x^*: solution to SLP within ε of optimality
Initialization: $\overline{Z} \leftarrow +\infty, Err \leftarrow +\infty, i \leftarrow 1$
while $Err \geq \varepsilon$ **do**
  $\hat{x}, \hat{\phi} \leftarrow \arg\min\{cx + \phi\}$
  $\underline{Z} = c\hat{x} + \hat{\phi}$
  $\hat{y}^\omega, \hat{\pi}^\omega \leftarrow \arg\min\{p^\omega f^\omega y^\omega\}$
  $\hat{Z} \leftarrow c\hat{x} + \sum_{\omega \in \Omega} p^\omega f^\omega \hat{y}^\omega$
  if $\hat{Z} \leq \overline{Z}$ **then**
    $\overline{Z} \leftarrow \hat{Z}$
    $x^* \leftarrow \hat{x}$
  end if
  Augment the set of cuts with $-G^i x + \phi \geq g^i$
  $Err \leftarrow \frac{\overline{Z} - \underline{Z}}{\min(|\overline{Z}|, |\underline{Z}|)}$
  $i \leftarrow i + 1$
end while

Fig. 2 Standard Benders decomposition algorithm

$$G^i = \sum_{\omega \in \Omega} p^\omega \pi^\omega B^\omega \tag{54}$$

$$g^i = \sum_{\omega \in \Omega} p^\omega \pi^\omega d^\omega \tag{55}$$

For more details and other forms of BD algorithms, please see reference Conejo et al. (2006).

3.4.2 Horizontal Decomposition

Progressive Hedging (PH) is aimed at decomposing a two-stage stochastic resource allocation problem *horizontally* by solving the problem for each scenario separately and adding *non-anticipativity* constraints to couple the first stage decision variables (standard PH) (Rockafellar and Wets 1991). The PH method for mixed-integer problems is a heuristic method that finds an upper bound answer for the non-convex optimization problem; however, Gade et al. (2016) developed a method to also calculate a lower bound for results of the PH algorithm in order to quantify the quality of results. One drawback of standard PH algorithm is that for problems with a large number of scenarios and integer variables, it may need a large number of iterations to satisfy non-anticipativity constraints (and sometimes it may never converge if no heuristic action is taken inside the algorithm).

For the standard PH algorithm, the TEP problem (1) can be rewritten as the following so-called *scenario* formulation:

$$Standard\ PH = \min_{x,y} \sum_{\omega \in \Omega} p^\omega (c\boldsymbol{x}^\omega + f^\omega y^\omega) \tag{56}$$

$$\text{st.}\quad A\boldsymbol{x}^\omega = b, \quad \forall \omega \in \Omega \tag{57}$$

$$-B^\omega \boldsymbol{x}^\omega + D^\omega y^\omega = d^\omega, \quad \forall \omega \in \Omega \tag{58}$$

$$\boldsymbol{x}^\omega \geq 0, \ y^\omega \geq 0, \quad \forall \omega \in \Omega \tag{59}$$

$$\boldsymbol{x}^1 = \cdots = \boldsymbol{x}^s \tag{60}$$

A copy of decision variable vector $\boldsymbol{x}^\omega$ is created for each scenario ω in Ω that allows solution of the TEP problem for each scenario independently, and non-anticipativity constraints (60) are added to couple first stage solutions and guarantee that the final expansion plan does not depend on scenarios.

Instead of decomposing the problem for each individual scenario, it is possible to use bundles of scenarios ($\mathscr{B} = \{\mathscr{B}_1, \ldots, \mathscr{B}_b\}$) for decomposition. Equations (56)–(60) can be rewritten for bundled PH as follows:

$$Bundled\ PH = \min_{x,y} \sum_{\mathscr{B}} [p^{\mathscr{B}_i}(c\boldsymbol{x}^{\mathscr{B}_i}) + \sum_{\mathscr{B}_i} Pu^{\omega} f^{\omega} y^{\omega}] \tag{61}$$

$$\text{st.} \quad A\boldsymbol{x}^{\mathscr{B}_i} \quad = b, \quad \forall \mathscr{B}_i \in \mathscr{B} \tag{62}$$

$$- B^{\omega}\boldsymbol{x}^{\mathscr{B}_i} + D^{\omega}y^{\omega} = d^{\omega}, \quad \forall \mathscr{B}_i \in \mathscr{B}, \forall \omega \in \Omega \tag{63}$$

$$\boldsymbol{x}^{\mathscr{B}_i} \geq 0, \; y^{\omega} \geq 0, \quad \forall \mathscr{B}_i \in \mathscr{B}, \forall \omega \in \Omega \tag{64}$$

$$\boldsymbol{x}^{\mathscr{B}_1} = \cdots = \boldsymbol{x}^{\mathscr{B}_b} \tag{65}$$

In this case, a copy of decision variable vector $\boldsymbol{x}^{\mathscr{B}_i}$ is created for all $\mathscr{B}_i$s in $\mathscr{B}$. Non-anticipativity constraints (65) are explicitly modeled for scenario bundles, and they are implicitly modeled for scenarios within each bundle (κ scenarios in each bundle already have the same first stage decision variable $\boldsymbol{x}^{\mathscr{B}_i}$). Therefore, a bundled PH will have fewer non-anticipativity constraints compared to a standard PH ($|\mathscr{B}| < |\Omega|$), which usually reduces the number of iterations for convergence.

Through an iterative process, PH will converge to a unique answer for the first stage decision variables by appropriately penalizing deviations of non-anticipative variables from their *mean* values. The PH algorithm with bundled scenarios is shown in Fig. 3. In the first line, the initial value of the iteration counter (υ) and multiplier vector ($\boldsymbol{W}^{\upsilon}_{\mathscr{B}_i}$) is set. From line 2–4, the TEP optimization problem for each bundle is solved separately (and can be parallelized). In line 5, the weighted sum of individual expansion plans ($\boldsymbol{x}^{\mathscr{B}_i,\upsilon}$s) is calculated. Line 6 calculates the deviation (Err) from averaged expansion plan ($\hat{\boldsymbol{x}}^{\upsilon}$). Lines 7–15 cover the main iterative part of the bundled PH algorithm. In line 8, the value of counter is updated. Line 9 updates the value of multiplier vector by using penalty vector $\boldsymbol{\rho}$. Lines 10–12 solve an updated TEP formulation with multiplier and penalizing deviation from average value of first stage decision variables. This optimization problem is solved for each bundle inde-

1: Initialization: $\upsilon \leftarrow 1, W^{\upsilon}_{\mathscr{B}_i} \leftarrow 0 \; \forall \mathscr{B}_i \in \mathscr{B}$
2: **for** $\forall \mathscr{B}_i \in \mathscr{B}$ **do**
3: $\quad x^{\mathscr{B}_i,\upsilon} \leftarrow argmin\, \zeta^{\intercal} x^{\mathscr{B}_i} + \sum_{\omega \in \mathscr{B}_i} Pu^{\omega} Q(x^{\mathscr{B}_i}, \xi^{\omega})$
4: **end for**
5: Aggregation: $\hat{x}^{\upsilon} \leftarrow \sum_{\mathscr{B}} P_{\mathscr{B}_i} x^{\mathscr{B}_i,\upsilon}$
6: $Err \leftarrow \sum P_{\mathscr{B}_i} \|x^{\mathscr{B}_i,\upsilon} - \hat{x}^{\upsilon}\|$
7: **while** $Err \geq \varepsilon$ **do**
8: $\quad \upsilon \leftarrow \upsilon + 1$
9: $\quad W^{\upsilon}_{\mathscr{B}_i} \leftarrow W^{\upsilon-1}_{\mathscr{B}_i} + \rho^{\intercal}(x^{\mathscr{B}_i,\upsilon-1} - \hat{x}^{\upsilon-1})$
10: $\quad$ **for** $\forall \mathscr{B}_i \in \mathscr{B}$ **do**
11: $\quad\quad x^{\mathscr{B}_i,\upsilon} \leftarrow argmin\, \zeta^{\intercal} x^{\mathscr{B}_i} + \sum_{\omega \in \mathscr{B}_i} Pu^{\omega} Q(x^{\mathscr{B}_i}, \xi^{\omega}) + W^{\upsilon}_{\mathscr{B}_i}{}^{\intercal} x^{\mathscr{B}_i} + \frac{\rho^{\intercal}}{2}(x^{\mathscr{B}_i} - \hat{x}^{\upsilon-1})^2$
12: $\quad$ **end for**
13: $\quad$ Aggregation: $\hat{x}^{\upsilon} \leftarrow \sum_{\mathscr{B}} P_{\mathscr{B}_i} x^{\mathscr{B}_i,\upsilon}$
14: $\quad Err \leftarrow \sum P_{\mathscr{B}_i} \|x^{\mathscr{B}_i,\upsilon} - \hat{x}^{\upsilon}\|$
15: **end while**

Fig. 3 Progressive hedging algorithm with bundled scenarios

pendently, so they can be solved in parallel. Lines 13 and 14 update the calculated average value for $\boldsymbol{x}$ and Err, respectively.

Stochastic unit commitment (Ryan et al. 2013), and transmission planning (Majidi-Qadikolai and Baldick 2018; Munoz and Watson 2015) are examples of PH algorithm application in power system. Crainic et al. (2014) used PH for commodity network design, and in (Escudero et al. 2012), PH algorithm is used for solving multi-stage stochastic mixed-integer problems.

3.4.3 Hybrid Decomposition

Hybrid decomposition uses both horizontal and vertical decomposition techniques to solve a large-scale stochastic optimization problem (Majidi-Qadikolai and Baldick 2018). It applies PH decomposition to horizontally decompose the original problem first, and then BD is used to vertically decompose each subproblem.

In PH algorithm (Fig. 3), extensive form of the problem is solved in lines 3 and 11. However, for very large-scale problems, solving the extensive form of these subproblems can also be computationally expensive. In the hybrid method, optimization subproblems in lines 3 and 11 of Fig. 3 will be solved using the BD algorithm. It divides the original problem into smaller subproblems to keep the original problem computationally tractable, and both PH and BD simulations can be distributed between multiple machines and be solved in parallel (see Sects. 4.3.5 and 4.3.6 for more discussion).

4 A Generalized Framework for Stochastic TEP Studies

A generalized decomposition framework for solving stochastic TEP studies for networks with different sizes, proposed in (Majidi-Qadikolai and Baldick 2018), is reviewed in this section. This framework is scalable, configurable, and easily maintainable.

4.1 Framework Overview

The framework is designed to be flexible and configurable for different problem sizes on different machines. It can be configured to solve a problem in extensive form (EF), or using PH, BD, and hybrid techniques (by setting its parameters) that provides more flexibility from the modeling perspective. The proposed framework can be summarized as follows:

Phase 0: Data preparation

Step 1: Input data and setting parameters
Input data includes the base network, scenarios, and candidate lines list. In this step, the planner configures the framework by setting its parameters; i.e. the number of scenarios in each bundle (κ) and the type of decomposition technique that should be used (PH, BD or Hybrid) for phases I and II. Settings for phase II can be modified later in step 4 if it is necessary.

Phase I: TEP without contingency analysis

Step 2: Scenario bundling
In this step, OPF for the base (existing) network is solved and calculated load shedding and wind curtailment will be used to develop an attribute for scenario bundling. After developing appropriate criteria, bundles of scenarios are formed (see subsection 4.2).

Step 3: Solving TEP
In this step, based on inputs from step 1 and bundles from step 2, TEP for normal operation states is solved. This step can be parallelized.

Phase II: TEP with contingency analysis
This phase is run if contingency analysis should be integrated in the TEP process.

Step 4: Scenario Bundling
Based on parameter settings, the scenario bundling method can be used to bundle scenarios.

Step 5: Solving TEP with contingency analysis
In this step, TEP with contingency analysis is solved. Either PH, BD, or hybrid may be used for solving this large-scale optimization problem. This step can be parallelized if PH and/or BD are selected as the solving algorithm. The contingency constraint reduction technique developed in (Majidi-Qadikolai and Baldick 2016a, b), can be used for solving TEP for each subproblem in this step.

Phase III: Quantifying the quality of results
If PH or hybrid is selected for phase I and/or II, then it will be necessary to find optimality gap to quantify the quality of results.

Step 6: Calculating a lower bound answer
In this step, the proposed lower bound formulation for PH in (Gade et al. 2016) is used to calculate a lower bound.

Step 7: Calculate optimality gap
The optimality gap (ε) can be calculated using the upper bound from step 5 (or step 3 in case of TEP without contingency analysis) and the lower bound from step 6. The selected plan is $\varepsilon - suboptimal$.

The framework is summarized in the flowchart in Fig. 4.

Fig. 4 Flowchart of the generalized framework

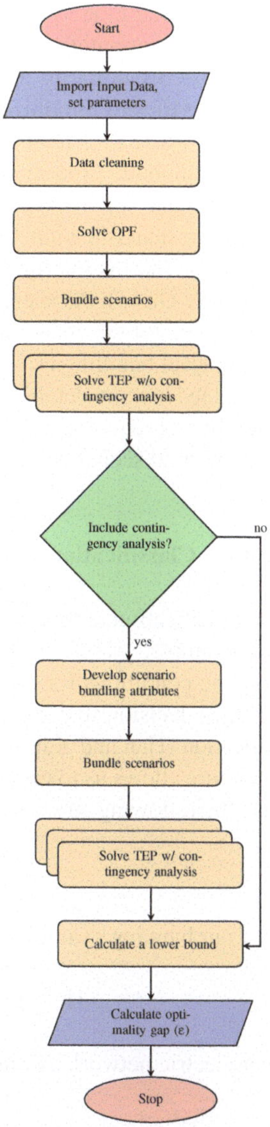

4.2 *Scenario Bundling*

The main purpose of scenario bundling is to create heterogeneous groups of scenarios with minimum dissimilarity *between* the groups collectively (based on selected attributes/criteria) and with relatively the same computational burden. Having similar bundles will improve the performance of PH algorithm by facilitating convergence of non-anticipativity constraints, as for a set of identical groups of scenarios, PH only needs one iteration to converge (although the choice of bundling does not necessarily reduce computational time). In contrast to clustering in which the objective is to minimize dissimilarity *within* groups (by forming homogeneous groups), scenario bundling tries to minimize dissimilarity *between* groups (see Majidi-Qadikolai and Baldick (2018) for mathematical formulation). As finding such a grouping can be computationally expensive, Majidi-Qadikolai and Baldick (2018) developed a heuristic method to solve this problem faster. This method bundles scenarios through three steps, i.e., classification, clustering, and grouping into bundles. The clustering step divides scenarios into multiple classes based on defined criteria (it is computationally very cheap). As scenarios in each class are clustered separately, the computational time for the clustering step is reduced. The grouping step allows integration of group level bundling criteria while forming heterogeneous bundles. These steps are explained in more detail in the following subsections. It should be noted that scenario bundling is required only if $1 < \kappa < |\Omega|$, where κ is the size of each bundle, Ω is the set of all scenarios, and $|\Omega|$ represents the size of this set.

4.2.1 Classification

In classification, a model or classifier is constructed to predict class labels such as, for example, "safe" or "risky" for bank loan application, or "light" and "heavy" loading conditions for electric networks. There are different classification methods such as decision tree induction, Bayes classification methods, and rule-based classification (Han and Kamber 2011). The rule-based method is used here, because its structure allows us to easily integrate expert knowledge into the bundling process. It has the following structure:

$$\textbf{IF} \;\; Condition \;\; \textbf{THEN} \;\; Conclusion \tag{66}$$

For our banking example, it can be written as

$$\textbf{IF} \;\; age \leq 25 \; AND \; student \;\; \textbf{THEN} \;\; Safe$$

For electric network example, we can have

$$\textbf{IF} \; average\,line\,loading \geq 50\% \; \textbf{THEN} \; Heavily\,loaded\,network$$

Rule-based classification will partition the original scenario set Ω into a finite number of non-empty classes $\mathscr{I} = \{\mathscr{I}_1, \ldots, \mathscr{I}_q\}$.

Different classification rules can be defined depending on the purpose of a study. For numerical analysis in Sect. 5, the number of important lines for contingency analysis (ICLs) can be used as a classifier in step 4. It might be necessary to adjust the number of scenarios in classes (those that are close to boundaries) for feasibility of the clustering step. Classification is an optional part of the bundling process, and if there is no classifier, then there will be only one class that includes all scenarios ($\mathscr{I} = \{\mathscr{I}_1\}$).

4.2.2 Clustering

Clustering is the process of grouping a set of objects in a way that objects within a cluster have the highest similarity. In this step, scenarios in each class ($\mathscr{I}_i$) are clustered based on selected attribute/developed criteria, and form the set $\mathscr{S}^i = \{\mathscr{S}_1^i, \ldots, \mathscr{S}_c^i\}$. Without loss of generality, scenarios are clustered in clusters with the same size, and the size of each cluster ($\mathscr{C}_s$) can be calculated from the following equation.

$$\mathscr{C}_s = \frac{|\Omega|}{\kappa} \tag{67}$$

where we assume that $|\Omega|$ is divisible by κ.

It is important to choose an attribute/criteria that is appropriate for the purpose of the study and provides insight for grouping phase. For example, for TEP without contingency analysis (step 3 of the framework), load shedding and wind curtailment penalties are major factors driving transmission expansion plans as they will be curtailed only if there is not enough transmission capacity to transfer their output (for wind) and/or supply them (for demand). Therefore, a weighted sum of load shedding and wind curtailment (LW) can be defined as a clustering attribute for this step. For phase II of the framework, TEP with contingency analysis is solved in step 5. As contingencies can have huge impact on selected transmission expansion plan (Majidi-Qadikolai and Baldick 2016b), important contingency list can be used to form an attribute for scenario clustering in this step.

Partitioning method is used to create clusters based on defined attributes. The objective of this clustering optimization problem is to minimize the distance between different attributes of objects (scenarios here) in a cluster. For step 2, scenarios with closest LW values are clustered together, and for step 4, scenarios with highest similarity in their important contingency lists will be clustered together (see Majidi-Qadikolai and Baldick (2018) for mathematical formulation).

4.2.3 Grouping into Bundles

In the last step, members of each cluster are distributed between groups (bundles) with the objective of minimizing dissimilarity *between* groups (by forming heterogeneous bundles). For the scenario set Ω, a bundle set $\mathscr{B} = \{\mathscr{B}_1, \ldots, \mathscr{B}_b\}$ of non-empty and mutually exclusive subsets ($\forall i \neq j,\ \mathscr{B}_i \bigcap \mathscr{B}_j = \emptyset$ and $\bigcup_j \mathscr{B}_j = \Omega$) is formed.

Scenarios in each cluster share similar characteristics (attributes used for classification and clustering). Therefore, one can form bundles of heterogeneous scenarios by randomly distributing members of each cluster between bundles. It is also possible to define new criteria for grouping in this step.

For example, for phase I of the framework, scenarios can be distributed between groups with the objective of minimizing the distance between aggregated LW values ($LW_{\mathscr{B}_i}$) between groups (bundles) because this attribute has a major impact on the TEP in step 3. For step 4 of the framework, total number of operational states (N_s) in each bundle can be used as a grouping attribute because it has a huge impact on computational time requirement for each bundle, and forming bundles with relatively the same computational burden will improve the performance of parallelizing in PH algorithm (see Sect. 5 for numerical results).

As a separate stochastic TEP is solved for each bundle in PH algorithm, the probability of each scenario should be updated based on Equations (68) and (69):

$$P_{\mathscr{B}_i} = \sum_{\omega \in \mathscr{B}_i} P^{\omega} \quad \forall \mathscr{B}_i \in \mathscr{B} \tag{68}$$

$$Pu^{\omega} = \frac{P^{\omega}}{P_{\mathscr{B}_i}} \quad \forall \omega \in \mathscr{B}_i, \forall \mathscr{B}_i \in \mathscr{B} \tag{69}$$

$$|\Omega| = \sum_{\mathscr{B}_i \in \mathscr{B}} |\mathscr{B}_i| \tag{70}$$

$$\sum_{\mathscr{B}_i \in \mathscr{B}} P_{\mathscr{B}_i} = 1 \tag{71}$$

where P^{ω} is the original probability of scenario ω, $P_{\mathscr{B}_i}$ is probability of bundle $\mathscr{B}_i$ in set of bundles $\mathscr{B}$, and Pu^{ω} is updated probability of scenario ω as a member of bundle $\mathscr{B}_i$. Equations (70) and (71) enforce scenario bundling to be mutually exclusive.

4.3 Model Performance Discussion

In this section, different factors affecting the performance of the framework are investigated.

4.3.1 Parameter Settings for the Framework

The size of each bundle (κ) and the choice of a decomposition method are set in step 1 in the framework (see Sect. 4.1). Table 1 shows different possible combinations for setting these two parameters. For the PH algorithm, by setting $\kappa = 1$ a standard PH is solved, $1 < \kappa < |\Omega|$ will result in a bundled PH, and $\kappa = |\Omega|$ is equivalent to solving the extensive form (EF) of the optimization problem. If BD is selected as the solving method, then for $1 \leq \kappa < |\Omega|$, the problem is solved separately for each bundle, and a heuristic method should be used to select a unique first stage answer. For $\kappa = |\Omega|$, a standard BD is solved. When hybrid method is selected, for $1 \leq \kappa < |\Omega|$, both PH

Table 1 Different parameter settings for the framework

	PH	BD	Hybrid
$\kappa = 1$	PH	Heuristic	Hybrid
$1 < \kappa < \|\Omega\|$	PH	Heuristic	Hybrid
$\kappa = \|\Omega\|$	EF	BD	BD

and BD are used for solving the problem in steps 3 and/or 5 in the framework. For $\kappa = |\Omega|$, hybrid method will be the same as BD method. These parameters can be set independently for phases I and II providing more flexibility, potentially improving the effectiveness of the framework.

4.3.2 Factors Affecting the Choice of Parameters

The size of the problem, the design of decomposition algorithms, existing hardware infrastructure, and solvers are critical for making a decision about setting parameters for the framework. These factors are briefly overviewed in the following.

- The size of the problem (d)
  The number of structural constraints (SC), Equations (5)–(8), continuous (CV) and binary (BV) decision variables are main factors for the size of the TEP optimization problem. For the extensive form of this TEP formulation from Sect. 3.1 (depending on the choice and design of decomposition algorithms, new variables and constraints may be added), these values can be calculated from the following equations:

$$d = \{SC, CV, BV\} \tag{72}$$

$$SC = (2 \times (|N_b| + |N_l|) \times |N_s^{\omega}| + |N_{wg}|) \times |\Omega| \tag{73}$$

$$CV = ((2 \times |N_b| + |N_l|) \times |N_s^{\omega}| + |N_g| + |N_{wg}|) \times |\Omega| \tag{74}$$

$$BV = |N_n| \tag{75}$$

  If no contingency constraint reduction technique is used, then $|N_s^{\omega}| = |N_l| + 1$ to model outage of each branch.
- Design of decomposition algorithms
  PH and BD are not black-box software packages with input and output vectors. These algorithms are designed based on specific needs and conditions. For BD, there are several different designs such as standard BD (Benders 1962), multi-cuts BD (Birge and Louveaux 1988), and nested BD (Roger Glassey 1973), and each design can be configured differently. For PH, either the standard form (Rockafellar and Wets 1991) or the bundled form (Wets 1989) might be used. Similar to BD, there are several internal settings for PH that can affect the performance of this algorithm.

- Existing hardware infrastructure
  The machine that is used to solve the TEP problem has an undeniable impact on the choice of a decomposition algorithm and the size of each bundle (κ). Machines with high computing power are usually capable of solving larger problems that make it possible to choose bundled PH with a large bundle size (κ). In the case of using multiple machines (or virtual machines for Cloud-based workstations), implemented parallel computation structure will be another key factor.
- Solvers
  The main feature of a solver that affects the choice of parameters for the framework is its capability to distribute computation burden over multiple cores of a CPU and use all computing power of the machine. GUROBI and CPLEX are examples of commercial solvers with this capability.

As discussed above, there are several factors that can affect hardware and software design of this framework. For a designed framework, running a few individual simulations can provide a relatively good insight about the performance of each module, and help on setting parameters for the framework.

4.3.3 PH Performance Improvement

Several heuristics such as finding appropriate values for $\boldsymbol{\rho}$, variable freezing, cyclic behavior detection, and terminating PH when the number of remaining unconverged variables is small can be used to improve the performance of the PH algorithm (Watson and Woodruff 2011). In the following, some of these heuristic methods are reviewed in detail.

- Choice of $\boldsymbol{\rho}$: A good approximation for $\boldsymbol{\rho}$ is important for the PH algorithm to perform well. As shown in Fig. 3, the value of multiplier vector ($\boldsymbol{W}^{\upsilon}_{\mathscr{B}_i}$) is updated using penalty vector $\boldsymbol{\rho}$, and an appropriate multiplier vector can affect the number of required iterations for PH convergence, and the quality of the lower bound answer (Gade et al. 2016). In Watson and Woodruff (2011), different heuristic methods for calculating effective values for $\boldsymbol{\rho}$ are proposed. Our experience with those methods shows that for the TEP problem using the following equation from Watson and Woodruff (2011) results in a better convergence rate.

$$\rho_l = \frac{\zeta_l}{x_l^{max} - x_l^{min} + 1} \tag{76}$$

  where ρ_l is the lth element of vector $\boldsymbol{\rho}$, and

$$x_l^{max} = \max_{\mathscr{B}_i \in \mathscr{B}} x_l^{\mathscr{B}_i} \tag{77}$$

$$x_l^{min} = \min_{\mathscr{B}_i \in \mathscr{B}} x_l^{\mathscr{B}_i} \tag{78}$$

For values of ρ_l close to the unit cost of its associated variable, the PH algorithm should have a better performance both from convergence speed and quality of results. Selecting higher values for ρ_l will increase convergence rate but may negatively affect the quality of results. On the other hand, very small values for ρ_l can improve the quality of results (by decreasing optimality gap), but can significantly increase the number of iterations and simulation time.

- Variable Freezing: To improve the convergence of PH algorithm, the *variable freezing* technique can be used. Based on this technique, first stage decision variables with values that did not change over the past ϑ iterations are frozen for future iterations. For example, for a case with 5 bundles and $\vartheta = 4$, the value of the decision variable x_l is frozen if for all 5 bundles during all 4 successive iterations $\upsilon + 1, \upsilon + 2, \upsilon + 3, \upsilon + \vartheta = \upsilon + 4$, its value did not change and was the same across all bundles ($x_l^{\upsilon+1,1} = \cdots = x_l^{\upsilon+4,5}$).
The impact of freezing variables can be investigated from two perspectives, namely simulation time and the selected plan.

  - Impact on simulation time
  By freezing binary variables, total number of binary decision variables is decreased as frozen variables have fixed values. It improves the performance of the algorithm by decreasing computational time for each iteration (as a TEP optimization problem with fewer binary variables will typically be solved faster) and reducing the number of iterations (as a PH problem with fewer non-anticipativity constraints will typically converge faster).
  - Impact on the selected plan
  When a decision variable is frozen, the implicit assumption is that its value will not change during subsequent iterations, but this assumption may not always be valid. Therefore, the selected plan might be negatively affected when variable freezing technique is used, especially for small values of ϑ like 1 or 2. By using more conservative values for ϑ, this effect can be mitigated.

The selected plan will be more sensitive to a small value for ϑ when there are several relatively similar candidate lines (in terms of cost and/or electric parameters) in a geographically limited area. For a large-scale network in which candidate lines are widely spread, a smaller value for ϑ can be selected.
Using the variable freezing technique may result in situations with only a very few unfrozen decision variables. Then PH can be terminated (to decrease the number of iterations), and the TEP with remaining binary variables solved in the extensive form or using a BD algorithm.

- Identical Parallel Candidate Lines: We have also noticed that having two (or more) identical parallel candidate lines can result in an unnecessary nonzero values of Err on lines 6 and/or 14 in PH algorithm (Fig. 3) when only one of those lines is selected as a part of expansion plan. We recommend to slightly modify the investment cost for otherwise identical lines to break the symmetry.

4.3.4 Optimality Gap

The optimality gap is used as a measure for quantifying the quality of results in an optimization-based TEP. Based on Table 1, the TEP problem is solved using one of these five methods, i.e., heuristic, extensive form (EF), PH, BD, and hybrid. For parameter settings that will result in a heuristic method, the optimality gap cannot be calculated to quantify the quality of results. For the EF method, the optimality gap of the final result will be less than or equal to the solver's setting for maximum optimality gap. For BD, achieving the optimality gap is set as the stopping criterion; therefore, for EF and BD methods, it is possible to guarantee a pre-defined optimality gap (assuming that the algorithm successfully terminates). On the other hand, for PH and hybrid methods, the optimality gap is calculated after the algorithm is terminated to quantify the quality of final results, and there is no guarantee that the final optimality gap will be less than or equal to a pre-defined threshold. As discussed in Sect. 4.3.3, using appropriate values for $\boldsymbol{\rho}$ and setting a conservative value for ϑ can improve the optimality gap of the PH algorithm.

4.3.5 Scalability and Maintainability

Scalability is one of the main features of this framework. Figure 5a shows the size of the EF of a stochastic TEP problem with security constraints. In this Fig., d^{ω} represents the size of the TEP problem for scenario ω ($d^{\omega} = \{SC^{\omega}, CV^{\omega}, BV^{\omega}\}$).

$$SC^{\omega} = 2 \times (|N_b| + |N_l|) \times |N_s^{\omega}| + |N_{wg}| \tag{79}$$

$$CV^{\omega} = (2 \times |N_b| + |N_l|) \times |N_s^{\omega}| + |N_g| + |N_{wg}| \tag{80}$$

$$BV^{\omega} = |N_n| \tag{81}$$

For a sample case with 6000 buses, 8000 existing branches, 500 conventional power plants, 100 wind farms, 100 candidate lines, and 10 scenarios, the size of the problem is $d^{\omega} = \{228.5M, 162.8M, 100\}$ when $|N_s^{\omega}| = 8101$ and $s = 10$ (M stands for million). Total size of the problem in Fig. 5a will be $d = \{2285M, 1628M, 100\}$. This problem is practically impossible to solve in the EF. There are constraint reduction techniques (Ardakani and Bouffard 2013; Madani et al. 2017; Majidi-Qadikolai and Baldick 2016a) that can be used to decrease the size of this problem. Let us assume using the VCL algorithm (Majidi-Qadikolai and Baldick 2016b) reduces the size of N_s^{ω} form 8101 to 50. The size of the EF of this problem will be $d = \{14M, 10M, 100\}$. Even after a massive problem size reduction, solving the EF of the problem still remains computationally extremely expensive.

The BD algorithm (shown in Fig. 5b) moves binary decision variables to the master problem and keeps all continuous variables in the subproblem. As the subproblem is a linear program, it is expected to be solved very fast; however, for the network in this example, the size of the subproblem will be $\{14M, 10M, 0\}$ which is not easy to solve especially when it should be solved at every BD algorithm iteration.

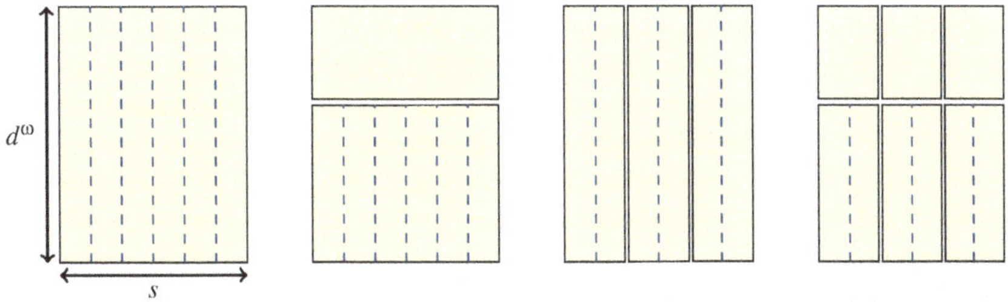

(a) Extensive Form (b) BD (c) Bundled PH (d) Hybrid

Fig. 5 Impact of different decomposition techniques, d^{ω}: size of the problem for scenario ω, s: the number of scenarios (6 for this example)

Figure 5c shows how bundled PH algorithm will decompose the problem. By creating bundles of two scenarios, the size of each subproblem for bundled PH will be $\{2.8M, 2.0M, 100\}$ (or $\{1.4M, 1.0M, 100\}$ for standard PH). Solving the extensive form of these subproblems might still be hard because of the large number of binary variables. In Fig. 5d, the hybrid method is used to decompose the problem both vertically and horizontally. By using this method, the size of each problem that needs to be solved in EF can be decreased up to $\{1.4M, 1.0M, 0\}$, which is a significant size reduction compared to $\{14M, 10M, 100\}$ for Fig. 5a.

The size of this case study may increase either by increasing the number of candidate lines or the number of scenarios. The BD feature of the hybrid method will keep us away from exponentially increasing computational time as a result of adding new binary variables, and the bundled PH feature will keep the size of each subproblem relatively unchanged even if the total number of scenarios is increased significantly (by increasing the number of bundles instead of increasing the size of each bundle). Therefore, the problem remains tractable, demonstrating the scalability of the proposed framework.

Another important feature of this framework (from practicality perspective) is its maintainability. Because it is module-based (BD algorithm, PH algorithm, bundling algorithm), each module can easily and (relatively) independently be upgraded as technology improves.

4.3.6 Parallelizing

With proper hardware, parallelizing decreases computational time for solving a series of independent simulations and improves scalability. Simulations in steps 3 and 5 in the framework can be parallelized, if PH, BD (with special configurations), or hybrid is selected to reduce elapsed time for solving TEP optimization problem by starting all simulations at the same time.

- PH algorithm: Based on PH algorithm for bundled scenarios shown in Fig. 3, lines 3 and 11 are run for each bundle (or each scenario in case of standard PH) independently. Therefore, we can parallelize both `for loops` (lines 2–4 and 10–12) in this algorithm and start all simulations in each `loop` at the same time to decrease computational time. It should be noted that lines 10–12 should be solved for each iteration of the PH algorithm, and decreasing computational time here can be rewarding from the performance improvement perspective. As shown in lines 5 and 13 in Fig. 3, the algorithm can proceed to the next step when all parallelized simulations are completed. In the bundling process, bundles should be developed that need relatively similar computational time, so that the framework can benefit the most from parallelizing.

- BD algorithm: For standard BD, in which one cut is sent to the master problem in each iteration, the subproblem is usually solved in extensive form. For multi-cuts BD (Birge and Louveaux 1988) and nested BD (Akbari et al. 2011; Khodaei et al. 2010; Roger Glassey 1973), it is possible to solve subproblems in parallel that will decrease computational time.
- Hybrid method: As hybrid algorithm uses both PH and BD to solve a problem, it can benefit from both vertical and horizontal decompositions and parallelize the problem-solving with both algorithms (if applicable). For example, by using bundled PH, the problem will be horizontally parallelized for each bundle $\mathscr{B}_i$. A nested BD can be used to solve each bundle, in which feasibility cuts for under contingency operation states can be created in parallel.

5 Case Study and Numerical Results

In this section, numerical analysis for three case studies from (Majidi-Qadikolai and Baldick 2016a, b, 2018; Majidi-Qadikolai et al. 2018) are presented. All simulations are done with a personal computer with 2.0-GHz CPU and 32 GB of RAM. MATLAB R2014a, YALMIP R20150626 package (Lofberg 2004), and GUROBI 5.6 (Gurobi Optimization, Inc 2014) are used as programming language, modeling tool and a solver respectively. To calculate the elapsed "Simulation Time," MATLAB built-in function `tic toc` is used. Steps 3 and 5 are parallelized using MATLAB built-in function `parfor` where PH is selected as a solving algorithm.

5.1 13-Bus Test System

This case study contains 13 buses, 33 existing lines, 16 power plants, 9 load centers, and 36 candidate lines with 100 scenarios to capture uncertainties in wind and load (Majidi-Qadikolai and Baldick 2016b) (shown in Fig. 6—See Appendix for details). A new line investment cost is assumed $\$1M$/mile, and load shedding is penalized at \$9000/MWh and \$500/MWh penalty for wind curtailment. This small case study with a large number of scenarios is used to demonstrate different steps of the framework. Table 2 shows developed case studies. The proposed method

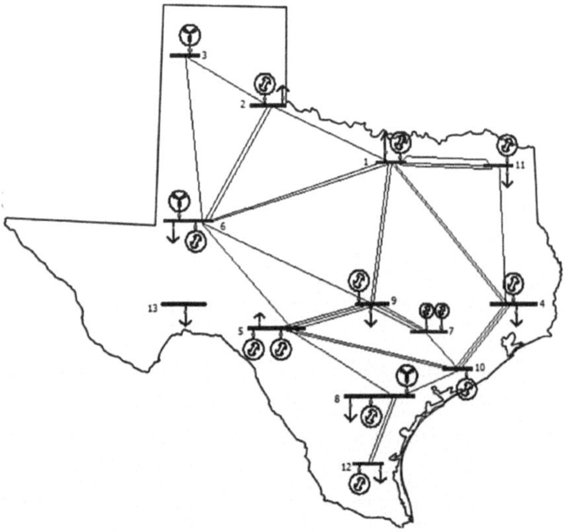

Fig. 6 13-bus system

Table 2 Case study definition

	Bundle size (κ)	Algorithm	Bundling Method
Case A	100	EF-Full	N/A
Case B	100	EF in (Majidi-Qadikolai and Baldick 2016a)	N/A
Case C	1	PH	N/A
Case D	20	PH	Random
Case E	20	PH	From Sect. 4.2

in (Majidi-Qadikolai and Baldick 2016a) is used for contingency constraint reduction for cases B–E.

In case A, the extensive form (EF) of two-stage stochastic TEP is solved without any constraint reduction. For case B, the proposed method in (Majidi-Qadikolai and Baldick 2016a) is used to reduce contingency constraints, and the EF of the reduced model is solved. Case C is a standard PH in which the size of bundles is set to 1. For case D, scenarios are bundled randomly with 5 bundles with size $\kappa = 20$ using MATLAB built-in function `randperm`. For case E, scenarios are bundled using the bundling method from Sect. 4.2.

5.1.1 PH Algorithm Settings

Values for ρ are calculated based on (76). Variables that are consistent across bundles and do not change over the most recent 4 iterations will be frozen at their values ($\vartheta = 4$). Moreover, if the number of remaining binary variables is less than or equal to 3, the PH algorithm is terminated, and the extensive form of the problem is solved for remaining decision variables. These settings are applied to cases C–E.

5.1.2 Model Performance Discussion

The simulation result for these five cases is summarized in Table 3. For case A, we were unable to get any results after 12 days. It shows that solving the EF of TEP with all constraints is not practical even for this small case study. For case B, the TEP optimization problem is solved in 25 min with 2.7% optimality gap. Standard PH in case C needs more than 2 hours to solve this problem, and the final result is 29.5%-suboptimal. It shows that the standard PH may not have a good performance when the number of scenarios is large. For Case D, bundling reduced computational time by 50% and optimality gap is dropped to 1.65%. For case E, computational time is reduced to 15 minutes, and the quality of results is significantly improved by decreasing optimality gap to 0.24%. The selected settings for framework for case E solves this problem more than 8 times faster than standard PH (case C) and 5 times faster than randomly bundled PH (case D). It also finds results with higher quality (optimality gap of 0.24% compared to 1.65% and 29.4% for randomly bundled PH and standard PH, respectively). From a computation time perspective, cases B and E are relatively similar, but the quantified quality of results is significantly different, and case E provides a better optimality gap in somewhat less time.

To further investigate the impact of parallelizing and variable freezing on computational time, we compared the performance of cases C–E under the following three alternatives:

Table 3 Summary of results for 13-bus system

	Case A	Case B	Case C	Case D	Case E
No. of added lines	–	16	21	17	16
Objective function ($b)	–	4.89	5.58	4.94	4.89
Simulation time (h)	288+	0.42	2.05	1.28	0.25
Optimality gap	–	2.7%	29.5%	1.65%	0.24%

Table 4 Impact of parallelizing and variable freezing on computational performance

		Alternative 1	Alternative 2	Alternative 3
Optimality gap	Case C	29.5%	0.85%	29.5%
	Case D	1.65%	0.13%	1.65%
	Case E	0.24%	0.12%	0.24%
Simulation time (h)	Case C	93.92	185.23	2.05
	Case D	7.38	132.97	1.28
	Case E	7.16	82.7	0.25

- *Alternative 1*: With variable freezing and without parallelizing
- *Alternative 2*: Without variable freezing and with parallelizing
- *Alternative 3*: With variable freezing and with parallelizing

Table 4 summarizes the impact of these two factors on optimality gap and computational time for cases C-E under these three alternatives.

The result from the second row shows that variable freezing may negatively affect the quality of results and increases the optimality gap (*Alternative 2*, in which variable freezing is ignored, has the lowest optimality gap). As expected, parallelizing will not affect the quality of results (similar optimality gaps for *Alternative 1* and *Alternative 3*). The third row in Table 4 shows the computational time for three alternatives. For *Alternative 1*, standard PH (Case C) is affected the most (compared to cases D and E) when parallelizing is not used because each iteration includes running TEP for all individual scenarios (simulation time increased from 2.05 to 93.92 hours). For bundled PH, both cases D and E could solve the problem in approximately the same time showing that when simulations are run sequentially (instead of in parallel), the impact of balancing computational burden between bundles (that will result in an earlier termination for a parallelized `for loop`) will be less effective. Variable freezing has a significant impact on computational time as it will decrease both the number of iterations and computational time for each iteration. Comparing the computational time and optimality gap for *Alternative 2* and *Alternative 3* shows the trade-off between quality of results and computational time. For example, for case E, the optimality gap is slightly increased from 0.12% to 0.24%; however, the computational time is decreased from 82.7 hours to 0.25 hours demonstrating the effectiveness of the heuristic methods used for PH performance improvements.

5.2 Reduced ERCOT System

A reduced ERCOT network is developed with 3179 buses, 474 generation units, 3598 load centers, 123 wind farms, and 4458 branches. All non-radial 138kV and 345kV lines in the ERCOT network are explicitly modeled. Generators and loads that were connected to lower voltage levels or radial network are moved to nearby

Table 5 Summary of results for reduced ERCOT system

	Case A	Case B	Case C	Case D	Case E
No. of added lines	–	–	6	9	4
Objective function ($b)	–	–	8.102	8.230	8.007
Simulation time (days)	15+	15	9.2	14.9	2.78
Optimality gap	–	–	3.1%	6.24%	0.97%

modeled buses. Ten different scenarios are developed to model load and wind uncertainties (using historical data) with 46 new lines as candidates for transmission expansion (Majidi-Qadikolai and Baldick 2018). Similar to the 13-bus system, five cases A–E are simulated to compare the results. As total number of scenarios is 10 for this case, κ is set to 10 for cases A and B. For phase I in case E, $\kappa = 5$ and for case D and phase II in case E, $\kappa = 2$. The proposed method in (Majidi-Qadikolai and Baldick 2016a) is used to solve TEP in lines 3 and 11 of the bundled PH algorithm (Fig. 3). The parameter ϑ is set to 3. Other parameters are set the same as the 13-bus system.

Numerical results are given in Table 5. We could not get a feasible solution for cases A and B after 15 days, demonstrating the need for decomposition-based methods for large-scale problems. As the number of scenarios is not large for this system, standard PH (case C) has a reasonable performance; however, the elapsed time of over a week may not be acceptable. For case D (randomly bundled scenarios), simulation is terminated manually after 14.9 days and a lower bound is calculated. The fifth column (case E) demonstrates the impact of the proper framework design/setting on improving quality of results (decreasing optimality gap from 6.24% to 0.97%) and reducing computational time (by more than 5.3 times) for solving this large-scale problem.

Results for this case demonstrates that bundling by itself may not necessarily improve the performance of PH without careful consideration of choice of bundles, because as explained in Sect. 3.4.2, each iteration for the PH algorithm is finished only when TEP for all bundles are completely solved. Because of this, randomly grouping scenarios may result in forming TEP subproblems with significantly different sizes (based on (73) and (74)) although the size of bundles (κ) is similar. This comparison also highlights the importance of the grouping step in the scenario bundling.

5.3 Full ERCOT System—High Load Growth Area Project

For this case study, a full ERCOT network model is used, and an area of the transmission system with high load growth at existing load centers is evaluated (Majidi-Qadikolai et al. 2018). In this project, a fast-growing load pocket in Central Texas is studied with the assumption that load growth will increase current load forecasts by 50% in the near-term horizon. Transmission planners may look at varying load assumptions as a sensitivity scenario to their base case studies. The sensitivity case studies are useful to anticipate what transmission infrastructure may be required if certain less anticipated conditions unfold. In this case study, the on-peak condition is evaluated. An initial list of candidate upgrades (including 23 lines and transformers) are identified by the transmission planning team. The area of study for this project and candidate options (doted lines) are shown in Fig. 7.

We have evaluated this project under the following conditions:

- Without low-cost option
- With low-cost option

"Low-cost option" refers to minor transmission upgrades that are not known/ included in the initial candidate lines list, but during practical TEP studies might be captured by planning experts and are added to their analysis (Majidi-Qadikolai et al. 2018).

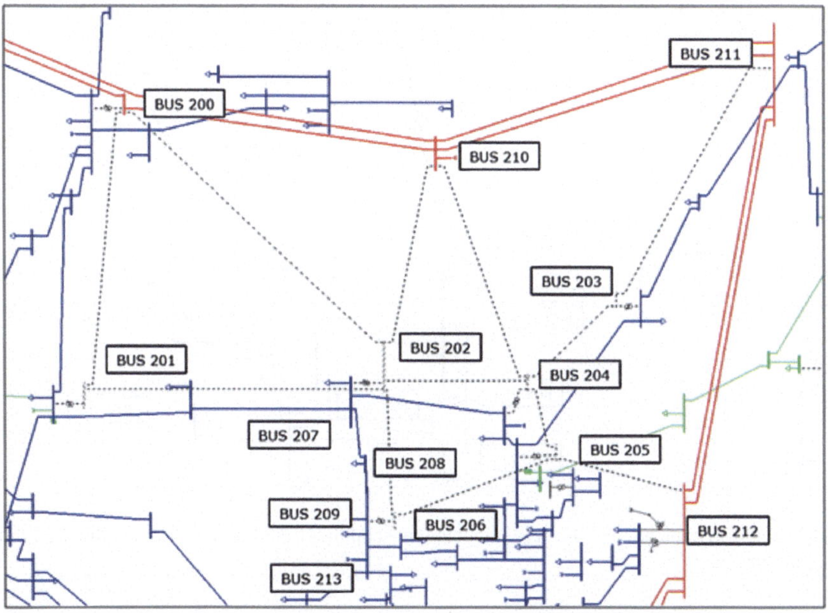

Fig. 7 Area of study for high area load growth project. Solid lines show existing branches (Red: 345 kV, Blue: 138 kV, Green: 69 kV). Dotted lines show candidate branches for expansion

Table 6 Summary of results for high load growth area project without low-cost options

	From	To	ID	Length (miles)
New line	202	210	1	22
New line	202	206	1	11.4
New transformer	202	207	1	NA
New transformer	206	209	1	NA
Total investment		105.4		(millions dollars)

5.3.1 Case A: Ignoring Low-Cost Options

If we ignore the possibility of "low-cost options" and use standard TEP optimization formulation (from Sect. 3.1) with preliminary candidate options, the TEP optimization tool selects two new lines and two transformers as the optimal expansion plan. The summary of results is shown in Table 6, and selected branches are highlighted (solid brown lines) in Fig. 8.

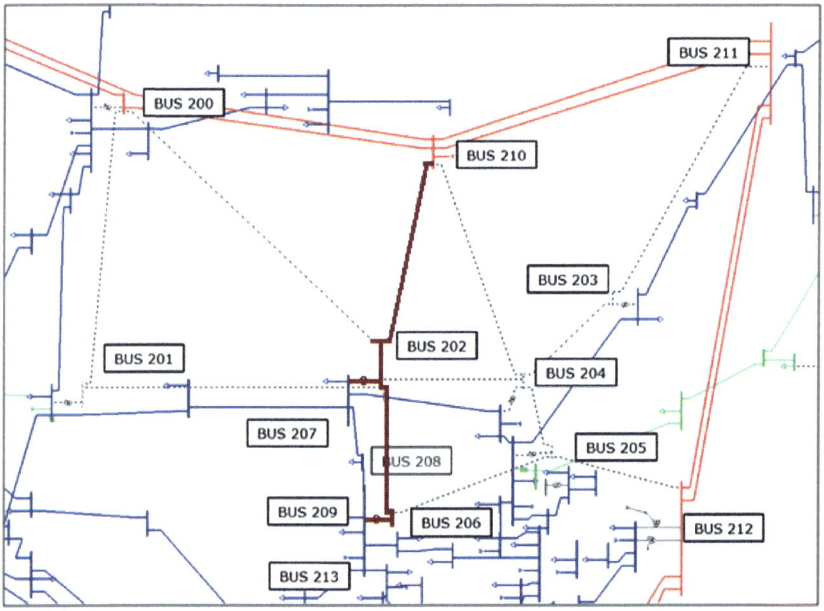

Fig. 8 Selected branches for case A (without low-cost options). New branches are highlighted with solid brown lines

Table 7 Summary of results for high load growth area project with low-cost options

	From	To	ID	Length (miles)
New line	202	210	1	22
New transformer	202	207	1	NA
Upgrade line	207	208	1	5.61
Upgrade line	209	208	1	1.96
Total investment		68.38		(millions dollars)

5.3.2 Case B: Integrating Low-Cost Options

For the second case, "low-cost options" feature is integrated into TEP formulation to capture potential upgrades in local area that might not be part of the initial candidate options. The summary of results for this case is provided in Table 7, and selected branches are highlighted in Fig. 9.

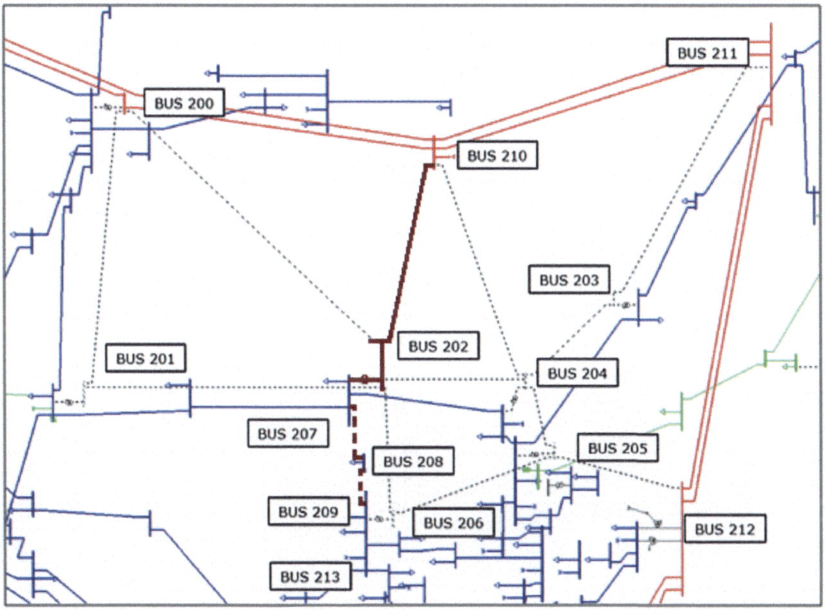

Fig. 9 Selected branches for case B (with low-cost options). New branches are highlighted with solid brown lines, and low-cost upgrades are highlighted with dotted brown lines

As shown in this Table, compared to case A, the TEP tool in this case suggested upgrading two existing lines as low-cost options instead of building one new transmission line and one new transformer. The planning engineers confirmed that these upgrades are practically possible, and they have added two new upgrades to the primary candidate list. Those upgrades are selected by TEP optimization tool when we rerun step 2. The selected plan in case B is around $35 million less expensive than the selected plan in case A (it is a fair comparison because in practice it is not possible to define all potential upgrades at the beginning of each study). Although the result of case A is optimal from mathematical perspective, the results of case B for this project demonstrates that a combination of advantages of optimization-based approaches and expertise of transmission engineers can lead to a better expansion plan.

Appendix

In this appendix, input date for 13-bus system is provided (Tables 8, 9, and 10).

Table 8 Load and generation data in (MW)

Bus	Gen	Load	Wind
1	21,374	19,519	0
2	2811	403	0
3	0	0	3000
4	24,292	20,895	0
5	8233	5066	0
6	6216	4509	4000
7	1208	0	0
8	5881	3755	1000
9	4657	7125	0
10	2750	0	0
11	3262	465	0
12	2503	2862	0
13	0	1000	0

Table 9 Existing transmission network data

From	To	Susceptance (P.U.)	Capacity (MW)
2	1	13.89	1000
1	4	8.20	625
1	4	8.20	625
1	6	8.85	812.5
6	1	8.85	912.5
1	9	11.11	875
1	9	11.11	937.5
1	11	15.87	1125
1	11	15.87	1125
1	11	15.87	1125
3	2	13.33	1062.5
2	6	12.35	1125
6	2	12.35	1125
3	6	9.26	875
4	10	27.78	1125
4	10	27.78	1125
4	10	27.78	1125
11	4	9.62	1000
6	5	8.55	937.5
8	5	15.87	812.5
9	5	25.00	1750
9	5	25.00	1750
5	9	25.00	1750
5	10	12.35	875
5	10	12.35	812.5
6	9	8.55	875
9	7	34.48	1250
9	7	34.48	1250
9	7	34.48	1250
7	10	22.22	1750
8	10	16.95	875
8	12	37.04	1312.5
8	12	37.04	1312.5

Table 10 Candidate lines

From	To	Susceptance (P.U.)	Capacity (MW)	Length (mile)
2	1	13.89	1000	144
2	1	13.89	1000	144
1	4	8.20	625	243
1	4	8.20	625	243
1	6	8.85	812.5	225
6	1	8.85	812.5	225
1	11	15.87	1125	126
3	2	13.33	1062.5	150
3	2	13.33	1062.5	150
2	6	12.35	1125	162
6	2	12.35	1125	162
3	6	9.26	875	216
3	6	9.26	875	216
4	10	27.78	1125	72
11	4	9.62	1000	207
6	5	8.55	937.5	234
6	5	8.55	937.5	234
8	5	15.87	812.5	126
9	5	25.00	1750	81
9	5	25.00	1750	81
6	9	8.55	875	234
6	9	8.55	875	234
7	10	22.22	1750	90
8	10	16.95	875	117
8	10	16.95	875	117
8	12	37.04	1312.5	108
8	12	37.04	1312.5	108
13	6	13.00	1125	173
13	5	20.05	1125	112.2
13	9	10.80	875	208.3
13	6	13.00	1125	173
13	5	20.05	1125	112.2
13	9	10.80	875	208.3
13	6	13.00	1125	173
13	5	20.05	1125	112.2
13	9	10.80	875	208.3

References

A. Abiri-Jahromi, F. Bouffard, On the loadability sets of power systems-part I: characterization. IEEE Trans. Power Syst. **32**(1), 137–145 (2017)

A. Abiri-Jahromi, F. Bouffard, On the loadability sets of power systems-part II: minimal representations. IEEE Trans. Power Syst. **32**(1), 146–156 (2017)

T. Akbari, A. Rahimikian, A. Kazemi, A multi-stage stochastic transmission expansion planning method. Energy Convers. Manage. **52**(8–9), 2844–2853 (2011)

N. Alguacil, A.L. Motto, A.J. Conejo, Transmission expansion planning: a mixed-integer LP approach. IEEE Trans. Power Syst. **18**(3), 1070–1077 (2003)

J. Alvarez Lopez, K. Ponnambalam, V.H. Quintana, Generation and transmission expansion under risk using stochastic programming. IEEE Trans. Power Syst. **22**(3), 1369–1378 (2007)

A.J. Ardakani, F. Bouffard, Identification of umbrella constraints in DC-based security-constrained optimal power flow. IEEE Trans. Power Syst. **28**(4), 3924–3934 (2013)

M. Armstrong, D.E.M. Sappington, *Recent Developments in the Theory of Regulation, volume 3 of Handbook of Industrial Organization*, Chap. 27 (Elsevier, Amsterdam, 2007), pp. 1557–1700

L. Bahiense, G.C. Oliveira, M. Pereira, S. Granville, A mixed integer disjunctive model for transmission network expansion. IEEE Trans. Power Syst. **16**(3), 560–565 (2001)

R. Baldick, K. Dixit, T.J. Overbye, Empirical analysis of the variation of distribution factors with loading, in *Power Engineering Society General Meeting, 2005*, vol. 1 (IEEE, New York, June 2005), pp. 221–229

J.F. Benders, Partitioning procedures for solving mixed-variables programming problems. Numer. Math. **3**, 238–252 (1962)

D. Bertsimas, D.B. Brown, C. Caramanis, Theory and applications of robust optimization. SIAM Review **53**(3), 464–501 (2011)

S. Binato, M.V.F. Pereira, S. Granville, A new Benders decomposition approach to solve power transmission network design problems. IEEE Trans. Power Syst. **16**(2), 235–240 (2001)

J.R. Birge, F.V. Louveaux, A multicut algorithm for two-stage stochastic linear programs. Eur. J. Oper. Res. **34**, 384–392 (1988)

G. Blackmon, *Incentive Regulation and the Regulation of Incentives*, 1st edn. (Springer, New York, NY, 1994)

M.O. Buygi, G. Balzer, H.M. Shanechi, M. Shahidehpour, Market-based transmission expansion planning. IEEE Trans. Power Syst. **19**(4), 2060–2067 (2004)

M. Carrion, J.M. Arroyo, N. Alguacil, Vulnerability-constrained transmission expansion planning: A stochastic programming approach. IEEE Trans. Power Syst. **22**(4), 1436–1445 (2007)

E.B. Cedeño, S. Arora, Performance comparison of transmission network expansion planning under deterministic and uncertain conditions. Int. J. Electr. Power Energy Syst. **33**(7), 1288–1295 (2011)

G. Chen, Z.Y. Dong, D.J. Hill, Transmission network expansion planning with wind energy integration: a stochastic programming model, in *Power and Energy Society General Meeting, 2012* (IEEE, New York, July 2012), pp. 1–10

A.J. Conejo, E. Castillo, R. Minguez, R. Garcia-Bertrand, *Decomposition Techniques in Mathematical Programming: Engineering and Science Applications* (Springer, Berlin, 2006)

T.G. Crainic, M. Hewitt, W. Rei, Scenario grouping in a progressive hedging-based meta-heuristic for stochastic network design. Comput. Oper. Res. **43**, 90–99 (2014)

Electric Reliability Council of Texas. ERCOT planning guide, section 4: Transmission planning criteria (2014)

ERCOT System Planning. 2014 Long-Term System Assessment for the ERCOT Region (2014)

ERCOT System Planning. 2016 Long-Term System Assessment for the ERCOT Region (2016)

L.F. Escudero, M.A. Garín, M. Merino, G. Pérez, An algorithmic framework for solving large-scale multistage stochastic mixed 0–1 problems with nonsymmetric scenario trees. Comput. Oper. Res. **39**(5), 1133–1144 (2012)

FERC. Federal energy regulatory commission (2019)

D. Gade, G. Hackebeil, S.M. Ryan, J.-P. Watson, R.J.-B. Wets, D.L. Woodruff, Obtaining lower bounds from the progressive hedging algorithm for stochastic mixed-integer programs. Math. Program. **157**(1), 47–67 (2016)

GAMS. Scenred: Scenario reduction algorithms (2002)

R. Garcia-Bertrand, R. Minguez, Dynamic robust transmission network expansion planning. IEEE Trans. Power Syst. **99**, 1 (2016)

L.L. Garver, Transmission network estimation using linear programming. IEEE Trans. Power Apparatus Syst. **PAS–89**(7), 1688–1697 (1970)

S. Granville, M. Pereira, G. Dantzig, B. Ivi-Itzhak, M. Avriel, A. Monticelli, L. Pinto, *Mathematical Decomposition Techniques for Expansion Planning Vol 2: Analysis of the Linearised Power Flow Model Using the Bender Decomposition Technique*. EPRI Technical Report EL-5299 (1988)

Gurobi Optimization, Inc. Gurobi optimizer reference manual (2014)

J. Han, M. Kamber, J. Pei, *Data Mining: Concepts and Techniques*, 3rd edn. (Elsevier, Amsterdam, 2011)

K.W. Hedman, R.P. O'Neill, E.B. Fisher, S.S. Oren, Optimal transmission switching with sensitivity analysis and extensions. IEEE Trans. Power Syst. **23**(3), 1469–1479 (2008)

M.R. Hesamzadeh, J. Rosellon, S.A. Gabriel, I. Vogelsang, A simple regulatory incentive mechanism applied to electricity transmission pricing and investment. Energy Econ. **75**, 423–439 (2018)

W. Hogan, J. Rosellon, I. Vogelsang, Toward a combined merchant-regulatory mechanism for electricity transmission expansion. J. Regul. Econ. **38**(2), 113–143 (2010)

R.A. Jabr, Robust transmission network expansion planning with uncertain renewable generation and loads. IEEE Trans. Power Syst. **28**(4), 4558–4567 (2013)

R. Jiang, J. Wang, M. Zhang, Y. Guan, Two-stage minimax regret robust unit commitment. IEEE Trans. Power Syst. **28**(3), 2271–2282 (2013)

S. Jin, S.M. Ryan, A tri-level model of centralized transmission and decentralized generation expansion planning for an electricity market: part I. IEEE Trans. Power Syst. **29**(1), 132–141 (2014)

P. Joskow, Incentive regulation in theory and practice: electricity transmission and distribution networks. Technical Report (2006)

P. Kall, S.W. Woodruff, *Stochastic Programming* (Wiley, Chichester, 1994)

A. Khodaei, M. Shahidehpour, S. Kamalinia, Transmission switching in expansion planning. IEEE Trans. Power Syst. **25**(3), 1722–1733 (2010)

G. Latorre-Bayona, I.J. Perez-Arriaga, Chopin, a heuristic model for long term transmission expansion planning. IEEE Trans. Power Syst. **9**(4), 1886–1894 (1994)

G. Latorre, R.D. Cruz, J.M. Areiza, A. Villegas, Classification of publications and models on transmission expansion planning. IEEE Trans. Power Syst. **18**(2), 938–946 (2003)

A.M. Leite da Silva, L.S. Rezende, L. A F Manso, G.J. Anders, Transmission expansion planning: a discussion on reliability and $N-1$ security criteria, in *2010 IEEE 11th International Conference on Probabilistic Methods Applied to Power Systems (PMAPS)*, pp. 244–251 (June 2010)

J. Lofberg, Yalmip : A toolbox for modeling and optimization in MATLAB, in *Proceedings of the CACSD Conference*, Taipei, Taiwan (2004)

R. Madani, J. Lavaei, R. Baldick, Constraint screening for security analysis of power networks. IEEE Trans. Power Syst. **32**(3), 1828–1838 (2017)

M. Majidi, S. Afsharnia, M.S. Ghazizadeh, A. Pazuki, A new method for optimal location of facts devices in deregulated electricity market, in *Electric Power Conference, 2008. EPEC 2008. IEEE Canada*, pp. 1–6 (Oct 2008)

M. Majidi, M.S. Ghazizadeh, S. Afsharnia, A novel approach to allocate transmission embedded cost based on mw-mile method under deregulated environment, in *Electric Power Conference, 2008. EPEC 2008. IEEE Canada*, pp. 1–6 (Oct 2008)

M. Majidi-Qadikolai, *Transmission Expansion Planning: Computational challenges toward real-size networks*. PhD Dissertation, The University of Texas at Austin (August 2017)

M. Majidi-Qadikolai, R. Baldick, Reducing the candidate line list for practical integration of switching into power system operation, in *22nd International Symposium on Mathematical Programming, (ISMP 2015)*, Pittsburgh, USA, (2015)
M. Majidi-Qadikolai, R. Baldick, Reducing the number of candidate lines for high level transmission capacity expansion planning under uncertainties, in *North American Power Symposium (NAPS), 2015*, pp. 1–6 (Oct 2015)
M. Majidi-Qadikolai, C. Urena, R. Baldick, Optimization-based approaches for practical transmission expansion planning studies: Ercot case studies, in *IEEE PES General Meeting 2018* (August 2018)
M. Majidi-Qadikolai, R. Baldick, Stochastic transmission capacity expansion planning with special scenario selection for integrating n-1 contingency analysis. IEEE Trans. Power Syst. **31**(6), 4901–4912 (2016)
M. Majidi-Qadikolai, R. Baldick, Integration of N-1 contingency analysis with systematic transmission capacity expansion planning: Ercot case study. IEEE Trans. Power Syst. **31**(3), 2234–2245 (2016)
M. Majidi-Qadikolai, R. Baldick, A generalized decomposition framework for large-scale transmission expansion planning. IEEE Trans. Power Syst. **33**(2), 1635–1649 (2018)
Market & Infrastructure Development, CAISO. California ISO 2016-2017 Regional Transmission Planning Process (2016)
Midcontinent ISO. MTEP16 MISO Transmission Expansion Plan (2016)
R. Minguez, R. Garcia-Bertrand, Robust transmission network expansion planning in energy systems: improving computational performance. Eur. J. Oper. Res. **248**(1), 21–32 (2016)
A. Monticelli, M.V.F. Pereira, S. Granville, Security-constrained optimal power flow with post-contingency corrective rescheduling. IEEE Trans. Power Syst. **2**(1), 175–180 (1987)
A. Monticelli, A. Santos, M.V.F. Pereira, S.H. Cunha, B.J. Parker, J.C.G. Praca, Interactive transmission network planning using a least-effort criterion. IEEE Power Eng. Rev. **PER–2**(10), 46–47 (1982)
R. Moreno, D. Pudjianto, G. Strbac, Transmission network investment with probabilistic security and corrective control. IEEE Trans. Power Syst. **28**(4), 3935–3944 (2013)
F.D. Munoz, J.-P. Watson, B.F. Hobbs, New bounding and decomposition approaches for MILP investment problems: multi-area transmission and generation planning under policy constraints, in *Sandia National Laboratories (SNL-NM)*, (SAND2014-4398J) (2014)
F.D. Munoz, B.F. Hobbs, J.L. Ho, S. Kasina, An engineering-economic approach to transmission planning under market and regulatory uncertainties: WECC case study. IEEE Trans. Power Syst. **29**(1), 307–317 (2014)
F.D. Munoz, E.E. Sauma, B.F. Hobbs, Approximations in power transmission planning: implications for the cost and performance of renewable portfolio standards. J. Regul. Econ. **43**(3), 305–338 (2013)
F. Munoz, J.-P. Watson, A scalable solution framework for stochastic transmission and generation planning problems. CMS **12**(4), 491–518 (2015)
F.D. Munoz, J.-P. Watson, B.F. Hobbs, Optimizing your options: extracting the full economic value of transmission when planning under uncertainty. Electr. J. **28**(5), 26–38 (2015)
NERC. Definition of adequate level of reliability (2007)
NERC. North american electric reliability corporation (2019)
NERC. Transmission system adequacy and security (2005)
R.P. O'Neill, E.A. Krall, K.W. Hedman, S.S. Oren, A model and approach for optimal power systems planning and investment. Math. Program. (2011)
T.J. Overbye, X. Cheng, Y. Sun, A comparison of the AC and DC power flow models for LMP calculations, in *Proceedings of the 37th Annual Hawaii International Conference on System Sciences*, p. 9 (Jan 2004)
H. Park, R. Baldick, Transmission planning under uncertainties of wind and load: sequential approximation approach. IEEE Trans. Power Syst. **28**(3), 2395–2402 (2013)

M.V.F. Pereira, L.M.V.G. Pinto, Application of sensitivity analysis of load supplying capability to interactive transmission expansion planning. IEEE Trans. Power Apparatus Syst. **PAS–104**(2), 381–389 (1985)

Potomac Economics. 2013 active constraint list. Personal Correspondence (2014)

R. Pringles, F. Olsina, F. Garcés, Real option valuation of power transmission investments by stochastic simulation. Energy Econ. **47**, 215–226 (2015)

F. Regairz et al., Market price signals and regulatory frameworks for coordination of transmission investments. CIGRE (2017)

R.T. Rockafellar, R.J-B. Wets, Scenario and policy aggregation in optimization under uncertainty. Math. Oper. Res. pp. 119–147 (1991)

C. Roger Glassey, Nested decomposition and multi-stage linear programs. Manage. Sci. **20**(3), 282–292 (1973)

R. Romero, A. Monticelli, A hierarchical decomposition approach for transmission network expansion planning. IEEE Trans. Power Syst. **9**(1), 373–380 (1994)

R. Romero, A. Monticelli, A zero-one implicit enumeration method for optimizing investments in transmission expansion planning. IEEE Trans. Power Syst. **9**(3), 1385–1391 (1994)

J. Rosellon, H. Weigt, A combined merchant-regulatory mechanism for electricity transmission expansion in europe. Energy J. **32**(1), 119–148 (2011)

A.M. Rudkevich, A nodal capacity market for co-optimization of generation and transmission expansion, in *2012 50th Annual Allerton Conference on Communication, Control, and Computing (Allerton)*, pp. 1080–1088 (October 2012)

P.A. Ruiz, A. Rudkevich, M.C. Caramanis, E. Goldis, E. Ntakou, C.R. Philbrick, Reduced MIP formulation for transmission topology control, in *2012 50th Annual Allerton Conference on Communication, Control, and Computing (Allerton)*, pp. 1073–1079 (October 2012a)

C. Ruiz, A.J. Conejo, Robust transmission expansion planning. Eur. J. Oper. Res. **242**(2), 390–401 (2015)

P.A. Ruiz, J. Contreras, An effective transmission network expansion cost allocation based on game theory. IEEE Trans. Power Syst. **22**(1), 136–144 (2007)

P.A. Ruiz, J.M. Foster, A. Rudkevich, M.C. Caramanis, Tractable transmission topology control using sensitivity analysis. IEEE Trans. Power Syst. **27**(3), 1550–1559 (2012b)

S.M. Ryan, R.J.B. Wets, D.L. Woodruff, C. Silva-Monroy, J.P. Watson, Toward scalable, parallel progressive hedging for stochastic unit commitment, in *2013 IEEE Power Energy Society General Meeting*, pp. 1–5 (July 2013)

D.E.M. Sappington, D.S. Sibley, Regulating without cost information: the incremental surplus subsidy scheme. Int. Econ. Rev. **29**(2), 297–306 (1988)

Y. Tohidi, M.R. Hesamzadeh, Multi-regional transmission planning as a non-cooperative decision-making. IEEE Trans. Power Syst. **29**(6), 2662–2671 (2014)

Y. Tohidi, M.R. Hesamzadeh, F. Regairaz, Sequential coordination of transmission expansion planning with strategic generation investments. IEEE Trans. Power Syst. **32**(4), 2521–2534 (2017a)

Y. Tohidi, L. Olmos, M. Rivier, M.R. Hesamzadeh, Coordination of generation and transmission development through generation transmission charges—a game theoretical approach. IEEE Trans. Power Syst. **32**(2), 1103–1114 (2017b)

D. Van Hertem, J. Verboomen, K. Purchala, R. Belmans, W.L. Kling, Usefulness of DC power flow for active power flow analysis with flow controlling devices, in *The 8th IEE International Conference on AC and DC Power Transmission, 2006. ACDC 2006* (IET, 2006), pp. 58–62

R.V. Villanasa, *Transmission network planning using linear and mixed linear integer programming*. Ph.D. thesis, Ressenlaer Polytechnic Institute (1984)

R. Villasana, L.L. Garver, S.J. Salon, Transmission network planning using linear programming. IEEE Trans. Power Apparatus Syst. **PAS–104**(2), 349–356 (1985)

I. Vogelsang, Price regulation for independent transmission companies. J. Regul. Econ. **20**(2), 141–165 (2001)

I. Vogelsang, J. Finsinger, A regulatory adjustment process for optimal pricing by multiproduct monopoly firms. Bell J. Econ. **10**(1), 157–171 (1979)

J.-P. Watson, D.L. Woodruff, Progressive hedging innovations for a class of stochastic mixed-integer resource allocation problems. CMS **8**(4), 355–370 (2011)

R.J.B. Wets, The aggregation principle in scenario analysis and stochastic optimization, in *Algorithms and Model Formulations in Mathematical*, ed. by S. Wallace (Springer, Berlin, 1989), pp. 91–113

J. Yen, Y. Yan, J. Contreras, P.-C. Ma, F.W. Felix, Multi-agent approach to the planning of power transmission expansion. Decis. Support Syst. **28**(3), 279–290 (2000)

Yu. Ermoliev and R.J.-B. Wets, *Numerical Techniques for Stochastic Optimization*. Springer series in computational mathematics (1988)

B. Zeng, L. Zhao, Solving two-stage robust optimization problems using a column-and-constraint generation method. Oper. Res. Lett. **41**, 457–461 (2013)

H. Zhang, V. Vittal, G. Heydt, An aggregated multi-cut decomposition algorithm for two-stage transmission expansion planning problems, in *Power Energy Society General Meeting, 2015* (IEEE, New York, July 2015), pp. 1–5

H. Zhang, V. Vittal, G.T. Heydt, J. Quintero, A mixed-integer linear programming approach for multi-stage security-constrained transmission expansion planning. IEEE Trans. Power Syst. **27**(2), 1125–1133 (2012)

O. Ziaee, O. Alizadeh Mousavi, F. Choobineh, Co-optimization of transmission expansion planning and TCSC placement considering the correlation between wind and demand scenarios. IEEE Trans. Power Syst. **33**(1), 206–215 (2018)

Regulated Expansion of the Power Transmission Grid

Thomas-Olivier Léautier

1 Introduction

Increasing the transfer capacity of the power transmission grid is essential to support the transition to low carbon electricity. Increasing the production from renewable energy sources (RES) over the next 30 years is an essential component of the energy transition. Since RES are located differently from existing power plants, increasing their production will profoundly transform the power flows on the transmission grid, hence will render necessarily a significant increase in the grid's capacity. For example, energy from onshore wind turbines located in West Texas must be transported to demand centers closer to the coast and from onshore and offshore wind turbines located in the North of Germany to demand centers in the South.

Capacity of the transmission grid can be increased by (i) investing in new assets, building new transmission lines, or increasing the size of the cables on existing corridors and/or (ii) operating existing assets better. For example, by measuring the system's condition in real time and adapting usage rules, the system operator can increase flow on existing lines. Building new transmission lines in the twenty-first century is more difficult than in twentieth century, due to local opposition (Not In My Backyard, NIMBY). This raises the cost of new construction and also raises the value of better operation of existing assets.

Toulouse School of Economics and EDF. This Chapter is extracted from Chap. 6 of Léautier (2019), which was written prior to my joining EDF on a sabbatical from TSE. The views presented here are therefore exclusively mine, and cannot be attributed to EDF. I am grateful to Ingo Vogelsang and the editors of this book for their insightful comments, and to MIT Press for allowing me to reproduce part of a Chapter for this text.

T.-O. Léautier (✉)
Toulouse School of Economics, Toulouse, France
e-mail: thomas.leautier@tse-fr.eu

M. R. Hesamzadeh et al. (eds.), *Transmission Network Investment in Liberalized Power Markets*, Lecture Notes in Energy 79,
https://doi.org/10.1007/978-3-030-47929-9_3

Therefore, it is essential that regulated owners and operators of transmission lines receive adequate incentives to expand the grid capacity.[1]

This chapter first presents the microeconomics analysis of transmission expansion, which produces three clear results:

1. Leaving the congestion rent generated by locational marginal pricing (sometimes called the merchandizing surplus) to the grid owner leads to suboptimal grid expansion. This is a classical result in economics: A monopoly maximizes his profits (a rectangle) not the net surplus (a triangle).
2. Making the monopoly responsible for the congestion cost (properly calculated) at the margin induces him to expand the grid optimally. This incentive regulatory mechanism was actually implemented in England and Wales for more than 10 years.
3. Regulated revenues that induce optimal expansion do not cover the fixed costs of the grid. This again is a standard result when increasing returns to scale exists (which is the case with fixed costs).

The second result deserves additional comments. The properly calculated congestion cost is the total loss of net surplus: Congestion forces the system operator to modify consumption and production compared to the unconstrained schedule. For example, some producers must produce more (they are constrained-on), some less (they are constrained-off). This creates a loss for them compared to the unconstrained schedule. Similarly, some consumers must reduce their consumption (constrained-off), while others must increase it (constrained-on). Again, this creates a net surplus loss for them compared to the unconstrained schedule. As will be discussed in more details later in this chapter, the cost of congestion is the sum of these net surplus losses.

A transmission system owner and operator who is made responsible for this congestion cost will aim to minimize the net surplus lost from congestion, i.e., to maximize net surplus. This provides the incentive for optimal expansion, which is the second result.

This chapter also discusses governance issues. In my opinion, these are more important, as they severely reduce the number of countries where an efficient incentive regulatory mechanism can be implemented. In the USA, ownership and operations of the transmission grid are separated, making it nigh impossible to implement an incentive regulatory mechanism: Transmission asset owners have no control over operating decisions, hence will vigorously object to being incentivized on a metric outside of their control; transmission system operators do not control maintenance of the transmission network, hence will similarly vigorously object to being incentivized on a metric outside of their control. Furthermore, since they are not-for-profit entities with very few assets, incentive regulation may simply not be feasible.

In Europe, transmission assets are usually owned and operated by a single entity, which solves the problem above. However, congestion occurs mostly at the borders between countries. The multiplicity of national transmission assets owners and oper-

---

[1] This chapter focusses on regulated expansion of the power grid, while merchant transmission is discussed in depth in Chap. 13 of this book, written by Papadaskalopoulos et al.

ators makes it nigh impossible to implement such an incentive regulatory mechanism across multiple countries.

The efficient incentive regulatory mechanism could be implemented today in electrical islands, for example, New Zealand, Texas, and Great Britain.

This chapter is structured as follows. Section 2 discusses the main results in "words", while the rest of the chapter proves these results using the appropriate equations. Using a simple two-market example, Sect. 3 introduces the two approaches to congestion management, and Sect. 4 derives an optimal incentive regulatory mechanism. Section 5 extends these result to a general network. Section 6 concludes by offering the author's perspective on the likelihood of a favorable change in governance in the USA and in Europe.

2 The Story in Words

2.1 A New Set of Problems

When the power industry was restructured, policy makers had to develop rules for transmission access, pricing, and expansion. This was a new set of problems. Historically, transmission access and pricing were non-issues for vertically integrated regional monopolies. Expansion followed engineering/economic analyses conducted by the utility: The transmission grid was built to transport electricity from the production centers (usually in the countryside, or near mines) to the consumption centers (usually towns). Its cost was considered part of the cost of developing new generation facilities and was included in the bundled regulated rate.

Access
Ensuring non-discriminatory access to the transmission grid for all market participants in the restructured power industry is of course essential.

In the USA, policy makers rapidly converged on wholesale and retail competition. In its landmark 1996 Order 888, the Federal Energy Regulatory Commission required all transmission asset owners and operators file an Open Access Transmission Tariff (OATT). In Europe, policy makers considered two alternative industry structures: (i) full wholesale and retail competition, and (ii) the single buyer model, in which local distribution and retail monopolies source electric power to meet their customers' needs from a competitive wholesale market (structured around a spot market or through long-term Power Purchasing Agreements). They eventually opted for full competition, hence regulators required OATTs.

In addition to OATTs, some form of vertical separation emerged as an important structural measure. In most countries, transmission assets ownership and operations were handed over to an independent company, separate from all other market participants, called a Transco, or a transmission system operator (TSO) in Europe. In the USA, electric power utilities retained ownership of the transmission assets while operations of transmission systems were handed over to independent not-for-profit entities called independent system operators (ISO). In France and Germany, transmission assets ownership and operations were entrusted to independent separate companies, yet still fully owned by the electric power utility, sometimes called legally separated transmission system operators (LTSO). The limited number of complaints by market participants suggests that, even without full vertical separation, fair access to the transmission grid has been secured.

Pricing

Prices users pay to access and use the transmission grid aim to achieve three objectives: (i) elicit the right production and consumption decisions in real time, (ii) elicit the right investment decisions, and (iii) cover the full costs of the network.

Transmission pricing is a complex issue. The difficulty arises from the laws of physics that rule power flows on a grid, transmission capacity limits, and from the cost structure of a transmission grid.

Laws of physics

One does not really "move electric power," as one moves a crate of tomatoes. Rather, one creates an electric current (technically an electromagnetic wave), which moves on the power grid following the laws of physics, not the law of economics.

Two implications of the laws of physics matter for pricing. First, energy losses: Electric current heats up conductors. This heat is dissipated in the atmosphere, hence is lost. A producer needs to produce around 105 megawatt-hours for his client to consume 100 megawatt-hours. Second, loop flows: Energy sold by a French producer to a German consumer travels from France to Germany, but also through Belgium, the Netherlands, Switzerland, and other countries.

Transmission capacity limits

Power flows on transmission lines are limited. Thus, a power market can be viewed as a series of "power islands" linked by bridges of limited capacity. When the traffic is low, power flows freely. When the power flow from France to Great Britain for example is equal to the capacity of the interconnection, the latter is congested, and it is impossible to increase exports from France into Great Britain.

Transmission capacity limits arise for two reasons. First, there are thermal limits: If power flowing on a line is too high, the line heats up and may break. Alternatively, the line sags and may touch the trees, which would produce a short circuit.

Second, there are operating limits. If a power plant or another line on the network fails, power flows are instantaneously rearranged, following the laws of physics. The operating limit on each line is such that, in the event of one (or more) failure on the system, the resulting flow on this line does not exceed the physical limit. This is called the (N-1) criterion, or the single contingency rule: The system is operated to withstand the loss of one major component. Some system operators use a (N-2) criterion and operate their system to withstand the loss of two major components.

Operating limits are often much lower than thermal limits. For example, in the early 2000s, the thermal capacity of the interconnection linking Quebec to New England was 2,000 MW, while its operating limit was hovering around 1,200 to 1,300 MW. This was economically costly for Hydro Quebec, which used the interconnection to export cheap electricity into New England, and for New England Load

Serving Entities, who bought it. More recently, in Europe, the Agency for Cooperation of Energy Regulators (ACER) found that most interconnections between European countries operated well below their thermal limit. For example, on average over 2015, the operating limit on the Belgium to the Netherlands interconnection was around 25% of its thermal limit.

Operating limits need not be identical in both directions, since the resulting flows in case of a failure are one-directional. For example, in 2015, the interconnection between Switzerland and France was operated at less than 20% of thermal capacity in the direction from Switzerland to France and at more than 40% in the reverse direction.

This operating practice was legitimate in the 1950s, but its cost is prohibitive today: On the specific example from Belgium to the Netherlands interconnection, only a quarter of the invested capital is used (on average). Transmission system operators (*TSOs*) and asset owners are installing measurement devices and developing algorithms to manage the operating limits dynamically, i.e., to meet dynamically the (N-1) criterion. In addition, recent analysis (Ovaere 2017) suggests that the (N-1) criterion itself could be made dynamic. This is an exciting development that will increase the usage of the grid. It does not modify the economic analysis presented in this chapter.

Cost structure

The variable cost of producing electric power is approximately proportional to output, and the fixed cost is approximately proportional to its installed capacity. Thus, power generation exhibits constant return to scale, at least to a reasonable first approximation.

The cost of moving electric power on one transmission line is not so simple. The variable cost of transmitting energy when the line is not congested is the cost of transmission losses. Under a reasonable approximation, transmission losses are proportional to the square of the energy flow.

The fixed cost of transmitting energy is the cost of developing, building, and operating a transmission line. It is not proportional to the capacity of the line. It can be approximated as a fixed part, independent of the capacity of the line, and a part proportional to the capacity of the line. A transmission network therefore exhibits strong returns to scale: It is much cheaper to build a transmission line of capacity 200 MW than to build two parallel transmission lines of capacity 100 MW.

2.2 *Locational Marginal Prices*

The solution

Fortunately (and somewhat surprisingly), engineers and economists have developed a simple and elegant solution to the problem of transmission pricing, called locational marginal prices.

The first seminal contribution was produced by Schweppe et al. (1988), who derive the optimal spot prices for electricity, including energy losses, loop flows, and congestion. This work generalizes peak load pricing derived in Boiteux (1949) to include spatial differentiation. At every point of the grid (called a node), electricity price is determined by the balance of local supply and demand, hence is equal to the marginal cost of the last megawatt-hour produced and/or the value of the last megawatt-hour consumed at this node. In addition, electricity prices at different nodes are related: The price consumers pay and producers receive is the price at a reference node plus their consumption's marginal contribution to losses and to congestion. Under a reasonable approximation of the equations governing power flows, these last terms have an extremely simple expression. These prices are called nodal prices, or locational marginal prices (LMPs).

Like Boiteux (1949), Fred Schweppe and its colleagues had a vertically integrated utility in mind when they derived the optimal spot prices. Hogan (1992) made another seminal contribution by showing how Schweppe's analysis can be used to solve the transmission pricing problem in restructured electricity markets. This article is perhaps the most influential academic contribution to the design of electricity markets.

Bill Hogan's intuition is that transmission does not need to be *explicitly* priced, rather the cost of moving power from node A to node B is *implicitly* defined as the difference between Schweppe's nodal prices at nodes B and A. If the network is not congested, the cost of transmission is simply the marginal cost of losses. Otherwise, the cost of transmission also includes the marginal contribution to congestion. Bill Hogan found a simple and elegant solution to an apparently intractable problem. This transmission pricing approach is called nodal pricing or locational marginal pricing.

Implicitly pricing transmission leaves market participants exposed to the difference in nodal prices, which is known only ex post. Bill Hogan's second contribution was to propose the creation of Financial Transmission Rights (FTRs), which grant market participants the difference between nodal prices at two points of the grid. If a producer located in Upstate New York wants to sell its power to a consumer in Long Island, he can purchase a FTR between these two points, hence lock in his profit margin.

As usual in economics, if competition is perfect, setting transmission price at the correct marginal cost for a given network generates incentives for optimal consumption and production in the short term, and optimal investment in generation assets and consumption centers in the long term.

2.3 *Regulated Transmission Expansion*

Underlying economics

LMPs signal the marginal value of transmission capacity. Formally, the analysis of transmission grid expansion is similar to the analysis of generation expansion. In the latter, generation capacity has value only at peak, when demand is equal to generation capacity. Optimal generation capacity equalizes expected operating margin of the plant during on-peak hours with the fixed cost of capacity. Since (i) entry in generation is (reasonably) easy, and (ii) the cost of generation capacity is approximately proportional to the size of an asset, the competitive equilibrium reaches the optimum.

Similarly, the capacity on a transmission line has value only when the line is congested. In the simplest case of two markets linked by an interconnection, the marginal value of transmission capacity is the expected price difference between the two markets. In a more complex network, the marginal value of capacity on a transmission line is a linear combination of LMPs. Optimal transmission capacity equalizes this marginal value with the marginal cost of capacity.

Unfortunately, the parallel with generation expansion stops here. First, there is no free entry in transmission: The number of suitable transmission sites is physically limited, hence competition among transmission providers is necessarily imperfect. Second, since the fixed cost of transmission capacity is significant, the congestion revenues at the optimal transmission grid do not cover the fixed capacity costs.

Incentives mechanism

In OECD countries, increasing the capacity of (portions of) existing grids to accommodate RES will be the large majority of transmission expansion projects. As previously mentioned, these will include a mix of "smarter" operating practices, software that will increase the rated capacity of existing assets, and new physical assets.

Economic analysis, formally presented in Sects. 4 for a two-market network, and 5 for the general case, delivers a good piece of news: If, as part of the incentives included in the regulatory contract, a for-profit transmission asset owner and operator is made responsible for the congestion cost (suitably defined), she faces incentives to operate and expand the grid optimally. A version of this incentive mechanism was successfully implemented in England and Wales from 1990 to 2006.

2.4 *Fixed Cost Recovery*

Finally, if the variable cost of transmission capacity is approximately proportional to the capacity, the congestion rent at the optimal capacity covers exactly the variable cost of transmission capacity, but not the fixed cost of the grid. Therefore, additional revenues must be raised to cover the full cost of the grid.

Historically, the total cost to be covered was spread across consumers and producers, through a mix of usage charge (expressed in \$/MWh) and access charges (expressed in \$/MW of peak demand or peak injection). These different approaches had different incentive properties, for example an access charge should lead to reduction in peak use of the grid, while a usage charge should lead to a reduction in average use. However, distortions were considered, rightly or wrongly, to be of secondary importance.

Matters are different today, due to the deployment of decentralized generation technologies, and the future deployment of localized storage technologies. Consider a residential user, paying 60 \$/MWh for network charges (transmission and distribution), and 80 \$/MWh for energy (suppose retailer's cost and margin are zero). Our user can install a solar panel on his roof, which produces at 100 \$/MWh. Since the electricity produced by the panel costs 100 \$/MWh, more than the wholesale spot price (80 \$/MWh), installing this solar panel is economically inefficient. Our user, however, compares the cost of consuming electricity he produces (100 \$/MWh) to the cost of electricity he purchases from the grid: 140 \$/MWh. He then rationally installs the panel.

The story does not stop here. The total network cost is unchanged, but the volume drawn from the network is reduced. Thus, the network charge per unit increases. This then encourages other users to go "off-grid". The cycle continues.

This outcome is economically inefficient. Redesigning transmission and distribution rate structures is a priority for regulators and policy makers in multiple countries and a promising area of research. The challenge is to charge users for the true cost they impose on the network. It has two components: designing the rate and implementing it. The first issue is economic, the second political. Both are intimately linked.

Most practitioners advocate rebalancing the rate structure to increase the fixed part and reduce the variable part of the tariff. This may solve the utilities problem in the short term, but I do not believe it will be sufficient: Changing the structure of the distribution rate creates winners and losers. The latter will be vocally against any change and likely to block it.

A more fruitful approach is to leverage the insights of cooperative game theory to design a fair hence politically acceptable rate structure. This issue has been extensively studied in a (somewhat esoteric) branch of the academic literature, where it is known as the cost-allocation problem (Young 1994). It involves sophisticated and highly abstract mathematics. Applying the insights of cooperative game theory to a power or distribution network requires a robust understanding of the cost structure of these networks, in particular the true cost one user imposes on the network, which varies with the user's profile and location. To my best knowledge, no utility has this information today.

This makes for a hard problem to solve. However, this added complexity is everything but superfluous: Cooperative game theory is precisely the approach required to design a tariff structure that is politically implementable. This issue constitutes an excellent field for further research and will doubtless occupy numerous scholars and regulators.[2]

2.5 Governance Issues

Readers should be encouraged by the previous discussion: a regulatory contract leading the transmission grid operator and owner to operate in the short term and expand in the medium to long term the grid optimally is feasible. Moving to a

[2]Nicolas Astier, a graduate student at the Toulouse School of Economics (TSE), working with Michel Lebreton, also at TSE has started this arduous journey (Astier 2017, Chap. 3).

rate structure leading to efficient fixed cost recovery is challenging, but can also be achieved.

However, the current governance of the electric grid in most countries makes such a contract nigh impossible to implement.

Such an incentive mechanism cannot be implemented in the USA, since ISOs are separate from asset owners. Since the former is not for-profit entities, it is almost impossible to subject them to incentive regulation. Since the latter does not operate their assets, it is almost impossible to make them responsible for an outcome (at least partly) outside of their control. They could claim to have expanded the grid optimally, but the ISO operated it suboptimally. Regulators in the USA have devised other measures to encourage transmission expansion, such as "bonified" allowed rates of return.

In Europe and other parts of the world where a single entity (called a TSO in Europe, a Transco in the USA) owns and operates the grid, an incentive regulation contract to minimize congestion is feasible. The challenge is then the "boundary problem": How do we reduce congestion at the boundary between TSOs? Because power flows cannot be controlled (the infamous loop flows), this is a serious problem, as illustrated by examples in Scandinavia and between Germany and Poland.

Therefore, a change in governance is required to see optimal expansion of the power grid.

3 Short-Term Congestion Management

Two approaches exist to manage congestion: nodal pricing and countertrading. They are presented successively in this section. First, we set the problem up.

3.1 Setup

Uncertainty

Electric power demand varies from one hour to the next. Electricity demand is higher during the day than at night, and on weekday than on weekends. In Northern Europe and Canada, electric heating leads to higher demand in the winter than in the summer. In addition, for a given hour, demand also varies randomly. For example, I expect demand tomorrow at 5 PM to be higher than tomorrow at 3 AM, due to a "structural" variation, the evening peak hour. However, I do not know exactly today the demand at 5 PM tomorrow, which depends on the temperature, which is a random variation.

To capture these two sources of variation, we introduce the notion of state of world, which corresponds to a particular realization of demand for a particular hour. The number of possible states of the world is infinite, and these are indexed by $\theta \in [0, +\infty)$. $F(\theta)$ and $f(\theta) = F^{'}(\theta)$ are the ex ante cumulative and probability density functions of state θ. The probability distribution $F(\theta)$ can be understood (and computed) by estimating the distribution of possible demand realization for every hour of the year. This produces a (possibly infinite) set of hourly demand realizations. The probability distribution $F(\theta)$ is the distribution of this set.

This approach has two limitations. First, it blurs the distinction between structural and random variations. This is acceptable for the simple analysis we present, but would be inappropriate for more sophisticated analyses, for example including intertemporal linkages or pricing demand volatility. Second, uncertainty affects both demand and supply conditions. Here, only demand uncertainty is explicitly modeled, since including production uncertainty does not modify the economic insights.

Industry structure
A large number of electricity producers compete and sell power in a wholesale spot market. Demand varies across states of the world, hence so does the wholesale spot price. A large number of retailers purchase on the wholesale market and resell it to customers at the wholesale spot price, and/or large customers source themselves directly on the wholesale spot market. Generation and retail are perfectly competitive. A fraction of customers face a constant price. This reduces the price elasticity of demand, but does not alter the economic intuition.

Network structure
Consider two markets indexed by $m = 1, 2$. To simplify the analysis, suppose that (i) each market is a single point on the network, called a node, and (ii) thermal losses are negligible. For $m = 1, 2$, denote $Q_m^s(\theta)$ and $Q_m^d(\theta)$ the aggregate quantities produced and consumed in market m in state of the world θ. The net export from market m is $Q_m(\theta) = Q_m^s(\theta) - Q_m^d(\theta)$. Since losses are negligible, net exports from market 1 must be equal to net imports into market 2: $Q_1(\theta) = -Q_2(\theta)$. Electrical engineers call $Q_m(\theta)$ the net injection at node m into the grid, and the set $\left\{Q_m^s(\theta), Q_m^d(\theta)\right\}_{m=1,2}$ a dispatch.

The marginal cost of production in market m is increasing and denoted $c_m(Q)$ and is independent of the state of the world θ. The value of consuming a marginal megawatt-hour is $P_m(Q, \theta)$ in state of the world θ. Market 1 is "cheaper" than market 2. For example, the marginal cost and marginal value of any quantity for any $Q \geq 0$ are higher in market 2: $c_2(Q) > c_1(Q)$ and $P_2(Q, \theta) > P_1(Q, \theta)$.[3]

3.2 Interconnection Unconstrained

Suppose first no congestion occurs. At the equilibrium, prices in each market are equal, and equal to the marginal cost of the last megawatt-hour produced, and the value of the last megawatt-hour consumed. This situation is presented on Fig. 1, which also illustrates energy balance on a power network: Demand is not equal to supply in each market. Rather, total demand is equal to total supply, and exports from market 1 are equal to imports into market 2.

[3]This assumption is of course unrealistic. It is used to simplify the exposition in this Section. All results hold without it.

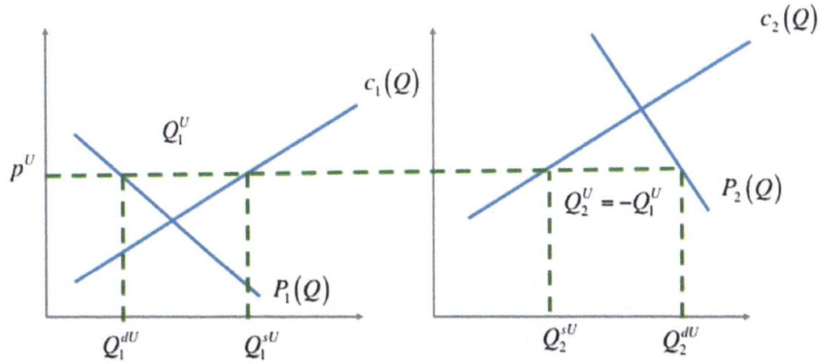

Fig. 1 Optimal production and consumption if marginal costs are continuously increasing and the interconnection is unconstrained. The unconstrained price is equal to the marginal cost of the last MWh produced and the value of the last MWh consumed in each market

3.3 *Locational Marginal Pricing*

Suppose now the interconnection is congested. Exports from cheaper producers located in market 1 are constrained by the interconnection capacity Φ^+. More expensive producers located in market 2 are therefore called on to produce. Price in each market is equal to the marginal cost of the last megawatt-hour produced and the value of the last megawatt-hour consumed in this market. The size of the interconnection determines the export (and import) volume hence the difference between nodal prices. The congestion rent received by the market operator is the surface of a rectangle: the interconnection capacity Φ^+ times the difference in nodal prices. This situation is presented on Fig. 2.

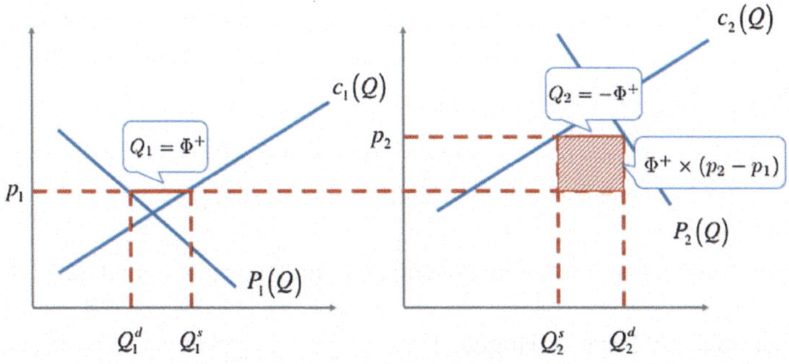

Fig. 2 Optimal production and consumption if marginal costs are continuously increasing and congestion is managed through nodal pricing. Price in each market is equal to the marginal cost of the last Megawatt-hour produced and the value of the last Megawatt-hour consumed in this market. The size of the interconnection determines the export (and import) volume, hence the difference between nodal prices. The congestion rent received by the market operator is the surface of a rectangle: the interconnection capacity Φ^+ times the difference in nodal prices

Merchandizing surplus

In a centralized market, producers in each market receive, and consumers in each market pay, the nodal price for all megawatt-hours produced and sold. The central market operator (i.e., the independent system operator in the USA, or the transmission system owner in Europe) buys and sells power at the nodal prices at each node. He collects a positive merchandizing surplus:

$$MS = \mathbb{E}\left[p_1(\theta) \times \left(Q_1^d(\theta) - Q_1^s(\theta)\right) + p_2(\theta) \times \left(Q_2^d(\theta) - Q_2^s(\theta)\right)\right].$$

Since

$$Q_2^d(\theta) - Q_2^s(\theta) = -\left(Q_1^d(\theta) - Q_1^s(\theta)\right) = \varphi(\theta),$$

we have

$$MS = \mathbb{E}\left[(p_2(\theta) - p_1(\theta)) \times \varphi(\theta)\right].$$

Since prices differ only when the interconnection is congested, i.e., $\varphi(\theta) = \Phi^+$, we have

$$MS = \mathbb{E}\left[p_2(\theta) - p_1(\theta)\right] \times \Phi^+.$$

The merchandizing surplus is equal to the expected difference in locational marginal prices times the interconnection capacity Φ^+. It is therefore also called the congestion rent. It is represented by the shaded rectangle in Fig. 2.

The merchandizing surplus is always non-negative. As discussed next, this merchandizing surplus can be used to provide market participants with insurance against congestion costs.

Financial Transmission Rights (FTRs)

Consider a producer in market 1, selling to a customer in market 2, at price p. If the line is congested, the producer cannot sell directly in market 2. Instead, he produces at cost c and sells to the market operator at $p_1(\theta)$ in market 1 and purchases from the market operator at $p_2(\theta)$ and sells to its customer at p in market 2. His profit per unit is thus

$$\begin{aligned}\pi(\theta) &= p_1(\theta) - c + p - p_2(\theta)\\ &= p - c + (p_1(\theta) - p_2(\theta)).\end{aligned}$$

The producer is thus exposed to the difference in nodal prices, called the "basis risk" by traders.

In his seminal 1992 article, Hogan (1992) suggests market participants can hedge this uncertainty by selling (or buying) financial forward products that pay the difference between nodal prices for every unit of hedge purchased. These Financial Transmission Rights (*FTRs*) are auctioned by the system operator and perfectly transferable.

If he owns a *FTR* that pays $(p_2(\theta) - p_1(\theta))$, our producer's unit profit becomes

$$\pi(\theta) = p - c + (p_1(\theta) - p_2(\theta)) + (p_2(\theta) - p_1(\theta)) = p - c.$$

Profit is thus insured against fluctuations in nodal prices.

The market operator pays *FTR* owners the price difference associated with their *FTR*. As previously mentioned, he receives the merchandizing surplus from purchasing and selling power at different prices, equal to the difference in nodal prices times the interconnection capacity. Thus, in the simple two node case, the market operator can pay exactly as many *FTRs* as there is capacity on the line. Hogan (1992) and Bushnell and Stoft (1996) generalize the result, and show that, if the sum of *FTRs* auctioned is feasible (i.e., if it is consistent with the grid configuration), the latter always exceeds the former: The market operator can always cover *FTR* payments.

Thus, the market operator auctions off *FTRs* forward, and market participants purchase *FTRs* at their expected value. When the market is run, congestion appears (or not), and the market operator pays exactly the congestion amount.

The auction proceeds go to the transmission asset owners, as part of their regulated revenues. The incentive properties of *FTR* payments are discussed later.

Rosellon and Kristiansen (2013), which gather contributions from leading academics, provide a recent and in-depth coverage of *FTRs*.

Nodal pricing provides a simple and elegant solution to two problems: congestion management and pricing of transmission services. It does so by applying to transmission pricing the same insight that gave rise to peak load pricing: When a facility is used below its capacity, the price is the variable cost of production, the cost of fuel in the case of a power plant, the marginal losses in the case of an interconnection. In our example, the latter is neglected, hence the price is zero. If the interconnection is not congested, nodal prices are equal, hence the price of transmission service is zero. On the other hand, when a facility is used at capacity, the price exceeds the variable cost. When the interconnection is congested, nodal prices are different, and the price of transmission service is no longer zero.

Power markets in the USA have all converged toward nodal pricing, which has helped them manage congestion efficiently.

3.4 Countertrading

"Countertrading" is another approach to manage congestion, used in particular in Nordic countries. It yields the same production and consumption as nodal prices, but different transfers.[4]

The market operator receives all offers (supply and demand). He first computes the unconstrained dispatch, ignoring transmission constraints. The resulting price is $p^U(\theta)$ in state of the world θ. In our example, since generation capacity in market 1 is assumed to exceed total demand, the unconstrained price is $p^U(\theta) = c_1(Q^s(\theta))$.

[4]An alternative to countertrading is redispatching: The former refers to a market-based mechanism for congestion management, while the latter refers to a non-market mechanism for congestion management. In redispatching, the TSO minimizes the total regulation cost. In the countertrading mechanism, the TSO receives up-regulation offers and down-regulation bids, and it minimizes the reported up-regulation cost minus the reported down-regulation saving.

Second, the market operator compares this dispatch with the interconnection capacity. If the resulting flows are lower than the interconnection capacity, all offers are accepted, no congestion occurs, hence no redispatching is required.

If the power flows exceed the interconnection capacity, the market operator must modify the transactions to adjust the flows. He purchases power from the producers at node 2 (and the consumers, if possible), in order to increase production at node 2 (and reduce demand), hence reduce imports. At the same time, he reduces production in market 1 (and increase demand, if possible), in order to reduce exports by the same amount as imports. The market operator adjusts until the power flow is exactly equal to the capacity of the line. The dispatch is thus exactly identical to the nodal pricing one, as represented on Fig. 2.

Compensation to constrained-off and -on producers
However, transfers are different. Under nodal pricing, all producers sell and all customers/retailers buy at p_1 in market 1, while consumers/retailers purchase at $p_2 > p_1$ in market 2. As illustrated in Fig. 2, the operator receives a positive surplus, equal to the congestion rent.

Under countertrading, the market operator purchases at price p^U actual production $Q_1^s(\Phi^+, \theta)$ and $Q_2^s(\Phi^+, \theta)$, sells at price p^U the volume $\left(Q_1^d(\Phi^+, \theta) + Q_2^d(\Phi^+, \theta)\right)$, and compensates constrained-on and -off producers and consumers for their adjustments.

Figure 3 illustrates the compensation to constrained-off and -on producers. Market 1 is export-constrained, i.e., $Q_1^s(\Phi^+, \theta) < Q_1^{sU}(\theta)$. For $x \in \left[Q_1^s(\Phi^+, \theta), Q_1^{sU}(\theta)\right]$, constrained-off producers are paid the net operating profit they would have received: $\left(p^U - c_1(x)\right)$. Market 2 is import-constrained, i.e., $Q_2^s(\Phi^+, \theta) > Q_2^{sU}(\theta)$. For $x \in \left[Q_2^{sU}(\theta), Q_2^s(\Phi^+, \theta)\right]$, constrained-on producers are paid their cost $c_2(x)$, hence their constrained-on payment is $\left(c_2(x) - p^U\right)$. Thus, compensation to constrained-off and -on producers in state θ is

$$\int_{Q_1^s(\Phi^+,\theta)}^{Q_1^{sU}(\theta)} \left(p^U - c_1(x)\right) dx + \int_{Q_2^{sU}(\theta)}^{Q_2^s(\Phi^+,\theta)} \left(c_2(x) - p^U\right) dx$$

$$= \sum_{m=1}^{2} \int_{Q_m^{sU}(\theta)}^{Q_m^s(\Phi^+,\theta)} \left(c_m(x) - p^U\right) dx.$$

It is equal to the surface of the two triangles on Fig. 3.

Compensation to constrained-on and -off consumers
Figure 4 illustrates the compensation to constrained-on and -off consumers. Market 1 is export-constrained, i.e., $Q_1^d(\Phi^+, \theta) > Q_1^{dU}(\theta)$. For $x \in \left[Q_1^{dU}(\theta), Q_1^d(\Phi^+, \theta)\right]$, constrained-on consumers are compensated for their surplus loss from consuming, hence are paid $\left(p^U - P_1(x, \theta)\right)$. Market 2 is import-constrained, i.e., $Q_2^d(\Phi^+, \theta) < Q_2^{dU}(\theta)$. For $x \in \left[Q_2^d(\Phi^+, \theta), Q_2^{dU}(\theta)\right]$, constrained-off consu-

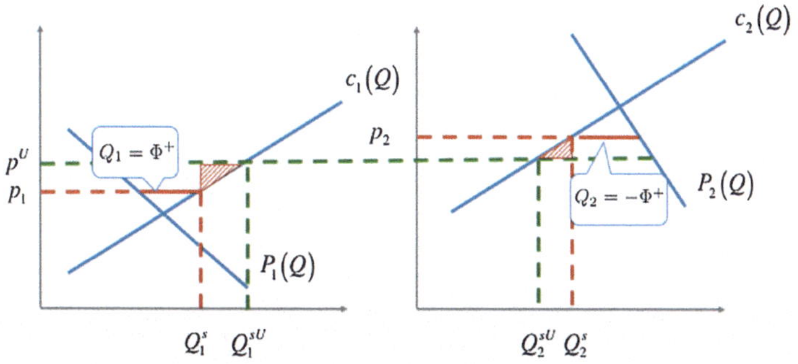

Fig. 3 Payments to constrained-on and -off producers if marginal costs are continuously increasing and congestion is managed through countertrading

mers receive the net surplus they would have derived from consuming: $(P_2(x, \theta) - p^U)$. Thus, compensation to constrained-on and -off consumers in state θ is

$$\int_{Q_1^{dU}(\theta)}^{Q_1^d(\Phi^+,\theta)} \left(p^U - P_1(x,\theta)\right) dx + \int_{Q_2^d(\Phi^+,\theta)}^{Q_2^{dU}(\theta)} \left(P_2(x,\theta) - p^U\right) dx$$

$$= \sum_{m=1}^{2} \int_{Q_m^d(\Phi^+,\theta)}^{Q_m^{dU}(\theta)} \left(P_m(x,\theta) - p^U\right) dx.$$

It is equal to the surface of the two triangles on Fig. 4.

Market operator net profit

In state θ, the market operator's payment to producers, including compensation to constrained-off and -on producers is

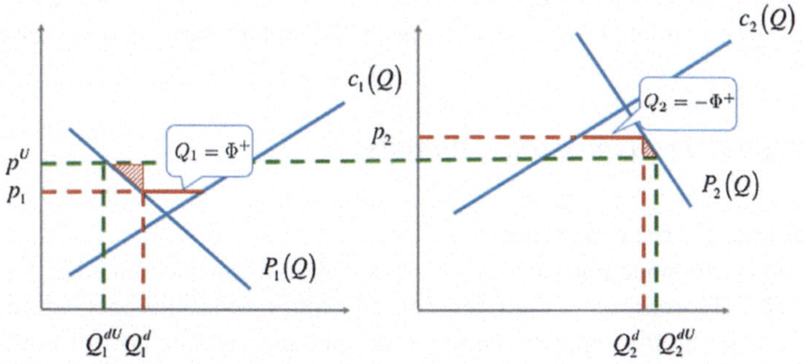

Fig. 4 Compensation to constrained-on and -off consumers if congestion is managed through countertrading

$$C\left(\Phi^{+},\theta\right)=p^{U}\left(Q_{1}^{s}\left(\Phi^{+},\theta\right)+Q_{2}^{s}\left(\Phi^{+},\theta\right)\right)+\sum_{m=1}^{2}\int\limits_{Q_{m}^{sU}(\theta)}^{Q_{m}^{s}(\Phi^{+},\theta)}\left(c_{m}\left(x\right)-p^{U}\right)dx.$$

Her revenues from electricity sale net of the compensation to constrained-on and off consumers are

$$R\left(\Phi^{+},\theta\right)=p^{U}\left(Q_{1}^{d}\left(\Phi^{+},\theta\right)+Q_{2}^{d}\left(\Phi^{+},\theta\right)\right)-\sum_{m=1}^{2}\int\limits_{Q_{m}^{d}(\Phi^{+},\theta)}^{Q_{m}^{dU}(\theta)}\left(P_{m}\left(x,\theta\right)-p^{U}\right)dx.$$

Her expected net profit is

$$\begin{aligned}
&\mathbb{E}\left[R\left(\Phi^{+},\theta\right)-C\left(\Phi^{+},\theta\right)\right]\\
&=-\mathbb{E}\left[\begin{array}{c}p^{U}\left(Q_{1}\left(\Phi^{+},\theta\right)+Q_{2}\left(\Phi^{+},\theta\right)\right)+\\ \sum_{m=1}^{2}\left[\int\limits_{Q_{m}^{sU}(\theta)}^{Q_{m}^{s}(\Phi^{+},\theta)}\left(c_{m}\left(x\right)-p^{U}\right)dx+\int\limits_{Q_{m}^{d}(\Phi^{+},\theta)}^{Q_{m}^{dU}(\theta)}\left(P_{m}\left(x,\theta\right)-p^{U}\right)dx\right]\end{array}\right]\\
&=-\mathbb{E}\left[\sum_{m=1}^{2}\left[\int\limits_{Q_{m}^{sU}(\theta)}^{Q_{m}^{s}(\Phi^{+},\theta)}\left(c_{m}\left(x\right)-p^{U}\right)dx+\int\limits_{Q_{m}^{d}(\Phi^{+},\theta)}^{Q_{m}^{dU}(\theta)}\left(P_{m}\left(x,\theta\right)-p^{U}\right)dx\right]\right].
\end{aligned}$$

since $Q_1\left(\Phi^{+},\theta\right)+Q_2\left(\Phi^{+},\theta\right)=0$. The market operator expected profit is negative and equal to minus the redispatching cost. It is covered by all consumers, through an uplift charge.

4 Transmission Grid Expansion Under Different Regimes

This section first characterizes the optimal transmission capacity, then examines the incentives produced by the congestion rent. Finally, it derives the incentive contract that induces the regulated monopoly to optimally expand transmission capacity.

4.1 Optimal Transmission Capacity

Marginal cost of grid expansion

For a single interconnection, the cost of capacity can be approximated as $\Gamma\left(\Phi^{+}\right)=\Gamma_0+\gamma\times\Phi^{+}$: Γ_0 represents the fixed cost of building the interconnection, which does not depend on the capacity, and γ represents the variable cost of building the interconnection, which increases approximately linearly with the capacity of the interconnection.

In a large network, matters are much more complex. First, increasing the size of a line (or building a parallel new line) increases the transfer capacity of that line (or interface between two regions), but also impacts other lines. Therefore, different solutions are available to increase capacity on a given interface, including increasing capacity on other interfaces. The resulting cost function is the minimum overall feasible solutions (Boyer et al. 2006 provide a very clear example of a natural gas network).

Second, as indicated by Hogan et al. (2010), the resulting cost function does not always possess "nice" mathematical properties. In some instances, increasing capacity on one interconnection reduces capacity on another. Suppose for example line *A* is the largest contingency setting the operating limit of line *B*. Increasing the capacity of line *A* increases the flow on line B if line *A* was to fail. Thus, the operating limit on line *B* has to be reduced. In this case, there may be no optimal grid, or multiple optima.

Since this chapter focusses on incentives generated by different remuneration mechanisms, it assumes that the cost function is such that a unique optimal grid exists. It is the case for a single interconnection linking two markets. Readers should be aware, however, that this property may not always hold for meshed networks.

Marginal value of the interconnection capacity
Increasing the interconnection capacity by 1 megawatt has no value when the interconnection is not congested. This is the equivalent of the observation that the marginal value of generation capacity is zero when demand is lower than capacity.

When the interconnection is congested and technology 2 is producing, it enables the substitution of 1 megawatt-hour produced with technology 1 for one megawatt-hour produced with technology 2, thus saving $\left(c_2\left(Q_2^S(\theta)\right)-c_1\left(Q_1^S(\theta)\right)\right)$. The marginal value of transmission capacity can be expressed using LMPs:

$$\Lambda\left(\Phi^+\right)=\mathbb{E}\left[p_2(\theta)-p_1(\theta)\right].$$

The marginal value of the interconnection is the expected difference in LMPs and is decreasing as the capacity of the line increases.

Optimal interconnection capacity
At the optimum, this marginal value is equal to the marginal cost of capacity on this interconnection:

$$\mathbb{E}\left[p_2(\theta)-p_1(\theta)\right]=\gamma. \tag{1}$$

Since the marginal cost of capacity is positive, Eq. (1) implies that interconnection must be congested in some states of the world. An interconnection that could never be congested in any state of the world would be too large. This does not mean that every line must be congested a few hours every year. It means that an interconnection that would never be congested in all the scenario studied would be oversized.

This result differs from current planning practice in two aspects. First, transmission grids within a utility service territory were often built never to be congested. In many cases, the transmission grid was built to transport electricity from the production centers to the consumption centers. The power flows were highly predictable, and often one-directional. Since the cost of transmission was (in most instances) a fraction of the cost of generation, interconnection size was simply the size of the power plant, and the cost of building the interconnection was included in the total cost of generation.

In the future, matters will likely be different. We expect renewables and decentralized production to progressively represent a large share of the generation mix. Power flows will thus become more variable, and interconnections will be used in both directions. Furthermore, building transmission lines is proving extremely challenging, hence costly. Thus, we will have to accept some lines are congested for some period of time and use relationship (1) to optimize the network.

Second, transmission planners often compute separately economic benefits and reliability benefits of an interconnection. Equation (1) provides a unified treatment of these two sources of value. Economic benefit is the substitution of cheap for dear power, as was discussed earlier. Reliability benefit is the reduction in the probability of having to curtail customers in case of tension on the system. As discussed in Leautier (2019, Chap. 3), curtailment (for example in market 2) would result in the price being set at the value of lost load (VoLL) in market 2. Thus, equation (1) captures this benefit by recognizing that the price $p_2(\theta)$ may be equal to the VoLL in some states of the world.

Fixed costs recovery

As discussed earlier, electricity generation exhibits approximately constant returns to scale: If prices are set to cover marginal costs, they also cover average cost. This is not the case for networks, which exhibit increasing returns to scale, hence covering marginal cost does not guarantee average cost coverage (e.g., Bell et al. 2011). In the two-market example, the congestion rent at the optimal capacity covers exactly the variable cost of the grid:

$$\Lambda\left(\Phi^{+}\right) = \gamma \Leftrightarrow \Lambda\left(\Phi^{+}\right) \times \Phi^{+} = \gamma \times \Phi^{+}.$$

This leaves the entire fixed cost Γ_0 uncovered. (Perez-Arriaga et.al 1995) show that the congestion rents recover only approximately 25% of total costs of a representative transmission grid. An additional charge must be levied to cover the remaining costs.

4.2 *The Congestion Rent Does Not Induce Optimal Grid Expansion*

Leaving the congestion rent to a monopoly grid company does not provide optimal incentives. If he receives the entire congestion rent

$$R\left(\Phi^{+}\right) = \Lambda\left(\Phi^{+}\right) \times \Phi^{+},$$

he finds himself in the situation of a classical monopoly. He maximizes his rent by setting the marginal revenue equal to the marginal cost:

$$\Lambda\left(\Phi^{+}\right)+\frac{d\Lambda\left(\Phi^{+}\right)}{d\Phi^{+}}\times\Phi^{+}=\gamma.$$

As we have seen, $\frac{d\Lambda(\Phi^{+})}{d\Phi^{+}}<0$: increasing the capacity on a line reduces its value. Thus, if the transmission company increases the interconnection, it captures the expected price difference on the marginal capacity, but also reduces the expected price difference, hence the rent, on all inframarginal units.

The result extends to a Transco owning and operating multiple lines. If he is granted the congestion rent, he will behave as a multiproduct monopolist and invest suboptimally.

Thus, leaving the congestion rent to the monopolist was quickly abandoned as an approach to induce optimal investment. However, transmission asset owners still receive the congestion rent through the auction of *FTRs* in the USA. In Europe, transmission asset owners receive the congestion rent associated with cross-border interconnections. In both cases, this rent is deducted from their required revenues so as not to provide incentives to increase congestion. Allocating the congestion rent to a fund to finance network expansion has been discussed as an alternative, but this has not been pursued.

4.3 Optimal Regulation of a Monopoly Transmission Company

The regulatory contract aims to induce (i) the optimal cost reduction/rent extraction trade-off and (ii) the optimal grid expansion.

Optimal cost reduction/rent extraction trade-off

This is the core of modern regulation theory, exposed for example by Laffont and Tirole (1993). The starting point is *information asymmetry* between the regulator (the principal) and the regulated firm (the agent). The latter has better information than the former concerning its cost, and the intensity of its cost reducing activities.

Historically, the regulator would cover the firm's reported costs, an approach known as cost of service regulation. In that case, the firm had no incentives to reduce costs. Since the regulator did not know the firm's cost-saving potential, there was widespread suspicion of inefficiency. This situation was captured by John Hicks' famous observation:

> The best of all monopoly profits is a quiet life.

A completely opposite approach is to fix the regulated firm's average price at a given level for the entire regulatory period, for example 5 years. If the firm reduces its costs below the fixed level, it is allowed to keep the savings realized, hence captures significant profit. If it fails to do so, it may incur a loss. Thus, the firm faces strong incentives to reduce costs.

Laffont and Tirole (1993) formally study this problem. Building on previous academic analyses (e.g., Baron and Myerson 1982), they propose a very simple representation of the situation. The variable cost for a firm to produce output q is $C(q)$ defined by

$$C(q) = (\beta - e)\, q,$$

where β is the natural cost efficiency of the firm and e a cost reducing effort, both unobservable to the regulator. The regulator observes the cost $C(q)$ in the regulated firm's financial statements. However, if the regulator observes low-cost $C(q)$, he does not know whether the firm is naturally efficient (low β), or exerts a very high-cost reducing effort (high e). Reducing cost generates private disutility for the managers of the firm, who no longer enjoy Hicks' "quiet life". Since effort is unobservable, the regulator cannot force managers to exert a certain level, rather she needs to induce them, by leaving them an information rent.

Using this model, Laffont and Tirole (1993) find that the regulator should propose a menu of (linear) contracts, and let the regulated firm choose its preferred contract within this menu. A naturally efficient firm (low β), optimally, chooses a contract close to a fixed-price contract and exerts high-cost reducing effort e, while a naturally inefficient firm chooses a contract close to a cost-plus contract and exerts low-cost reducing effort.

This model is extremely simple, yet extremely powerful. It has exerted a profound influence on academic economists and policy makers and figured prominently in the citation for Jean Tirole's Nobel prize in 2014.

Optimal expansion

Laffont and Tirole (1993) do not discuss in great depth how to induce the regulated firm to produce the optimal output. They prove a dichotomy property, i.e., they derive a sufficient condition on the cost function $C(q)$ for the output choice to be treated independently from cost reduction/rent extraction problem.

Leautier (2000) proves that if demand is completely price-inelastic (which was a reasonable representation of power markets in the late 1990s), the redispatching cost is an efficient congestion metric, i.e., that having the regulated firm responsible for the redispatching cost at the margin gives it incentives to optimally expand the grid. As will be shown below, this result also holds if demand is price responsive. The proof is slightly technical, but the intuition is simple: As shown in Figs. 3 and 4, the redispatching cost is a triangle. The marginal redispatching cost is therefore the vertical side of the triangle, i.e., no inframarginal effect occurs. This argument is formalized below:

Lemma 1 *The marginal redispatching cost is equal to the marginal value of transmission capacity:*

$$\frac{d}{d\Phi^{+}}\mathbb{E}\left[R\left(\Phi^{+},\theta\right)-C\left(\Phi^{+},\theta\right)\right]=\Lambda\left(\Phi^{+}\right).$$

Proof From equation

$$\begin{aligned}
\frac{d}{d\Phi^{+}}\mathbb{E}\left[R\left(\Phi^{+},\theta\right)-C\left(\Phi^{+},\theta\right)\right] &= -\mathbb{E}\left[\left(\begin{array}{c}\left(p^{U}-p_{1}\right)\frac{dQ_{1}^{d}}{d\Phi^{+}}-\left(p_{2}-p^{U}\right)\frac{dQ_{2}^{d}}{d\Phi^{+}}\\ -\left(p^{U}-p_{1}\right)\frac{dQ_{1}^{s}}{d\Phi^{+}}+\left(p_{2}-p^{U}\right)\frac{dQ_{2}^{s}}{d\Phi^{+}}\end{array}\right)\right]\\
&= -\mathbb{E}\left[\left(p^{U}-p_{1}\right)\frac{dQ_{1}}{d\Phi^{+}}+\left(p^{U}-p_{2}\right)\frac{dQ_{2}}{d\Phi^{+}}\right]\\
&= \mathbb{E}\left[-\sum_{m=1}^{2}p_{m}\left(\Phi^{+},\theta\right)\frac{dQ_{m}}{d\Phi^{+}}+p^{U}\sum_{m=1}^{2}\frac{dQ_{m}}{d\Phi^{+}}\right]\\
&= -\mathbb{E}\left[\sum_{m=1}^{2}p_{m}\left(\Phi^{+},\theta\right)\frac{dQ_{m}}{d\Phi^{+}}\right].
\end{aligned}$$

When the interconnection is not congested, $\frac{dQ_1}{d\Phi^+}=\frac{dQ_2}{d\Phi^+}=0$. When the interconnection is congested, $Q_1=\Phi^+=-Q_2$, hence $\frac{dQ_1}{d\Phi^+}=1=-\frac{dQ_2}{d\Phi^+}$. Therefore,

$$\frac{d}{d\Phi^{+}}\mathbb{E}\left[R\left(\Phi^{+},\theta\right)-C\left(\Phi^{+},\theta\right)\right]=\mathbb{E}\left[\left(p_{2}\left(\Phi^{+},\theta\right)-p_{1}\left(\Phi^{+},\theta\right)\right)^{+}\right]=\Lambda\left(\Phi^{+}\right),$$

which proves the Lemma.

Lemma 1 enables us to prove the main result of this section:

Proposition 1 *If a for-profit transmission asset owner and operator are responsible for the redispatching cost on the interconnection, he selects the optimal interconnection capacity.*

Proof The Transco profit is

$$\Pi\left(\Phi^{+}\right)=\mathbb{E}\left[R\left(\Phi^{+},\theta\right)-C\left(\Phi^{+},\theta\right)\right]-\left(F+\gamma\Phi^{+}\right).$$

The Transco chooses Φ^+ to maximize its profit. The first-order condition is

$$\frac{d\Pi}{d\Phi^{+}}=0\Leftrightarrow\frac{d}{d\Phi^{+}}\mathbb{E}\left[R\left(\Phi^{+},\theta\right)-C\left(\Phi^{+},\theta\right)\right]=\Lambda\left(\Phi^{+}\right)=\gamma,$$

which characterizes the optimal interconnection capacity.er condition defines the unique maximum.

Section 5 shows Proposition 1 holds for a general N-node network. This result is particularly powerful in the case of a Transco, which owns and operates the transmission grid. Increasing the capacity transfer on the grid can be achieved by physical investment, but also by improving the operating procedures, for example monitoring asset conditions in real time.

Transcos may object to being exposed to the full redispatching cost, hence incentive regulation may expose them only to the deviation from a target, with a cap on potential gains and penalties. For example, if the target is $500 millions in congestion cost, the Transco receives (or pay) the difference between the target and the actual cost of congestion, up to $100 millions.[5] This provides incentives for efficient management of congestion, while capping the potential gains and liabilities.

This approach was implemented in England and Wales in the 1990s and led to significant reduction in congestion cost. It stopped in 2006 when National Grid Company, the Transco in England and Wales, could no longer beat the target and had to pay penalties.

The problem is more challenging if one firm owns the grid while a not-for-profit entity is responsible for its operation, as is the case in the USA. There, financial incentives cannot be included in the regulatory compact, and other approaches must be implemented.

However, even with a Transco, implementing Proposition 1 may not be straightforward: If a market is organized around nodal prices (as it should), a separate computation is required to compute the congestion cost. As we have seen before, both approaches to manage congestion require the same inputs (costs and demand curves at every node, available transmission capacities) and produce the same production and consumption plan. Still, the regulator should perform a separate analysis to compute the redispatching cost.

To address this issue, Bill Hogan, Juan Rosellon, and Ingo Vogelsang have attempted in a series of articles to design a mechanism relies on *FTRs*. This interesting stream of research is summarized in Hogan et al. (2010). A Transco is a multiproduct monopoly, which, if left on its own, would likely produce less output than socially optimal to increase its profit. This problem has been extensively studied in the late 1970s and 1980s (e.g., by Vogelsang and Finsinger 1979, and Sappington and Sibley 1988). Therefore, Bill Hogan and his co-authors hoped to be able to transpose these results to a Transco. This has proven to be harder task than expected. One of the issues is that the cost function for the production of *FTRs* (which is derived from the cost function of the expansion of the interconnections on the network, but different) is not well behaved: For example, increasing the amount of *FTRs* available between two nodes may reduce the amount of available *FTR*s. (The interested reader is referred to Rosellon et al. (2012) for a richer discussion of transmission cost functions).

Vogelsang in Chap. 12 of this book describes how a slightly different variation of the initial mechanism can be applied to an transmission asset owner, named H-R-G-V mechanism after the initials of the authors developing it. As it turns out, the H-R-G-V mechanism is extremely close to the mechanism described in Proposition 1, with two differences. First, the H-R-G-V mechanism is not concerned by the rent extraction versus cost minimization trade-off and focusses only on investment incentives. This difference is not as significant as it may seem: The dichotomy property discussed by Laffont and Tirole (1993) enables us to separate the determination of optimal investment incentives from the rent extraction versus cost minimization trade-off. Proposition 1 focusses solely on the former.

---

[5]The cap, expressed as a maximum gain/loss, may be determined in relation to the return on the regulated asset base.

Second, the H-R-G-V mechanism provides a payment to the transmission asset owner, while Proposition 1's mechanism provides an incentive payment/penalty. As Vogelsang observes in Chap. 12 of this book, the decision by the regulator in England & Wales to stop implementing Proposition 1's mechanism after National Grid paid penalties for a few years "may show the potential downsides of too strict rent extraction policies," hence weakens the applicability of the mechanism. While this point is legitimate, I am more optimistic: Social pressure, in particular in England & Wales but in other countries as well, forces regulators to be more demanding. A well-calibrated incentive mechanism is feasible.

My concern with the applicability H-R-G-V mechanism is the governance of the transmission system. The H-R-G-V incentive mechanism applies to the transmission asset owner and takes the output of the ISO dispatch as given. As previously discussed, I am uncertain of the feasibility of such a mechanism: Why would the transmission asset owner accept a compensation that depends on the performance of the ISO, who has significant discretion in the dispatch process? Further research is required to resolve this issue.

5 Extension to a General Network

We now extend the derivation of nodal prices and the previous results to a general network, following Schweppe et al. (1988). This presentation should be sufficient for most economists. Readers interested in a discussion of the physics underlying this discussion are referred to the Appendix D in Schweppe et al. (1988).

The main difference with the two-market network is a phenomenon called "loop flows". Power flows on a network do not follow contractual arrangements. Rather they obey Kirchhoff's circuit laws, derived in 1845 by German physicist Gustav Kirchhoff. A megawatt-hour produced at point A to be consumed at point B does not solely "travel" on the line from A to B. Rather, it "travels" on the entire transmission grid, following the paths of least resistance. The power flows thus trace loops on the grid. These loop flows produce surprising results and have profound implication for market design. Market designers have attempted to ignore them at their peril! However, including loop flows does not fundamentally alter the economic intuition.

5.1 Setup

Notation

Consider a power network consisting of M markets, linked by L interconnections. For every market $m = 1, ..., M$, $Q_m^s(\theta)$, $Q_m^d(\theta)$, and $Q_m(\theta) = Q_m^s(\theta) - Q_m^d(\theta)$ are the production, demand, and net injection in state of the world θ. For every interconnection, $l = 1, ..., L$, $\varphi_l(\theta)$ is the power flowing on the interconnection in state of the world θ. Without loss of generality, we assume $\varphi_l(\theta)$ measures the flow from the lowest number node to the highest number node. Φ_l^+ et Φ_l^- are the maximum power flow in each direction, with the convention that Φ_l^+ corresponds to $\varphi_l(\theta) > 0$. The vector of oriented flows is $\varphi(\boldsymbol{\theta}) \in \mathbb{R}^L$.

Energy balance
In every state θ, energy balances require that total production is equal to total demand plus thermal losses $\tilde{LO}(\varphi(\boldsymbol{\theta}))$:

$$\sum_{m=1}^{M} Q_m^s(\theta) = \sum_{m=1}^{M} Q_m^d(\theta) + \tilde{LO}(\varphi(\boldsymbol{\theta})).$$

As will be shown below, the M net injections are linked by the above equation, thus only $(M-1)$ net injections are independent. We thus define a reference node (e.g., node 1), and write the flow equations as a function of the vector $\underline{\mathbf{Q}}(\theta) \in \mathbb{R}^{M-1}$ of the $(M-1)$ remaining net injections.

DC load flow approximation
Electric power is in fact a wave. It is represented by two dimensions, an amplitude and a phase angle, or normal and reactive power. The Direct current (DC) load flow approximation assumes that the phase angles at the extremities of all lines are very close. Neglecting second-order terms yields a linear relationship between the vector of flux and the vector of net injections

$$\varphi(\boldsymbol{\theta}) = \mathbf{H} \cdot \underline{\mathbf{Q}}(\theta)$$

where $\mathbf{H} \in \mathbb{R}^L \times \mathbb{R}^{M-1}$ is the admittance transfer matrix.

Under the DC load flow approximation, thermal losses are a quadratic function of the flows, hence of net injections

$$\tilde{LO}(\varphi(\boldsymbol{\theta})) = \underline{\mathbf{Q}}^T(\theta) \cdot \mathbf{B} \cdot \underline{\mathbf{Q}}(\theta) = LO\left(\underline{\mathbf{Q}}(\theta)\right)$$

where the matrix $\mathbf{B} \in \mathbb{R}^{M-1} \times \mathbb{R}^{M-1}$ is symmetric. Global energy balance is thus

$$\sum_{m=1}^{M} Q_m^s(\theta) = \sum_{m=1}^{M} Q_m^d(\theta) + LO\left(\underline{\mathbf{Q}}(\theta)\right).$$

Denote $\mu_e(\theta)$ the Lagrange multiplier associated with this constraint. $\mu_e(\theta)$ represents the value of a marginal megawatt-hour in state θ.

Maximum flow on each interconnection
Since every line has a maximum transfer capacity

$$-\Phi_l^- \leq \varphi_l(\theta) \leq \Phi_l^+ \Leftrightarrow -\Phi_l^- \leq \sum_{m=2}^{M} H_{lm} Q_m(\theta) \leq \Phi_l^+.$$

At the optimal dispatch, only one constraint is binding. Without loss of generality, denote Φ_l the binding constraint in state θ and $\eta_l(\theta)$ the associated Lagrange multiplier.

As previously mentioned, capacity on interconnections is increasingly determined dynamically. The transmission capacity constraint would then be written as $-\Phi_l^-(\theta) \leq \varphi_l(\theta) \leq \Phi_l^+(\theta)$. The economic intuition would be unchanged.

5.2 *Characterization of the Optimal Dispatch*

Optimization problem

Introduce $C_m(Q)$ and $c_m(Q)$, respectively, as the total cost and marginal cost of producing quantity Q.

Schweppe et al. (1988) state the optimization problem using $U(Q, \theta)$, and the utility consumers derive in state θ from quantity Q. The marginal utility is equal to the value of the marginal megawatt-hour: $\frac{\partial U(Q,\theta)}{\partial Q} = P(Q, \theta)$.

The optimization problem is thus to chose $\left\{Q_m^s(\theta), Q_m^d(\theta)\right\}_{\{m,\theta\}}$, the production and consumption at each node m and in each state θ, to maximize the net surplus, subject to the constraints imposed by the laws of physics:

$$\begin{gathered} max_{Q_m^s(\theta), Q_m^d(\theta)} \mathbb{E}\left[\sum_{m=1}^M \left\{U_m\left(Q_m^d(\theta), \theta\right) - C_m\left(Q_m^s(\theta)\right)\right\}\right] \\ st: \begin{cases} \sum_{m=1}^M Q_m^s(\theta) = \sum_{m=1}^M Q_m^d(\theta) + LO\left(\mathbf{Q}(\theta)\right) \;\forall\theta \;(\mu_e(\theta)) \\ \sum_{m=2}^M H_{lm} Q_m(\theta) \leq \Phi_l \;\forall(l, \theta) \qquad\qquad (\eta_l(\theta)) \end{cases}. \end{gathered}$$

Locational marginal prices

The Lagrangian of the optimization program is

$$\mathcal{L} = \mathbb{E}\left[\begin{array}{c} \sum_{m=1}^M \left\{U_m\left(Q_m^d(\theta), \theta\right) - C_m\left(Q_m^s(\theta)\right)\right\} \\ +\mu_e(\theta)\left(\sum_{m=1}^M Q_m^s(\theta) - \sum_{m=1}^M Q_m^d(\theta) - LO\left(\mathbf{Q}(\theta)\right)\right) \\ +\sum_{l=1}^L \eta_l(\theta)\left(\Phi_l - \sum_{m=2}^M H_{lm} Q_m(\theta)\right) \end{array}\right].$$

The first-order derivatives are

$$\frac{\partial \mathcal{L}}{\partial Q_m^d(\theta)} = P_m\left(Q_m^d(\theta), \theta\right) - \mu_e(\theta)\left(1 - \frac{\partial LO}{\partial Q_m}\right) + \sum_{l=1}^L \eta_l(\theta) H_{lm}$$

for every m such that $Q_m^d(\theta) > 0$, and

$$\frac{\partial \mathcal{L}}{\partial Q_m^s(\theta)} = -c_m\left(Q_m^s(\theta)\right) + \mu_e(\theta)\left(1 - \frac{\partial LO}{\partial Q_m}\right) - \sum_{l=1}^L \eta_l(\theta) H_{lm}$$

for every m such that $Q_m^s(\theta) > 0$, with the convention that $\frac{\partial LO}{\partial Q_1} = H_{l1} = 0$.

The optimal prices $p_m(\theta)$ are defined for all $m \geq 1$ by the local equilibrium conditions

$$p_m(\theta) = \begin{cases} P_m\left(Q_m^d(\theta), \theta\right) \text{ if } Q_m^d(\theta) > 0 \\ c_m\left(Q_m^s(\theta)\right) \quad \text{if } Q_m^s(\theta) > 0 \end{cases}, \tag{2}$$

and the global equilibrium conditions

$$\begin{cases} p_1(\theta) = \mu_e(\theta) \\ p_m(\theta) = \mu_e(\theta)\left(1 - \frac{\partial LO}{\partial Q_m}\right) - \sum_{l=1}^{L} H_{lm}\eta_l(\theta) \text{ for } m > 1 \end{cases}. \tag{3}$$

The local equilibrium conditions (2) impose that price at a node m is equal to the marginal surplus from consumption and/or the marginal cost of production.

The global equilibrium conditions (3) take into account the network externalities: thermal losses and congestion. They are defined relative to the reference node, i.e., no externality is included in the price at node 1, which is the global reference price $\mu_e(\theta)$.

Consider the consumer's perspective for a market $m > 1$ where $Q_m^d > 0$. She pays the global reference price $\mu_e(\theta)$, plus her marginal contribution to thermal losses $\frac{\partial LO}{\partial Q_m^d} = -\frac{\partial LO}{\partial Q_m}$, where losses are valued at the global reference price, plus for every line her contribution to congestion on this line $\frac{\partial \varphi_l}{\partial Q_m^d} = -\frac{\partial \varphi_l}{\partial Q_m} = -H_{lm}$, valued at the (virtual) value of the congestion $\eta_l(\theta)$. Symmetrically, if $Q_m^s > 0$, a producer sells her energy at the reference price $\mu_e(\theta)$, minus her marginal contribution to thermal losses $\frac{\partial LO}{\partial Q_m^s} = \frac{\partial LO}{\partial Q_m}$ valued at $\mu_e(\theta)$, minus her marginal contribution to congestion on every line l $\frac{\partial \varphi_l}{\partial Q_m^s} = \frac{\partial \varphi_l}{\partial Q_m} = H_{lm}$ valued at $\eta_l(\theta)$.

Every market participant thus faces the marginal externalities created by her decisions. Under perfect competition, nodal pricing decentralizes the social optimum. Conversely, if market participants face prices other than nodal prices, net surplus is reduced.

5.3 *Merchandizing Surplus*

The merchandizing surplus in state θ is

$$MS(\theta) = \sum_{m=1}^{M}\left(Q_m^d(\theta) - Q_m^s(\theta)\right)p_m(\theta) = -\sum_{m=1}^{M} Q_m(\theta)\, p_m(\theta).$$

Schweppe et al. (1988) show that

$$MS(\theta) = \mu_e(\theta)\, LO\left(\underline{\mathbf{Q}}(\theta)\right) + \sum_{l=1}^{L} \eta_l(\theta)\, \Phi_l.$$

The proof proceeds as follows. Substituting the *LMPs* given by equations (3) in the expression of the merchandizing surplus yields

$$MS(\theta) = -\sum_{m=1}^{M} Q_m(\theta)\left[\mu_e(\theta)\left(1-\frac{\partial LO}{\partial Q_m}\right)-\sum_{l=1}^{L} H_{lm}\eta_l(\theta)\right]$$
$$= -\mu_e(\theta)\sum_{m=1}^{M} Q_m(\theta)\left(1-\frac{\partial LO}{\partial Q_m}\right)+\sum_{m=1}^{M} Q_m(\theta)\sum_{l=1}^{L} H_{lm}\eta_l(\theta).$$

Consider the first term on the right-hand side. Observe that (i) the energy balance yields $\sum_{m=1}^{M} Q_m(\theta) = LO\left(\underline{\mathbf{Q}}(\theta)\right)$, and (ii) since $LO\left(\underline{\mathbf{Q}}(\theta)\right)$ is a quadratic function of net injections, $LO\left(\underline{\mathbf{Q}}(\theta)\right) = \frac{1}{2}\sum_{m=1}^{M} Q_m(\theta)\frac{\partial LO}{\partial Q_m}$.

Therefore,

$$\sum_{m=1}^{M} Q_m(\theta)\left(1-\frac{\partial LO}{\partial Q_m}\right) = LO\left(\underline{\mathbf{Q}}(\theta)\right) - 2LO\left(\underline{\mathbf{Q}}(\theta)\right) = -LO\left(\underline{\mathbf{Q}}(\theta)\right).$$

Consider now the second term on the right-hand side. Inverting the order of summations yields

$$\sum_{m=1}^{M} Q_m(\theta)\sum_{l=1}^{L} H_{lm}\eta_l(\theta) = \sum_{l=1}^{L}\eta_l(\theta)\sum_{m=1}^{M} H_{lm}Q_m(\theta) = \sum_{l=1}^{L}\eta_l(\theta)\varphi_l(\theta) = \sum_{l=1}^{L}\eta_l(\theta)\Phi_l,$$

since $\eta_l(\theta) > 0 \Leftrightarrow \varphi_l(\theta) = \Phi_l$.

Putting the two pieces together proves the result.

Locational marginal pricing thus generates a positive surplus for the market maker, equal to the value of the losses (valued at $\mu_e(\theta)$), plus value of the congestion on the network.

5.4 Optimal Transmission Capacity

We first suppose that increasing capacity on one line has no impact on other lines' capacities, i.e., $\frac{\partial \Phi_j}{\partial \Phi_l} = 0$ for $j \neq l$. Relaxing this assumption will be discussed at the end of this section. The optimum is defined by

$$max_{Q^s_m(\theta),Q^d_m(\theta),\Phi_l}\mathbb{E}\left[\sum_{m=1}^{M}\left\{U_m\left(Q^d_m(\theta),\theta\right)-c_mQ^s_m(\theta)\right\}\right]-\sum_{m=1}^{M} r_mk_m - \Gamma(\Phi)$$
$$st: \begin{cases} Q^s_m(\theta) \leq k_m\ \forall m & (\lambda_m(\theta)) \\ \sum_{m=1}^{M} Q^s_m(\theta) = \sum_{m=1}^{M} Q^d_m(\theta) + LO\left(\underline{\mathbf{Q}}(\theta)\right) & (\mu_e(\theta)) \\ \sum_{l=1}^{m} H_{lm}Q_m(\theta) \leq \Phi_l\ \forall l & (\eta_l(\theta)) \end{cases}.$$

Differentiating the Lagrangian with respect to Φ_l, necessary conditions satisfied by the optimal grid are

$$\Lambda_l(\Phi) = \mathbb{E}[\eta_l(\theta)] = \frac{\partial \Gamma(\Phi^*)}{\partial \Phi_l} \ \forall l. \tag{4}$$

As mentioned previously, the cost function $\Gamma(\Phi)$ may not be well behaved. The system of equations (4) may not admit a solution or admit multiple solutions.

5.5 Optimal Regulatory Contract

Lemma 1 extends to a general network. Therefore, so does Proposition 1, assuming the system of equations (4) admits a unique solution.

Redispatch cost
The logic of the two-market network applies. If the power flows exceed the interconnection capacity, the market operator must modify the transactions to adjust the flows. At import-constrained nodes, she purchases power from the producers (who are constrained-on) and consumers (who are constrained-off) in order to increase production hence reduces imports. Simultaneously, at export-constrained nodes, she reduces production and increases demand. The market operator adjusts until the power flow on congested interconnections are exactly equal to the capacity of these lines. The dispatch is thus exactly identical to the nodal pricing one.

We now compute the redispatching cost. Consider an export-constrained node, characterized by $Q_m^s(\mathbf{\Phi}) < Q_m^{sU}$ and $Q_m^d(\mathbf{\Phi}) > Q_m^{dU}$. For $x \in \left[Q_m^{dU}, Q_m^d(\mathbf{\Phi})\right]$, constrained-on consumers are compensated for their surplus loss from consuming, hence are paid $\left(p^U - P_m(x,\theta)\right)$. For $x \in \left[Q_m^s(\mathbf{\Phi}), Q_m^{sU}\right]$, constrained-off producers are paid the net operating profit they would have received: $\left(p^U - c_m(x)\right)$. The total redispatching cost at export-constrained nodes in state θ is

$$\int_{Q_m^{dU}}^{Q_m^d(\mathbf{\Phi})} \left(p^U - P_m(x,\theta)\right) dx + \int_{Q_m^s(\mathbf{\Phi})}^{Q_m^{sU}} \left(p^U - c_m(x)\right) dx.$$

Consider an import-constrained node, characterized by $Q_m^s(\mathbf{\Phi}) > Q_m^{sU}$ and $Q_m^d(\mathbf{\Phi}) < Q_m^{sU}$. For $x \in \left[Q_m^{sU}, Q_m^s(\mathbf{\Phi})\right]$, constrained-on producers are paid their cost $c_m(x)$, hence their constrained-on payment is $\left(c_m(x) - p^U\right)$. For $x \in \left[Q_m^d(\mathbf{\Phi}), Q_m^{dU}\right]$, constrained-off consumers receive the net surplus they would have derived from consuming: $\left(P_m(x,\theta) - p^U\right)$. The total redispatching cost at import-constrained nodes in state θ is

$$\int_{Q_m^{sU}}^{Q_m^s(\mathbf{\Phi})} \left(c_m(x) - p^U(\theta)\right) dx + \int_{Q_m^d(\mathbf{\Phi})}^{Q_m^{dU}} \left(P_m(x,\theta) - p^U(\theta)\right) dx.$$

Comparing the expressions above, we observe redispatching costs are formally identical at export- and import-constrained nodes.

In state θ, the market operator purchases and sells the constrained quantities $Q^s_m(\Phi, \grave{}\,)$ and $Q^d_m(\Phi, \grave{}\,)$ at the unconstrained price $p^U(\theta)$ and pays the redispatching cost. Observing that $\sum_{m=1}^{M} Q_m(\theta) = LO\left(\underline{\mathbf{Q}}(\theta)\right)$, her net profit is

$$\Pi^{CT}(\Phi) = -\mathbb{E}\left[\begin{array}{c} p^U LO\left(\underline{\mathbf{Q}}(\theta)\right) + \\ \sum_{m=1}^{M}\left(\int_{Q^{sU}_m}^{Q^s_m(\mathbf{\Phi})}\left(c_m(x) - p^U(\theta)\right)dx + \int_{Q^d_m(\mathbf{\Phi})}^{Q^{dU}_m}\left(P_m(x,\theta) - p^U(\theta)\right)dx\right)\end{array}\right].$$

The market operator pays the redispatch cost and the transmission losses, valued at the unconstrained price.

Marginal redispatching cost

Lemma 2 *Suppose* $\frac{\partial H_{jm}}{\partial \Phi_l} = 0$ *and* $\frac{\partial \Phi_j}{\partial \Phi_l} = 0$ *for* $j \neq l$. *The marginal redispatching cost with respect to capacity on line* l *is equal to the marginal value of line* l*:*

$$\frac{\partial \Pi^{CT}}{\partial \Phi_l} = \Lambda_l(\Phi).$$

Proof Since $LO\left(\underline{\mathbf{Q}}(\theta)\right) = \sum_{m=1}^{M} Q_m(\theta)$, differentiation with respect to Φ_l yields:

$$\begin{aligned}\frac{\partial \Pi^{CT}}{\partial \Phi_l} &= -\mathbb{E}\left[p^U \sum_{m=1}^{M}\frac{\partial Q_m}{\partial \Phi_l} + \sum_{m=1}^{M}\left(\left(p_m - p^U\right)\frac{\partial Q^s_m}{\partial \Phi_l} - \left(p_m - p^U\right)\frac{\partial Q^d_m}{\partial \Phi_l}\right)\right] \\ &= -\mathbb{E}\left[p^U \sum_{m=1}^{M}\frac{\partial Q_m}{\partial \Phi_l} + \sum_{m=1}^{M}\left(p_m - p^U\right)\frac{\partial Q_m}{\partial \Phi_l}\right] \\ &= -\mathbb{E}\left[\sum_{m=1}^{M} p_m \frac{\partial Q_m}{\partial \Phi_l}\right].\end{aligned}$$

The next step follows the proof of the expression of the merchandizing surplus. Substituting the *LMPs* given by equations (3) in the expression of the marginal redispatching cost yields

$$\begin{aligned}\frac{\partial \Pi^{CT}}{\partial \Phi_l} &= -\mathbb{E}\left[\sum_{m=1}^{M}\left[\mu_e(\theta)\left(1 - \frac{\partial LO}{\partial Q_m}\right) - \sum_{j=1}^{L} H_{jm}\eta_j(\theta)\right]\frac{\partial Q_m}{\partial \Phi_l}\right] \\ &= -\mathbb{E}\left[\mu_e(\theta)\sum_{m=1}^{M}\left(1 - \frac{\partial LO}{\partial Q_m}\right)\frac{\partial Q_m}{\partial \Phi_l} - \sum_{m=1}^{M}\sum_{j=1}^{L} H_{jm}\eta_j(\theta)\frac{\partial Q_m}{\partial \Phi_l}\right].\end{aligned}$$

Consider the first term on the right-hand side. Since $\sum_{m=1}^{M} Q_m(\theta) = LO\left(\mathbf{Q}(\theta)\right)$, $\sum_{m=1}^{M} \frac{\partial Q_m}{\partial \Phi_l} = \frac{\partial LO(\mathbf{Q})}{\partial \Phi_l}$, and by construction $\frac{\partial LO(\mathbf{Q})}{\partial \Phi_l} = \sum_{m=1}^{M} \frac{\partial LO}{\partial Q_m}\frac{\partial Q_m}{\partial \Phi_l}$. Therefore,

$$\sum_{m=1}^{M}\left(1 - \frac{\partial LO}{\partial Q_m}\right)\frac{\partial Q_m}{\partial \Phi_l} = \sum_{m=1}^{M}\frac{\partial Q_m}{\partial \Phi_l} - \sum_{m=1}^{M}\frac{\partial LO}{\partial Q_m}\frac{\partial Q_m}{\partial \Phi_l} = \frac{\partial LO(\mathbf{Q})}{\partial \Phi_l} - \frac{\partial LO(\mathbf{Q})}{\partial \Phi_l} = 0.$$

Consider now the second term on the right-hand side. Inverting the order of summations yields

$$\sum_{m=1}^{M}\sum_{j=1}^{L} H_{jm}\eta_j(\theta)\frac{\partial Q_m}{\partial \Phi_l} = \sum_{j=1}^{L}\eta_j(\theta)\sum_{m=1}^{M} H_{jm}\frac{\partial Q_m}{\partial \Phi_l}.$$

Since $\eta_j(\theta) > 0 \Leftrightarrow \varphi_j(\theta) = \Phi_j$, we limit the first sum to lines such that $\varphi_j(\theta) = \sum_{m=1}^{M} H_{jm}Q_m = \Phi_j$. Assuming $\frac{\partial H_{jm}}{\partial \Phi_l} = 0$,

$$\sum_{m=1}^{M} H_{jm}\frac{\partial Q_m}{\partial \Phi_l} = \frac{\partial\left(\sum_{m=1}^{M} H_{jm}Q_m\right)}{\partial \Phi_l} = \frac{\partial \Phi_j}{\partial \Phi_l}.$$

Thus, if $\frac{\partial \Phi_j}{\partial \Phi_l} = 0$ for $j \neq l$,

$$\sum_{j=1}^{L}\eta_j(\theta)\frac{\partial \Phi_j}{\partial \Phi_l} = \eta_l(\theta) \Rightarrow \mathbb{E}\left[\sum_{j=1}^{L}\eta_j(\theta)\frac{\partial \Phi_j}{\partial \Phi_l}\right] = \Lambda_l(\Phi).$$

Putting the two pieces together and taking the expectation prove the result.

Lemma 2 is sufficient to extend Proposition 1 to a general N-node network:

Proposition 2 *If a for-profit transmission asset owner and operator are responsible for the redispatching cost on the grid, he selects the optimal capacity on every line l.*

Extensions

Leautier (2000) shows that Lemma 2 continues to hold if we assume $\frac{\partial H_{jm}}{\partial \Phi_l} \neq 0$, which is realistic. Physically increasing the capacity on an interconnection is likely to modify its admittance, which then modifies the admittance transfer matrix. If this occurs, new terms are added, but the economic intuition is unchanged.

As mentioned earlier, increasing the transfer capacity on line l may modify the $(N-1)$ contingency on line j, hence impact the transfer capacity on line j. Full treatment of this issue is beyond the scope of this chapter. The main issue is likely to be that the cost function may no longer be well behaved. Assuming the cost function remains well behaved, the economic intuition and the main results hold.

Suppose for example transfer capacity on line 2 is impacted by transfer capacity on line 1 and another decision X: $\Phi_2(\Phi_1, X)$. Transfer capacities on all other lines are independent.

Differentiating the Lagrangian of the optimal grid problem with respect to Φ_1 and X yields the two first-order conditions:

$$\mathbb{E}\left[\eta_1(\Phi,\theta)+\eta_2(\Phi,\theta)\frac{\partial\Phi_2}{\partial\Phi_1}\right]=\frac{\partial\Gamma}{\partial\Phi_1}+\frac{\partial\Gamma}{\partial\Phi_2}\frac{\partial\Phi_2}{\partial\Phi_1},$$

and

$$\mathbb{E}\left[\eta_2(\Phi,\theta)\frac{\partial\Phi_2}{\partial X}\right]=\frac{\partial\Gamma}{\partial\Phi_2}\frac{\partial\Phi_2}{\partial X}.$$

Suppose this system admits a unique solution, which is a maximum.

Lemma 2's derivations yield

$$\frac{\partial\Pi^{CT}}{\partial\Phi_1}=\mathbb{E}\left[\sum_{j=1}^{L}\eta_j(\theta)\frac{\partial\Phi_j}{\partial\Phi_1}\right]=\mathbb{E}\left[\eta_1(\theta)+\eta_2(\Phi,\theta)\frac{\partial\Phi_2}{\partial\Phi_1}\right]$$

and

$$\frac{\partial\Pi^{CT}}{\partial X}=\mathbb{E}\left[\sum_{j=1}^{L}\eta_j(\theta)\frac{\partial\Phi_j}{\partial X}\right]=\mathbb{E}\left[\eta_2(\Phi,\theta)\frac{\partial\Phi_2}{\partial X}\right].$$

Proposition 2 thus continues to hold.

Suppose now the transfer capacity varies across states of the world. For example $\Phi_l=\Phi_l(X_l,\theta)$, where X_l is an observable measure (e.g., voltage of the line), or possibly a vector of observable measures. The cost function is then $\Gamma(\mathbf{X})$, and equations (4) become

$$\mathbb{E}\left[\eta_l(\theta)\frac{\partial\Phi_l}{\partial X_l}\right]=\frac{\partial\Gamma(\mathbf{X})}{\partial X_l}.$$

Proposition 2 thus continues to hold.

6 Concluding Observations

After reading this chapter, readers are (hopefully) convinced that designing a regulatory mechanism to induce a monopoly transmission grid owner and operator to optimally increase the capacity of the grid is straightforward technically: Compute the redispatch cost and make the company responsible for it at the margin, as was done for ten years in England and Wales in the 1990s.

As is often the case when it comes to policy making in the electricity industry, the challenge arises from the political economy, not the microeconomics. In the USA, transmission operation and ownership are separated, which make it nigh impossible to implement such a regulatory mechanism. In Europe, the transmission grid is fragmented among national operators. Even if they built the optimal national grid, there would be no mechanism to induce optimal expansion between countries.

The political economy of moving to Transcos in the USA, where ISOs have been operating for more than twenty years, is challenging: Established institutional arrangements are difficult to transform. Power utilities are hesitant to part with the stable and regulated cash flows provided by transmission assets. As long as they own the grid, they will not be able to operate it, to protect free access. Infrastructure funds could step in, but they would need to roll up transmission assets owned by numerous companies in multiple states. While this outcome is possible (and desirable), it seems highly unlikely, barring a significant change in public policy.

In Europe, the odds are slightly better than in the USA, although still low. Existing TSOs will want to merge to grow (some already have). Attracted by the savings potential and the gains from incentive regulation, they may be able to convince national governments, which are also often their owners that a transnational Transco does not pose a threat to national security of supply. The European Commission may seize the opportunity.

References

N. Astier, *Essays on the Economics of Modern Electricity Markets*. Ph.D. thesis, Toulouse School of Economics (June 2017)

K. Bell, R. Green, I. Kockar, G. Ault, J. McDonald, Academic review of transmission charging arrangements. Report produced on behalf of the Gas and Electricity Markets Authority, Project TransmiT (2011)

M. Boiteux, La tarification des demandes en pointe: application de la theorie de la vente au cout marginal. Revue Generale de l'Electricite pp. 321–40 (August 1949)

J. Bushnell, S. Stoft, Electric grid investment under a contract network regime. J. Regul. Econ. **10**, 61–79 (1996)

W. Hogan, Contract networks for electric power transmission. J. Regul. Econ. **4**, 211–242 (1992)

W. Hogan, J. Rosellon, I. Vogelsang, Towards a combined merchant-regulatory mechanism for electricity transmission expansion. J. Regul. Econ. **38**(2), 113–143 (2010)

T.-O. Leautier, *Imperfect Market and Imperfect Competition: An Introduction to the Microeconomics and Political Economy of Power Markets* (MIT Press, Cambridge and London, 2019)

T.-O. Leautier, Regulation of an electric power transmission company. Energy J. **21**(4), 61–92 (2000)

M. Ovaere, *Economics of Power Transmission Reliability*. Ph.D. thesis, KU Leven (2017)

J.I. Perez-Arriaga, F.J. Rubio, J.F. Puerta Gutierrez, Marginal pricing of transmission services: An analysis of cost recovery. IEEE Trans. Power Syst. **10**, 546–553 (1995)

J. Rosellon, T. Kristiansen (eds.), *Financial Transmission Rights: Analysis, Experiences and Prospects* (Springer, London, 2013)

J. Rosellon, I. Vogelsang, H. Weigt, Long-run cost functions for electricity transmission. Energy J. **33**, 131–160 (2012)

F. Schweppe, M. Caramanis, R. Tabors, R. Bohn, *Spot Pricing of Electricity* (Kluwer, Norwell, MA, 1988)

H.P. Young, *Handbook of Game Theory with economic applications*, Vol. 2, chapter Cost allocation (Elsevier Science, Amsterdam, 1994), pp. 1195–1235

Transmission Planning and Operation in the Wholesale Market Regime

Frank A. Wolak

1 Introduction

The wholesale market regime implies a dramatically different role for transmission planning and operation regulation relative to the former vertically-integrated monopoly regime. The monopoly supplier has the potential to capture any economies to scope between planning and operating the transmission network and building and operating all of the generation units connected to this transmission network because a single vertically-integrated firm performs all of these tasks. In contrast, suppliers in the wholesale market regime are typically financially independent of the transmission network owner and system operator and therefore condition their entry and operating decisions on the configuration of the transmission network.

This difference in the incentives generation unit owners face for locating and operating their units in the wholesale market regime versus the vertically-integrated monopoly regime has wide-ranging implications for the design and operation of the transmission network in the two regimes. The purpose of this paper is to explore these implications in order to adapt the transmission planning and operation regulatory process fully to the wholesale market regime, particularly one with a significant amount of intermittent renewables.

In the wholesale market regime, the configuration of the transmission network determines the extent of competition that suppliers face, with a more extensive transmission network facing suppliers with greater competition, which increases their unilateral incentive to submit offer prices closer to their marginal cost of production. This logic implies that a supplier owning low-cost generation capacity in a portion of the grid with limited transmission interconnection capacity to the remainder of the

F. A. Wolak (✉)
Program on Energy and Sustainable Development and Department of Economics, Stanford University, Stanford, CA, USA
e-mail: wolak@stanford.edu

M. R. Hesamzadeh et al. (eds.), *Transmission Network Investment in Liberalized Power Markets*, Lecture Notes in Energy 79,
https://doi.org/10.1007/978-3-030-47929-9_4

grid may find it expected profit-maximizing to withhold output from this capacity in order to raise the price it receives for the energy this capacity does supply. In contrast, a vertically-integrated, output price-regulated monopoly has little incentive to withhold output from low-cost units because this would only increase the total cost of serving demand throughout its service area with no corresponding revenue increase because the monopoly's revenues are the product of an output price set by the regulator and the realized demand at that price.

Because the configuration of the transmission network impacts the extent of competition that a supplier faces in the wholesale market regime, the transmission network configuration that leads to the lowest average price of electricity delivered to final consumers in the wholesale market regime is likely to require more capacity than the configuration of the transmission network that achieves this same outcome in the vertically-integrated monopoly regime. Below, I present two simple models to illustrate this point. This logic implies different measures of grid reliability in the two regimes—engineering reliability in the vertically-integrated monopoly regime and economic reliability in the wholesale market regimes.

The divergent roles of the transmission network in the two regimes arise from these two definitions of grid reliability. In the vertically-integrated monopoly regime, changes in the configuration of the transmission network can improve the performance of an *imperfectly regulated* vertically-integrated monopoly. In the wholesale market regime, changes in the configuration of the transmission network can improve the performance of an *imperfectly competitive* wholesale electricity market.

For each environment, it is only possible to obtain second-best solutions for the industry in the sense of Lipsey and Lancaster (1956), because certain features of the economic environment make it is impossible to implement the least-cost or first-best solution. For the vertically-integrated monopoly, a second-best solution is only possible because of the asymmetric information problem between the regulator and the monopolist. For the wholesale market regime, a second-best solution is only possible because large suppliers have the ability and incentive to exercise unilateral market power. Many studies have documented the fact that suppliers in wholesale electricity markets have the ability and incentive to exercise unilateral market power.[1]

This logic also implies that the "second-best" the optimal configuration of the transmission network depends on how transmission congestion is managed in the short-term market. Specifically, a single-pricing-zone wholesale market with a pay-as-bid mechanism for managing congestion implies a different least-cost-of-delivered-electricity transmission network configuration from a locational marginal pricing (LMP) wholesale market that integrates congestion management into the market mechanism.[2] A multiple-pricing-zone wholesale market design implies a

[1]For example, see Wolfram (1999) and Wolak and Patrick (2001) for the case of the England and Wales, Wolak (2000) for the case of Australia, Borenstein et al. (2002) and Wolak (2003b) for the case California, Bushnell et al. (2008b) for the PJM Interconnection, New England ISO, and California, and Wolak (2009) and McRae and Wolak (2014) for New Zealand.

[2]See Schweppe et al. (2013) for a detailed discussion of locational marginal pricing.

distinctly different least-cost-of-delivered electricity transmission network configuration from the ones that are least-cost for the single-zone or LMP market design.

The need to match the transmission planning and regulatory process to the wholesale market design requires a substantially more sophisticated transmission network planning process than the vertically-integrated monopoly regime. The transmission planner must recognize the fact that generation unit owners and load-serving entities will account for the current and future configuration of the transmission network in making their expected profit-maximizing entry and operating decisions. To this end, I outline a general forward-looking framework for evaluating transmission network expansions in the wholesale market regime. In the language of game theory, the transmission network planner should behave as a Stackelberg leader taking into account the best-reply entry and operating decisions of generation unit owners and load-serving entities in its planning and construction decisions.

The regulator's role of protecting consumers from retail electricity prices that reflect the exercise of unilateral market power in the wholesale market regime also implies a different criterion for measuring the economic benefits of a transmission expansion in the wholesale market regime versus the vertically-integrated monopoly regime. As Joskow (1974) notes, United States regulators set the vertically-integrated monopolist's output price to allow it the opportunity to recover all prudently incurred costs necessary to serve demand, including an adequate return on capital invested. Any transmission upgrade that reduces the total cost of serving demand throughout the utility's service territory more than the cost of the upgrade will allow the regulator to reduce the regulated price and should therefore be undertaken.

For the wholesale market regime, the regulator cannot ensure that generation unit owners receive output prices that only allow them to recover their prudently incurred costs. In the wholesale market regime, the regulator can only set prices for use of the transmission network and determine whether to allow revenue recovery for a transmission expansion. The extent of competition faced by each supplier determines whether the market price only recovers the incurred cost of that supplier. Because the configuration of the transmission network impacts the ability of suppliers to exercise unilateral market power which, in turn, impacts the wholesale electricity price, the regulator should account for the impact of the configuration of the transmission network on the ability and incentive of suppliers to exercise unilateral market power in deciding whether to approve a transmission expansion.

This logic implies an additional source of economic benefits from transmission expansions in the wholesale market regime besides simply reducing the production costs associated with serving demand. A transmission upgrade typically increases the extent of competition supplier's face, which then causes these suppliers to offer to supply energy at prices closer to their marginal cost of production. Lower offer prices lead to lower wholesale energy prices and lower total wholesale energy payments by electricity consumers. Consequently, in the wholesale market regime, if the total surplus increase to electricity consumers from an upgrade is less than the cost of the upgrade then the upgrade should be undertaken because it increases net surplus to electricity consumers.[3] A consumer benefits focus on regulating transmission planning and pricing in the wholesale market regime also argues for a competitive procurement process for an expansion determined by the planner. Joskow (2019) describes a recent experience with competitive transmission expansion procurement processes in the United States.

Because the wholesale market regime involves the risk that suppliers can exercise a substantial amount of unilateral market power in a relatively short period of time, there is likely to be significantly more uncertainty in the realized economic benefits of a transmission expansion.[4] Consequently, presenting a single point estimate of the economic benefits of a transmission expansion or a small number of scenario-based estimates, as is typically the case in the vertically-integrated monopoly regime, does not convey sufficient information about the range and likelihood of specific values of the realized benefits of transmission expansion in the wholesale market regime. This logic argues for presenting an estimated distribution of the realized economic benefits so that the regulator can fully assess the insurance value of the upgrade. To this end, I propose a general methodology for computing the distribution of realized economic benefits from an upgrade in the wholesale market regime.

Transmission network expansions can provide insurance against wholesale market outcomes that result in the exercise of significant unilateral market power or periods of local supply scarcity. It may, therefore, be prudent for a regulator to insure against these extreme market outcomes in the form of a transmission expansion, even if the expected realized benefit from an upgrade is less than the cost. The potential for large economic losses is the same reason that consumers purchase insurance against damages to their homes and car. For the same reason that consumers do not view the money for insurance against these large economic losses as wasted if this damage does not occur, regulators may feel the same way about insurance against market outcomes where a substantial amount of unilateral market power is exercised or local scarcity conditions arise. The insurance value of a transmission expansions is also likely to be even greater as the share of intermittent renewable generation in a region increases, because of the need to import more energy from neighboring control areas when within-region renewable energy production is low.

Both the competitiveness benefits and insurance value of the transmission expansions in the wholesale market regime argue for a transmission planning process over the entire interconnected transmission network because upgrading one link of an interconnected transmission network can change transfer capacities between many other parts of the transmission network. This logic implies the need for an inter-regional transmission network planning process to account for these economic

[3]The change in the variable cost of serving demand is not the relevant criteria in the wholesale market regime. Many upgrades that significantly reduce the ability of suppliers to exercise unilateral market power and therefore yield total surplus increases for consumers that are greater than the cost of the upgrade would not be undertaken by this criteria.

[4]Examples of wholesale markets where substantial amounts of unilateral market power was exercised are: California, Borenstein et al. (2002); New Zealand, Wolak (2009); and Colombia, McRae and Wolak (2016).

benefits, which is a substantial change relative to the state-level planning process in the United States and country-level planning process in other parts of the world.

Transmission expansions to support the deployment of renewable resources are also impacted by the paradigm shift in measuring the economic benefits of transmission expansions in the wholesale market regime. The location of rich sources of renewable resources is typically well-known and the only way for major load centers to access these resources is through transmission network interconnections between these renewable resource locations and major load centers. A forward-looking transmission planning process in a region with ambitious renewable energy goals should build transmission network interconnections between these regions and major load centers with sufficient interconnection capacity for the load centers to access these resources, anticipating the expected profit-maximizing entry decisions of investors in new renewable generation capacity. Again this planning process should take place over the geographic scope of the interconnected transmission network, not just at the state-level or country-level.

A final issue introduced by the wholesale market regime is the increased economic benefits associated with coordinating the planning of the transmission network over the largest possible geographic region and with the planning of the natural gas transmission network. The location of the natural gas transmission network capacity will influence the expected profit-maximizing location decisions of natural gas-fired generation unit owners. Consequently, a forward-looking and coordinated electricity transmission and natural gas transmission network planning process has the potential to increase the competitiveness of both wholesale electricity and natural gas markets.

The remainder of the paper proceeds as follows. The next section explains the distinction between engineering reliability and economic reliability. Section 3 demonstrates why these two criteria imply different optimal configurations of the transmission network. Specifically, the wholesale market regime typically requires more transmission capacity than the same industry structure under the vertically-integrated monopoly regime. Section 4 discusses the consequences of transmission planners and regulators continuing to rely on transmission expansion assessment methodologies from the former vertically-integrated monopoly regime. In particular, it becomes increasingly difficult to protect consumers from wholesale prices that reflect the exercise of unilateral market power, which reduces the likelihood consumers will realize benefits from electricity industry restructuring. Section 5 proposes a general transmission planning process for the wholesale market regime that is forward-looking, anticipating the profit-maximizing entry and operating decisions of generation unit owners. This planning process assumes a distribution of the future system conditions that impact the realized economic benefits from a transmission expansion. Section 6 argues that the wholesale market regime requires a substantially more sophisticated economic modeling framework for transmission planning relative to the vertically-integrated monopoly regime. This process should be broader in geographic scope and be coordinated with the input fuel infrastructure planning process and the location of renewable energy resources. This section also discusses the viability of a pure market-based approach to transmission planning and expansion where builders of transmission infrastructure finance projects from locational price differences. Section 7 argues that the distribution of the realized economic benefits from most transmission expansions in the wholesale market regime is significantly more positively skewed. This logic implies that transmission expansions provide insurance against these rare, but very costly events, which implies that a single point estimate for the economic benefits of a transmission expansion may not be as informative as the distribution of these realized benefits.

2 Grid Reliability in the Vertically-Integrated Monopoly Versus Wholesale Market Regime

The transmission planning process in an electricity supply industry with a formal wholesale market is fundamentally different from the process that existed when the industry was a vertically-integrated geographic monopoly that built and operated the transmission network, the portfolio of generation facilities, and the local distribution networks for a given geographic region. This difference is the result of the incentives faced by generation unit owners in the wholesale market regime versus the vertically-integrated monopoly regime.

A crucial determinant of the reliability of the transmission network in the vertically-integrated geographic monopoly regime is that a single firm owns, or at least controls, the operation of all generation resources available to serve final consumers in that geographic region. The second feature is the fact that the relevant regulatory authority prospectively sets the output price of the vertically-integrated monopoly and requires it to serve all demand at this price, which effectively makes the monopoly's total revenues invariant to how it serves this demand.

Consequently, the geographic monopoly market structure combined with retail price regulation eliminates many of the incentives for inefficient operation by generation unit owners that can arise in the wholesale market regime.[5] Because the revenues received by the vertically-integrated monopoly are largely independent of how it operates its generation units, a profit-maximizing monopolist has an incentive to operate its available generation units to minimize the cost of serving demand. This logic implies that in the vertically-integrated monopoly regime a transmission network is deemed to be reliable if there is sufficient capacity for the firm that owns and operates all of the generation units in the region and has interconnections with neighboring control areas to maintain a pre-specified level of reliability of the supply of electricity to final consumers.

This definition of reliability is based purely on engineering criterion because it assumes the transmission network, the fleet of generation units, and portfolio of supply contracts from outside of the control area are all owned and operated by the

[5]The vertically-integrated monopoly regime creates other sources of inefficiencies in generation investment and system expansion and operation not present in the wholesale market regime. Wolak (2014) describes the causes and consequences of these inefficiencies in the vertically-integrated and wholesale market regimes.

monopolist to serve demand in real-time. A grid that satisfies these criteria is said to meet the engineering standard for reliability.

In the wholesale market regime, the generation segment of the industry is open to competition, and the transmission network is operated as an open access facility for all generation unit owners and retailers. The regulator has a limited ability to specify where new generation facilities will be built or how new and existing generation facilities will be operated. Privately-owned generation unit owners are likely to build new facilities at the most profitable locations, which could be near a major load center and/or on the constrained side of congested transmission paths. Moreover, generation unit owners will offer their facilities into the short-term wholesale market and operate them to maximize the return to their shareholders from this investment taking into account the configuration of the transmission network and the mechanism used to price congestion in the transmission network.

For this reason, transmission planning and operation is now a crucial component of the wholesale market regime regulatory process because the configuration of the transmission network impacts the competitiveness of the wholesale electricity market. The regulator can protect electricity consumers from prices that reflect the exercise of significant unilateral market power through transmission expansions that increase the number of independent generation unit owners able compete to supply energy at each location in the transmission network. These upgrades make it less likely that a generation unit owner will find it unilaterally profit-maximizing to withhold output and congest the transmission network in order to increase the price it receives for the output that it sells.

Although many wholesale electricity markets, particularly those in the United States, have local market power mitigation mechanisms in place to limit the ability of suppliers to take advantage of the configuration of the transmission network in order to raise the price they receive for their output, these local market power mitigation mechanisms do not completely eliminate the incentive or ability of suppliers to exercise unilateral market power. Consequently, there is still likely to be a role for transmission expansions to increase the extent of competition suppliers face and thereby limit their ability and incentive to exercise unilateral market power.

The new role of the transmission planning and operation process in limiting the ability and incentive of market participants to exercise unilateral market power suggests a new definition of reliability for the wholesale market regime. An economically reliable transmission network has sufficient capacity to all locations in the transmission network so that suppliers at those locations face significant competition from enough independent suppliers to cause them to offer to supply energy at close to their marginal cost the vast majority of the hours of the year.

In the language of Wolak (2000), an economically reliable transmission network is one that faces all suppliers with very elastic residual demand curves the vast majority of hours of the year. As shown by Wolak (2000) for the case of Australia and McRae and Wolak (2014) for the case of New Zealand, the residual demand curve a supplier faces determines its ability to exercise unilateral market power in a formal wholesale market. The more firms that can compete to sell electricity at a supplier's location in the transmission network, the flatter is the residual demand curve that supplier faces. Increasing the capacity of a transmission network will typically increase the number of competitors a supplier faces and thereby flatten the distribution of residual demand curves that supplier faces.

Because of the role that transmission upgrades play in reducing the ability and incentive of suppliers to exercise unilateral market power in the wholesale market regime, the relevant planning standard in this regime is economic reliability. The first few years of operation in all of the restructured markets in the United States demonstrated that transmission networks that met the engineering reliability standards were insufficient to operate single-zone and multi-zone wholesale electricity markets. These markets experienced levels of transmission congestion not experienced in the former vertically-integrated regime, and this led to significant transmission network expansions and a shift to LMP market designs that price all transmission network and other relevant operating constraints in the day-ahead and real-time short-term markets.

3 Optimal Configuration of Transmission Network in the Vertically-Integrated Monopoly Versus Wholesale Market Regime

Because transmission expansions in the wholesale market regime limit the ability and incentive of suppliers to withhold output to raise wholesale electricity prices–an action that the vertically-integrated monopolist has little incentive to undertake–the optimal configuration of the transmission network in the wholesale market regime is likely to require more capacity than the optimal configuration in the vertically-integrated monopoly regime.

3.1 Second-Best Solutions for Monopoly and Wholesale Market Regimes

The first step in this argument must recognize that there is no single optimal transmission network configuration for both regimes. For the case of the vertically-integrated monopoly regime, the fact that the monopolist knows more about how to produce its output than the regulator implies the existence of informational asymmetries between the firm and regulator. As discussed in Wolak (1994), the regulator can only know the monopolist's incurred cost of producing its output, it can never know the least-cost way to produce the monopolist's output. This implies that transmission expansions in the vertically-integrated monopoly regime can only improve the performance of an imperfectly price-regulated monopoly. The regulator can only determine if a transmission expansion is likely to reduce the monopolist's incurred cost of serving demand more than the incurred cost of the transmission network expansion. Because of the informational asymmetries between the regulator and monopolist concerning the monopolist's cost of production, the regulator can never determine the least-cost configuration of the monopolist's transmission network.

For the case of the wholesale market regime, finding the least-cost transmission network configuration is impossible because the variable costs of generation units are not known by the transmission planner or system operator. In this regime, generation units are called upon to supply electricity based on offer prices, not variable costs. Even if the regulator knew each generation unit owner's minimum cost of production, it is extremely unlikely that all suppliers would find it unilaterally profit-maximizing to submit their minimum cost of supplying electricity as their offer price during all hours of the year. In all offer-based wholesale electricity markets, some generation unit owners have the ability and incentive to exercise unilateral market power during a number of hours of the year. This means that the resulting dispatch of generation units would not be the least-cost, and the transmission network that is optimal for the least-cost dispatch of generation units would not be the least cost to consumers for the case that suppliers exercised unilateral market power during those hours of the year. Consequently, at best, transmission expansions in the wholesale market regime can only improve the performance of an imperfectly competitive wholesale electricity market.

Therefore the optimal configuration of the transmission network in both regimes necessarily implies solving for a "second-best" transmission network configuration in the sense of Lipsey and Lancaster (1956). In the case of the vertically-integrated monopoly regime, the informational asymmetries about the monopolist's production process and the demand it faces between the firm and the regulator are the constraint that implies an optimal "second-best" solution. In the case of the wholesale market regime, the fact that wholesale electricity markets are not perfectly competitive and suppliers exercise unilateral market power implies an optimal "second-best" solution in the wholesale market regime.

3.2 Why Wholesale Market Regime Is Likely to Require More Transmission Capacity?

This section presents two simple models that illustrate the economic forces that imply the "second-best" optimal amount of transmission capacity for a region in the wholesale market regime is typically larger than that it is for the same region in the vertically-integrated monopoly regime. The first model focuses on the mechanism that more transmission capacity allows lower cost sources of electricity to supply more energy to final consumers. The second model focuses on the mechanism that more transmission capacity faces suppliers with the ability and incentive to exercise unilateral market power with a flatter residual demand curve.

For the first model suppose the wholesale market is composed of N identical firms, each with cost function $C(q, T) = c(T)q + F$, where q is the firm's output level, T is the amount of transmission capacity in the region, F is the fixed cost of production for the firm, and $c(T)$ is marginal cost of production, which is a decreasing function of T, $\mathrm{d}c(T)/\mathrm{d}T < 0$. This assumption implies that output level q can be supplied at a lower marginal cost with a larger value of T. For the case of the vertically-integrated monopoly regime, assume that $\mathrm{CI}(q, T) = \mathrm{ci}(T)q + \mathrm{FI}$, is the cost function for the vertically-integrated monopoly, where $\mathrm{ci}(T)$ is the incurred marginal cost of the vertically-integrated monopoly and FI is its incurred fixed cost. Let $P(q)$ equal the inverse demand curve for electricity and $\mathrm{TC}(T)$ is the total cost of transmission capacity T, where $\mathrm{dTC}(T)/\mathrm{d}T > 0$ and $d^2\mathrm{TC}(T)/\mathrm{d}T^2 > 0$, which implies that the total cost of transmission capacity is increasing at an increasing rate in T.

For the case of the vertically-integrated monopoly regime, the regulator is assumed to set the output price to maximize the sum of consumer and producer surplus subject to the monopolist recovering its incurred costs. This yields the following constrained optimization problem:

$$\mathrm{Max}_{\{q,T\}} \int_0^q P(s)\mathrm{d}s - \mathrm{CI}(q, T) - \mathrm{TC}(T) \text{ subject to } P(q)q - \mathrm{CI}(q, T) - \mathrm{TC}(T) = 0$$

An equivalent form of this problem maximizes consumer surplus subject to the monopolist recovering its incurred costs:

$$\mathrm{Max}_{\{q,T\}} \int_0^q P(s)\mathrm{d}s - P(q)q \text{ subject to } P(q)q - \mathrm{CI}(q, T) - \mathrm{TC}(T) = 0$$

The solution to either of these problems satisfies the following first-order conditions:

$$\begin{aligned} &(P - \mathrm{ci}(T))/P = -k/\varepsilon, \\ &-q(\mathrm{dci}(T)/\mathrm{d}T) = \mathrm{dTC}(T)/\mathrm{d}T, \\ &\text{and } P(q)q - \mathrm{CI}(q, T) - \mathrm{TC}(T) = 0, \end{aligned}$$

where $1 > k > 0$ and ε is the own-price elasticity of demand for electricity. The first equation is the standard Ramsey-pricing result that requires marking up the output price above marginal cost in order to recover sufficient revenues to cover the monopolist's fixed costs, FI, and $\mathrm{TC}(T)$. The second equation sets the marginal generation cost reduction equal to the marginal transmission cost increase from a one-unit change in T. The third equation requires that total revenues equal total incurred costs. Note that k lies in the interval (0,1) if unrestricted monopoly pricing would recover more revenues than the firm's incurred costs, a likely outcome in the electricity supply industry.

For the case of the wholesale market regime, I still assume that the regulator would like to maximize consumer surplus. However, the regulator can no longer set the output price to achieve this outcome. Competition among suppliers in the wholesale market sets the market-clearing price. I assume price is set by quantity-setting competition among the N producers. As shown in Waterson (1984), equilibrium in this market implies the following relationship between the market price, the marginal cost of each firm, the own-price elasticity of the market demand, and the number of firms in the market: $(P - c(T))/P = -1/(N\varepsilon)$, with each firm producing $q_i = q/N$.

This logic implies that the regulator knows that once the value of T is set, the value of q will be determined by quantity-setting competition among the N suppliers. Applying the implicit function theorem to $(P(q(T)) - c(T)) = -P(q(T))/(N\varepsilon)$ and making the simplifying assumption that ε, the price elasticity of demand is constant $(P(q) = Aq^{1/\varepsilon})$, implies

$$\frac{\mathrm{d}q(T)}{\mathrm{d}T} = \frac{\frac{\mathrm{d}c(T)}{\mathrm{d}T}}{P'(q(T))\left(1 + \frac{1}{N\varepsilon}\right)}$$

and $\mathrm{d}q(T)/\mathrm{d}T > 0$, because $\mathrm{d}c(T)/\mathrm{d}T < 0$, $P'(q) < 0$ and $(1 + 1/(N\varepsilon)) > 0$ in order for a quantity-setting oligopoly equilibrium to exist.

Similar to the vertically-integrated monopoly regime, the regulator chooses the transmission network capacity to maximize consumer surplus less than the cost of the transmission grid. Different from the vertically-integrated monopoly regime, the regulator must respect the constraint that industry output and the market-clearing price are determined from quantity-setting competition. This yields the following optimization problem for the regulator:

$$\mathrm{Max}_{\{T\}} \int_0^{q(T)} P(s)\mathrm{d}s - P(q(T))q(T) - \mathrm{TC}(T).$$

Using the above definition of $\mathrm{d}q(T)/\mathrm{d}T$, the first-order condition in T reduces to:

$$-q\frac{\frac{\mathrm{d}c(T)}{\mathrm{d}T}}{\left(1 + \frac{1}{N\varepsilon}\right)} = \frac{\mathrm{dTC}(T)}{\mathrm{d}T}.$$

Note that because $(1 + 1/(N\varepsilon)) < 1$, if $\mathrm{ci}(T) = c(T)$ for all T (the vertically-integrated monopoly's incurred marginal cost is equal to the marginal cost of each of the N symmetric firms in the wholesale market regime), the optimal value of T under the wholesale market regime is greater than the optimal value of T under the vertically-integrated monopoly regime. This result follows from the fact that the second derivative of $\mathrm{TC}(T)$ is positive.

The first-order condition for T for the wholesale market regime illustrates the competitiveness benefits of transmission upgrades in this regime, because more transmission capacity reduces marginal cost of supplying output for each of N the firms, which lowers the output price paid by consumers. Therefore, for the same value of T, the marginal benefit, $\mathrm{MB}(T)$, of an additional unit of transmission capacity is larger under the wholesale market regime than the vertically-integrated monopoly regime. Specifically,

$$\mathrm{MB}(T)|_{\text{Wholesale}} = -q\frac{\frac{\mathrm{d}c(T)}{\mathrm{d}T}}{\left(1+\frac{1}{N\varepsilon}\right)} > \mathrm{MB}(T)|_{\text{Vertically-Integrated}} = -q\frac{\mathrm{dci}(T)}{\mathrm{d}T}.$$

A second model introduces an additional channel through which the competitiveness benefits can be realized. The vertically-integrated monopoly solution is the same as above, but for the wholesale market regime assume that the monopoly is divested in such a way that $K < N$ firms own a sufficient amount of generation capacity to be able to set quantity strategically and the remaining $N - K$ firms behave as a price-takers. Let SO(p, T) equal the supply curve of these-price-taking firms and assume that $\partial \mathrm{SO}(p, T)/\partial p > 0$, $\partial \mathrm{SO}(p, T)/\partial T > 0$ and $\partial^2 \mathrm{SO}(p, T)/\partial T/\partial p > 0$, which implies that the supply curve is increasing in price, increasing in the amount of transmission capacity into regions where the K strategic firms compete, and increases in T increase the output responsiveness to the market price of the price-taking firms.

The remaining K firms are symmetric with a marginal cost equal to $c(T)$, where $\mathrm{d}c(T)/\mathrm{d}T < 0$ implies that more transmission capacity increases their ability to sell output from lower marginal cost units. Define $\mathrm{DR}(p, T) = D(p) - \mathrm{SO}(p, T)$ as residual demand faced by these K strategic firms, where $D(p)$ is the demand function associated with the inverse demand curve, $P(q)$. The first-order conditions for the symmetric quantity-setting competition equilibrium between the K strategic firms facing the residual demand curve, DR(p, T) is equal to:

$$(P - c(T))/P = -1/(K\eta(P, T)),$$

where $\eta(P, T) = (P/\mathrm{DR}(P, T)) * (\partial \mathrm{DR}(P, T)/\partial p)^{-1}$, is the price elasticity of this residual demand curve. The regulator knows that once T is chosen, competition among the K strategic firms and the $N - K$ price-taking firms yields a market-clearing price that solves the equation

$$P(T) - c(T) = -P(T)/(K\eta(P(T), T)).$$

Applying the implicit function theorem to this equation yields:

$$\frac{\mathrm{d}P(T)}{\mathrm{d}T} = \frac{\mathrm{d}c(T)/\mathrm{d}T + (\partial\eta(P, T)/\partial T)P(T)/(K[\eta(P(T), T)]^2)}{1 + \frac{1}{K\eta(P(T), T)} - \{P(T)/(K[\eta(P(T), T)]^2)\}\partial\eta(P, T)/\partial P}.$$

The regulator's problem for setting the optimal transmission network capacity then becomes:

$$\mathrm{Max}_{\{T\}} \int_0^{D(P(T))} P(s)\mathrm{d}s - D(P(T))P(T) - \mathrm{TC}(T)$$

Using the above expression for dP(T)/dT and the fact that $D(P(T)) = q$, the first-order condition in T reduces to

$$\frac{dTC(T)}{dT} = -q\left\{\frac{dc(T)/dT + (\partial\eta(P,T)/\partial T)P(T)/(K[\eta(P(T),T)]^2)}{1 + \frac{1}{K\eta(P(T),T)} - \{P(T)/(K[\eta(P(T),T)]^2)\}\partial\eta(P,T)/\partial P}\right\}$$

It can be shown that our assumptions on SO(p, T) implies, $\frac{\partial\eta(P,T)}{\partial T} < 0$, meaning that increasing T makes the residual demand curve facing the duopolists more price responsive. For simplicity, if we assume that $\partial\eta(P,T)/\partial P = 0$, elasticity of the residual demand curve facing the duopolists does not change as the price changes, then the first-order condition simplifies to:

$$\frac{\mathrm{dTC}(T)}{\mathrm{d}T} = -q\frac{\mathrm{d}c(T)}{\mathrm{d}T}\left\{\frac{1 + \frac{\partial\eta(P,T)/\partial T)P(T)/(K[\eta(P(T),T)]^2)}{\mathrm{d}c(T)/\mathrm{d}T}}{1 + \frac{1}{K\eta(P(T),T)}}\right\}.$$

The term $\left\{\frac{1 + \frac{\partial\eta(P,T)/\partial T)P(T)/(K[\eta(P(T),T)]^2)}{\mathrm{d}c(T)/\mathrm{d}T}}{1 + \frac{1}{K\eta(P(T),T)}}\right\}$ is greater than one because the numerator is greater than one and the denominator is less than one. This implies that if $c(T) = ci(T)$, the optimal transmission capacity for the wholesale market regime is greater than optimal capacity in the vertically-integrated monopoly regime.

These two examples demonstrate that as long as there is imperfect competition in the wholesale electricity market, there will be a difference between the "optimal second-best" transmission network configuration in the wholesale market regime and the vertically-integrated monopoly regime. That is because increasing transmission capacity increases the extent of competition suppliers with the ability to exercise unilateral market power face, a source of consumer surplus increase not present in the vertically-integrated monopoly regime.

4 Consequences of Continuing to Rely on Methodologies from the Vertically-Integrated Monopoly Regime

The analysis of the previous section demonstrates that an important difference between transmission planning in the vertically-integrated monopoly regime and the wholesale market regime is that the regulator controls the firm's price and output level. The classical regulatory bargain is that if the regulator sets a price that allows the firm an opportunity to recover its costs, the firm must satisfy all demand at the regulated price. The configuration of the transmission network impacts the regulated firm's incurred cost of supplying this output, so it is optimal to invest in transmission capacity until the marginal benefit of lower production costs to serve demand from an additional unit of transmission capacity, $-q\mathrm{d}ci(T)/\mathrm{d}T$, equals the marginal cost of an additional unit of transmission capacity, $\mathrm{dTC}(T)/\mathrm{d}T$.

In the wholesale market regime, output prices are determined by the competition between imperfectly competitive suppliers. The regulator only knows that once the capacity of transmission network is chosen, firms will make their entry decisions and set their output levels to maximize profits given this transmission capacity. Therefore, the "optimal second-best" transmission capacity should maximize consumer surplus less the cost of this transmission capacity accounting for the fact that all suppliers will maximize profits given this choice of transmission capacity. Specifically, the transmission planner chooses the value of the transmission capacity accounting for the expected profit-maximizing responses of all suppliers to this choice. This outcome will typically result in the regulator selecting a larger value of transmission capacity because of the improvements in market performance, as measured by market prices closer to marginal cost, resulting from the additional transmission capacity.

The different industry structures and the desire of the regulator to protect consumers from prices that reflect the exercise of market power imply different approaches to valuing transmission network investments. In case of the vertically-integrated utility, the regulator prospectively sets an output price to recover all of the costs—generation, transmission, distribution, and retailing—that the utility incurs to serve its customers. In addition, because the regulator sets the utility's output price, or more generally, the utility's revenue, to recover the utility's total production costs, the relevant welfare criteria for transmission planning is to maximize consumer surplus subject to the utility receiving an output price that allows it an opportunity to recover its incurred cost. This objective implies that if the incurred cost of serving load is reduced more than the cost of the transmission expansion, this upgrade should be undertaken.

The major challenge facing the regulator in the vertically-integrated regime is making the firm's incurred cost of production equal to the minimum cost of producing its output. Because of the asymmetric information problem between the firm and the regulator, solving this problem will result in the regulated firm earning some informational rents. This means that the price paid by consumers will be above that necessary to recover the minimum cost of producing the firm's output because the regulator is legally bound (at least in the United States) to set a price that allows the monopoly the opportunity to recover all prudently incurred costs associated with serving demand.[6]

For the wholesale market regime, the regulator provides no guarantee of cost recovery for the suppliers and has a limited ability to prevent suppliers from earning revenues substantially in excess of their production costs, including an adequate return on capital invested. The regulator can only set the market rules and participate in the transmission planning process. Transmission prices remain regulated in the wholesale market regime in the sense that the prices charged to consumers must allow the transmission network owner the opportunity to recover its costs. For this reason, the relevant welfare criterion is consumer surplus net of the cost the transmission network, because transmission costs must be recovered regardless of market outcomes in the wholesale market (assuming the transmission network is prudently operated).

---

[6]Wolak (1994) provides an estimate of the magnitude of these information rents for the case of regulated water utilities in California.

Although the regulator cannot set the price paid to suppliers, its choice of the capacity of the transmission network does impact equilibrium outcomes in the wholesale market, as the examples in the previous section demonstrated. This logic implies that the transmission planning process now serves a regulatory function in the sense of protecting consumers from the exercise of unilateral market power. In particular, if a transmission network upgrade increases consumer surplus (because increased competition in the wholesale electricity market) more than the cost of the transmission upgrade, then the transmission upgrade should be undertaken in the wholesale market regime.

In this regime, the impact of a transmission network upgrade on the cost that suppliers incur in producing their output, is largely irrelevant to the valuation of an upgrade.[7] This is because wholesale prices are based on offers to supply energy into the short-term market, not the cost of supplying this energy. As Wolak (2000, 2003c, 2007) demonstrates, the expected profit-maximizing offers of a supplier with the ability to exercise unilateral market power depend on its cost of producing output and the extent of competition the supplier faces. A supplier's offer curve can be vastly different from its marginal cost curve if it does not face sufficient competition. Although there are explicit forms of regulatory intervention into market mechanisms such as market power mitigation mechanisms, offer caps, and price caps to limit the ability of suppliers to exercise unilateral market power in the wholesale market regime, these mechanisms do not completely eliminate the exercise of unilateral market power or the ability of transmission expansions to limit the ability of suppliers to exercise unilateral market power.

The following chain of logic determines how the benefits of transmission expansions should be assessed in the wholesale market regime. The benefits of an upgrade depend on its impact on wholesale market prices. Market prices depend on the offers suppliers submit into the short-term market and these offers depend on the configuration of the transmission network. Consequently, the planning process should be forward-looking in the sense of anticipating the expected profit-maximizing responses of market participants to the capacity and configuration of the transmission network, because this impacts the expected profit-maximizing behavior of suppliers with the ability to exercise unilateral market power.

Taking this argument further, in the wholesale market regime, the entry decisions of market participants depend on the characteristics of the transmission network. By recognizing and anticipating the profit-maximizing entry as well as the offer behavior response of generation unit owners to a given transmission upgrade, greater system-wide benefits from all transmission expansions can be realized.

[7] Further, evidence for the irrelevance of a supplier's production costs to valuing transmission expansions in the wholesale market regime is the fact that these costs are largely unobservable in the wholesale market regime. Suppliers do not make detailed accounting costs filings with the regulator, as is the case in the vertically-integrated monopoly regime. Moreover, the goal of a wholesale market regime is to make the market sufficiently competitive that suppliers find it unilaterally expected profit-maximizing to submit offers into the short-term market close to their minimum marginal cost of production. Unfortunately, this goal has proven difficult, if not impossible, to obtain during all hours of the year in any wholesale electricity market.

There is a considerable first-mover benefit that electricity consumers receive from a transmission expansion policy that leads to new generation entry decisions, because transmission projects typically take significantly longer to plan, site, and construct than most new generation investments. For this reason, transmission expansions should lead rather than follow generation entry decisions.

Consider a wholesale electricity market contemplating a change in the transmission network configuration. If the planner chooses the new transmission configuration taking into account how this configuration will impact the future entry and operating decisions of generation unit owners, it can make any amount of spending on transmission investments more effective at reducing the ability of suppliers to exercise unilateral market power in the short-term market. The frequency of abnormally high market prices, out-of-merit energy costs, and other reliability costs can be significantly reduced if the transmission planning process is forward-looking and anticipates where new entry is likely to take place and how suppliers will operate given the configuration of the transmission network.

In contrast, a transmission expansion policy that responds to new generation investment decisions puts the planning and construction process in a continual game of catch-up with the entry decisions of new generation unit owners, because of the longer time it takes to plan, site and construct transmission facilities versus generation units. Such a policy would very likely result in higher average retail electricity prices to consumers because it would preclude consideration of many transmission expansions that provide access to low-cost distant generation in favor of the construction of generation units local to load centers.

Because the planning process must address current conditions in the transmission network before they create significant reliability problems, many longer horizon transmission expansions must be removed from consideration in a planning process that is not forward-looking. Only new local generation units can be considered given the short time horizon available to address the reliability concern. This implies that the wholesale market will have to rely increasingly on local market power mitigation mechanisms and other regulatory interventions to prevent generation unit owners from exercising the local market power associated with their location in the transmission network. These suppliers face inadequate competition for their output given the limited amount of transmission capacity into these load centers. Consequently, the regulatory interventions necessitated by a transmission expansion policy that responds to generation entry decisions severely limits the benefits accruing to consumers from wholesale electricity competition.

Any uncertainty about proposed transmission upgrades becoming a reality provides an opportunity for existing suppliers to exercise unilateral market power in the forward market. In the above example, if market participants do not believe that a proposed transmission upgrade will take place within two years, distant electricity suppliers are no longer credible competitors to the suppliers near the load center served by the retailer. Because of this delay, retailers can expect to pay a higher price for a forward contract for electricity that begins delivery two years in the future because of the reduced level of competition faced by suppliers providing this energy.

A transmission expansion policy that serves the interests of electricity consumers should create a level playing field for all generation sources to compete to provide the lowest-priced electricity at all delivery horizons. For example, distant generation from oil sands cogeneration, coal, or nuclear units can compete with natural gas-fired generation units located close to the major load centers only if there is sufficient transmission capacity to allow this to occur. Because there is considerable uncertainty over future fossil fuel prices and the price of GHG emissions, a transmission policy that allows all electricity generation technologies to interconnect and compete to supply energy to the major load centers will ensure that electricity consumers have access to the full range of available sources of electricity at all delivery horizons.

An efficient transmission expansion policy maximizes the competitiveness of the forward market for energy at all delivery horizons. However, different from the short-term market for energy, the actual transmission capacity does not need to exist when a forward contract is negotiated in order to discipline the behavior of suppliers in the forward market. Participants must only be confident that the capacity will exist when it is necessary for the seller of the forward contract to deliver energy. For example, suppose that a retailer is negotiating a forward contract for energy to be delivered several years in the future. A distant source of energy will discipline the offers of suppliers near the load center served by the retailer if all parties believe that the transmission capacity between this distant source of energy and the load center will exist when the contract begins delivering energy. For example, if the retailer is negotiating a contract that will begin delivery in two years, then it is only necessary that all market participants believe that the transmission capacity will be operating within two years. This logic emphasizes that the benefits of a forward-looking transmission expansion policy accrue to purchasers of electricity at all horizons to delivery when upgrades take place on time and according to plan.

The sequence of events that arise from transmission investments following generation investments is broadly consistent with outcomes in a number of United States wholesale electricity markets that do not have forward-looking transmission expansion policies. Transmission expansions are undertaken largely in response to new generation entry decisions rather than in anticipation of these entry decisions. These wholesale markets have experienced increasing amounts of transmission congestion, with growth rates in excess of the rate of growth of system load. The frequency and incidence of local market power problems have necessitated increasing reliance on local market power mitigation mechanisms, which typically set market-clearing prices based on loosely regulated variable costs of production of the mitigated generation units. The lack of a forward-looking transmission expansion policy that recognizes that the transmission network configuration plays a major role in allowing suppliers to exercise unilateral market power in wholesale electricity markets in many parts of the United States. Many transmission projects that would satisfy the regulatory test for in approval wholesale market regime will not be approved using a planning methodology designed for the wholesale market regime, thereby limiting the potential benefits consumers can realize from electricity industry restructuring.

5 A Methodology for Evaluating Transmission Expansions in the Wholesale Market Regime

Because the benefits of a transmission expansion depend on market prices and market prices are driven by inputs costs and the extent of competition suppliers face, there is considerably more uncertainty in the distribution realized benefits of transmission expansions in the wholesale market regime relative to the vertically-integrated monopoly regime, where the major source of uncertainty is the future cost of producing energy by the vertically-integrated monopoly. This section outlines a forward-looking methodology for evaluating transmission expansions in the vertically-integrated regime that accounts for the increased uncertainty in the realized benefits of these projects in the wholesale market regime.

5.1 Modeling Challenges in the Wholesale Market Regime

The ideal methodology for evaluating transmission upgrades is an equilibrium model with multiple strategic generation owners bidding for the right to supply electricity each hour of the day through a transmission network model that reflects the physical realities of system operation under any potential realization of future system conditions such as demand, input prices, hydrology and other factors that impact supplier behavior. With this methodology, market outcomes could be simulated with and without the proposed transmission upgrade for a large number of realizations from the distribution of future system conditions. Any decision criteria for determining whether to go forward with a proposed transmission upgrade will be a function of the distributions of market outcomes with and without the upgrades. With this modeling tool in hand, the planner/regulator could evaluate the viability of any potential transmission upgrade.

However, given the current state of economic theory and computing power, this methodology cannot be implemented without making significant modeling compromises. Specifically, even solving for the equilibrium day-ahead bidding/scheduling/congestion management strategy for only two firms owning multiple generating facilities in accordance with the California market rules (specifically, ten price and quantity bid increments for each generating facility, each of which can change on an hourly basis each day) is an extremely complex problem, even for the case in which there is no underlying transmission network model constraining the set of feasible production levels of generation unit owners. The strategy space for each player is enormous. In the day-ahead energy scheduling and congestion management process, this means setting the values of more than 500 parameters each day for each generating unit. A firm that owns 8 units, which is similar to the number owned by several California market participants, would have a 4000-dimensional strategy space.

Wolak (2000) has implemented a procedure for computing the expected profit-maximizing price and quantity offers of a single supplier in the Australian electricity market given the offer behavior of its competitors and the distribution of system demand. Computing this best-reply bidding strategy for a single day requires solving a roughly thousand parameter nonlinear programming problem subject to linear equality constraints. Computing an equilibrium with two firms setting-expected profit-maximizing offer curves would require solving a massive nonlinear complementary problem involving thousands of choice variables.

Determining the equilibrium strategies of firms operating in a wholesale electricity market in a manner that reflects the actual market rules and the actual size of each firm's strategy space increases the computational complexity to the point of being impossible to solve in a reasonable period of time. This conclusion is valid without attempting to account for the configuration of the transmission network in the wholesale market model. In general, computing a Nash equilibrium requires solving an extremely large nonlinear complementary problem subject to equilibrium constraints. Any attempt to account for the constraints on generation unit owner behavior implied by the physical configuration of the network massively increases computational complexity.

Because the purpose of the proposed methodology is to assess the benefits of transmission upgrades, adding a realistic network model is essential to achieving that goal. Unfortunately, firms competing through a transmission network with finite capacity can create discontinuities in the profit function of one firm with respect to the strategies of other firms, even in a two-node network model with two suppliers, as shown by Borenstein et al. (2000). This property implies that small changes in the behavior of one firm can lead to large changes in the best-reply of the other firm, which makes computing equilibrium strategies using standard techniques impossible. Both the enormity of the strategy space and the complications introduced by having firms compete through a realistic transmission network make solving the ideal model virtually impossible given the current state of computing power and solution methods.

Consequently, in order to make progress on this question, some economic modeling compromises must be made. Taking stock of what is actually feasible computationally and what is available in terms of historical data on the performance of a wholesale market available from the system operator, the following simplification seems to balance the goals of realism and tractability. All system operators have network models available that can compute market outcomes given the bids and schedules submitted by all market participants. These system operators also have a number of years of data available on offer behavior as a function of market conditions. The proposed simplified methodology is to use the current model of the transmission network with and without the upgrade and data on historical offer behavior and system conditions to analyze the potential benefits of a transmission upgrade.

5.2 A General Forward-Looking Methodology

An outline of this methodology follows. Let θ_d denote the firm's action choice for day *d*. This *K*-dimensional vector is composed of all of the parameters that a supplier submits to the system operator expressing its willingness to sell energy from each unit it owns each hour of the day. This vector is composed of the start-up cost, no-load cost, and the price and quantity parameters of the energy offer curve for each generation unit owned by the firm, assuming each of these offer parameters exists for the market under consideration. As noted above, the value of *K* can easily be in the thousands. Let Ω_d denote the set of variables known to the firm at the start of day d that it conditions its offers on. These variables could include: the load forecasts for all hours of the day, the temperature forecasts for the day at various locations in the control area, the demand for operating reserves, the price of natural gas and other input fuels, measures of water availability for hydroelectric units, and the amount of generating capacity owned by other firms within some radius of the plants owned by this firm, and most importantly for our purposes, the amount of available transmission capacity at various interfaces in the control area.

Let $\pi(\theta_d|\Omega_d)$ denote the realized profits of the firm for day d given Ω_d. To compute this magnitude the supplier solves following optimization problem in θ_d for day d given the set of conditioning variables, Ω_d,

$$\max_{\{\theta_d\}} E(\pi(\theta_d|\Omega_d) \text{ subject to } h(\theta_d) \leq 0,$$

where $h(\theta)$ is vector-valued function defining the set of technological and market rule constraints that restrict the values of θ_d that the firm can choose. The solution to this problem yields the expected profit-maximizing value θ_d as a function of the variables in Ω_d. Re-write the optimal value of θ_d as the vector-valued function $f(\Omega_d)$. Because both θ_d and Ω_d are observable, we can approximate the function $f(\Omega_d)$, as a very high-order polynomial in the elements of Ω_d using stochastic function approximation techniques. Modern machine learning techniques such as the Lasso (Tibshirani 1996) or Random Forests (Breiman 2001) could be employed for this task. This function $f(\Omega_d)$ could be estimated for each market participant in the control area using a large sample of data from the operation of the wholesale market.

Given estimates of $f_j(\Omega_d)$ for each firm *j* with the ability to exercise unilateral market power in the wholesale market during any hour of the year, implement the proposed transmission upgrade and compute new values of θ_{dj} for strategic market participant *j* using this function. Using the estimated functions $f_j(.)$ for each strategic player, compute $f_j(\Omega_d(\text{proposed}))$, where Ω_d(proposed) is the value of Ω_d with the transmission capacity after the proposed upgrade is in place. Then feed the values of θ_{dj} for all strategic market participants implied by Ω_d(proposed) into the market model to compute new market-clearing prices and quantities. This yields the counter-factual market outcomes to compare to the baseline market outcomes without the transmission upgrade. To get baseline market outcomes, for system conditions Ω_d, compute $f_j(\Omega_d(\text{actual}))$, where $\Omega_d(\text{actual})$ is the value of Ω_d with the transmission capacity before the proposed upgrade.

To account for the uncertainty in future load growth, low water conditions, input fuel prices, the entry of new generation, and other elements of Ω_d, an estimate of the joint distribution of the elements is necessary. Given this distribution, values of $f_j(\Omega_s(\text{proposed}))$ and $f_j(\Omega_s(\text{actual}))$ can be computed for each strategic market participant for each draw of the vector of future system conditions, Ω_s, from this distribution. The realized values of market outcomes can then be computed for both the proposed and actual configuration of the transmission grid in the future for each of these realizations of Ω_s.

Let $M(\Omega_s(\text{proposed}))$ equal the vector of market outcomes—locational prices and production levels and demands—for future system conditions realization Ω_s with the proposed upgrade in place and $M(\Omega_s(\text{actual}))$ equal the vector of market outcomes for these future system conditions without the upgrade. Let $B(M(\Omega_s(\text{proposed})), M(\Omega_s(\text{actual})))$ be the function that maps these two vectors of market outcomes into a measure of the economic benefits of the upgrade for future system condition Ω_s.

This process gives rise to a distribution of economic benefits of the upgrade driven by future system conditions and the predictive relationship between system conditions and offers submitted by suppliers based on historical data. This approach to assessing the benefits of transmission expansion in a wholesale market regime has been applied to the California ISO's proposed Path 26 upgrade, in Awad et al. (2010). An important outcome of this analysis is an estimate of the distribution of future economic benefits of the upgrade. Although the expected value of these future benefits exceeds the expected cost of the project, the distribution of the benefits is very positively skewed, indicating the realized benefits of the upgrade can be extremely large under certain future system conditions. The mapping from system conditions to benefits, $B(M(\Omega_s(\text{proposed})), M(\Omega_s(\text{actual})))$, provides valuable information to the decision-makers because it identifies what values of the elements of the vector of future system condition, Ω_s, yield large realized benefits from the upgrade.

Another important outcome from the Awad et al. (2010) analysis is that although the upgrade under consideration allowed more presumably low-cost generation to serve load in Southern California, the major source of economic benefits from the upgrade was the reduction in the amount of the unilateral market power that was exercised as a result of the transmission network expansion. Suppliers near the major population centers in Southern California would face greater competition as a result of the upgrade because $f_j(\Omega_s(\text{proposed}))$ predicted values for the offers of local strategic suppliers closer to their marginal costs, which led to lower prices in that region.

Wolak (2015) applies a version of this methodology to assess the competitiveness benefits of the transmission expansion policy that exists in the Alberta wholesale electricity market. This analysis also found that the reduction in the ability of strategic suppliers to exercise unilateral market power was the source of the vast majority of the economic benefits associated with eliminating transmission congestion in the

Alberta market. The expected economic benefits associated with Alberta's transmission expansion policy were also found to be significantly larger with a larger share of intermittent wind generation in the system.

Hesamzedeh et al. (2010a, b) formulate an economic model to quantify how transmission network changes impact the ability of strategic suppliers to exercise unilateral market power. Hesaamzedeh et al. (2010c) construct an equilibrium model of competition between strategic generation unit owners and use it to quantify both the economic efficiency improvements and the competitiveness benefits of transmission expansions. The authors simplify the process of computing equilibrium outcomes with and without the transmission upgrade by restricting the strategic players to a finite number of actions. They employ the extremal-Nash equilibrium concept of Hesamzedeh and Bigger (2012) to compute the equilibrium with and without the transmission upgrade equilibria because their game typically has many Nash equilibria. Hesamzadeh et al. (2011) extend the authors' earlier transmission expansion modeling framework to account for the fact that expansions also allow the deferral generation capacity investments because more energy from distant locations can be used to serve demand. Their model decomposes the economic benefits of transmission expansions into efficiency benefits (lower dispatch costs), competitiveness benefits (more competitive behavior by suppliers), and deferral benefits (deferral of generation capacity investments).

These analyses emphasize the importance of accounting for the competitiveness benefits in measuring the economic benefits of transmission expansions in the wholesale market regime. Many consumer welfare-improving expansions for the wholesale market regime are likely to fail the traditional dispatch cost reduction test used in the former vertically-monopoly regime, which implies that consumers are ultimately paying more electricity than necessary. Consequently, in order for consumers to realize the full economic benefits of the electricity industry restructuring the transmission planning process must recognize this new source of economic benefits from transmission capacity in the wholesale market regime.

5.3 Implementing a Forward-Looking Transmission Planning Process

A credible estimate of the distribution of realized economic benefits from a transmission expansion requires credible estimates of the joint distribution of future system conditions. Estimates of the joint distribution of future demand conditions, input fossil fuel prices, hydrological conditions, and new generation capacity entry decisions and locations are essential to providing a forward-looking assessment of the distribution of economic benefits of a transmission expansion. Under certain realizations of future system conditions, a proposed upgrade may have very small economic benefits, but for other realizations, it may have very large economic benefits, so it is important to know the probabilities associated with each of these outcomes.

Unfortunately, it is extremely difficult, if not impossible to obtain an estimate of the joint distribution of all of the elements of vector of future system conditions, including future generation entry decisions. At best it is possible to estimate marginal distributions of these magnitudes. For example, historical data could be used to simulate the marginal distributions of future load growth, future hydrological conditions, or future input fossil fuel prices. However, as the dimension of the vector of future system conditions grows, estimating its joint distribution becomes increasingly challenging.

One approach to addressing this problem is to use information on the marginal distribution of each dimension of the vector of future system conditions to constrain the unknown joint distribution of future system conditions. Consider the following example. Suppose the vector of future system conditions Ω has three dimensions. Let the unknown joint probability that Ω takes on the specific value Ω_{ijk}, equal ρ_{ijk}. Suppose there are I realizations of the first dimension, J realizations of the second dimension, and K realizations of the third dimension of Ω_{ijk} and the marginal probabilities of each realization of each dimension are known. By the properties of joint and marginal probabilities, the following equalities hold:

$$\rho_i = \sum_{j=1}^{J}\sum_{k=1}^{K}\rho_{ijk} \text{ for } i = 1, 2, \ldots, I, \quad \rho_j = \sum_{i=1}^{I}\sum_{k=1}^{K}\rho_{ijk} \text{for } j = 1, 2, \ldots, J$$

$$\rho_k = \sum_{i=1}^{I}\sum_{j=1}^{J}\rho_{ijk} \text{ for } k = 1, 2, \ldots, K \text{ and } 1 = \sum_{i=1}^{I}\sum_{j=1}^{J}\sum_{k=1}^{K}\rho_{ijk}$$

The realized value of the benefits of the upgrade could be computed for each value Ω_{ijk} for all possible values i, j, and k. The analyst could then compute the distribution of realized economic benefits from the upgrade by choosing the unknown elements of ρ_{ijk} to maximize the expected value of the upgrade subject to the four sets of linear constraints given above for the known marginal distributions, $\rho_i(i = 1, 2, \ldots, I)$, ρ_j $(j = 1, 2, \ldots, J)$, and $\rho_k(k = 1, 2, \ldots, K)$, of each element of the vector of future system conditions. The same joint density could be computed for the ρ_{ijk} that minimizes the expected value of the distribution of economic benefits. These two estimated distributions of the future economic benefits provide the regulatory process with valuable information about what specific realizations of Ω_{ijk} and the associated value of ρ_{ijk} lead to the extreme high and low realizations of the future economic benefits from the upgrade. For an illustration of this approach applied to a transmission upgrade in the California ISO control area, see Awad et al. (2010).

Hesamzedeh et al. (2010a, b) formulate the transmission network expansion problem as a single leader and multiple follower game between the single transmission planner and multiple strategic generation unit owners. The transmission network owner explicitly recognizes the strategic use of the transmission network configuration by generation unit owners to maximize the profits earned in the short-term energy market. Hesamzedeh and Yazdani (2014) formulate this leader–follower game between the transmission planner and generation unit owners with the short-term energy market between quantity-setting generation unit owners. Tohidi et al. (2017b) extend this leader–follower approach to modeling transmission network expansions to account for both the strategic entry and operating decisions of generation unit owners. Because the configuration of the transmission network impacts generation unit entry decisions, Tohidi et al. (2017a) attempt to achieve more efficient transmission and generation expansion in the wholesale market regime through the use of locational transmission network changes. These charges capture the impact of incremental generation unit investments on transmission network costs.

All of these forward-looking approaches to modeling transmission network expansions described above explicitly account for the expected profit-maximizing strategic response of generation unit entry and operating decisions in the transmission planning process in order to maximize the economic benefits consumer receive from transmission expansions in the wholesale market regime.

5.4 Modeling Policy-Driven Future Entry Decisions

Renewable energy goals are likely to be achieved at significantly lower costs to consumers with a forward-looking transmission planning process. One element of the vector of future system conditions could be the extent and rate at which renewable energy goals are met. For example, if a region has aggressive renewable energy goals and a marginal probability distribution associated with these goals being met, under the realizations where these goals are met, the benefits of a substantial transmission expansion into a region with rich renewable resources could have substantial realized economic benefits. An expansion policy that is not forward-looking might instead choose a smaller expansion that subsequently forecloses significant new generation investments into this region because of the high cost of adding incremental transmission capacity into this region.

A forward-looking transmission expansion policy is also the least-cost way to ensure that renewable energy can compete to be part of the total generation mix. The cost of the transmission interconnection facilities for the typical wind or solar project is a much larger fraction of the cost of constructing the generation facility because these generation units tend to be located far from major load centers. In addition, because there are likely to be many individual renewable generation projects at a single remote location, the size of the interconnection facility needed to serve all of these projects is substantially larger than the interconnection facility needed to serve any single renewable resource project at that location. For example, a location may have the potential to support 1000 MW of wind resources, but the average size of the wind projects at this location may be 100 MW. Because of economies to scale in constructing transmission interconnection facilities, it may be much cheaper from a discounted present value of the dollar per MW cost perspective to construct interconnection facilities with the capacity to serve the 1000 MW wind generation potential that exists at this location rather than builds only the capacity needed to serve the initial 100 MW project and then add more interconnection capacity as more wind generation capacity enters at this location.

If the costs of coordinating all of the expected renewable resource suppliers at a remote location in order to construct the single large interconnection facility are sufficiently high, then renewable resources owners may instead choose to construct these interconnection facilities sequentially as each new facility begins producing. This sequential construction of the necessary interconnection facilities will result in a total cost for interconnecting all of the eventual renewable suppliers at that location that is larger than the cost of the single interconnection facility built to serve all of these suppliers at the time the first supplier begins producing. However, if the total costs of such a large interconnection facility were charged to the first entrant, it may be so high as to prevent development at all. A forward-looking transmission policy will ensure that positive net benefit facilities will be constructed despite the fact that no individual renewable electricity supplier would find unilaterally profit-maximizing to construct it.

Finally, because the entry decisions of suppliers, the ability of suppliers to exercise unilateral market power as well as uncertainty in future system conditions and future input fuel prices, demand growth, hydrological conditions, and future renewable energy goals impact the realized economic benefits of a transmission expansion, the traditional small-number-of-future-scenarios approach to quantifying benefits of transmission upgrades is likely to provide a very incomplete estimate of the distribution of future benefits. A full characterization of the distribution of future realized benefits is likely to lead to more informed transmission planning decisions.

6 Increased Sophistication of Transmission Planning Process

As should be clear from the previous sections, the sophistication of the economic modeling required to assess the benefits of transmission expansions in the wholesale market regime is much greater than that required for the vertically-integrated regime. In the vertically-integrated regime, there no need to model the strategic response of electricity suppliers to the transmission network expansion. There is also no need to account for strategic entry and exit decisions and locations of generation units in response to network expansion. Finally, there is no need to model the strategic response of suppliers to load growth, input fuel prices, hydrological conditions, and other future system conditions.

6.1 The Downside of Open Access

The need for a sophisticated transmission planning process is greater in the wholesale market regime because no single entity has a financial interest in finding the least-cost combination of transmission and generation capacity to meet load throughout the entire wholesale market.[8] Under the vertically-integrated monopoly regime, the monopolist had little incentive to take actions to increase the total cost of meeting its load obligation by operating expensive local generation units because it had a legal obligation to serve all demand in its service territory at a regulated retail price. The combination of a fixed retail price and the obligation to serve all demand at that price gave the vertically-integrated monopolist a strong incentive to find the least-cost mix of generation and transmission investments to meet these load obligations and a strong incentive to operate its fleet of generation units in a least-cost manner.

As discussed above, in the wholesale market regime, a generation unit owner that faces insufficient competition from other suppliers has an incentive to take advantage of its location in the transmission network to increase the price that it is paid to supply electricity by changing its offer price or the amount of energy it makes available to the short-term market. Moreover, a supplier may also have an incentive to construct new generation capacity in locations where it can take advantage of its favorable location in the transmission network to raise wholesale prices through its offer price and capacity availability decisions. All of these factors imply significant benefits to consumers from a transmission policy that attempts to find the "optimal second-best" configuration of the transmission network.

6.2 The Form of Congestion Management Matters for Benefits Measurement

The specific mechanism used to manage and price transmission congestion must be modeled in order to determine the economic benefits of transmission expansions in the wholesale market regime. That is because how congestion is managed and priced impacts how suppliers behave in the wholesale market regime and ultimately market-clearing prices and the amount consumers pay for wholesale electricity. For example, offers that are expected profit-maximizing for suppliers in a single-zone or multi-zone market may no longer be expected profit-maximizing in the LMP market design. Performing an assessment of the economic benefits of a transmission expansion using an LMP market design when the actual market sets a single market-wide price or prices in a small number of zones is likely to lead to extremely inaccurate estimates of the economic benefits on an upgrade. For example, Bushnell, Hobbs,

[8]In the United States markets the Independent System Operator (ISO) is only charged with operating the transmission network, although it is a major participant in the transmission planning process. United States ISOs are non-profit entities that do not receive a direct financial benefit from finding the least-cost mix of transmission and generation capacity.

and Wolak (2008a) note how to offer behavior in the California ISO's zonal market would change as a result of the shift to an LMP market design from zonal market design.

This logic implies that different congestion management mechanisms are likely to have different "optimal second-best" amounts of transmission capacity. For example, a single zonal price model implicitly assumes that all generation units in the control area are able to compete against each other to supply electricity during all hours of the year. This logic implies that optimal amount of transmission capacity for a single-zone market is likely to larger than the optimal amount of transmission capacity for a multi-zone market that only assumes that all generation units in each zone are able to compete against each other to supply electricity during all hours of the year.[9]

Even the local market power mitigation employed for the same market design will impact the "optimal second-best" transmission capacity. There is some degree of substitutability between the stringency of the market power mitigation mechanism and transmission expansions in limiting the ability and incentive of suppliers to exercise unilateral market power. Consequently, the distribution of economic benefits of a given transmission upgrade will also depend on the form of the local market power mitigation mechanism employed.

6.3 Expanded Geographic and Industry Scope

The geographic scope of the planning process is another dimension along which the sophistication of the process should increase relative to the vertically-integrated monopoly regime. Because most formal wholesale electricity markets were formed from joining the service territories of multiple vertically-integrated utilities, the geographic scope of the transmission planning process must expand to account for this fact. Because of the looped nature of many transmission networks, expanding capacity in one geographic area can significantly alter the available transmission capacity in other geographic regions. The benefits and costs of an upgrade should, therefore, be accounted for in the transmission planning and expansion process for the entire region.

In the former vertically-integrated monopoly regime, regulators typically only counted benefits from a transmission expansion that accrued to the utility undertaking the expansion. If an expansion by one utility benefitted a neighboring utility, these economic benefits were not typically counted in the transmission planning process for that utility. While there may have been some logic to this approach to benefits assessment in the vertically-integrated monopoly regime, this approach makes very little sense in the wholesale market regime.

---

[9]A major reason for the abandonment of zonal market designs in all wholesale markets in the United States and the increasing challenges faced by zonal markets in Europe is the failure of the transmission planning and expansion process to make these implicit assumptions into reality through forward-looking transmission expansions.

There is even an argument for expanding this economic benefits calculation to include neighboring control areas, assuming there is a way for the region undertaking the investment to capture these economic benefits. This could be possible through some cost-sharing agreement with the neighboring control area negotiated before the upgrade takes place. Tohidi and Hesamzedeh (2014) model multi-regional transmission planning as a non-cooperative game between neighboring control areas that only care about the economic surplus in their control area versus a cooperative regional transmission planning process where the planner cares about the total economic surplus in both areas. The authors use their modeling results to argue that there are significant economic benefits from regional coordination of transmission planning processes. Tohida et al. (2018) employ a modified Benders decomposition to solve this game incorporating a transmission network investment risk based on the probability of a supply shortfall.

By the same logic that transmission network expansions enhance the competitiveness of wholesale electricity markets, natural gas transmission and distribution network expansions can enhance the competitiveness of wholesale natural gas and electricity markets. If expanding a gas transmission line reduces the frequency of gas curtailments and short-term natural gas price spikes, this will provide lower and less volatile natural gas prices to electricity generation unit owners, which should, in turn, increase the extent of competition to supply electricity.

Expanding natural gas pipeline capacity near locations with significant interconnection capacity for new natural gas-fired generation capacity will facilitate new entry of generation capacity and increase the competitiveness of the wholesale electricity market. For these reasons, there is a clear consumer benefit in terms of protecting consumers from the exercise of unilateral market power in the natural gas and wholesale electricity market from coordinating the natural gas and electricity transmission planning process.

An additional source of economic benefits from coordinating these two planning processes arises in wholesale markets with significant renewable energy goals. The cost of storing renewable electricity as hydrogen or natural gas is facilitated by the proximity of renewable generation capacity to the natural gas network. This will reduce the cost of injecting hydrogen or natural gas produced from renewable energy into the natural gas network.

6.4 The Viability of Market-Based Transmission Expansions

A distinguishing feature of a looped transmission network is that expanding one link can provide economic benefits to users of virtually all of the links in the transmission network. For this reason, it is generally impossible for the entity undertaking a transmission upgrade to capture all or even a significant fraction of the benefits of that upgrade. This logic has important implications for market-based mechanisms for funding transmission expansions. Specifically, relying on the revenues earned from locational price differences to fund transmission expansions is likely to lead to very limited transmission expansions and high levels of congestion in the transmission network.

One approach that has been proposed to fund transmission expansions is what has been called the "merchant transmission model" where an investor constructs a transmission line in exchange for the receiving the difference between the prices at the source and the sink of the transmission link times the capacity of the transmission line each trading period.[10] For example, if the price at the sink of the transmission line is $80/MWh and the price at the source is $50/MWh, then the owner would receive $30/MWh times the capacity of the transmission link. The merchant transmission model assumes that these locational price differences provide the economic signals necessary for fund transmission expansions.

There is virtually no empirical evidence to support the viability of the merchant transmission model, except in very rare circumstances.[11] As Joskow (2019) notes, competition to supply transmission capacity typically takes place after the regulatory process has decided to undertake a transmission expansion project. Because the locational price difference between two points in the transmission network typically captures a small portion of the benefits of the transmission upgrade, there have been few, if any, financially viable merchant transmission projects in any wholesale market. Virtually all transmission expansions are the result of a formal transmission planning process and are funded through a single system-wide transmission tariff.

7 The Insurance Value of Transmission Expansions

Future system conditions are the major driver of the realized benefits of any transmission upgrade. There are many sources of uncertainty that impact future system conditions. Market prices depend on many unknown factors such as input fossil fuel prices, the amount of entry by new generation unit owners, the level of load growth, and the outages of generation units and transmission facilities. In hydroelectric-dominated systems, water levels are a crucial determinant of wholesale electricity prices. Another source of short-term price uncertainty is the amount of fixed-price forward market obligations sold by suppliers. To compute an accurate estimate of the expected benefits of a proposed upgrade, the analyst must account for the full range of uncertainty in each of these dimensions of future system conditions. Otherwise, the expected benefits of a transmission upgrade under the wholesale market regime will be dramatically underestimated. This logic also emphasizes that transmission upgrades have a substantial insurance value, particularly under the wholesale market regime.

[10] Joskow (2019) discusses the economic viability of this merchant transmission investment model.

[11] The few examples of viable merchant transmission projects are direct current (DC) lines from a remote location to a generation load pocket, rather than upgrading or building a link in a looped alternating current (AC) high voltage network.

Transmission upgrades can significantly reduce the likelihood of system conditions that produce extreme prices. For example, large interconnections between California and neighboring control areas can substantially reduce the probability of extreme prices in California. For example, if a temporary shortfall in natural gas availability in California causes electricity prices to rise significantly, a large interconnection with the Pacific Northwest allows hydroelectric energy to substitute for expensive natural gas-fired electricity. A large interconnection with the Desert Southwest could allow coal-fired energy to displace expensive natural gas-fired energy in California. Under normal conditions for natural gas availability in California, this interconnection may not be fully utilized, but it does provide insurance against this and other potential supply uncertainties within the state.

Because the impact of physical constraints on system conditions are often exacerbated by the strategic behavior of suppliers, the insurance value of transmission expansions is likely to be even larger under the wholesale market regime than under the vertically-integrated monopoly regime. For example, there are many examples of from hydroelectric-dominated wholesale markets around the world of fossil fuel suppliers taking advantage of low water conditions and submitting much higher offer prices because they know that hydroelectric suppliers must conserve water rather than compete vigorously to supply electricity to the short-term market.[12] Similar logic applies in a natural-gas-dominated market such as California. If the price of natural gas rises substantially, then out-of-state coal-fired generation unit owners could submit higher offer prices because they face less competition at their former offer prices from the natural gas-fired generation unit owners. However, if there is substantial interconnection capacity with neighboring control areas, the coal-fired suppliers will still face competition from coal-fired suppliers in other control areas and will be unable to raise wholesale prices in California.

The events of June 2000 to June 2001 in the California electricity market provide a vivid illustration of the extent to which extreme events can drive the benefits of a transmission expansion.[13] Specifically, had there been significant transmission capacity available to transfer electricity from the Eastern Interconnection to the Western Interconnection, it is unlikely that the enormous increase in electricity prices in the Western US would have occurred during this time period. This transmission capacity could have allowed consumers in the Western US to avoid paying prices that were orders of magnitude higher than prices in the Eastern US during this time period. In addition, this interconnection would have also eliminated the need for the State of California to sign long-term forward contracts during the winter of 2001 at prices more than double wholesale prices during first two years of operation of the California market in order to commit suppliers to the California market during the summer of 2001 onwards. A very conservative estimate of the realized discounted present value of the benefits of this interconnection to consumers in the

[12] Wolak (2009) describes the case of New Zealand and McRae and Wolak (2016) the case of Colombia.

[13] Wolak (2003a) provides a diagnosis of the causes and consequences of the California electricity crisis and Borenstein et al. (2002) assess its economic efficiency consequences.

Western US (because it would have prevented the events of June 2000 to June 2001 from occurring in the Western US) is on the order of 30 billion dollars.[14]

The substantial economic harm caused by a sustained period of extreme wholesale electricity prices argues in favor of incorporating some degree of risk aversion into the process used to assess the distribution of net benefits from a transmission expansion. For example, electricity consumers are likely to prefer a transmission expansion project that has produces market outcomes with a certain \$1 million net benefit relative to a competing reliability project that has a −\$100 million net benefit and a \$102 million net benefit each with equal probability, despite the fact that both projects have the same expected net benefit. A transmission expansion project that increases the number of distant suppliers that can sell energy into the market has a much more certain net benefit distribution than a demand response or local generation project that does not increase the number of new suppliers able to sell energy into the wholesale market. Consequently, if risk aversion is an important concern, then the transmission planning process should guard against under-investment in the transmission network rather than over-investment in the transmission network.

Over-investment (relative to an expected net economic benefit criterion) in the transmission network protects against rare, but extremely costly market outcomes. Specifically, even though consumers will be asked to pay for more transmission capacity in all future states of the world, the upgrade will eliminate the realization of a market outcome that is extremely costly to consumers. Under-investment in the transmission network subjects consumers to the prospect of extremely costly market outcomes in exchange for slightly lower transmission charges in all states of the world. If consumers are risk-averse then they should prefer an outcome that slightly over-invests in transmission capacity relative to one that slightly under-invests in transmission capacity, even if consumers expect to pay the same price for retail electricity under both scenarios.

The argument for a transmission planning process that treats over-investment in the transmission network as less harmful to consumers than the under-investment is strengthened by the fact that less than 10% of the average retail price of electricity in most jurisdictions pays for the transmission network. This percentage is unlikely to increase because of expectations of increasing fossil fuel prices and a positive price for greenhouse gas emissions. These two factors imply that consumers can realize even greater economic benefits from a wholesale electricity market that faces all suppliers with the maximum amount of competition and allows consumers to have access to the lowest-cost sources of electricity for as many hours of the year as possible. This set of circumstances can only exist if there is a forward-looking transmission policy that plans, sites, and builds transmission facilities in anticipation of generation unit entry and operating decisions.

---

[14] Wolak et al. (2004) provide this conservative estimate of the cost of the California electricity crisis.

8 Conclusion

The current regulatory structure in the United States governing transmission planning and expansions is poorly suited to the wholesale market regime that serves the vast majority of electricity consumers in the United States. The foregone benefits to United States electricity consumers associated with the current regulatory framework governing transmission planning and expansions are substantial and are very likely to become much larger as the electricity supply industry transitions to low-carbon energy sources. A coordinated transmission planning and expansion process tailored to the wholesale market regime can significantly increase the economic benefits electricity consumers realize from all money spent on transmission expansions and substantially increase the rate at which low-carbon electricity sources are able to interconnection and sell electricity to final consumers and the ultimate benefits realized from electricity industry restructuring.

References

M.L. Awad, K.E. Casey, A.S. Geevarghese, J.C. Miller, A. Farrokh Rahimi, A.Y. Sheffrin, M.Z. Eric Toolson, G. Drayton, B.F. Hobbs, F.A. Wolak, Using market simulations for economic assessment of transmission upgrades: application of the California ISO approach. Restruct. Electr. Power Syst. 241–270 (2010)

S. Borenstein, J. Bushnell, S. Stoft, The competitive effects of transmission capacity in a deregulated electricity industry. RAND J. Econ. (2000)

S. Borenstein, J.B. Bushnell, F.A. Wolak, Measuring market inefficiencies in California's restructured wholesale electricity market. Am. Econ. Rev. **92**(5), 1376–1405 (2002)

L. Breiman, Random forests. Mach. Learn. **45**(1), 5–32 (2001)

J. Bushnell, B. Hobbs, F. Wolak, Final opinion on 'The DEC Bidding Activity Rule under MRTU' (2008a). Available at https://www.caiso.com/Documents/MSCFinalOpiniononDECBiddingActivityRuleunderMRTU.pdf

J.B. Bushnell, E.T. Mansur, C. Saravia, Vertical arrangements, market structure, and competition: an analysis of restructured US electricity markets. Am. Econ. Rev. **98**(1), 237–266 (2008b)

M.R. Hesamzadeh, D.R. Biggar, Computation of extremal-Nash equilibria in a wholesale power market using a single-stage MILP. IEEE Trans. Power Syst. **27**(3), 1706–1707 (2012)

M.R. Hesamzadeh, M. Yazdani, Transmission capacity expansion in imperfectly competitive power markets. IEEE Trans. Power Syst. **29**(1), 62–71 (2014)

M.R. Hesamzadeh, N. Hosseinzadeh, P.J. Wolfs, Transmission system augmentation based on the concepts of quantity withheld and monopoly rent for reducing market power. IEEE Trans. Power Syst. **25**(1), 167–180 (2010a)

M.R. Hesamzadeh, N. Hosseinzadeh, P.J. Wolfs, A leader–followers model of transmission augmentation for considering strategic behaviours of generating companies in energy markets. Int. J. Electr. Power Energy Syst. **32**(5), 358–367 (2010b)

M.R. Hesamzadeh, D.R. Biggar, N. Hosseinzadeh, P.J. Wolfs, Economic transmission augmentation with explicit modeling of the competition benefit. IEEE Trans. Power Syst. **25**(3), 1714–1721 (2010c)

M.R. Hesamzadeh, D.R. Biggar, N. Hosseinzadeh, P.J. Wolfs, Transmission augmentation with mathematical modeling of market power and strategic generation expansion—Parts I and II. IEEE Trans. Power Syst. **26**(4), 2040–2056 (2011)

P.L. Joskow, Inflation and environmental concern: Structural change in the process of public utility price regulation. J. Law Econ. **17**(2), 291–327 (1974)
P.L. Joskow, Competition for electric transmission projects in the U.S.: FERC Order 1000, MIT Working Paper (2019)
R.G. Lipsey, K. Lancaster, The general theory of second best. Rev. Econ. Stud. **24**(1), 11–32 (1956)
S.D. McRae, F.A. Wolak, How do firms exercise unilateral market power? Evidence from a bid-based wholesale electricity market, in *Manufacturing Markets: Legal, Political and Economic Dynamics*, ed. by E. Brousseau, J.-M. Glachant Cambridge University Press, 2014), pp. 390–420
S.D. McRae, F.A. Wolak, Diagnosing the causes of the recent El Nino event and recommendations for reform (2016). Available at http://www.stanford.edu/~wolak
F.C. Schweppe, M.C. Caramanis, R.D. Tabors, R.E. Bohn, *Spot Pricing of Electricity*. (Springer Science & Business Media, 2013)
R. Tibshirani, Regression shrinkage and selection via the lasso. J. Roy. Stat. Soc.: Ser. B (Methodol.) **58**(1), 267–288 (1996)
Y. Tohidi, M.R. Hesamzadeh, Multi-regional transmission planning as a non-cooperative decision-making. IEEE Trans. Power Syst. **29**(6), 2662–2671 (2014)
Y. Tohidi, L. Olmos, M. Rivier, M.R. Hesamzadeh, Coordination of generation and transmission development through generation transmission charges—a game theoretical approach. IEEE Trans. Power Syst. **32**(2), 1103–1114 (2017a)
Y. Tohidi, M.R. Hesamzadeh, F. Regairaz, Sequential coordination of transmission expansion planning with strategic generation investments. IEEE Trans. Power Syst. **32**(4), 2521–2534 (2017b)
Y. Tohidi, M.R. Hesamzadeh, F. Regairaz, Modified Benders decomposition for solving transmission investment game with risk measure. IEEE Trans. Power Syst. **33**(2), 1936–1947 (2017)
M. Waterson, *Economic theory of the industry* (Cambridge University Press, Cambridge, 1984)
F.A. Wolak, An econometric analysis of the asymmetric information, regulator-utility interaction. Annales d'Economie et de Statistique 13–69 (1994)
F.A. Wolak, An empirical analysis of the impact of hedge contracts on bidding behavior in a competitive electricity market. Int. Econ. J. **14**(2), 1–39 (2000)
F.A. Wolak, Diagnosing the California Electricity Crisis. Electr. J. August/September, 11–37 (2003a)
F.A. Wolak, Measuring unilateral market power in wholesale electricity markets: the California market, 1998–2000. Am. Econ. Rev. **93**(2), 425–430 (2003b)
F.A. Wolak, Identification and estimation of cost functions using observed bid data, in *Advances in Economics and Econometrics: Theory and Applications, Eighth World Congress*, vol. 2 (Cambridge University Press, 2003c), p. 133
F.A. Wolak, Quantifying the supply-side benefits from forward contracting in wholesale electricity markets. J. Appl. Econ. **22**(7), 1179–1209 (2007)
F.A Wolak, An assessment of the performance of the New Zealand wholesale electricity market. *Report for the New Zealand Commerce Commission* (2009)
F.A Wolak, Regulating competition in wholesale electricity supply, in *Economic Regulation and Its Reform: What Have We Learned*? (University of Chicago Press, 2014), pp. 195–289
F.A Wolak, Measuring the competitiveness benefits of a transmission investment policy: The case of the Alberta electricity market. Energy Policy **85**, 426–444 (2015)
F.A. Wolak, R.H. Patrick, *The Impact of Market Rules and Market Structure on the Price Determination Process in the England and Wales Electricity Market*. No. w8248. (National Bureau of Economic Research, 2001)
F.A. Wolak, B. Barber, J. Bushnell, B. Hobbs, Comments on the California ISO's Transmission Expansion Assessment Methodology (TEAM) (2004). Available at http://www.caiso.com/docs/2004/06/01/200406011457422435.pdf
C.D. Wolfram, Measuring duopoly power in the British electricity spot market. Am. Econ. Rev. **89**(4), 805–826 (1999)

Cost Allocation Issues in Transmission Network Investment

Michel Rivier and Luis Olmos

1 Why Does Transmission Network Cost Allocation Matter?

Transmission pricing has the primary purpose of recovering the cost of building, operating and maintaining the grid while also promoting the overall efficiency of power supply. Under appropriate pricing, transmission grid activities should be sustainable (i.e., profitable enough to attract investment) and the pricing signals should promote an efficient response by the transmission network users. Marginal prices are well known to promote efficiency. Under ideal conditions, including the continuity of network investments and the absence of economies of scale, long-term energy (marginal) prices would coincide with short-term ones—the Locational Marginal Prices (LMPs)—and their application would result in the exact recovery of the cost of the grid. However, conditions in real-life systems are far from being ideal. As a consequence of this, long-term energy prices do not exist in many cases, while the application of short-term ones results in net revenues for the system that fall very short of those needed to pay the grid (Pérez-Arriaga et al. 1995). Empiric assessments show that no more than 25% of total grid costs could be expected to be recovered through network incomes resulting from LMPs.

The application of short-term marginal energy prices (or LMPs, as they are known in many jurisdictions) implicitly allocates a fraction of the cost of the grid that is smaller than those costs directly attributable to specific network users. This is due to the fact that changes in the decisions made by individual network users result in changes in the cost of development of the grid, representing those costs attributable to the former, that largely exceed the fraction of these costs assigned through LMPs.

M. Rivier · L. Olmos (✉)
Comillas Pontifical University, Madrid, Spain
e-mail: Luis.Olmos@iit.comillas.edu

M. R. Hesamzadeh et al. (eds.), *Transmission Network Investment in Liberalized Power Markets*, Lecture Notes in Energy 79,
https://doi.org/10.1007/978-3-030-47929-9_5

Then, complementary charges applied to recover the remaining fraction of the grid costs should be designed so as to allocate those attributable, or assignable, network costs not assigned through LMPs to those network users that cause them, i.e., those whose decisions make the system incur these costs (Rubio-Odériz and Pérez-Arriaga 2000). Besides, as previously stated, complementary charges should result in the recovery of the remaining network costs that are reasonably incurred. This chapter is focused on discussing the design of efficient complementary charges, the options that may exist for this, and the interaction between transmission network cost allocation and other aspects of the functioning of the transmission activity.

1.1 Relationship with Other Aspects of Transmission Regulation

The allocation of the cost of transmission investments is central to the development of the grid, especially in regional markets comprising several systems. Only if the cost allocated to each system within a regional market (national one, or each State, in some markets) is commensurate with the benefits that this system expects to obtain will these systems agree with the cost allocation, and therefore, with the undertaking of the corresponding investment, see (Olmos et al. 2018). At the same time, the composition of the set of reinforcements to the grid, identified by planning authorities or private promoters as being optimal from their point of view, and being promoted by them and eventually built, conditions the set of benefits produced by each individual reinforcement and the distribution of these benefits to the network users and stakeholders in general. This is due to the fact that the benefits produced by the several network reinforcements planned are mutually interdependent (Bañez et al. 2017a).

Analogously, given that the benefits produced by network investments depend on the solution implemented for the dispatch and operation of the system and the market, these will influence both the allocation of the cost of these investments that should be carried out and the decision on which investments to undertake. Network expansion planning, obviously, will affect the grid development and the result of the market and system operation processes. Once network expansion planning and cost allocation are satisfactorily dealt with, getting the approval of parties to these investments and recovering their cost should be easy, since there would be a commitment from the systems to pay each a fraction of this cost.

But there is one last problem that should be considered before achieving the construction of a new transmission asset. Local stakeholders and authorities in the area where this asset is to be located may oppose its construction, since they may perceive that this project may create some local negative impact on the environment, as well as some negative economic impact on land and business owners. Overcoming the resistance of local pressure groups may also relate to the distribution to be carried

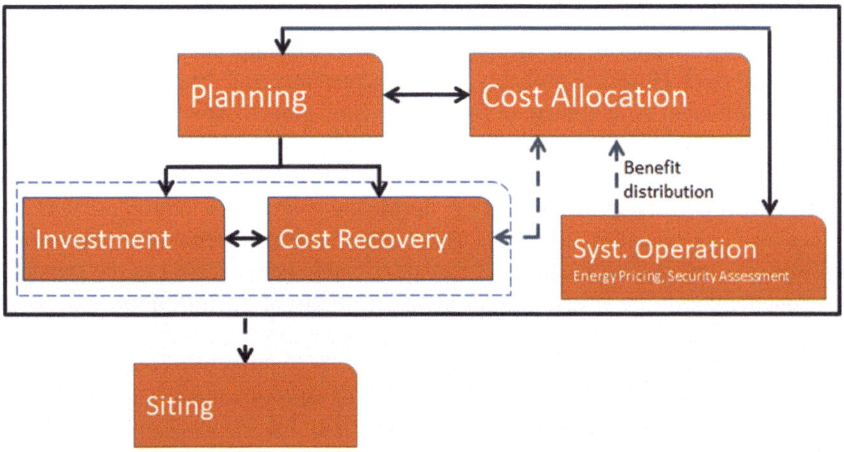

Fig. 1 Relationship between the transmission network cost allocation and other aspects of the regulation of the functioning of the transmission activity

out of the benefits and costs caused by the reinforcement. Thus, one way forward, for example, could be the payment of compensations to local parties affected.

All these relationships of interdependence among the several aspects of transmission regulation are graphically depicted in Fig. 1.

1.2 Cost Allocation for Regulated Investments: Completing the Recovery of the Allowed Regulated Revenue

Regulated investments are those defined and promoted by an entity looking after the interests of the whole electricity system, normally the central network planning authority, which often coincides with the System Operator. Regulated network investments are selected as those best serving the interests of the network users as a whole. The electricity transmission activity is widely considered a natural monopoly, since the most efficient solution for the system is to have a single network, i.e., a set of interconnected assets which all the agents can access to exchange electricity and other related products (Rivier et al. 2013). Given that a single, central, grid infrastructure exists to be used by all the system stakeholders, the management of the grid confers the ability to significantly affect the functioning of the system, regarding the system security or its efficiency. Due to this fact, the entity in charge of identifying the reinforcements to promote and eventually undertake should not normally be one that has economic interests in other activities within the system. Otherwise, this entity could use his dominant position in the electricity transmission activity to favor its other businesses. This does not mean that all the reinforcements to the grid should be built as regulated investments. As discussed in chapter "Competition for

Electric Transmission Projects in the USA: FERC Order 1000," there is room in the system for investments driven by private promoters to complement those promoted as regulated ones. However, probably, only regulated investments can be relied upon to drive a sufficient development of the transmission grid (Joskow and Tirole 2005).

Given that transmission expansion and operation planning confers large market power, the remuneration scheme applied to regulated investments cannot result in incentives for the planning authorities to decide the investments to promote in order to increase their revenues at the expense of other system stakeholders (generators, consumers or transmission companies). Additionally, being central to the satisfactory functioning of the system, the economic viability of the transmission activity should be guaranteed, to the extent possible. This leads the remuneration of regulated investments to be defined to allow the owner of these facilities to recover the cost of these investments and achieve a reasonable rate of return on them. Then, the regulated electricity transmission service should be a low risk one whose profitability is high enough to attract a sufficient number of companies (Rivier et al. 2013).

The revenues of investors in regulated network reinforcements, which are defined by regulation, are normally termed Allowed Regulated Revenue of these investments. As argued above, the net revenues resulting from the application of efficient energy prices (marginal ones exhibiting locational and temporal differentiation), generally, do not manage to recover the full cost of the socially efficient network investments (those to be promoted as regulated investments). Then, additional charges need to be applied on network users to complete the recovery of the Allowed Regulated Revenues of the aforementioned investments. These are normally called Complementary Transmission Charges, or transmission tariffs (Rivier et al. 2013). Besides completing the collection of the Allowed Regulated Revenue of regulated investments, transmission tariffs should provide incentives driving an efficient development of the system and not interfering with the efficient operation of the system, as discussed in Sect. 2. A multiplicity of methods have been proposed for the computation of complementary transmission charges. Some of them have been implemented while others not. Section 2 discusses the objectives to be achieved through the application of transmission charges and to what extent each of the main types of charging methods achieves these objectives.

1.3 Cost Allocation for Investments Promoted by Private Parties

As mentioned above, regulated transmission investments may be complemented by those driven by private parties. There may be situations where transmission investment needs cannot be covered by the regulated development of the grid. In many systems, the owner (and investor) of regulated transmission facilities is the SO. In other systems, at least part of these facilities are owned by transmission companies. The access to funds of transmission companies, or even TSOs, may be limited by their financing capability, especially for systems in developing countries where generators and, especially, consumers may not be able to pay the regulated transmission charges. Additionally, private promoters, having access to different information from that considered by planning authorities, may be able to identify relevant investment needs overlooked by the latter.

Private network investments may be promoted by merchant entrepreneurs, willing to make profits out of the commercial exploitation of the transmission capacity they build, or by network users, willing to increase their market profits by improving their access to these markets be means of the new transmission capacity whose construction they promote. In the following paragraphs the allocation of the cost of these investments is analyzed.

1.3.1 Merchant Investments

Merchant investments are promoted by private entrepreneurs willing to make a profit out of the commercial exploitation of these facilities. Entrepreneurs promoting merchant investments aim to maximize their commercial profits. Traditionally, the revenues of merchant owners have amounted to the market value of this capacity. This market value, corresponding to the congestion rents merchant investments could produce, or, equivalently, the revenues from the sale of rights over these congestion rents, tend to differ significantly from the social benefit produced by the corresponding reinforcements. Thus, private promoters focus on congestion rent maximizing investments, which do not need to be the social welfare maximizing ones. In fact, the incentives to maximize the congestion rents drive promoters to pursue underdeveloped reinforcements where congestion is relevant. Thus, most of the socially beneficial reinforcements are not attractive for merchant entrepreneurs, while those tended to be built by them have less capacity than what is socially optimal. Because of this, only a small subset of the beneficial reinforcements can be thought to be built by merchant promoters.

The cost of the former type of merchant investments is afforded by the merchant promoter in the first place. Given that the promoter aims to make a profit out of his investment, the congestion rents, or transmission rights over them, to be paid by network users are expected to more than recover the original cost of the investment. Therefore, normally, these investments should end up being paid from the net revenues resulting from the application of nodal, or zonal, short-term energy prices on both ends of the corresponding facilities. Since short-term marginal energy prices do not coincide with long-term ones internalizing the network development costs, the allocation of the cost of these lines to network users cannot be deemed to be fully efficient. However, to the extent that short-term energy prices are related to long-term marginal ones, one can argue that cost-causality, or benefit allocation, principles are partly driving the allocation of the cost of these reinforcements. This can be intuitively corroborated by the fact that the energy price differences creating the congestion rents paying these investments are due to the fact that generation injecting electric energy in the merchant asset is selling it at a lower price than that paid by the consumers retrieving energy from this asset. Then, generation on the exporting side and demand on importing side of the merchant line could be deemed to be those paying the cost of this line. These two groups of agents are the same ones deemed to be benefiting from the existence of a merchant line connecting two nodes or areas.

Recently, merchant promoters in many jurisdictions have been given the option to negotiate with the future users of the facilities they promote the charges they would have to pay to access the capacity of these facilities. This is aimed at allowing the merchant promoter to extract part of the benefits obtained by network users from the transmission project the former undertakes, which would result in a partial alignment of the private incentives of the promoter when selecting the network investment to undertake with the social interest. This negotiation normally takes place before the construction of the merchant asset. The negotiation process needs to take the form of a competitive process, usually an auction. Network users willing to pay access charges are within those that will benefit from the construction of this asset (again, generators on the exporting side and consumers on the importing one). Then, one can conclude that the network users paying the cost of this type of merchant investments will be part of those benefiting from the investment. However, there may probably be some network users benefiting from the construction of this asset (for example, because the energy prices they earn are more favorable after the construction of the merchant project) that will not pay access charges. Then, some level of free riding will normally occur.

These two promotion schemes may be combined with the option for the merchant promoter to modulate the amount of capacity, within that built, that is made available to the network users. This is aimed at making profitable from the outset some investments that, being larger than what is optimal from the private investor's point of view, may never be built, or may be built with some delay, otherwise. This feature of merchant investments should not alter the considerations on cost allocation that have just been made.

The several types of merchant investments that exist are discussed in Part 3 of the book.

1.3.2 Investments by Associations of Network Users

Network users may be interested in promoting the construction of certain network facilities themselves when these would be highly beneficial for the aforementioned users and they have some doubts over the capability of the central planner, or the intention of merchant investors, to promote and achieve the construction of these facilities. However, free riding may be a serious problem blocking the undertaking of many transmission projects by associations of network users. Gathering all, or, at least, a majority of the network users benefiting from a new project within the association promoting it may prove to be very challenging in many cases, especially those of transmission projects whose benefits would be widely spread over a large number of network users in the system.

Network users creating an association to build a reinforcement would certainly benefit from it. However, as aforementioned, there may be other users also benefiting from that reinforcement that do not belong to the association. Thus, despite the fact that the Association of network users will always have to pay, at least, a fraction of the reinforcement, for some projects of this type, another fraction of the cost of the reinforcement may be allocated to third parties, not within the association, that are also expected to benefit from this reinforcement. Normally, in these cases, the fraction of the cost of the reinforcement to be paid by third parties through regulated charges is related to the ratio of the benefits from the line perceived by third parties not promoting the reinforcement to the total social benefit created by the reinforcement (CRIE 2005).

Network charging schemes in place in the system considered are used to allocate to third parties that do not belong to the promoting association the allowed regulated revenue of this type of reinforcements, or fraction of the cost of these reinforcements to be collected from regulated network charges. The fraction of the cost of these projects to be paid by the promoting coalition is allocated to the members of the coalition according to the criteria they have privately defined, though these criteria should normally be based on the expected distribution among the coalition members of the benefits produced by the reinforcement.

In some systems or regions, there may be coalitions of network users promoting reinforcements that, if approved by regulatory authorities, end up being fully paid from regulated network charges. In this case, the efficiency of the allocation of the cost of these reinforcements would fully depend on that of the transmission network charging scheme in place.

2 Guidelines to Complete the Recovery of the Allowed Regulated Revenue

This section aims to present and discuss the main principles to consider for cost allocation of electricity transmission network infrastructures. In a national context, the allocation is performed by means of the design of efficient grid charges to be applied to the different users (electricity market agents) of the transmission network. In a supranational or regional context, the discussion may be rather related to discuss the criteria to allocate the cost of transmission interconnections (or transmission facilities relevant at supranational level) to the different countries involved in the regional market. The latter will be addressed in Sect. 5.

As discussed later on, the physical and economic characteristics of electricity transmission networks render the efficient allocation of the cost of these facilities among their users a very challenging task. Actually, still nowadays and despite academic research efforts and years of applying various different charging methods in systems around the world, none indisputable methodology exists. Second-best solutions are commonly used both at conceptual and practical levels.

In those cases, it is of paramount importance to properly understand the limitations of the different approaches proposed so far and the difficulties to properly implement them. But a proper assessment of cost allocation methodologies should be based, first, on a comprehensive discussion of the main principles that should ideally drive the cost allocation solution.

Next Sect. 2.1, addresses the conceptual principles of transmission network cost allocation, while Sect. 2.2 reviews some practical implementation issues in the design of transmission charges, generally common to all kind of transmission network cost allocation methodologies. Then, Sect. 3 is entirely devoted to the "beneficiary-pays" approach as one of the soundest cost allocation methods, and Sect. 4 reviews other alternative methods that have been implemented in real regulatory frameworks or proposed in the literature.

2.1 Fundamentals of Transmission Network Charges Design: Main Principles. Discussing the Relationship Between Cost-Causality and Benefit-Driven Cost Allocation

Allocating the costs of the transmission network among all network users should follow economic principles oriented to produce an efficient usage of the network, both regarding the short-term (operation) and the long-term (investment) decision making. As previously discussed, nodal energy prices (LMPs) do actually provide those efficient economic signals in the short-term (operation). Should economies of scale not exist and a perfectly adapted network be possible, net incomes from applying nodal energy prices would be enough to fully recover network investment costs, so that the issue of allocating network infrastructures costs to market agents would have been solved in an irrefutable and efficient way, complying also with the revenue reconciliation goal. But, as previously explained, this is not by far the case, and the design of a proper transmission network cost allocation methodology to allocate the remaining fraction of the network cost -or the full network cost whenever LMPs are not in place- still remains an issue.

As previously mentioned, the physical and economic characteristics of electricity transmission grids prevent setting up a theoretically fully sound methodology to address the design of grid charges. However the experience learned through many years of applying different approaches to different power systems enables us to propose a set of principles any grid charge design should comply with, irrespective of the specific methodology adopted (Pérez-Arriaga and Smeers 2003; MIT 2011).

Apply, as close as possible, the "cost-causality" economic principle

According to the economic theory, network users should be charged the fraction of the grid costs they are responsible for, which should result in efficient grid charges. This is known as the "cost-causality" principle. Moreover, theoretically speaking, the optimal grid economic signal to be sent to network users corresponds to the long-term marginal impact of their investment and operation decisions on network costs (Olmos and Pérez-Arriaga 2009; Pérez-Arriaga and Smeers 2003). Given that these long-term marginal costs actually do not exist due to the discrete nature of network investments, grid charges to be paid by agents should reflect the long-term incremental network costs they cause. Unfortunately, computing the latter is intractable in practical terms at least at the transmission network level.

Thus, basing transmission charges on the drivers of network investment decision making could be of help. Decision makers apply certain criteria to come up with the decision to reinforce or extend the network. Therefore, allocating network costs in proportion to the contribution of each network user to the decision to carry out each reinforcement, according to those criteria, can be understood as a "cost-causality" approach. In the case of transmission networks, as discussed in previous chapters, conducting a cost-benefit analysis (CBA) is deemed to be the soundest approach, from an economic point of view, to decide on the reinforcement of the network. The "beneficiary-pays" approach, which has inspired transmission pricing regimes since 30 years ago (Read and Sell 1989), is intimately linked to network expansion CBA. This methodology proposes to allocate the cost of network reinforcements in proportion to the benefits that each network user obtains from the existence of that infrastructure.[1] The rationality behind this approach and its link to the "cost-causality" principle is further discussed in Sect. 3.1.

Network transmission charges should be independent of commercial transactions
In a mature and well-functioning electricity market, where decisions made by market players are rationale from an economic point of view and these have access to full information, the actual production and consumption of each agent will not depend on the specific commercial transactions they have agreed on. Whatever bilateral agreements may be in place, energy spot markets (day-ahead market or balancing markets) will rationally lead the cheapest units in the system to produce power and the most valuable demand to be satisfied. Indeed, for instance, a generator will resort, whenever the spot market energy price is cheaper than its production costs, to buying energy at the spot market to comply with its commercial agreements rather than producing it itself. Thus a cheaper unit, located somewhere else, will be producing instead of a more expensive one, regardless of the commercial transactions the market agents could have agreed with third parties.

Therefore, the net injections/withdrawals at each node of the transmission grid will be independent of agents' commercial agreements, while the physical distribution of energy flows through the whole transmission network should depend only on those net injections/withdrawals. Thus, the benefit (or use) each agent draws from

[1] Instead, long term incremental cost pricing is better suited to address the computation of distribution network charges applied on end consumers, due to the usually radial configuration of distribution grids. This usually results in network charges with a capacity structure (per MW ones) that are based on the expected contribution of network users to the simultaneous peak load, see (MIT 2016). In that case, the drivers of distribution network reinforcements usually are the peak loading of network facilities. Therefore, the allocation of network costs follows the very same principle as that used in the "beneficiary-pays" approach for transmission networks.

(makes of) the grid will be independent of agents' commercial agreements. Instead, transmission charges should depend on the location of the agents in the network and on the temporal patterns of their power injections (for generators) and withdrawals (for loads), since the benefit the obtain form the grid (or their use of it) does depend on those aspects (Pérez-Arriaga and Smeers 2003).

This principle applies in regional markets. At first glance, it may lead to counter-intuitive conclusions. Indeed, this principle means that a producer located in system A that comes up with a trading agreement with a consumer entity in system B to sell energy, should pay the same transmission charge as if, instead, it were contracted to supply a neighboring load sited within its own A system. Nevertheless, it is straightforward to conclude, applying the same reasoning as above, that, if there is open network access and no barriers to inter-regional trade, the decentralized interaction between the regions and their agents should approximate the ideal outcome of an inter-regional efficient generation dispatch, regardless of who trades with whom, inside or outside each region.

Recognized in both the US and EU as undesirable—(FERC Order No. 888 Open Access) and (European Commission 2011) respectively—the opposite results in "pancaking" or piling up of transmission charges, whereby network users are required to pay accumulating fees for every system through which their power is deemed, by contract, to pass between the buyer and seller, regardless of actual power flows. Pancaking makes transmission charges depend on the number of administrative borders between buyer and seller. Such pricing tends to stifle trade and to prevent buyers from accessing low cost sellers.

Network transmission charges should be established ex ante

For the network transmission charges to be efficient locational economic signals, they should encourage potential new generators to site at locations that will increase as little as possible the need for network reinforcements. The first principle discussed above focuses on this goal. But, in practical terms, transmission charges, if updated annually, become unpredictable, failing to provide the investors with a low financial risk to decide the most convenient location.

Provided new generators will not be able to change any longer their location once they construct their facilities, it is certainly more relevant to estimate ex ante, as accurately as possible, the transmission cost they should be charged, for the generator to make an informed investment decision, than to update this charge annually to fit better the real benefit (or use) drawn (made) by the generator from the use of the network. Indeed, computing transmission charges ex-post, based on the real benefits finally provided by the reinforcements once they are built, would not allow network users to decide on their investments taking into account the cost of the network reinforcements associated with the former, which would prevent the investment decisions by network users from being made considering the associated network development costs and, therefore, would not prevent inefficient network reinforcements from taking place. These transmission charges would be inefficient as far as they would not achieve a reduction in the network development costs, see (Olmos and Pérez-Arriaga 2009). Only if investors in new generation (or demand) facilities have certainty, before undertaking the corresponding investments, about the specific level of the network charges they would have to pay for these facilities, will they internalize the corresponding costs in their investment decisions, increasing the efficiency of them.

It is therefore sensible to set ex ante, for a time horizon long enough (10 years could be a reasonable trade-off period), the transmission charge to be paid by a new generator requesting connection to a specific point of the grid. It could be arguable if this discussion should also be extended to consumers, but no significant impact is expected on the siting decisions of consumers, at least the small ones since transmission is a minor component of their total electricity payment, which normally is not a major ingredient of the consumers' budget. At the end of this period, transmission charges could be updated to reconcile, from then on, the actual benefits obtained by agents and the expected ones used to compute charges.

2.2 Some Practical Implementation Issues in the Design of Transmission Charges

Transmission charges deem, first, to recover the cost of the transmission network infrastructure and, second, to send as efficient as possible economic signals to the users of the network. Some practical implementation issues related to these objectives are highlighted next.

Tariff structure

Once it has been determined, according to whatever selected methodology, the fraction of the network development costs not recovered from congestion rents that each network user has to pay, the specific format of implementation of the complementary charge used to recover this fraction of network costs matters. Tariffs for the application of complementary charges could be designed in terms of a volumetric charge (€/MWh), a capacity charge (€/MW), a lump-sum (€) or a combination of them. The choice of the format may have implications on the short- and long-term behavior of the network users. Indeed, a volumetric charge will be internalized by agents as an additional variable cost, influencing the result of the dispatch and the energy prices in the system (the short-term marginal cost of electricity). Indeed, the merit order of units could be distorted according to the transmission tariff (complementary charge) applied upon them, leading to an inefficient dispatch. Instead, a capacity charge (€/MW) would be considered as an additional cost by investors in new generation or demand facilities only depending on the size of their investment. If structured as a capacity charge, its level should be made dependent on the amount of capacity built so as to represent the amount of costs caused by the new generator or load facilities for each possible capacity of these. In general terms, transmission charges should be structured as fixed ones (Olmos and Pérez-Arriaga 2009) avoiding volumetric ones.[2]

Revenue reconciliation path

Any kind of cost allocation method may face a temporal mismatch between the incomes resulting from transmission charges and the distribution of payments due to network owners. For instance, when applying the "beneficiary-pays" methodology described in the next section, the chronological distribution of benefits may not (usually will not) coincide with the chronological distribution of allowed regulated revenues of transmission facilities. Although it is not a critical problem, the design of transmission charges should take this into account.

Transmission charges applied on generators

Many regulatory practical schemes apply transmission charges exclusively to the demand. One of the most common arguments for it is that consumers will end up paying all costs, no matter the scheme adopted, since generators will pass on to the demand all charges levied on them.

In principle, both the generators and the consumers are responsible for the cost of (or benefit from) network services, so that it is indisputable that they should both be charged transmission costs. It is true that under a very competitive generation regime, producers will pass-through a large part of those kind of costs whatever the format of the charge is, either in the short-term through their market bids if the format of the transmission charge is €/MWh, or in the long term by postponing investment decisions till the energy prices increase enough as to recover the cost of those charges, if the format is rather a lump-sum or a €/MW one. But this is not commonly the case and the generators often enjoy opportunities that restrict to some degree competition. In those cases, generators can be charged transmission costs without any anticipated pass-through to consumers.

Moreover, transmission charges are aimed at providing efficient locational signal for siting of new generation facilities, in order to minimize total network investment costs. At least the relative values among transmission charges over the different grid locations should matter to guide the siting of new entrants. Obviously, this cannot been addressed if no transmission charges are applied to generators. This could become even more important in a context of large penetration of renewable generation, which frequently require costly transmission investments and has multiple possibilities for siting.

Socialization of part of the grid costs

The physical and technical characteristics of the transmission network (mainly, as previously mentioned, the large economies of scale and the pronounced discrete

[2]Note that probably for distribution network charges the volumetric format may be the more appropriate ones, provided they are time dependent (higher in periods with peak loading of network facilities), since the distortion in wholesale market results will be low and instead a proxy of the long term marginal cost is achieved (see footnote 1).

nature of investments) results in significant, although economically justified, overcapacity. This is particularly true for the first years of new transmission facilities. Therefore, market agents will generally only make use of a portion of the network facilities.[3]

This indisputable fact raises the pertinent question of whether network users should be charged the entire cost of network facilities, or just the part of the cost corresponding to the maximum loading level of each facility. The authors in Olmos and Pérez-Arriaga (2009) largely discuss this issue. They propose using cost causality, or a proxy to it, to charge the network users only a fraction of the cost of each facility corresponding to the ratio of the loading rate of this facility to that of other facilities in the system playing a similar function, obviously limiting to 100% the fraction of the cost of the facility to be charged according to cost causality. This fraction can be deemed to correspond to the one that generators and consumers are actually using, and benefiting from. The remaining part of the cost of the facility should probably be socialized to all the network users. Notice that a reinforcement could be devised larger than strictly needed as for the present situation, anticipating future needs and taking full advantage of economies of scale. In this case, it does not seem reasonable to charge specific network users for a part of the facility that is expected to be used in the future.

3 Benefit-Driven Network Cost Allocation: The "Beneficiary-Pays" Approach

This section is devoted to analyzing the cost allocation methodology driven by the benefits each network users gets from the network. First, in Sect. 3.1, its theoretical foundation and link to economic principles is discussed. Then, in Sect. 3.2, the main open issues and implementation difficulties of such a methodology are reviewed. See also Vogelsang's chapter in this book (chapter "A Simple Merchant-Regulatory Incentive Mechanism Applied to Electricity Transmission Pricing and Investment: The Case of H-R-G-V") for complementary and detailed additional information on the "beneficiary-pays" approach.

3.1 Benefit-Driven Cost Allocation and the "Cost-Causality" Principle

The "beneficiary-pays" methodology proposes to allocate the cost of network reinforcements in proportion to the benefits each agent gets from the existence of the network infrastructure. A network reinforcement does impact the benefits of the

[3]Obviously this is not always the case since network congestion does exist even in properly expanded transmission systems.

different agents in the system as far as it may change the pattern of dispatch of some generation units (maybe because it relieves a congestion) and it also may impact the energy prices at each node of the grid. This will modify the incomes and/or production or consumption patterns of agents, resulting in a different benefit for them.

The "beneficiary-pays" approach is actually very much linked to the "cost-causality" principle. Indeed, since transmission facilities should be built when the aggregated benefits of the additional investment to the network users exceed the incurred costs, the responsibility of system stakeholders in network investments tends to be closely connected to the benefits obtained by these stakeholders from the former investments. Then, the benefits to be produced by these reinforcements are the main reason for undertaking those (Rivier et al. 2013). To the extent that the benefits earned by the network users drive the network investments, and if the resulting complementary transmission charges are applied at a time when they can affect the (long term) decisions by the network users resulting in these network investments, these network charges should be efficient. Network investments, whose cost exceed that recovered from LMPs in the form of congestion rents, are realized in most cases so as to provide specific individual network users with certain amount of expected benefits. Then, the benefits expected to be earned by specific individual users justify a major fraction of network investments. This involves that the corresponding fraction of the network development costs should be directly assigned to the aforementioned network users, i.e., this fraction of the network costs should be deemed directly assignable costs, as opposed to common costs. When they install some new facilities (generation or consumption ones), the network users should be made to pay those network investments undertaken in order to provide some benefits to these facilities. Then, these users should internalize the corresponding network development costs in their investment decision, which is efficient and could potentially lead to savings in network development costs. The same can be said about the decisions by the existing network users about the decommissioning of their facilities. Complementary charges computed in the aforementioned way would act as economic signals coordinating the development of the grid with that of demand and generation.

This is not, strictly speaking, equivalent to the use of the long-term marginal impact of agents' investment and operation decisions on network costs, but it actually complies with the "cost-causality" principle. Indeed, one can imagine some case examples where two agents that make use of the network, located at the same network bus, may have a similar long-term marginal impact on network costs, although the benefits they get from a marginal network investment are rather different. However, if a network reinforcement is undertaken, it is because the net benefits the agents, all together, get from it, are larger than the reinforcement investment costs. Thus, agents are responsible for that cost ("cost causality") in proportion to the benefits they get from the reinforcement.

The reader should note that applying the beneficiaries pay principle to allocate part of the cost of the grid largely differs from applying Ramsey pricing. Despite not being inefficient in conceptual terms,[4] the usual application of Ramsey pricing in transmission cost allocation, built on nodal prices (LMP) as the relevant marginal costs, fails to efficiently allocate the entire cost of construction of the network when, as in reality, LMP largely differ from long-term marginal prices.

Note that this implies that both generators and consumers should be charged because both (in general) benefit from the expansion of the transmission network. When the benefits of a transmission project are very widely distributed, which if often the case for some reinforcements that aim at upgrading the security of the whole system, it might be reasonable to "socialize" these costs, i.e., to apply a flat charge to recover the project costs.

A relevant practical additional advantage of the beneficiaries pay approach is that, if implemented, no stakeholder will be charged more than the benefits it obtains from network reinforcements. Indeed, being the cost of the network reinforcement lower than the overall profits it generates (otherwise the reinforcement should not be approved), allocating that cost to agents in proportion to their benefits will ensure no agent is charged more than the benefits it gets from the reinforcement.

Thus, the "beneficiary-pays" methodology has been adopted as principle in several regions in the world. For instance the regulation in place in USA (FERC 2010) and the EU (European Union 2013; Agency for the Cooperation of Energy Regulators 2013), establishes that the allocation of the cost of new transmission network investments to network users should be driven by the benefits that the latter obtain from the former.

Note that the beneficiaries pay principle can be deemed to be closely linked to Activity Based Costing. By linking the allocation of the cost of the network investments to the benefits that these are deemed to bring about to the production and consumption of electricity by the network users, i.e., the operation benefits produced by network investments, the application of this principle establishes a clear relationship between the allocation of the network development costs incurred and the activities of production or consumption of electricity (an intermediate activity, in the latter case).

This being said, the practical application of this methodology is not exempt from difficulties, for which no orthodox solutions that are sound from an economic point of view, or at least uncontroversial solutions, have still been provided.

3.2 *Main Open Issues and Implementation Difficulties*

This section addresses three of the main hurdles faced when computing the benefits of transmission reinforcements, namely the need to estimate the future evolution of the system to properly estimate the agents' benefits from reinforcements of the network, the so-called counterfactual problem, and the difficulties to actually compute the benefits of individual network projects within a full network expansion plan.

[4]By definition, Ramsey pricing is always second best when marginal cost pricing is not enough to recover full costs, and therefore cannot be "inefficient".

3.2.1 Availability of the Information Required to Compute Benefits

Computing the net benefits of individual stakeholders from the reinforcements of the network conceptually consists in estimating the difference in their benefits between the situation where the reinforcement is in place and the situation where the reinforcement is not undertaken. These are usually called the "with and without" scenarios. Regarding the generators, their benefit results from the difference between their incomes from selling energy and their costs of producing that energy. For consumers instead, benefits result from the difference between the utility value of the energy consumed and the payments for buying that energy. Some obvious difficulties arise when trying to compute individual benefits in practical terms.

First, a proper computation requires knowing the actual production costs of generators and the utility value of consumers. This is confidential information only deemed to be revealed in some compulsory one-shot organized markets (the offers and bids in those markets are supposed to mirror those values). Normally, there is neither access to these data, nor easy way of estimating them, especially regarding the consumption utility value.

Fortunately, some proxies to the benefit changes due to the network reinforcements can be adopted. All generation technologies are well known, and, although their actual production costs will depend on the price of their private fuel purchase agreements, a good approximation of their production costs could be available. Then, the generation dispatch results (pattern of production) and the resulting energy prices (single or nodal energy prices) can be also estimated for both the "with and without" scenarios. The computation of the impact of reinforcements on the incomes and costs, and, therefore, on the benefits, of generators, directly follow. Regarding consumers, assuming a quite inelastic demand, the pattern of consumption will barely be modified. Then, only the change in the energy price (which could be computed) will affect its benefits. There is the need, however, to have in mind that, if demand becomes more active, this assumption will be much more questionable. It is worth noting that this difficulty does not apply exclusively to the "beneficiary-pays" cost allocation process, but also to the proper CBA conducted to make the decision to undertake the network reinforcement. Indeed, as discussed previously, both issues, investment decision drivers and cost allocation drivers, are tightly linked.

Second, network assets lasts many years and the benefits individual stakeholders get from a network reinforcement should be computed for the entire life of this reinforcement. Indeed, the actual share of benefits resulting from the reinforcement may vary along time quite drastically depending on the future evolution of the system. This is a major difficulty. Two approaches may be adopted. One consists in repeating annually the computation of the agents' benefits "with and without" the reinforcement, updating yearly the cost allocation of that reinforcement and adapting it to the actual system evolution. Although it may be deemed to be fairer, it fails to comply with the principle of providing stable ex ante transmission charges, at least to new entrants. Also, after several years, the "without" situation to be analyzed may not make sense at all, since part of the future network could have been expanded based on that reinforcement. This problem is very much linked to the counterfactual issue discussed later on.

The other approach involves estimating the individual benefits produced by the reinforcement over its whole life (due to the application of the discount rate, in practical terms, considering a shorter horizon should be enough). This requires assessing the future evolution of the system, usually by means of a set of possible future scenarios. This obviously makes it complex in terms of defining future scenarios for which data should be collected (fuel price evolutions, demand evolutions, etc.). On the other hand, and again, this is the exercise the system planning and regulatory authorities should have carried out to decide on the expansion of the network. Thus, again, one process could feed the other.

3.2.2 Counterfactual

While carrying out the CBA of a network expansion project, a conceptual problem arise which has still not been properly addressed. When computing the benefits created by the project, making use of the "with and without" situations, one may argue that, in several situations, the "without" scenario may not make much sense. Indeed, if this reinforcement were not be built, eventually, alternative solutions would have been adopted. This is particularly true for large projects. This is called the counterfactual problem, that involves defining which should be the alternative situation to be analyzed when considering the "without" scenario for computing the benefits of a piece of network infrastructure.

A very illustrative example is that of trying to apply the "with and without" assessment of the functioning of the system to an old, backbone line of an existing power system. Without that line, the dispatch and energy price results obtained make no sense at all (for instance leading to lot of unserved energy). If this line had not been constructed, other solutions would have been adopted, including implementing other network topologies, or other siting for generators.

Another illustrative example could be that of undertaking the CBA of a network development to connect a set of off-shore wind farms to the main in-land network. Could it be argued that the "without" situation makes sense, with those wind farms fully isolated in the middle of the sea? Or should we assume that the counterfactual situation would rather consider those very same agents investing in in-land wind farms somewhere else, so that the benefits they are expected to get from the expansion project are not computed by comparing their production and incomes just "with" and "without" the line, but "with" the line and in "a counterfactual" reasonable alternative situation.

All the cost allocation methods based on the "beneficiary-pays" approach share this very same problem, and, as far as we know, no sound solution has been provided to it. Thus, it remains an open issue nowadays. Usually the "with" and "without" situations need to be defined only when allocating the cost of new transmission expansion projects, since these are the projects whose cost is allocated according to the "beneficiary-pays" approach. Other more classical cost allocation approaches (see Sect. 4) are employed to allocate the cost of the previously existing grid. This is relevant, since defining the "without" scenario for the latter assets may prove to be very challenging.

3.2.3 Computing the Benefits of Individual Projects Within an Expansion Plan

Network reinforcements are usually not analyzed in an isolated way. Most of them are part of a wider transmission network expansion plan devised by planners in a system or region. The expansion plan embracing a certain time horizon is made up of a set of individual expansion projects. Although some of them may be quite independent from the others, many of them are actually interrelated. The benefits they produce are linked. Thus, the benefits of the overall expansion plan are not the result of adding up the benefits of the individual projects in the plan computed considering that only each of these projects is undertake at each time.

Thus, one should, in principle, devise a set of transmission charges by allocating the cost of the full expansion plan proportionally to the overall benefits each agent obtains from the former. However, it may be of interest and sometimes required to breakdown the benefits of the whole expansion plan into the benefits of individual expansion projects. In practical terms, even if a full expansion plan has been approved, not all the projects within an expansion plan are finally built, normally, while only the cost of those projects actually undertaken should be allocated to agents. Indeed, some projects may face fierce opposition from local authorities and interest groups or may face some funding difficulties, who could delay the construction of these projects or block them definitely.

If some of the projects within a plan are not built, network charges should only achieve the recovery of the costs of those projects undertaken. The only way to achieve this is separately allocating the cost of each expansion project (part of the whole expansion plan) according to the benefits this individual project produces. Note that computing the benefits and costs of each individual project in the plan would allow ranking the projects in the plan in order to give some of them priority if, for instance, budget constraints exist.

Besides, individual projects within an expansion plan are, generally, finally approved individually, after getting all the permits and funds required. But there is no certainty of which specific reinforcements within the remainder of the plan are finally going to be built in the considered time horizon, and which others will be temporally or definitely blocked. So, even if finally the whole expansion plan turns out to be built, the time scheduling for the construction of the different individual projects within the plan is uncertain, and the cost allocation for the first ones built should be computed without the others being in place, and not having the uncertainty about whether these will be built.

Thus, in this case, probably the most sensible way to proceed is separately allocating the cost of each project assuming that the remainder of projects in the plan are going to be built. This requires computing the benefits produced by each individual project in the context of the plan. However, as mentioned previously, the benefits they produce are interdependent and cannot be assessed considering each project in isolation. Incrementally computing the benefits each individual project brings out, assuming a certain order of deployment of the projects, would result in the benefits of individual projects adding up to the total benefit produced by the whole plan. However, the benefits associated with each individual project will largely depend on the assumed order of deployment of projects and, therefore, the resulting cost allocation, considering a certain order, should be deemed arbitrary.

Some simplified approaches to address this problem have been considered, for instance, in Europe to compute the benefits of regional transmission expansion projects, namely the TOOT (Take Out One at the Time) and the PINT (Put IN one at the Time) methods (ENTSO-e 2013). The TOOT method assumes that the project whose benefits are being computed is the last one to be deployed within the expansion plan. Thus, the benefits are computed assuming the remainder of the plan is already in place. On the contrary, the PINT method assumes that each project whose benefits are assessed is the first one to be deployed (Bañez et al. 2017a, b). Thus, its benefits are computed assuming the remainder of the plan has not been deployed at all. In both cases, as the method is applied to each individual project, no specific order of deployment of the rest is assumed, which reduces the level of arbitrariness of the method. However the sum of benefits produced by all individual projects largely fails to match the benefit produced by the whole expansion plan. Then, proportional scaling must be used.

Other families of methods have been proposed in the literature trying to better take into account the interactions occurring among individual projects without having to assume a specific order of deployment of them. Although they are deemed to be more consistent, they are much more complex. The best-known approaches are those based on the Shapley and the Aumann-Shapley concepts. The Shapley method computes the incremental benefits produced by each project as the average of those created by the project over all the possible orderings of deployment of the projects in the plan (Hasan et al. 2014). The Aumann-Shapley method is an extension of the Shapley one devised to prevent results from being dependent on the size of the project, partially preventing perverse incentives for project promoters to merge their projects with others to get their projects more highly ranked in priority rankings. It divides each expansion project into elemental subprojects, all of the same size, and computes the benefits produced by the project by adding up those assigned to the subprojects when deploying them in all the possible orders (Bañez et al. 2017a, b).

3.2.4 Treatment of the Negative Benefits Produced by Projects

A transmission expansion project may induce both positive and negative benefits for different network users. For instance, a project reinforcing the link between two areas with a large energy price difference will usually produce positive benefits for the generators in the low price area and the consumers in the high price area, and negative benefits for the generators in the high price area and the consumers in the low price area, since the energy prices of both areas will tend to converge (the one in the low price area will increase, while that in the high price area will decrease).

For the reinforcement to take place, the net benefit produced by the project (positive ones minus negative ones) should exceed the total cost of the line. Although some authors argue that not considering negative benefits would make the network development more dynamic and agile, since a larger set of network expansion projects will be approved, this would result in projects producing a net benefit that is lower than their cost being approved, which is not efficient from a social point of view.

Regarding the cost allocation of the project, one could easily agree that only network users positively affected by the reinforcement should be charged part of its costs. Those network users disadvantaged by the project should, instead, not be charged at all. The question is rather if the latter should be compensated for the loss of benefits they incur. Doing so, these agents will be shielded from taking wrong siting decisions, which seems not to be an appropriate signal. Market agents should not be protected from the losses incurred as a result of the increase in the level of market competition created by network investments.

However, it may be possible that some stakeholders, for one or another reason, have a veto power over the project (Coase 1990). This could be, for instance, the case of transmission expansion plans at regional level where countries or States involved may have the final decision to make on the construction of network facilities within their administrative territory. Those reinforcements, although socially beneficial for the whole region, will never be undertaken if they produce negative benefits for some of these countries, unless they are compensated for the losses they incur. In these cases, for the project to go ahead, those stakeholders being positively affected by it will have to pay larger charges allowing compensations to be paid to those other stakeholders being negatively affected. The reader should be aware that, even in those cases, no agent will be charged more than the benefit it gets form the project, since total net benefits (positive ones minus negative ones) exceed the cost of the project.

4 Other Approaches to Transmission Network Cost Allocation

The "beneficiary-pays" methodology to set transmission charges, although adopted in guidelines for grid charging both in Europe and America, has not been yet extensively implemented due to the previously discussed difficulties.[5] Then, a set of alternative approaches have been actually followed to address the transmission charges design. Most of them lack a solid economic background, although some try to resort to some proxy of the cost-reflectivity principle. They are next briefly reviewed and analyzed, being grouped in two families, the network usage-based ones and the non-locational ones.

[5]Despite these difficulties, Argentina (1992) and California (1998) adopted network charging regulatory approaches inspired in the "beneficiary pays" concept.

4.1 Network Cost Allocation Methods Based on Usage

Due to the fact that determining long-term marginal, or incremental, costs, or the benefits produced by network investments, is challenging, many regulatory frameworks for cost allocation resort to making use of a proxy of the former deemed easier to compute: the transmission network usage made by each agent. These frameworks, or methods, are commonly referred to as usage-based network cost allocation methods. Although in some cases the benefits that network users obtain from the grid are largely different from the usage they make of this grid, considering this proxy may be acceptable in general terms. Indeed, it makes sense that the larger is the use of a transmission asset by a market agent, the larger the latter benefits from the line or the larger it will be responsible for further investments in that transmission link if reinforcements are needed in the future.

Although easier to address, quantifying the usage of the grid by a network user is unfortunately also challenging, since network flows result from the interaction of all the power injections by generators and the power withdrawals by consumers. Network flows are not traceable from one point of injection to the ones of withdrawal. The responsibility of a certain agent in a certain flow of a certain line cannot unambiguously defined. Thus, several proposals exist to compute this. The choice made is not irrelevant, since they usually provide quite different values. In fact, only some of the usage-based methods can actually be considered an acceptable proxy of the "Beneficiaries pay" methodology or the "cost-causality" principle, in our opinion, namely, the average participation (AP) and Aumann-Shapley (AS) methods.

Next, some of the more popular usage-based methods are described. A detailed critical assessment of some usage-based methods can be found in (Rubio-Odériz and Pérez-Arriaga 2000).

Contract path

This method, widely used in the past to charge bilateral trading arrangements (known as "wheeling") in a context previous to the creation of organized wholesale markets, assumes that the energy flow associated with bilateral trade arrangements follows a predetermined path, the contract path, from the point of injection to the point of withdrawal. The transmission charges are then determined as a fraction of the cost of the lines where the transaction "flows" according to this path.

This method has some clear drawbacks. First, it violates the principle discussed earlier stating that network transmission charges should be independent of commercial transactions. Second the path followed by the energy flow between the two grid connection points involved in the bilateral trade agreement cannot be unambiguously identified. Actually, this path usually comes out of an agreement between the seller, the buyer and the transmission company. Third, as previously discussed, energy flows are not traceable in a grid, but depend on the full set of power injections and withdrawals in the whole grid. The utilization of the grid by such a bilateral arrangement is an entelechy. Fourth, when applied to a regional context, this method leads to the "pancaking" phenomenon discussed previously, since a commercial agreement between a seller and a buyer located in different systems will entail paying a transmission charge for each system supposedly crossed by the flows created by this transaction, making the transmission charges dependent on the existing political borders, which makes no sense at all. Fifth, the method is hardly applicable to energy trading arrangements based on organized market pools where bilateral agreements, simply, do not fit.

MW-mile (MW-km)

This method also applies to bilateral trading agreements and, therefore, is subject to the very same critics as the "contract path" one. It is, actually, a more sophisticated version of the latter, trying to provide a more precise measure of the network usage by bilateral transactions, taking into account not only the amount of power transmitted (MW) but also the length of each line used. Note that the latter was also implicitly taken into account in the "contract path" method, since the cost of a longer line would obviously be larger than that of a shorter line.

The computational procedure to follow is as follows. First, transmission flows (MW) for a baseline case, representative of system operation, are identified and the flow carried by each line is multiplied by the length of the line to determine the usage made of it. Adding these all together provides the total MW-mile usage associated with the baseline case. Then, for each one of the bilateral transactions to be charged, that particular bilateral transaction is removed from the baseline case and the new total MW-mile usage of the grid is computed. The difference between the two values is the usage of the network attributable to this transaction. These estimates of the use of the grid by each bilateral transaction are finally used as proportional factors to allocate the total cost of the network.

Marginal participations (MP)

This method aims to determine the marginal usage of the network by each player in the system. That is, it computes the change in the flows of all the grid facilities for a marginal increase in the power injected/withdrawn by each generator/consumer. This is computed for all agents and a set of representative scenarios. Based on this, the usage of each network facility by each agent is identified as the product of the marginal increase of the flow in the line, the power generated or consumed by the agent, and the duration of the scenario. This usage metric is then used to proportionally allocate the individual cost of each particular facility.

Although the methodology seems sensible, it has a severe drawback. Indeed, the computation of the incremental flows, as described above, requires the arbitrary choice of a slack/swing/reference bus in the system, and the results prove to be largely dependent on this choice. Transmission charges based upon such a methodology could be easily disputable. Distributed slack nodes have been proposed to mitigate this problem, but, still, the arbitrariness on the selection of such a node remains.

Only in systems with a very dominant load center, chosen as the slack node, may this method provide sensible results. For instance, this is the case of the Argentinean and Chilean electricity systems, where variations of this method (known as "areas de influencia") have been implemented. In both cases, Buenos Aires and Santiago de Chile—where a large fraction of the total demand of the country is located—are selected as slack nodes. As a consequence, the demand pay very low transmission charges while generators pay higher charges the farther they are located from the capital. A similar method, known as CRNP (cost-reflective network pricing) is in place in Australia, while a more sophisticated version of it is applied in the Single Electricity Market of Ireland.

The investment cost related pricing (ICRP)

This method is actually similar to that of the marginal participations (MP) one, in the sense that the responsibility of each agent in the grid investments costs is assumed to be associated with the marginal increase in the use made of the grid, measured in a peak load scenario. Nevertheless, MP is usually applied to allocate the cost of the existing network so that grid charges are computed multiplying this unit usage factors by the annualized unit actually incurred network costs, while ICRP aims at providing a proxy of the long-term marginal network cost with respect to the power injected/retrieved by each generator/consumer, using therefore updated unit replacement network costs instead of the actually incurred ones. The resulting locational unit network charges are applied to network users and lead to the recovery of the cost of a portion of the network. The remaining network cost is usually socialized to grid users through a residual charge.

This method has been applied in the UK[6] and Colombia. Analogously to the case of MP, this method requires the somehow arbitrary choice of a slack bus for computing the marginal usage of the network. A virtual slack node is chosen so that the proportion of charges applied to generators and consumers corresponds to a pre-set value.

Average participations (AP)

Based on common sense, the method applies a simple heuristic rule to "trace downstream" the power injections into the grid (by generators) and to "trace upstream" the power withdrawals from the grid (by consumers). Based on the actual pattern of flows observed in the grid, the power flows entering a node are split among the outflows proportionally to the size of the latter. For this, only lines where the flow is coming out of the bus are considered. Then the process is repeated for each bus reached by the power injected by an agent until reaching the demands supplied with this power. A similar process is followed for power withdrawals.

Although this method could be perceived as too heuristic and therefore lacking a solid theoretical foundation, actually the only hypothesis is resorts to seems very reasonable, that is that a fraction of a flow bifurcates at each node in the same proportion as the total flow. Moreover this method deals with the real observed flows which are values that can be measured and quantified. The consequence is that, actually, this method provides in most of the cases quite sensible results. Although this is not a solid argument to support its application from an economic or technical viewpoint, it is relevant in many regards.

---

[6]In the UK, the method differentiates conventional and renewable generation, taking into account the average load factor for the latter.

This method allocates the flow carried by each line to certain injections, on the one hand, and certain withdrawals, on the other. A strict application of those results will lead to a 50%-50% allocation of the overall total network costs to generators and consumers. However the method is fully flexible in this regard since the "trace downstream" (from generators) and the "trace upstream" (from consumers) computations are performed on a separate basis and can be associated with a different proportion of the cost of the transmission assets. Therefore the method adapts easily to any policy-driven or negotiated-driven decision making on the transmission cost to be shared by generators and consumers.

This method was first proposed and applied in New Zealand's Trans Power, see (Read and Sell 1989). Another reference to this is (Bialek 1996).

Aumann-Shapley (AS)

Although not yet adopted by any existing regulatory layout, the Aumann-Shapley approach in Junqueira et al. (2007) provides also reasonable cost allocation results. These are close to those produced by the Average Participation method. In fact, both approaches have some similar features. The Aumann-Shapley cooperative game can be formulated having network users as the players in the game. In this case, the overall use made of the network is allocated to these players. Alternatively, Aumann-Shapley can be formulated having individual transmission expansion projects as the players in the game, in order to allocate to these projects the benefits produced by the expansion of the grid as a whole (Bañez et al. 2017a).

According to AS, the use made of the grid, and each transmission asset, by each generator, respectively each consumer, in each operation situation is determined as its average incremental contribution to the grid usage, and to the flow in each asset, when considering all the possible orderings of deployment of the power injected into, respectively withdrawn from, the grid by the generators, respectively the consumers, in the system in this operation situation. In order to consider all the possible orderings of injecting power into, or withdrawing power from, the grid by the network users, the injections and withdrawals by these users in each situation are divided into elemental power injections or withdrawals, all of the same size. Then, these elemental power injections and withdrawals are considered in all the possible orders, and, for each order, the incremental contribution of each elemental power injection or withdrawal to the flow in each network element is computed trying to minimize it. Then, the use made of the grid by each network user for this ordering is computed as the sum of the incremental contributions to the use of the network made by all the elemental power injections or withdrawals corresponding to this network user. Lastly, the average incremental contribution of each generator or load to the overall usage made of the grid in a certain operation situation is computed as the average of the network, and individual line, usage attributed to this generator or load over all the orderings of deployment of power injections, or withdrawals, in the system that have been considered.

4.2 Tariffs Without a Locational Component

Most of the regulatory approaches to transmission charging applied in real life do not include any locational term. Although not desirable in general, this could be acceptable when the weight of the transmission network in total electricity supply costs is low, and the transmission grid is already well-developed and fully meshed and it does not require major reinforcements in the foreseeable future. In this case, providing more sophisticated signals may not be worth the effort required for this. Besides, as discussed, these signals are not straightforward to compute and could be contestable. However, the current trends toward the integration of large amounts of wind and solar generation in regional grids may advise including locational signals in transmission charges.

Postage stamp
This method involves applying a uniform charge per MW connected to, or per MWh injected into or withdrawn from, the grid (as a "postage stamp" tariff) to all the network users regardless of their actual location. It is often applied only to consumers, although some systems do allocate an arbitrary part of the total network cost also to generators. It is, by large, the most popular method. Most of the European countries, many US systems, and many countries in the world, apply this approach.

Ramsey pricing
This method is aimed at applying transmission charges interfering the least possible with the market and system operation, and the investment decisions made by market players. No locational signal is still provided, but a Ramsey criterion (a well-known second best approach to allocate costs) is followed. The Ramsey criterion seeks to allocate the transmission costs to network users in inverse proportion to their price elasticity. According to this criterion, the less elastic the decisions made by agents are to a change in the transmission charge, the higher are the charges these agents are required to pay. Under this criterion, those generators whose operation decisions are most price elastic are almost exempt from paying transmission charges, while other generators whose residual demand is inelastic and domestic consumers pay the highest charges, and industrial consumers pay intermediate charges. Although not explicitly assumed in their methodology, most of the countries applying postage stamp charges, are implicitly applying a Ramsey criterion to allocate costs to groups of agents.

5 Transmission Network Cost Allocation in Regional Power Systems

New issues appear when trying to extend the principles of transmission pricing from a single power system to a multiple—usually termed regional—system. First, we discuss the new requirements. Then, we describe and evaluate the solutions adopted in three mature and quite different regional markets: The Internal Electricity Market (IEM) of the European Union, the US Regional Markets, and the Regional Electricity Market in Central America (MER for its name in Spanish).

5.1 The Need for Coordination and Orthodoxy in Inter-system and Intra-system Transmission Pricing

Regional power systems require adequate interconnections to be able to exploit the advantages of joint operation and planning. Developing new transmission reinforcements of regional scope requires to agree on the inter-system allocation of their cost. The individual systems will only approve of a new transmission investment if they perceive that the expected future benefit from the project exceeds the cost to be born. Therefore, application of the principle of allocation of the cost of these transmission investments according to some measure of the expected benefits that each system is expected to obtain is, in the regional context, even more justified than in the single system context. Violation of this principle would probably lead some systems to oppose the construction of some investments and to miss the opportunity of building network reinforcements that are highly beneficial for the region and whose benefits are spread across several systems. This would, in turn, significantly affect the operation of the power systems in the region, since the fact that the grid is underdeveloped may largely constrain the power exchanges among these systems.

A frequent mistake in inter-system transmission cost allocation stems from violating the principle that transmission pricing should be independent of commercial transactions. Charging the costs of interconnections only to those agents involved in cross-border commercial transactions is fatally wrong and seriously deters inter-system trade, since the cost of the cross-border network assets must be more widely shared among all agents that benefit from these transactions, which is a much wider group than those that have signed cross-border contracts.

Once the costs of the regional transmission network have been allocated to the different participating systems, each one of them must decide how to price transmission internally. The individual systems may want to have autonomy in establishing their intra-system transmission pricing scheme. However, some minimum coordination level is advisable, since, otherwise, economic efficiency will suffer. Intra-system transmission pricing may affect the investment and operation decisions made by the network users in a region. The systems may choose among the different methods presented in the previous sections. The resulting charges will affect the decisions made by the network users on whether, when and where to install new generation and load. Additionally, if these charges are poorly designed—e.g. it is decided to apply a volumetric transmission charge in $/kWh—the level of these charges will have a distortionary impact on the economic dispatch of the power plants in the region, as their effective variable costs will be modified.

Consequently, if the set of rules applied to determine the contribution of individual agents within each system to the cost of each regional network reinforcement are not efficient and homogeneous across all the systems, some systems may arbitrarily become more attractive for the installation of new generation or load than others. The locational signals provided through network charges to the network users of a certain type in a system would, then, mainly depend on whether the authorities within this system have decided to apply more favorable rules for this specific group of network users than the authorities in other systems.

5.2 *Case Examples of Cost Allocation Approaches in Regional Markets*

Within the regional markets in the world, there are three ones considered to be paradigmatic due to their distinct features, also with respect to the solution adopted for the allocation of the cost of the regional transmission grid: the IEM of the European Union, the regional markets that have developed in the USA, and the MER in Central America. In the following paragraphs, we describe and assess the coordination scheme applied within each of them. In order to do so, we start by describing some general features of these markets that condition the cost allocation scheme adopted in each case.

5.2.1 European Scheme

The Internal Electricity Market (IEM) of the European Union is based on the coordination of the national systems and the markets existing in this region. The national authorities within each country and the existing markets are deeply rooted in their respective systems. Therefore, the regional and national authorities decided to organize the main aspects of the functioning of the IEM through the cooperation among the already existing systems and markets. Market Operators are undergoing a process of concentration, i.e., they are merging to create larger ones. Eventually, this could lead to setting up a single Market Operator all over Europe. However, it is highly unlikely that national authorities and System Operators will transfer large amounts of decision making power to regional institutions in the short to medium term future.

Regarding the allocation of the cost of the transmission grid, first, there is the need to highlight the fact that short-term energy prices within each country are the same for all the network users, with some exceptions, like some countries in the Nordic Electricity Market, Nordel. Then, only those transmission assets crossing a political border (or a bidding zone one in the aforementioned few cases where some locational differentiation in energy prices exists within a country) produce some rents out of the application of energy prices that can be used to recover part of the cost of these assets. For the rest of assets, the complementary charges must allocate their whole cost.

As for the design and implementation of complementary charges in the IEM, instead of computing pan-European transmission charges, the authorities have decided to have only in place national transmission charges, whose level should be adapted to take into account the use that the agents within each national system are making of the grid of other national systems. This has been implemented through a system of compensations among countries associated with the cross-border use of national transmission grids. For each country, a compensation is computed aiming to reflect the use that generators and consumers in other countries are making of the grid of the former. The compensations owed to all the countries make the global compensation fund. Lastly, the contributions of countries to the global compensation fund are determined, allegedly, according to the responsibility of each country in cross-border flows. Thus, as a result of the implementation of this scheme, for each country, a compensation to receive for the external use made of its grid, and one to be paid corresponding to is contribution to the global compensation fund, are computed. National authorities are free to decide on how these two should be reflected in national transmission charges, which are the only ones paid by network users. Part A of the Annex to the Regulation (EU) 838/2010 sets out guidelines on the definition of the Inter-TSO Compensation (ITC) mechanism.

The network charging scheme applied aims to provide a solution to the problem of allocating the regional network development costs across countries, while, at the same time, being the least intrusive possible regarding the interaction with national schemes. Thus, it is not aimed at providing locational signals to individual network users; i.e., it does not address the problem of driving efficient long- and short-term decisions by individual generators and consumers. In this regard, the national and regional authorities have agreed to achieve some harmonization of national transmission charges in order not to distort competition in the European market. Harmonization efforts have focused on the level of transmission charges within each country that should be levied on generation. Regulation 838/2010 sets the range where transmission charges for the use of the grid paid by generation should lie. Surprisingly, the limits set vary across countries, which may create distortions in the competition taking place among the generators from these countries. In addition to that, there is significant heterogeneity across countries regarding a number of charging aspects. A list of them follows:

- The structure of the charges: the fraction of these charges applied as a fixed charge, on the basis of the energy produced or consumed, or based on the nominal capacity of generators or capacity contracted by consumers;
- The nature of connection charges: whether deep charges or shallow ones are applied;
- The locational content of charges: whether use of the system charges have some locational content, and how this locational component of charges is computed.

All these aspects may affect the competitive position of agents (mainly generators and large consumers) in the electricity sector and the economic activities they are involved in.

Regarding the computation of compensations among countries and the schemes applied within each country to modify the national charges according to the net compensation to be paid by this country (the net amount resulting from deducting the compensation owed to the country for the external use of its grid from the contribution of the country to the global compensation fund), there is also a long way to go to implement an efficient solution. The inter-TSO compensation scheme applied is far from computing a sensible estimate of the external use that is being made of the grid of each country, while electrical usage normally does not accurately represent the benefits that the agents in other countries are making of the grid of the one considered. The contributions of countries to the global compensation fund result from the application of a simplistic method and do not reflect the responsibility of each country in the use of the grid of others. Lastly, most countries are not reflecting in an economically meaningful way the compensation they have to pay in the local transmission charges applied on consumers and generators. Hence the locational signals that these compensations may convey are largely lost. For a sensible way to compute locational signals at country and agent level in a European context, see (Olmos and Pérez-Arriaga 2007a, b). As a result of all this, there is a debate in Europe over the convenience to modify the current ITC mechanism. Thus, the Agency for the Cooperation of Energy Regulators (ACER), in its Recommendation No 05/2013, proposes allocating the cost of new transmission assets based on cross-border cost allocation agreements among national authorities, leaving the ITC mechanism to allocate only the cost of the infrastructure already existing at the end of the year 2015. Cross-border cost allocation agreements among countries are expected to be based on the expectations of the countries about the benefits that each of them would obtain from the new assets to be built.

Within the IEM, the regulated transmission network investments coexist with others promoted by merchant entrepreneurs (merchant investments). Most of these are subject to open access, earning congestion rents resulting from the system operation. However, there are some for which access to their capacity is negotiated between the owner and prospective users.

5.2.2 USA Scheme

Within the USA, as aforementioned, regional markets, the so-called RTO ones, have developed over the last two decades. These are fully integrated markets that have applied fully integrated transmission pricing solutions. However, the level of coordination and solutions implemented at inter-regional level to allocate the cost of transmission projects affecting, and benefiting, several regions, may also be a matter of discussion. Coordination among regions in this, and many other, regards is fairly limited. Thus, two levels may be distinguished when discussing transmission pricing within the USA: the regional one and the inter-regional one.

At the regional level, most regional markets have implemented energy prices including some sort of locational signals. Some have applied nodal pricing, like PJM, while other have opted for zonal energy pricing, like for the California Independent System Operator (ISO). Thus, congestion rents resulting from the application of energy prices can be used to recover part of the cost of the grid, though, as argued above, the majority of it remains to be collected through the application of complementary transmission charges.

The federal energy regulator (FERC) has authority, for reasonableness and fairness, over the allocation of the cost of regulated network assets, and the network costs of interstate commerce in general. However, States can also block line construction, or make it difficult, when they disagree with the allocation of the cost of these lines. Originally, the Standard Market Design (SMD), a blueprint for the design of regional markets by FERC (2002), stated that regional access tariffs should adopt the form of postage stamp charges or license plate rates that vary across areas. Consumers should be in charge of paying them. However, SMD recommendations have not become rules to be enforced. Sometime after, FERC order 1000, FERC (2012), required the cost of the regulated transmission projects to be allocated proportionally to the benefits expected to be reaped by network users from the former. In the case of reliability projects, this dictates charging those agents causing the corresponding constrain violation or, alternatively, for those projects expected to improve supply conditions for large parts of the system, the socialization of their cost. Federal regulation also mandates that the cost allocation schemes applied are transparent and do not prevent, or hamper, the undertaking of transmission projects producing large benefits.

The requirement to allocate regulated transmission costs according to the benefits created by the corresponding projects has led the authorities in most regions to implement charging methods that consider a proxy to the benefits these projects produce. For example, in many regions, connection facilities are to be paid by the network users installing in a certain node, or area. Besides, there are also many regions where flow-based mechanisms are employed to determine the use that generators and loads are making of transmission assets and, based on this, how much they should contribute to the recovery of the cost of these assets. In the later cases, network usage is being deemed a suitable proxy to determine the distribution of the benefits produced by the considered assets, being network usage computed in different ways in different regions. On the other hand, network charging methods requiring the computation of the benefits produced by transmission projects are rarely applied, see (PJM 2010). The charging schemes applied within each region to allocate the cost of different types of transmission projects, according to their function of the benefits they produce, may be different. The entities in charge of developing and approving the charging methodology for the transmission network in a region are the planning authorities within it. If the authorities within a region do not define the scheme to allocate the cost of regulated transmission projects, FERC is allowed to set one.

According to what has just been discussed, one can easily conclude that different regions tend to apply different transmission network cost allocation schemes. This can potentially distort competition among agents involved in reliability, economic, or RES energy based power exchanges among regions. Because of the existence of power exchanges across regions, there are transmission projects that benefit agents located in several regions. This is clear for those projects interconnecting two or more regions. Then, the cost of these projects must be allocated to (the network users within) several regions. According to FERC regulation, this should be based on the application of a common method agreed by all the affected regions, at least in the case of projects crossing several regions. However, nothing is said about the specific features that this method should have, nor about how the fraction of the cost of a project to be paid by each region should be allocated internally to the network users within the region. Then, despite the allocation of the costs of transmission investments within each region could be based on sensible rules monitored by FERC, there may be difficulties to efficiently allocate the cost of inter-regional projects. This may jeopardize the undertaking of these projects. Besides, while the efficiency of locational signals and the level of interference of transmission charges with the operation of the system within each region depend on the features of the specific solution adopted in the region, the lack of harmonization across regions of the cost allocation schemes applied negatively affects the efficiency of the locational signals conveyed at country level (across neighboring regions), as well as the creation of a level playing field for the competition of generators and consumers from different regions.

As in the IEM in Europe, merchant investments of the several types discussed in Sect. 1.3.1 coexist with the regulated investments.

5.2.3 Central American Scheme

The Regional Electricity Market (MER) in Central America is a seventh market superimposed on the six national ones within this region. Regional authorities and institutions have jurisdiction over the functioning of the MER. This concerns the planning of the expansion and development of the regional transmission grid, the dispatch of regional energy transactions, the pricing of the energy involved in these transactions (according to a system of nodal prices applied on the nodes of the regional transmission grid), as well the transmission capacity used by them (including the regional contracts), and the definition of regional transmission charges and their implementation at country level. However, national authorities have the power to decide on how the transmission charges computed for the generation and load in each country should be finally applied to the individual network users within their territory. For more information on the basic design and rules of the Regional Electricity Market in Central America, see the regulation (Reglamento del Mercado Eléctrico Regional, or RMER for its name in Spanish) developed by the regional regulator CRIE.

Thus, the national authorities are the ones eventually setting the locational signals sent to individual network users driving their investment and operation decisions. Little progress has been made regarding the definition of enforceable rules that national transmission charging methods should comply with, nor with the harmonization of the level, structure, and cost components included in these transmission charges. Therefore, the economic signals conveyed through these charges are unlikely to drive an efficient distribution of generation and demand in the region and may interfere with the regional system and market operation.

The rules originally developed in RMER to allocate the cost of regional transmission assets to the regional and national transactions, and eventually to countries, were deemed to be too complex. Then, the regional regulator (CRIE) developed some interim regulation on the method to apply to compute transmission charges, the eventual transmission payments to be made by network users, and the invoicing, and settlement of these payments for part of the regional transmission grid (RTR for its name in Spanish), the so-called SIEPAC line, see (CRIE 2011). While the original regulation was complex and inefficient, the remuneration currently applied to allocate to the network users, and eventually to the countries in the region, the cost of the assets belonging to the SIEPAC line is far too simple to reflect the benefits that each user and country is expected to obtain from these assets. Basically, the net revenues resulting from the application of nodal prices to regional transactions (congestion rents) are used to pay part of the cost of the SIEPAC line, which represents the backbone of the RTR. The rest of the cost of the SIEPAC line is allocated to the countries based on the location of the corresponding reinforcements and the demand within each country. Thus, the fraction of the cost of the interconnectors to be paid from complementary charges is allocated to the countries in the form of postage stamp charges to be applied on all the demand within the region. Thus, the allocation of this cost to the countries is proportional to the demand within each country. On the other hand, the cost of non-interconnector transmission assets within the SIEPAC line to be recovered through complementary charges is to be paid fully by the country where these assets are located. According to the proposal in the regional regulation, this cost should be levied on network users in the country as a postage stamp charge to be paid by the demand. Fortunately, regional authorities have authority over the planning of the expansion of the regional grid and the approval of the regional reinforcements. This means that, even when the allocation of the cost of regional reinforcements to the countries in the region may not be efficient, regional authorities may be able to achieve the construction of these reinforcements despite the opposition by the national authorities of those countries that are badly treated in the cost allocation process. However, one main problem faced when trying to promote the construction of regulated transmission assets is the lack of funds that can be accessed by national and regional planning authorities. This may prevent both national and regional authorities to achieve the construction of those reinforcements that are perceived as necessary. Additionally, due to the aforementioned lack of funds, the development of the national transmission grids tends to be largely insufficient in some countries. Then, some regional transmission investments may end up being used to host flows created by national generation and demand, instead of allowing an increase in cross-border trade in the region.

There is also the possibility to combine the construction of regulated assets with the undertaking of investments at risk promoted by (associations of) network users in the region. These are especially well suited for the connection of large generation developments to the bulk regional transmission system. However, the potential lack of capacity in the RTR to transfer the energy produced by these generation facilities to other areas in the region where it can be consumed may deter these investments.

6 Conclusions

Transmission network pricing methods are overly simple in most power systems worldwide. In most cases these costs are socialized among the network users, adding flat volumetric charges ($/kWh) to the other components of the consumers' tariffs and frequently sparing the generators from any charges. Only a handful of countries for which the transmission costs represent a relevant part of total electricity supply costs (typically large countries) or for which the grid still requires large reinforcements have applied more elaborated pricing schemes that try to convey locational signals associated with some measure of cost causality.

However, the relevance of transmission pricing is rapidly increasing. Decarbonization of power systems requires installing large amounts of renewable generation, for which a large set of candidate sites is possible. This may require undertaking significant network reinforcements, depending on the location of the new investments. Forward-looking cost-reflective locational transmission charges will contribute to the efficient siting of network users. Regional electricity market integration around the world is also creating the need for large supranational transmission network infrastructures. Again, a sound cost allocation approach—one based on cost reflectiveness or, equivalently, on beneficiary pays—will be of essence. In a regional (multi-system) context, the individual systems will probably oppose the construction of those transmission assets for which they have to pay a fraction of their cost that is not commensurate with the benefits that they expect to obtain from these projects.

This chapter has presented the basic principles that should guide the design of complementary transmission charges: transmission charges should be forward-looking and cost-reflective, or, equivalently, based on the beneficiary-pays principle; they should not depend on commercial transactions among the agents of the system, either in a single or multiple-system context; and the charges should be set ex ante, at the time the generation or demand siting decisions are made. Proxies might have to be used to avoid unnecessary complexities when deemed reasonable.

Transmission-related locational signals start with the design of energy prices, when they have a locational component. However, it has been found that nodal short-term energy prices—also called locational marginal prices—grossly under-recover the cost of most transmission lines in actual power systems. Complementary charges are therefore needed to achieve the complete recovery of the regulated transmission cost. These charges must send forward-looking long-term locational signals to network users, while avoiding distortion of the short-term signals. In most power systems the short-term energy prices do not contain any locational differentiation and the "complementary" charge is the only transmission charge, besides any initial connection charge.

Merchant lines, that is, network investments promoted by private investors and not subject to regulated incomes, may run under different network charge schemes, often resulting from negotiation. The cost of those investments will then be charged to the users that benefit from them, though the private promoters run the risk of having finally to bear part of it.

A diversity of transmission pricing methods is applied around the world, in most cases flat volumetric charges only applicable to consumers or interconnection charges associated to the power being traded. As it has been explained, poorly designed transmission charges, in particular those applicable to new generators and to allocate the cost of interconnection network assets, will have to be urgently replaced by more cost-reflective schemes. This chapter has made the case that the "beneficiary-pays" approach, equivalent to responsibility in future network investments, makes economic and technical sense and it is the guide to be followed, in particular when transmission cost allocation matters. Beneficiary- pays is intrinsically linked to the widely accepted rule that justifies transmission network investment decisions (the Golden rule based on cost-benefit analyses). This explains why the beneficiary-pays principle has been adopted as a conceptual guideline for transmission network cost allocation in some of the most relevant regions of the world, such as the EU and US FERC regulations. However, many of the practical difficulties of computing the benefits of transmission investments and deriving transmission charges based on them have not been solved yet. This chapter discusses the theoretical foundations of the beneficiary-pays principle and reviews the main difficulties, implementation hurdles and open issues that still need to be addressed before successfully implementing charges based on this principle. Some of the most relevant aspects to explore include the definition of a counterfactual when assessing the benefits of a transmission project, or alternative situation to that where this investments project is undertaken; the computation of the benefits produced by each of the projects within an expansion plan; the collection of the information required to compute the future evolution of the system; and the treatment to be given to the negatives benefits obtained by some network users from the construction of a transmission project. Further research is still required to find a solution to the aforementioned problems.

Then, this chapter also reviews other methodologies for transmission network cost allocation in place around the world, highlighting the practical advantages of one of them, the average participation method (AP), which provides robust and sensible practical results wherever it has been applied, but also highlighting the fact that it lacks a solid supporting economic and technical foundation.

Finally, the allocation of the cost of transmission networks in a multi-system, or regional, context is discussed. The relevance of implementing efficient signals in this case is reviewed and the solutions adopted for cost allocation in three of the most relevant regions in the world, characterized by the existence of relevant regional markets, are described and briefly assessed. The regional markets explored in this regard correspond to the ones created in the USA and the interaction among them at national level, the Regional Electricity Market in Central America, and the Internal Electricity Market in the EU. A review of the solutions adopted in each of these cases allows one to conclude that, despite the fact that sound guiding principles are formally taken as a reference when devising these solutions in several cases, the specific cost allocation schemes adopted at regional level are far from being efficient. These tend to be simple schemes involving the socialization of transmission network costs, or schemes based on network usage that result in regional network charges, or an allocation of the regional grid cost, not reflecting the real usage of the grid by each agent, or system.

References

ACER, Recommendation of the Agency for the Cooperation of Energy regulators No. 05/2013 of 25 March 2013, on a new regulatory framework for the inter-transmission system operator compensation. A document by the Agency for the Cooperation of Energy regulators (2013), p. 7. Available at: https://acer.europa.eu/Official_documents/Acts_of_the_Agency/Recommendations/ACER%20Recommendation%2005–2013.pdf

Agency for the Cooperation of Energy Regulators, Recommendation No. 07/2013 regarding the cross-border cost allocation requests submitted in the framework of the first union list of electricity and gas projects of common interest. Staff report (2013)

F. Bañez, L. Olmos, A. Ramos, J.M. Latorre, Estimating the benefits of transmission expansion projects: an Aumann-Shapley approach. Energy **118**, 1044–1054 (2017a)

F. Bañez, L. Olmos, A. Ramos, J.M. Latorre, Beneficiaries of transmission expansion projects of an expansion plan: an Aumann-Shapley approach. Appl. Energy **195**, 382–401 (2017b)

J. Bialek, Tracing the flow of electricity. IEE Proc. Gener. Transm. Distrib. **143**(4), 313–320 (1996)

R.H. Coase, *The firm, the market and the law* (The University of Chicago Press, Chicago, 1990), p. 2002

Commission Regulation (EU), No 838/2010 of 23 September 2010 on laying down guidelines relating to the inter-transmission system operator compensation mechanism and a common regulatory approach to transmission charging. EU Commission, p. 7. Available at: https://publications.europa.eu/en/publication-detail/-/publication/50e99710-f80f-477a-a5b6-5c4b59ed4113/language-en

CRIE, Reglamento del Mercado Eléctrico Regional. A document developed by the Comisión Regional de la Interconexión Eléctrica (2005), p. 366. Available at: http://www.crie.org.gt/wp/wp-content/uploads/2013/12/rmer.pdf

CRIE, Metodología Inicial Transitoria de Cálculo, Conciliación, Facturación, y Liquidación del Peaje Operativo y cargo Complementario de los Cargos por Uso de la Red de Transmisión Regional para la Remuneración de la Línea SIEPAC. A document developed by the Comisión Regional de la Interconexión Eléctrica (2011), p 366

ENTSO-e, ENTSO-e Guideline for Cost Benefit Analysis of Grid Development Projects. A document by European Network of Transmission System Operators for Electricity (2013). Available at: https://www.entsoe.eu/Documents/SDC%20documents/TYNDP/ENTSO-E%20cost%20benefit%20analysis%20approved%20by%20the%20European%20Commission%20on%204%20February%202015.pdf

European Commission, COM, 658 final, 2011. Proposal for a Regulation of the European Parliament and of the Council on Guidelines for Trans-European Energy Infrastructure (2011)

European Union, Regulation (EU) No 347/2013 of the European Parliament and of the Council of 17 April 2013 on guidelines for trans-European energy infrastructure and repealing Decision No 1364/2006/EC and amending Regulations (EC) No713/2009, (EC) No 714/2009 and (EC) N, vol. 2013, no. 347, pp. 39–75 (2013)

FERC, Standard market design notice of proposed rulemaking. Docket No. RM01-12-000. Issued by the Federal Energy Regulatory Commission (2002)

FERC, Transmission planning and cost allocation by transmission owning and operating public utilities. Docket No. RM10-23-000 (2010). Available at: http://elibrary.ferc.gov/idmws/common/opennat.asp?fileID=12372947

FERC, Transmission planning and cost allocation by transmission owning and operating public utilities, Docket No. RM10-23-001; Order No. 1000-A. Issued by the Federal Energy Regulatory Commission (2012). Available at http://www.ferc.gov/industries/electric/indus-act/trans-plan.asp

K.N. Hasan, T.K. Saha, D. Chattopadhyay, M. Eghba, Benefit-based expansion cost allocation for large scale remote renewable power integration into the Australian grid. Appl. Energy **113**, 836–847 (2014)

P. Joskow, J. Tirole, F. Merchant transmisison investment. J. Ind. Econ. **LIII**(2), 233–264 (2005)

M. Junqueira, L.C. da Costa, L.A. Barroso, G.C. Oliveira, L.M. Thome, M.V. Pereira, An Aumann-Shapley approach to allocate transmission service cost among network users in electricity markets. IEEE Trans. Power Syst. **22**(4), 1532–1546 (2007)

MIT, The future of the electric grid. Massachusetts Institute of Technology (2011). Available at: http://web.mit.edu/mitei/research/studies/the-electric-grid-2011.shtml

MIT, Utility of the future. An MIT Energy Initiative response to an industry in transition. A report by the Massachusetts Institute of Technology Energy Initiative in collaboration with the Pontifical Comillas University (2016), pp. 382. Available at: energy.mit.edu/uof

L. Olmos, I.J. Pérez-Arriaga, Comparison of several inter-TSO compensation methods in the context of the internal electricity market of the European Union. Energy Policy **35**(4), 2379–2389 (2007a)

L. Olmos, I.J. Pérez-Arriaga, Evaluation of three methods proposed for the computation of inter-TSO payments in the Internal Electricity Market of the European Union. IEEE Trans. Power Syst. **22**(4), 1507–1522 (2007b)

L. Olmos, I.J. Pérez-Arriaga, A comprehensive approach for computation and implementation of efficient electricity transmission network charges. Energy Policy **37**(12), 5285–5295 (2009)

L. Olmos, M. Rivier, I.J. Pérez-Arriaga, Transmission expansion benefits: the key to redesigning the regulation of electricity transmission in a regional context. Econ. Energy Environ. Policy **7**(1), 47–62 (2018)

I.J. Pérez-Arriaga, F. Rubio-Odériz, J.F. Puerta Gutiérrez, J. Arcéluz Ogando, J. Marín, Marginal pricing of transmission services: an analysis of cost recovery. IEEE Trans. Power Syst. **10**(1), 65–72 (1995)

I.J. Pérez-Arriaga, Y. Smeers, Guidelines on tariff setting, in *Chapter 7 in the book 'Transport pricing of electricity networks'*, ed. by F. Lévêque (Kluwer Academic Publishers, New York, 2003)

PJM, A survey of transmission cost allocation issues, methods and practices. A report issued by the Pennsylvania-New Jersey-Maryland INterconnection (2010). Available at: http://ftp.pjm.com/~/media/documents/reports/20100310-transmission-allocation-cost-web.ashx

E.G. Read, D.P.M. Sell, A framework for transmission pricing, in *Principles for pricing electricity transmission*, ed. by A. Turner (Trans Power New Zealand, Wellington, 1989)

M. Rivier, I.J. Pérez-Arriaga, L. Olmos, Electricity transmission, in *Regulation of the power sector*, ed. by I.J. Pérez Arriaga (Springer, London, 2013), pp. 251–340

F. Rubio-Odériz, I.J. Pérez-Arriaga, Marginal pricing of transmission services: a comparative analysis of network cost allocation methods. IEEE Trans. Power Syst. **15**(1), 448–454 (2000)

Transmission Planning, Investment, and Cost Allocation in US ISO Markets

R. P. O'Neill

1 Introduction and Background

U.S. transmission investments by Commission-jurisdictional transmission providers increased from $2 billion/year in the 1990s to $20 billion/year in 2013–2017. In 2017, the US electric power system had annual revenues of over $400 billion. In ISOs, transmission investment decisions can change the entry decisions for generators. Even modest improvements in modeling and decision making can result in billions of dollars of cost savings. Such potential indicates the need for improvements to the decision process, modeling, and cost allocation in the electric power transmission planning.

For the first eight decades of the twentieth century, the US electric power system was characterized mostly by weakly interconnected vertical-integrated for-profit utilities that owned and controlled the generation, transmission, and distribution inside a franchised system boundary. Most of these utilities were cost-of-service regulated by the state of physical residence. Generally, planning consisted of forecasting load growth, deciding on the next generator to build and expanding transmission and distribution to reliably deliver the power to load. Load forecasts were based on forecasted economic growth. During this period, load growth, increasing economies of scale in generation, and other technological advances resulted in lower prices and a dominance of large nuclear and coal generators that required large amounts of rate-based capital.

In 1935, the Federal Power Act was amended to fill the regulatory 'gap' for transmission and wholesale sales in interstate commerce. Rates (aka prices) for transmission and wholesale sales are required to be just and reasonable and not

Views expressed do not necessarily represent the views of the Commission or the USA.

R. P. O'Neill (✉)
ARPA-E, Dept. of Energy, Washington, D.C., USA
e-mail: richard.oneill@hq.doe.gov

M. R. Hesamzadeh et al. (eds.), *Transmission Network Investment in Liberalized Power Markets*, Lecture Notes in Energy 79,
https://doi.org/10.1007/978-3-030-47929-9_6

unduly discriminatory. Federal Power Commission (later renamed as the Federal Energy Regulatory Commission or hereinafter simply the Commission) was given this responsibility. For the next five decades, cost-of-service regulation determined these prices.

Over time, reliable operations became more important and the interconnections between utilities in different states were used to increase reliability and to execute economic trades often based on prices known as 'split the savings.' To facilitate this trading, some utilities formed power pools.

In 1978, fearing shortages of natural gas, Congress passed the Powerplant and Industrial Fuel Use Act (FUA) that outlawed the use of natural gas in new generators. In addition, in 1978, the Natural Gas Policy Act (NGPA) promoted a pricing regime to increase supply and removed the barriers to intrastate and interstate trade. Over time, the assumptions of the NGPA and FUA that natural gas was in short supply proved incorrect. In 1987, the FUA was repealed. Over time, most sections of NGPA were repealed.

Also in 1978, to encourage new forms of generation, Congress passed the Public Utility Regulatory Policies Act (PURPA) that required utilities to the purchase of energy from certain sources including co-generation, wind and solar at the utilities 'avoided costs' (similar to a feed-in tariff). A few states set avoided-cost rates high enough to attract wind and solar facilities. In other states, industrial customers built co-generation. PURPA gave birth to independent power producers (IPPs).

In the late 1970s and early 1980s, the prime rate for capital rose significantly and load growth in many utilities was considerably lower than predicted. The result was very expensive excess capacity. During the 1980s, economies of scale for coal and nuclear generation stopped increasing. Some policy discussions raised the idea of generation competition instead of geographic franchised monopolies (for example, see Joskow and Schmalensee 1983). Some utilities saw their generation investment as not earning a reasonable return on equity and sold off some of their generation to independent power producers. The average price of coal plants was about 200% of book value and average price of nuclear plants was about 10% of book value. To encourage competition, the Commission required open access to the transmission system as a condition of mergers. 'Experts' testified that open access would cause instability and blackouts. This was proven incorrect by actual experience.

In 1996, the Commission's Order 888 required that all utilities provide open access to their transmission system. Utilities had the option of forming an independent system operator (ISO). ISOs were given the responsibility for operating day-ahead and real-time energy and ancillary service auction markets with market power mitigation. This market design regulated by the Commission produced just and reasonable prices. In addition, ISOs were given certain transmission planning responsibilities including generation interconnection.

After correcting some early mistakes, the ISO energy markets have performed remarkably well and improved over time as the modeling, software and the underlying hardware all increased in capability to produce more efficient results. Over the next two decades after Order 888, seven ISOs formed and grew in geographic size. Today, US ISO markets account for over two-thirds of generation and consumption. The

ISO energy markets are highly competitive. The efficient energy dispatch function remains an independent monopoly service provided by the ISO. In the transmission sector, competition to build produces cost savings. The ISO remains an independent monopoly service for planning the transmission system.

In the 2000s, concerns about climate change and cleaner energy increased. Governments around the world increased subsidies for renewable energy and imposed carbon taxes. The competition in generation, technology advances in natural gas production, and renewable subsidies brought new challenges. Lower ISO energy prices caused concerns about premature retirement of coal and nuclear generators. Some states established aggressive 'clean energy' standards to combat climate change and other environmental issues. New federal and state subsidies formed the incentives to build wind, solar, and geothermal generators.

In 2003, to further articulate the open access interconnection process, Order 2003 separated the transmission expansion process and generation interconnection process. The rule implicitly used a vertical-integrated utility model and explicitly excluded transmission service from the interconnection process.

In 2005, Energy Policy Act added Section 219 to the FPA stating in part 'The rule shall (1) promote reliable and economically efficient transmission and generation of electricity by promoting capital investment in the enlargement, improvement, maintenance, and operation of all facilities for the transmission of electric energy in interstate commerce, regardless of the ownership of the facilities.' This responsibility falls to the Commission. The states retained regulation of retail prices, distribution, and siting decisions for generation, transmission, and distribution.

Historically, reliability standards were guidelines and compliance was voluntary. Steps to formalize, standardize, and computerize reliability started after the 1965 Northeast Blackout. Generally, reliability was confined to a vertically integrated utility and was a weakly defined concept that often included considerable judgment. Due in part to the 2003 Northeast Blackout, EPAct 2005 gave the Commission formal authority to regulate and enforce reliability standards.

In 2007, Order 890 required greater consistency and transparency in the transmission planning process on both local and regional level, economic planning studies, and cost allocation. In 2011, Order 1000 required the transmission planning process to consider transmission needs driven by public policy requirements established by state or federal laws or regulations. The rule requires that each public utility transmission provider must participate in a regional transmission planning process that has a regional cost allocation method for new transmission facilities selected in the regional transmission plan. Costs allocation must be 'roughly commensurate' with estimated benefits.

Over time, experience with the process raised the need for a course correction. In 2014, Former FERC commissioner Clark (2018) argued that less benefit has come from Order 1000 than expected. Clark concludes that the Commission should tailor the rule for ISOs.

Many transmission projects (often labeled repair and replacement of existing facilities, asset management or supplemental) have limited or no ISO review for either benefits or costs. The project costs are placed into rate base. Over the last decade, PG&E spent over 60% of its annual capital additions on 'self-approved' projects and overall ISOs, about 47% of the projects receive limited review (Bone 2018). In 2019, Huntoon states that 'virtually none of the costs [capital spending on transmission] is supported by cost-benefit analysis.' Industrial customers state they are seeing transmission costs rise each year without any benefits to show for it (RTO insider July 4, 2016).

From 2013 to 2018, competition was limited to only 2% of total U.S. transmission investments. Nevertheless, competitive processes led to innovations in proposed solutions, low bids, cost caps, cost control measures, and innovative financial structuring. Winning bids for competitive process averaged 40% below initial cost estimates while non-competitive projects were completed at 34% above initial estimates (Pfeifenberger et al. 2018). Subsequently, a study commissioned by the utilities argued that these claims were incorrect (Nicholson et al. 2019).

In the future, the mix of assets will change the nature of power systems and the way we model them. Traditional expansion-planning models focus on peak periods and certain off-peak periods. The penetration of renewables brings reliability concerns making the traditional analytic assumptions no longer valid. For example, it will be increasingly difficult to predict peak or stressful operation periods. 'Off-peak' or low consumption periods may experience higher prices and scarcity due to the lack of wind and/or solar generation.

In addition, it appears there will be a proliferation of new smaller devices, for example, smaller generators and storage devises. There will also be at least 100 times more information about the power system from smart meter penetration and phasor measurement units (PMUs) allowing more price-responsive demand to achieve economic efficiency. The regulatory response and modeling often experience significant institutional inertia. Market responses are much faster.

Existing approaches to transmission planning and investment have implicit and explicit assumptions, and approximations that need to be re-examined in the context of a smarter grid and increased amounts of energy from wind and solar generators, batteries, and price-responsive demand. Some approximations and assumptions in current models were necessary to make the planning problem computationally practical decades ago. Other assumptions and approximations are made to simplify uncertainty, such as failure modes and demand growth. Still, other assumptions and approximations were made in order to harmonize planning and investment approaches with the market design *de jour*. Many of these assumptions and approximations are out of date and limit advancements in optimal planning and cost allocation of the electric grid.

With the advent of large amounts of wind and solar along with storage and more price-responsive demand, the current approaches need to be modified. Today, for computational and management reasons, reliability models are decomposed, compartmentalized, and reduced in size using a mixture of engineering judgment, experience, and less transparent modeling. Planning results are tested for adequate voltage stability, short circuits, transient stability, and various other aspects of reliability. Over time, more of the constraints have been and will be modeled explicitly over larger regions as the data, hardware, and software for solving the problem improve.

The important issues are finding efficient transmission expansions, siting, cost overruns, efficient rate design, beneficiaries-pay cost allocation, and risk allocation.

In Sect. 2, we present the necessary components of 'reliable and economically efficient' power systems. In Sect. 3, we examine the principal uncertainties in planning. In Sect. 4, we examine the transmission expansion models. In Sect. 5, we analyze the transmission competition processes. In Sect. 6, we examine the cost and transmission rights allocations. In Sect. 7, we examine the transmission expansion process. In Sect. 8, we conclude with recommendations.

2 'Reliable and Economically Efficient'

The FPA requires that the Commission promote 'reliable and economically efficient transmission and generation.' The Commission accomplishes this through a combination of competition and cost-based regulation. The transmission expansion process consists of reliability upgrades, economic expansions, public policy projects, interconnection, cost allocation, and transmission rights allocation.

2.1 Co-Optimization

In 2013, Liu et al. (2013) strongly recommended co-optimization (optimization of the entire system) for planning. Co-optimization has many dimensions. Currently, the transmission planning process is decomposed into many separate analyzes. Some issues get less attention. For example, until recently the fuel supply was not analyzed explicitly because it was assumed not be a constraint on the optimal transmission expansion. Some reliability issues are studied in isolation without fully examining the options or cost/benefit analysis.

Reliability is a process of creating rules and penalties for non-compliance to reduce the probability of cascading blackouts, serious equipment damage, and forced load curtailment. Cascading blackouts affect large geographic areas and their prevention is a club good for those areas. The focus of planning has been N-1 reliability, that is, the system operation must be stable and able to survive the failure of any one asset with a high probability. In some areas, this focus is N-2. Reliability includes other rules for situational awareness, vegetation management, for example, tree trimming, and operator training that are not discussed in this chapter.

Reliability engineers and economic planners differ significantly in education and orientation. Reliability engineers often ignore the benefit/cost of the reliability solution. Without strong regulatory oversight, a cost-of-service regulated transmission owner would choose the solution with the higher capital costs. With smart grid technologies, less expensive alternatives may be available. Future planning should consider cheaper alternatives like remedial action schemes (RAS) and price-responsive demand.

Economic planners focus on finding efficient expansions with reliability rules as constraints. They prefer price-responsive demand to balance and stabilize the system as the first choice. Consequently, the reliability projects and efficient planning often proceed separately.

Reliable and economically efficient are concepts that should not be separated. Most if not all projects have both economic and reliability effects. Reliability upgrades are almost always by definition highly beneficial because they reduce the probability of a costly cascading blackout or forced curtailments. Reliability is an economic issue disguised in engineering terms. The economic benefits of not having a cascading blackout can and have been quantified. Economic upgrades have reliability benefits and reliability upgrades have economic benefits. Consequentially, it is more efficient to analyze both reliable and economically efficient projects as economic projects.

In ISOs, interconnection for large generation without access to transmission makes little sense. Order 2003 requires an interconnection customer to pay for interconnection before knowing the costs or scope of its transmission service. It could be better to present a complete cost of market participation. The transmission expansion process should include the interconnection process to maximize the expected economic efficiency of future power systems.

2.2 Price-Responsive Demand

Almost all reliability planning explicitly or implicitly employs a value of lost load (VOLL) calculation in its process. The VOLL is calculated by taking a reliability metric, for example, 1 outage event in 10 years using the average cost of constructing and operating a CT. The average cost of the marginal CT is the implied VOLL. Table 1 presents some examples of implied VOLL under various assumptions. Depending on the metric and the assumption in the analysis, the VOLL is usually greater than \$2000/MWh and often much greater. Few would believe that that given the choice of consuming at \$2000/MWh or more (over 20 times more than current average prices) and voluntarily reducing consumption, many consumers would choose the latter. To a reliability engineer, load reduction looks like a remedial action scheme (RAS). To economists, load reduction is a normal reaction to market prices.

Price-responsive demand is explicitly bidding a demand function into the energy auction markets. Historically, it was not possible to signal and charge most consumers the actual cost of producing energy because the metering process was incapable of measuring consumption over intervals less than a month. With the advent of smart interval meters and the high-speed Internet, measuring consumption and responding to dynamic prices are no longer a technical problem. High renewable penetration has made time-of-use pricing much less efficient.

Table 1 Various VOLL assumptions

Value of service (VOLL) \$/MW-year	Net capital cost (net CONE) \$/MWH	Hours per outage event hours/event	Optimal LOLE events/year	Optimal nines
\$4000	\$120,000	5	6.0	2.5
\$4000	\$80,000	5	4.0	2.6
\$4000	\$40,000	5	2.0	2.9
\$2000	\$120,000	5	12.0	2.2
\$2000	\$80,000	5	8.0	2.3
\$2000	\$40,000	5	4.0	2.6
\$20,000	\$120,000	5	1.2	3.2
\$20,000	\$80,000	5	0.8	3.3
\$20,000	\$40,000	5	0.4	3.6

Source Astrape Consulting (2013, p. 29)

Price-responsive demand can resolve many reliability issues. Forced unexpected curtailment and voluntary reductions in consumption have different values. Price-responsive demand can shift demand to other periods acting like storage. It can forego voluntarily consumption reducing the peak in the energy market and saving money while increasing the efficiency of the market. Price-responsive demand does not need a capacity commitment because it can get off the system when prices are high and is in effect its own reserve. It can also be a reserve (ancillary service) in the energy market, for example, AGC.

The transmission expansion plan should maximize the expected economic efficiency of future power systems using a price-responsive demand curve that includes VOLL at the high end, but more price sensitivity at the lower end.

Price-responsive demand should be modeled comparably to generators. If the load chooses to be explicitly price-responsive (bid into the market), it should have comparable bidding parameters to generation and storage. For example, load can bid the value of consumption in a single period or can bid a single value for an entire eight-hour shift using minimum run parameters. Price-responsive demand needs no capacity commitments since demand will voluntarily curtail itself when the price is too high.

2.3 Market Power

Restriction of transmission access creates market power concern by creating barriers to entry for efficient generation. The game theoretic discussions can be found in Sauma and Oren (2007) and Kimbrough et al. (2014). Game theoretic analysis adds an additional computational burden to an already difficult problem. In addition, game theoretic approaches are often very complex and require many assumptions that move markets away from market efficiency. In the US ISO energy markets, to avoid market power issues, generator offers are mitigated if necessary and transmission markets are predominately, cost-of-service regulated. Some transmission projects are competitively procured.

2.4 *Siting and Eminent Domain*

States retain the rights to determine the generation resource mix and have siting authority inside their respective states. At the state level, there is an ongoing debate of the balance between markets, subsidies, environmental, and regulatory concerns.

3 Uncertainty

All forecasts are wrong. Some are useful (generally attributed to the statistician George Box). History shows that power system planning is subject to profound long-run uncertainties in policy, externalities, technology, fuel costs, load shape, and load growth. Uncertainties in planning require planners to develop scenario visions of the future. Traditional planning methods have typically applied simple and ad hoc methods to address power system uncertainties. Integration of large amounts of renewable and distributed resources presents additional challenges.

Perhaps the biggest issue affecting ISO transmission planning is the uncertainty over the generation mix. Shifts are occurring in the generation mix with reductions in coal and nuclear offset by increases in natural gas and renewables. These changes have been accompanied by lower energy prices that pressure some types of generation to exit the market. However, there has been substantial resistance from both generators and states for some generators to exit due to, consideration of market externalities, such as resilience, fuel security, jobs, and importance of plants to local communities. These issues are usually not addressed directly in planning models.

3.1 *Natural Gas Price Uncertainty*

Natural gas prices are very difficult to forecast and currently are the principal determinant of energy prices. Figure 1 shows the history of natural gas prices and two regressions (linear and cubic). In 2018, the price of natural gas in real terms was the essentially same as it was in 1978. The linear regression shows a price increase over time. The cubic fit shows a mild cyclic behavior. Both have large error bands. The historic tendency is to predict future prices using a depletion theory that requires future long-term price forecasts to increase with a static economic resource base. Cyclic prices could be the result of new technology stimulated by higher prices that

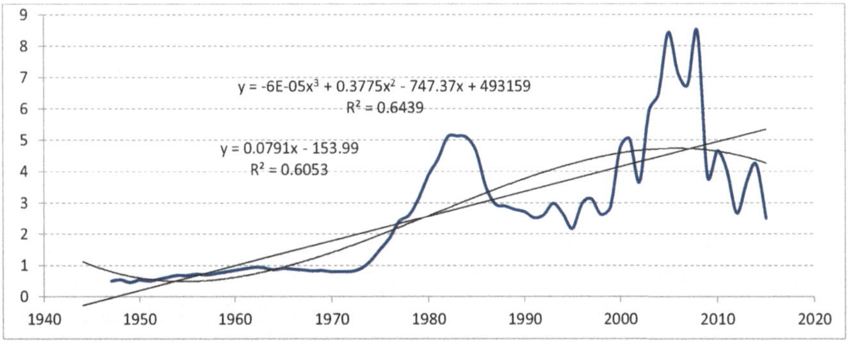

Data source: U.S. Energy Information Administration

Fig. 1 U.S. Natural Gas Wellhead Price (in $2006/MMBtu). *Data source* U.S. Energy Information Administration

increases the economic resource base. Under certain assumptions, using natural gas CCCT generation to charge EVs is more efficient and less polluting than gasoline vehicles.

3.2 *Weather Uncertainty*

Historically, the largest contingency was the largest generator on the system. In the near future, bad weather forecasts may be the largest contingency. Seventy percent of generator failures are due in part to weather. Fossil generator output is a function of temperature and the time since the last maintenance. Demand is a function of temperature and humidity. Transmission capability is due in part to weather. Solar output is a function of sunshine. Wind output is a nonlinear function of wind velocity and extreme cold temperatures. Hydro-output is a function of rain and snowfall. This creates difficulty in determining where and when the system is under the most stress. A cloudy and windless day or sequence of days requires significant amounts of storage discharge, fossil fuel generators, and/or price-responsive demand. A sunny and windy day may need little other generation with storage charging.

3.3 *Technology Innovation Uncertainty*

In the past, technology innovation has lowered the cost of coal generation. More recently, it has lowered the costs of wind and solar generation and the cost of natural gas. FACTS devices, better information and faster computers have increased the controllability of the transmission system and topology optimization (see O'Neill et al. 2005a, b; Fisher et al. 2008; Hedman et al. 2008, 2009, 2010). Smart meters have made price-responsive demand easier to implement. Technology innovations are difficult if not impossible to predict.

3.4 Risk Management

More complex risk-management techniques have been suggested such as value at risk and conditional value at risk. Most of these approaches are at the research stage of development for planning. Moreover, risk tolerance is both individual and systemic. The governments must decide what risks to socialize and what risks are privatized. Socialized risks can create moral hazards.

3.5 Summary

New uncertainties add more complexity to the process and a less predictable evolution of power systems. They raise questions of whether existing planning methods are adequate. As renewable penetration increases, flexibility for generators, load, and transmission becomes more important. Some ISOs recognize the need to incent operational flexibility. Therefore, co-optimization should include the ability to model flexibility (for example, ramping capability and operational range) in resource portfolios. Modeling operational reserve requirements and proper modeling of the costs of fossil fuel unit cycling need consideration. It is possible to develop co-optimization tools that handle uncertainty, but at a significant increase in computational burden and debate among the market participants. In ISOs, transmission expansions need greater transparency because they must pass market participants and Commission review.

4 Models

Models approximate reality. Models must tradeoff fidelity, detail, breath, and scope with the computational burden and cost of operation. Useful models must pass a benefit/cost test. New models must work against the institutional inertia—the traditional way of doing things. The result is a suite of models where each model focuses on a particular part of the process. This leads to iteration between high-level models with less detail and greater scope and more focused higher-fidelity models with greater detail and less scope. Some models test for reliable (feasible) solutions. Some search for economically efficient (optimal) solutions.

Lower-fidelity models are used to solve many rough-cut scenarios quickly in preliminary high-level analysis. Larger, higher-fidelity models are used to ensure the detailed or final decisions are consistent with lower-fidelity models.

Good approximations simplify the formulation to make it easier to solve while minimizing the impact on the optimal outcome. Today, approximations are mostly a mathematical art form passed from one generation of modelers to the next often with insufficient testing and documentation. They became part of 'good utility practice.' Weak approximations yield weak results that are harder to support.

4.1 Literature Review of Models

Many planning models use approximations to handle the magnitude of the problem and the computational difficulty presented by binary investment decisions. Garver (1970) and Villasana et al. (1985) presented linear programming approaches for finding feasible transmission network expansions given future loads and generation. Dusonchet and El-Abiad (1973) discussed the use of dynamic programming to deal with the size and complexity of a transmission planning optimization problem.

Romero and Monticelli (1994) proposed a method for solving network expansion-planning problems using mixed-integer programming, by relaxing the network problem to a transportation model and then successively introducing the complicating constraints. Baughman et al. (1995) discussed models for the inclusion of transmission expansion decisions. Gallego et al. (1996) presented a least-cost transmission expansion problem using simulated annealing. Gallego et al. (1998) presented a genetic algorithm approach for solving the transmission expansion problem. De la Torre et al. (2008) presented a mixed-integer program for long-term transmission investment planning in a competitive pool-based electricity market. Kazerooni and Mutale (2010) solve the N-1 security constrained transmission expansion optimization problem with environmental constraints. O'Neill et al. (2013) proposed a stochastic two-stage chance-constrained mixed-integer planning model. The objective of the model is to maximize the expected economic efficiency from investment.

Commercial models require higher documentation, verification, and transparency. Commercial modeling tools include production cost models that simulate operations, capacity expansion models, and reliability models. The model types and the issues they address are in Table 2.

ISOs use commercial models along with internal software. The ISO New England uses a high-level production cost model (http://www.iso-ne.com). The New York ISO uses ABB's Gridview, GE's MAPS (http://www.gepower.com) and Portfolio Ownership and Bid Evaluation (http://www.nyiso.com). PJM, Midwest ISO and SPP use Ventyx PROMOD (http://www.ventyx.com). California ISO uses ABB Gridview and PLEXOS (https://energyexemplar.com).

Table 2 Model type and reliability issue

Reliability issue	Model type		
	Generation and transmission capacity expansion	Production cost (unit commitment and dispatch)	Reliability (AC power flow, dynamic stability)
Generator adequacy (meet demand satisfying the loss of load probability	Often	Yes	No
Flexibility (adequate ramp rate and operating range)	Depends	Yes	No
Transmission adequacy (maintain thermal, voltage, and stability limits)	Mostly no	Partially	Yes
Generator contingencies (maintain reliability in a generator failure)	Mostly no	Somewhat	Yes
Transmission contingencies (maintain reliability in a transmission line failure)	Mostly no	Somewhat	Yes
Frequency stability (maintain frequency using inertia, primary frequency (governor) response, and regulating reserves	Mostly no	Somewhat	Yes
Voltage stability (maintain system voltage using reactive power)	No	No	Yes
Transient/rotor angle stability	No	No	Yes

Source Boyd (2016) modified

4.2 Hydro-Dominated Systems Models

Hydro-dominated systems have a different focus than non-hydro-dominated systems. For hydro-dominated systems, the main concern is a multiyear drought. In addition, hydro-generators are often a significant distance from load. Pereira and Granville (2001) explored a Benders decomposition approach to solving mixed-integer programming problems for the transmission expansion problem. Alguacil et al. (2003) proposed a mixed-integer programming formulation of the long-term transmission expansion problem with binary transmission investment decisions and applied it to a 46-node single period model of the Brazilian power system. Binato et al. (2005) presented a sigmoid function approach for binary investment variables in the optimal transmission expansion problem and tested it on a model of the south-eastern Brazilian system. In a market dispatch, the most important parameter is the opportunity cost of hydropower that changes based on the water levels in the reservoirs.

4.3 Production Cost Models

The current framework for production cost modeling involves simulations of the economic dispatch process for a chosen footprint and time horizon. The dispatch simulations may be performed with DC power flow or 'transportation-type' transmission constraints and with or without unit commitments, the introduction of binary decisions adds one or more orders of magnitude to the computations. While the current production cost modeling framework is useful for quantifying the economic effects of specific projects, it is weak as a tool for seeking the economically efficient set of projects from among a set of proposals or potential projects. For example, given a set of potential transmission and generation expansions, many production cost models do not give the option to find the economically optimal combination of projects under different scenarios. Such abilities may be useful in the context of analysis to support system-wide planning for the integration of renewable resources. Most optimal transmission expansion models do not incorporate transmission investments as binary decision variables. Co-optimized, stochastic models are mostly experimental and in limited use.

4.4 Reliability Models

Reliability models are necessary because the high-level models cannot adequately model reliability issues. Reliability models test the candidate transmission and generation expansions for reliability violations. Generally, they are high fidelity models with a narrow focus. They simulate dynamic events that occur in seconds not minutes. A transient stability model simulates whether generators remain synchronized after a contingency. AC power flow models check steady-state operational feasibility. Traditional reliability analysis focuses on periods of high load and whether the system remains stable after a power plant loss, a transmission line loss or power system instability. System dynamic models simulate dynamic events under fault conditions to examine transient stability. Network reliability models include GE's Positive Sequence Load Flow (PSLF), and Siemens' Power System Simulator for Engineering (PSSE).

4.5 Model Size and Approximations

Models can quickly balloon in size and computational complexity making it important to reduce its size without over-compromising fidelity. The ideal high-level planning model is a large, high fidelity, stochastic, mixed integer, AC power flow, and variable topology model. At this point, it is in the early research stage. The objective of the model is to maximize the expected market surplus (benefits to society) from new and existing investment. The approach advocated here integrates aspects from production cost modeling and investment models with large scale. It adds the capability to optimize transmission expansions over alternatives. Optimal topology including transmission switching is relevant because if a low capacity line in a circuit could block a valuable line then the low capacity line can be removed to improve the market performance. The model also recognizes generic generation investment alternatives and co-optimizes generation with transmission expansions with specified reliability levels and environmental goals.

The models are simplified in various ways. Simplifications include changing the granularity in topology, time step, number of periods and scenarios. In addition, some binary variables are converted to continuous variables. Table 3 presents the various degrees of fidelity and approximations. Planning and investment model can reasonably be given 10–50 times longer to solve than the day-ahead market models.

When high-level models are relaxed, this may create a need for additional intermediate models with more detail. One approximation or assumption may imply another. As the time step gets larger, for example, from one hour to one day, startup, and ramp rates issues fade in importance or disappear. Less granularity may remove the need to model the explicit probability of failure, unit commitment, minimum up and downtime constraints and ramp rate constraints. Approximations of this type may cause the model to lose some of the issues that new technology presents, for example, imposing a greater requirement on system ramp rate capabilities to respond to weather events, or near real-time decisions to start combustion turbines. Storage can be modeled as 'pumped' storage with time lags between charging and discharging or battery type without time lags.

Another approach is to model a typical and/or extreme weather day or week for selected seasons. Here, time granularity allows for commitment decisions. Sensitivity and scenario analysis can address many issues including sensitivity to data inputs, assumptions, and approximations. The list of possible sensitivities is large and can be computationally intense.

Table 3 Potential approximations and fidelity for high-level and intermediate models

Parameter/asset	Fidelity		
	High	Intermediate	Weak
Time period			
Year increment	1 year	5 years	10 years
Seasonal	Week	4 seasons	Peak annual
Daily	24 h	4 periods	Daily peak
Network topology			
Minimum voltage level	69 kV or lower	130 kV	225 kV
Geographic	Nodal	Balancing area	State level
Network equations	AC	DC	Transportation
Topology optimization	Optimal	Transmission switching	None
Max capacity	Flexible	Seasonal	Steady state
Generator			
Startup	Binary	Relaxed penalized binary	None
Minimum operating level	Binary	Relaxed penalized binary	None
Avoidable costs	Yes	Average costs	Marginal costs
Maximum operating level (generation and transmission)	Weather dependent	Steady state with moderate penalties for minor violations	Steady state with strong penalties for violations
Ramp rates	Yes	No	No
Minimum run time	Yes	No	No
Reliability	Full N-1	Sub-regional capacity set aside	ISO capacity set aside
Inelastic demand scenarios	5	3	1
Price-responsive demand	Like generators	Simple demand curve	None
Storage	Full arbitrage	Fixed	None
Relative computational difficulty	>1000	>50	1

5 Transmission Competition Models

In this section, we examine three ISO competition models: merchant transmission, competitive solicitation, and the sponsorship. Each has different positive aspects. To function properly, the rules must be firm, understood, and applied consistently.

5.1 The Merchant Transmission Model

In 2000, the Commission first granted negotiated rate authority to a merchant transmission project developer [see TransEnergie US Ltd., 91 FERC ¶ 61,230, at 61,838 (2000)]. A transparent open season process allocates some or all transmission capacity. Investors and their customers in a merchant transmission project assume the full market risk of the project. Currently, this process takes place outside the ISO transmission expansion process.

5.2 The Competitive Solicitation Model

In the competitive solicitation model, transmission planners with stakeholder input identify the efficiency-enhancing projects and then solicit bids from developers. The solicitation details should include who assumes the risks, what to build, and bidder qualifications. Market participants submit offers to build with their offer costs (that is, revenue requirements). The winning projects are eligible for regional cost allocation. CAISO, MISO. ERCOT, and SPP use this approach.

5.3 The Sponsorship Model

In the sponsorship model, transmission planners and stakeholders identify transmission needs and allow developers to propose potential solutions. The sponsorship model is performing well at finding innovative solutions. The choice of winning projects can be more subjective and subject to challenge. PJM, ISO-NE, and NYISO have used the sponsorship model.

5.4 Cost Caps

All projects in the transmission planning process should have cost caps and be evaluated at the cost caps. Cost caps for projects change the standard transmission development process by transferring some of the risk of overruns from ratepayers to the builder who is in the best position to control costs. Developers who fail to stay within their caps risk both the project and the offer cost recovery.

6 Cost and Transmission Rights Allocation

6.1 *Beneficiaries Pay*

Cost allocation occurs after each iteration of the optimal transmission plan. Cost allocation is a part of setting just and reasonable rates as the law requires. Conceptually, there is a general agreement and a circuit court decision (see *Illinois Commerce Commission, v. FERC*, U. S. Court of Appeals for the Seventh Circuit, August 6, 2009) that beneficiaries of transmission should pay for the transmission. The Commission requires that costs of transmission projects should be allocated to its beneficiaries 'roughly commensurate' to benefits. They also may receive the tradable associated transmission rights. There are significant disagreements on what beneficiaries-pay means, how much each market participant should pay and how the transmission rights are allocated. New projects must have a pre-construction benefit/cost ratio greater than one. If actual costs decrease the ratio below one, the additional costs of the projects should be based on rules set out when the project was authorized.

Some legacy approaches to cost allocation are license plate, postage stamp, highway/byway, distribution factor, and voltage level. Many do not pass the beneficiaries-pay cost allocation test. Beneficiaries often include generators, but generators are seldom allocated costs in the transmission expansion process. Order 1000 explicitly allows transmission expansion costs to be allocated to generation (Order 1000-A, 139 FERC ¶ 61,132 at P 680), but seldom does. Generally, transmission expansion costs are assigned to load regardless of the benefits to other market participants.

Benefits should be determined by the expected change in benefits or profits at the node due to the upgrade. When cost allocation disagreements occur, usually the strongest disagreements are in allocating costs to market participants not expected to benefit or not allocating cost to those who benefit (free riders). The 'Argentina' method where market participants vote on cost allocation based on proportion to their proposed cost allocation (see Littlechild and Ponzano 2007) as a method for allocation cost may be an appropriate approach to cost allocation. It could be binding or advisory.

Beneficiaries-pay cost allocation should be used for all projects including reliability and interconnection projects.

6.2 Theory of Cost Allocation

Some argue that transmission expansion is a public good. Since each transmission asset has a finite capacity and can become congested, it should not be characterized as a public good. When transmission assets become congested, they take on the characteristics of private good. Transmission should instead be characterized as a club good.

Cooperative game theory allows the participants to form into groups to cooperate and negotiate the cost allocation. Cooperative game theory contrasts with non-cooperative game theory where market participants are not allowed to communicate explicitly with each other. Markets are often analyzed under the non-cooperative game theory paradigm, for example, a Nash or perfect equilibrium as the model for deciding the optimal expansion. There is a vast literature on game theoretic cost allocation (see Young 1985, 1995). Many approaches are mathematically complex, others are computationally intensive and still others are both. Cost allocation using cooperative game theory includes the Shapley value, Nucleolus, and empty core models. If the market has an empty core or a free-rider problem, the market participants may not be able to agree on allocation rules and the Commission must impose them.

Projects may be complementary or mutually exclusive. For cost allocation in a multi-project environment, all projects should be taken as a whole. The value for all projects taken as a whole is not the sum of the individual value of each individual project.

6.3 Two-Node Example of Cost Allocation

We present a simple model of cost allocation. All costs and benefits are expected. To simplify the examples, we assume market participants are risk neutral. First, we calculate the difference in the expected costs of energy at each bus with and without the new investments. This is a relatively easy problem to solve since the investment decisions are fixed.

Let SB be the incremental efficiency gains or benefits from a set of transmission projects; let DTR be the new transmission rights created by the expansion and TTC be the total cost of the transmission expansion. Auction the DTR, receiving RTR. Let NR (net revenues) = RTR–TTC. If NR ≥ 0, no cost allocation is necessary.

Let B_i be the difference between the expected costs of energy under the expected optimal investment and the costs of energy under no investment for a market participant or defined group of market participants i. $B_i > 0$ corresponds to lower costs of energy consumption for market participants i or higher profits for production under the investment as compared to no investment.

Example 1 Add a line from 1 to 2 increasing capacity from q^{max} to q'^{max}

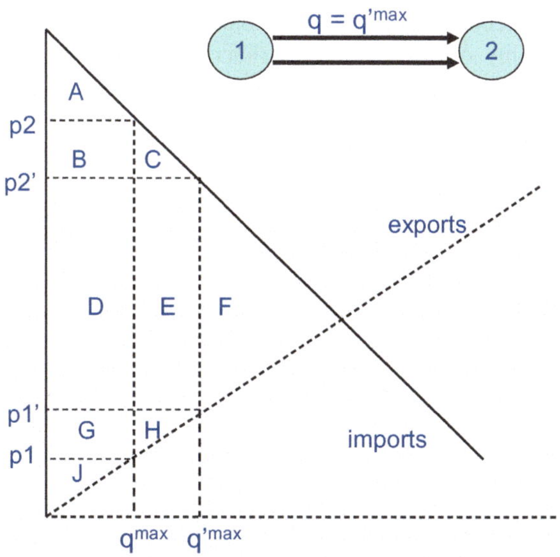

Example 1 presents the benefits of an expansion similar to Hogan (2010). The pre-expansion export transmission capacity is q^{max}, and the benefits to the import region 2 benefits are the area A. The FTR or flowgate benefits are the area D + B + G = $(p2 - p1)q^{max}$ and the benefits to the export region are the area J.

After the expansion, the value of transmission rights is D + E = $(p2' - p1')q'^{max}$.

The post-expansion capacity is q'^{max} with cost of TTC. After the expansion, the total value of transmission rights is D + E = $(p2' - p1')q'^{max}$. The efficiency criterion to build is C + E + H (benefits) > TTC (total costs). The incremental benefits to the import region are the area B + C, the incremental transmission right benefits are the area E, and the benefits to the export region are the area G + H. The existing transmission rights are diminished by B + G. B and G are called pecuniary benefits (aka business stealing) because they are transfers from transmission rights holders not efficiency gains to regions 1 and 2.

If E > TTC, a merchant transmission developer will build for transmission rights.

If C + E + H > TTC > E, merchant transmission will not build without support from regions 1 and 2. A cost allocation is: the total net benefits are TB = C + E + H. The import region 2 is willing to pay up to B + C, the transmission incremental right holders are willing to pay up to E, and import region 1 is willing to pay up to G + H. Since C + E + H > TTC, there is a cost allocation where all beneficiaries are better off.

Should winners compensate the losers? Losers in this example are original transmission rights holders. If B + D + G − (D + E) < 0, the value of transmission rights decrease. This value is transferred to regions 1 and 2. By the assumptions, there is enough value to compensate the loss.

6.4 Transmission Rights and Allocation

In ISOs, the fundamental unit of a transmission right is a flowgate right with no risk of becoming a liability. A financial transmission right (FTR) is the right or obligation to receive or pay the price difference between two nodes and is cashed out in the day-ahead market. An FTR is a portfolio of purchases and sales of flowgate rights. They are sold under projected day-ahead market topology in proportion to the distribution factors between the two nodes. The portfolio changes if the topology changes. Who should take the risk of topology changes? An FTR has the risk of becoming a liability if the nodal price differences are negative and an underfunding risk if the topology in the day-ahead market is different from the FTR auction topology assumption. Who should take the risk or get the reward of topology changes? The TO who changed the topology or the transmission rights holders.

6.5 Numerical Examples of Beneficiaries Pay

In the series of two-node examples below, we illustrate some properties of the beneficiaries-pay cost allocation and the allocation of transmission rights to beneficiaries. In these examples, the flowgate right on flowgate 12 and the FTR from node 1 to node 2 are the same. We illustrate with example how generators benefit, how load benefits, how FGR holders benefit, and how to allocate costs to multiple beneficiaries. It is a straightforward calculation to extend the examples to a reticulated network.

Base Case Table 4 has the energy market parameters for the base case. The cost of flowgate 12 upgrade is \$10/MW.

Table 4 Generators, load, and transmission parameters

Unit	Gen at node 1	Flowgate 12	Gen at node 2	Load at node 2
Network	①--②			
Minimum operating level (MW)	0	0	0	0
Maximum operating level (MW)	900	100	1200	1100
Marginal value (>0) or Marginal costs (<0) in \$/MWh	−10	0	−50	90

Table 5 Base case: economically efficient solution with market surplus of $48,000 and flowgate 12 capacity is 100 MW

Unit	Gen node 1	Flowgate 12	Gen node 2	Load node 2
Dispatch in MWh	100	100	1000	1100
Maximum operating level (MW)	900	100	1200	1100
LMP/flowgate marginal price in $/MWh	10	40	50	50
Revenue (≥0)/payment (≤0) in $	1000	4000	50000	−55000
Cost (≤0)/value (≥0) in $	−1000	0	−50000	99000
Profit (≥0)/benefit (≥0) in $	0	4000	0	44000

The auction market results without a transmission upgrade are in Table 5. The economically efficient solution has a market surplus of $48,000. The marginal flowgate value on flowgate 12 is $40/MWh.

Case 2 With an expansion of at a cost of $7000, the capacity of flowgate 12 is 800 MW. The market results for Case 2 are in Table 6. The market surplus without the cost of expansion increases from $48,000 in the pre-expansion base case to $76,000—an increase of $28,000. The net benefits of expansion netting out the expansion cost is $21,000 ($28,000 − $7000). The B/C is $28,000/$7000 = 4. The entity that paid for the upgrade receives 700 MW flowgate 12 rights. The net benefits of the expansion accrue to the flowgate rights holder. The flowgate 12 value increases from $4000 to $32,000 for net increased benefits of $28,000. There is no net benefit change for generators or load.

Case 3 With an expansion of 900 MW at a cost of $9000, the capacity of the transmission flowgate is 1000 MW. The market results for Case 3 are in Table 7 with a marginal flowgate value of $0/MWh and the LMPs are the same at both nodes. The market surplus without the cost of expansion increases from $48,000 in the pre-expansion base case to $80,000—an increase of $32,000. The benefits of the expansion accrue to the generator at node 1 whose profits increase from 0 to $36,000 compared to base case. The flowgate is decongested and loses $4000 in value from the expansion compared to the base case. The generator and load at node

Table 6 Case 2: economically efficient solution with market surplus of $76,000 and flowgate 12 capacity of 800 MW

Unit	Gen node 1	Flowgate 12	Gen node 2	Load node 2
Dispatch in MWh	800	800	300	1100
Maximum operating level (MW)	900	800	1200	1100
LMP/flowgate marginal price in $/MWh	10	40	50	50
Revenue (≥0)/payment (≤0) in $	8000	32000	15000	−55000
Cost (≤0)/value (≥0) in $	−8000	−7000	−15000	99000
Profit (≥0)/benefit (≥0) in $	0	25000	0	44000

Table 7 Case 3: economically efficient solution with market surplus of $80,000 and flowgate 12 capacity of 1000 MW

Unit	Gen node 1	Flowgate 12	Gen node 2	Load node 2
Dispatch in MWh	900	900	200	1100
Maximum operating level (MW)	900	1000	1200	1100
LMP/flowgate marginal price in $/MWh	50	0	50	50
Revenue (≥0)/payment (≤0) in $	45000	0	10000	55000
Cost (≤0)/value (≥0) in $	−9000	0	−10000	99000
Profit (≥0)/benefit (≥0) in $	36000	0	0	44000

2 do not benefit. The generator at node 1 (the only beneficiary of the upgrade) would pay $9000 for the upgrade and receive 900 MW flowgate rights on flowgate 12 as a future congestion hedge.

Case 4 We increase the flowgate 12 capacity by 901 MW to 1101 MW at a cost of $9010. In addition, we increase the generation capacity at node 1–1150. The market results for Case 4 are in Table 8. The market surplus without the cost of expansion increases from $48,000 in the pre-expansion base case to $88,000—an increase of $40,000. All benefits accrue to the load at node 2 whose benefits increase from $44,000 to $88,000. The marginal flowgate value is $0/MWh and the energy prices are the same at both nodes. All benefits of the expansion accrue to the load at node 2 whose benefits increase from $44,000 in the pre-expansion base case to $88,000. The generators do not benefit. The load pays $9010 for the upgrade and receives 901 MW flowgate rights on flowgate 12 as a hedge against future congestion.

Case 5 We increase the flowgate 12 capacity by 901 to 1101 MW at a cost of $9010. In addition, we increase the generation capacity at node 1–1150 and add a zero marginal cost generator with a capacity of 500 MW. The market results for Case 5 are in Table 9. The market surplus without the cost of expansion increases from $48,000 in the base case to $93,000—an increase of $45,000. The load at node 2 and the new generator at node 1 benefit. The load at node 2 benefits increase from $44,000 to $88,000. The generator at node 1 benefits is $5000. With a marginal

Table 8 Case 4: economically efficient solution with market surplus of $88,000. Flowgate 12 capacity of 1105 MW and generation capacity at node 1–1150

Unit	Gen node 1	Flowgate 12	Gen node 2	Load node 2
Dispatch in MWh	1100	1100	0	1100
Maximum operating level (MW)	1150	1101	1200	1100
LMP/flowgate marginal price in $/MWh	10	0	10	10
Revenue (≥0)/payment (≤0) in $	11000	0	0	11000
Cost (≤0)/value (≥0) in $	−11000	−9010	0	99000
Profit (≥0)/benefit (≥0) in $	0	0	0	88000

Table 9 Case 5: economically efficient solution with market surplus of $93,000 and flowgate 12 capacity of 1105 MW and generation capacity at node 1–1150

Unit	Node 1		–	Node 2	
	Gen 1	Gen 2	Flowgate 12	Gen1	Load
Dispatch in MWh	600	500	1100	0	1100
Maximum operating level (MW)	1150	500	1101		
LMP/flowgate marginal price in $/MWh	10	10	0	10	10
Revenue (≥0)/payment (≤0) in $	6000	5000	0	0	11000
Cost (≤0)/value (≥0) in $	−6000	0	−9010	0	99000
Profit (≥0)/benefit (≥0) in $	0	5000	0	0	88000

Table 10 Beneficiaries-pay cost allocation

	Incremental benefits	Share of benefits	Allocated costs in $	Allocated flowgate 12 rights in MW
Load	44000	0.898 (=44/49)	8091	809
Gen2	5000	0.102 (=5/49)	919	92
Total	49000	1	9010	901

flowgate value of $0/MWh and loses $4000, the energy prices are the same at both nodes.

The incremental benefits of the expansion that accrue to the load at node 2 are $44,000 compared to pre-expansion case. The new generator at node 1 benefits is $5000. The load and new generator at node 1 pay $9010 for the upgrade and receive flowgate rights on 901 MW upgrade in proportion to their benefits. The load and gen2 pay and receive flowgate rights in proportion to the benefits. The calculations are in Table 10

6.6 *Efficient Incentives*

Currently, ISOs have two dominant transmission rate designs: stated rates and formula rates. Stated rates are set in a rate case and stay in effect until another rate case is filed or the Commission finds them unjust and unreasonable and changes them. For stated rates, the TO can keep any profits it earns by reducing its average costs between rate cases. Formula rates are set in a rate case and are updated annually based on actual costs. The formula stays in effect until another rate case is filed or the Commission finds it unjust and unreasonable and changes them. It is unusual for the Commission to find either rate unjust and unreasonable.

In 2000, Léautier (2000) proposed a regulatory contract that induces network operators to optimally expand the grid. The proposed mechanism builds on a contract used in England and Wales. In 2009, Léautier and Thelen (2009) find that vertical separation is not sufficient to induce grid expansion and needs a well-designed incentive scheme.[1] Contemporaneously, Hogan et al. (2010) considered combining the merchant and regulatory approaches that rely on FTRs. They suggested benchmark or price regulation for monopoly transmission and practical incentive mechanisms on two-part tariffs. The basic idea is that, in order to promote expansion of transmission networks, the foregone congestion rents are compensated to the TRANSCO with an increase of the fixed part of the tariff. The overtime rebalancing of the fixed and variable parts of the two-part tariff also promotes convergence to an optimal social-welfare steady state. In 2018, Hesamzadeh et al. (2018) proposed an approach to optimal pricing/investment that combines the Hogan et al. (2010) approach with the Loeb and Magat (1979) subsidy approach and suggested ways to incorporate demand and cost functions changing over time. Also, recently, Vogelsang (2018) advocates the Hesamzadeh et al. (2018) as a mechanism that compares favorably to a central planning and stakeholder bargaining approaches.[2]

The Commission's principal focus is getting new transmission built that is reliable and economically efficient over large regions. The principal impediments are rights of ways and beneficiaries-pay cost allocation. ISOs have more than one TO and much of the literature assumes a single TO.

7 Transmission Expansion Process

The optimal transmission planning process needs high fidelity data, good expansion proposals, a good suite of models, reasonable assumptions about the future, transparency, and market participant involvement. In addition, due to the uncertainty and approximations, this process must be iterative.

7.1 Scenarios

Scenarios are the result of a vigorous transparent public debate. Scenarios need to focus on assumptions about technology, environment, input prices, government mandates, and the probability of each scenario. Technological innovation and scientific discoveries have perplexed prognosticators for centuries. The assumptions about technological innovation can radically change the model outcomes. Controversial but important scenario parameters include the future prices of coal, oil, natural gas, and carbon (or amount of carbon emissions permitted). EIA produces annual long-term

---

[1] See also Léautier's chapter in this book (chapter "Regulated Expansion of the Power Transmission Grid").

[2] See also Vogelsang's paper in this book (chapter "A Simple Merchant-Regulatory Incentive Mechanism Applied to Electricity Transmission Pricing and Investment: The Case of H-R-G-V").

forecasts that are generally considered the default assumptions in analysis. This is not because EIA necessarily gets it right, but because they are the least biased and have the best information base. Some policy objectives are exogenously determined and can be incorporated with constraints and modification of cost coefficients. Each environmental pollutant (for example, CO_2, SO_2, or NO_x) can be priced or constrained in any geographic region. Minimum resource portfolios (for example, wind, solar or geothermal) can be required for any geographic region. The models are used to guide the planning process. Transmission projects are chosen to maximize economic efficiency.

7.2 *Strawman Transmission Expansion and Interconnection Process for ISOs*

We present a strawman transmission expansion process that includes the interconnection process.

Step 1. Update system data including transmission topology, generator, load, and storage parameters

Step 2. Estimate future demand, asset costs and operating parameters, and fuel costs. Create future scenarios. Assign a probability to each scenario.

Step 3. Find the expected optimal ('reliable and economically efficient') topology. This is a complex optimization problem. All scenario project results including cost allocation using beneficiaries-pay approach are presented to stakeholders. New generators may drop out of the process.

Step 4. Assemble a set of transmission projects that could lead to an economically efficient result. Conduct a competition for new projects chosen by the process.

Step 5. Have identified beneficiaries vote on transmission cost allocation weighted by the proposed cost allocation. Consumers may reduce their share of the cost allocation by agreeing to be price-responsive demand.

Step 6. Coordinate expansion with neighboring system.

Step 7. If there is general agreement, file the results at the Commission. If not go to step 2 or submit the results to the Commission to resolve disagreements.

8 Summary, Conclusions, and Recommendations for Further Study

8.1 Summary

In this paper, we presented a process for transmission planning, raised questions for approximations and relaxations, and examined several approaches to allocation of transmission costs and rights. The Commission must approve expansions, cost allocation, and transmission rights awards. Each state holds the ultimate veto over transmission expansion in the state because it retains the eminent domain decision for most projects.

8.2 Recommendations for Study of Modeling Process, Cost and Transmission Rights Allocation

- Consider merging the interconnection with transmission planning processes to co-optimize and to clear up the inconsistencies and uncertainties, to lower transactions costs and to increase the expected economic efficiency.
- Promote greater transparency and participation of the market participants especially those who receive a cost allocation.
- Generation and load are treated comparably.
- Combine the analysis of reliability and economic projects.
- Encourage the industry to improve modeling capability.
- Expand the competition models to more projects.
- Beneficiaries pay should be the overarching cost allocation principle.
- Those who request a public policy upgrade should pay for it.
- Offer flowgate rights on the upgrades.

The transmission expansion is a complicated and complex process. It should be subject to continuous improvement and not be static.

Glossary

B/C Expected benefit/expected cost ratio
DFAX Distribution factor
EPAct Energy Policy Act
ERIS Energy resource interconnection service
FGR Flowgate right that entitles the holder to the marginal value of a flowgate (FMV)

Flowgate A transmission line or collection of tightly interconnected transmission assets

FMV Flowgate marginal value is the value of another unit of capacity on the flowgate

FTR Financial transmission right obligation to pay/receive the difference in nodal energy prices. It is a portfolio of purchases and sale of flowgate rights. The value of the portfolio is determined by the flowgate marginal values

FUA Powerplant and Industrial Fuel Use Act

IPP An independent power producer not owned by the interconnected utility

ISO Independent system operator (an RTO is also an ISO.)

LGIA Large generator interconnection agreement

NITS Network Integration Transmission Service allows a network customer to integrate and economically dispatch and regulate its current and planned network resources to serve its network load in a manner comparable to the way a transmission provider uses its transmission system to serve its native load customers. Order No. 676-H, 2014

NRIS Network resource interconnection service

OATT Open access transmission tariff

Option FTR Financial transmission right the right to receive flowgate rights. It is a portfolio of purchases and subsequent sale of flowgate rights

PMU Phasor measurement unit

Pseudo tie is a transmission service that allows the generator to be dispatched by the receiving BA. The energy transfer is updated in real time and included in the actual net interchange term like tie line in the affected BAs' control ACE equations or alternate (NERC)

PURPA Public Utility Regulatory Policy Act

RAS Remedial action scheme that generally relies on control mechanisms to satisfy reliability

Resource adequacy Occurs when all generators are available there is enough generation to serve forecasted non-price-responsive load and have sufficient reserves taking into account the transmission constraints and outages

RTO Regional transmission operator

Specific delivery is a contract between a generator and load that requires energy injected into the system to be delivered to the load. This is physically impossible except in simple systems. A milder form of contract requires the injections correspond to the withdrawals

Sunk cost is a cost that has already been incurred and has no value in an alternative use

TLR Transmission line loading relief

TO Transmission owner

VOLL Value of lost load

References

M. Alguacil, Conejo, Transmission expansion planning: a mixed-integer LP approach. Power Syst. IEEE Trans. **18**(3), 1070–1077 (2003)

Astrape Consulting, *The Economic Ramifications of Resource Adequacy*. For EISPC and NARUC Funded by the U.S. Department of Energy (2013)

L. Bahiense, G.C. Oliveira, M. Pereira, S. Granville, A mixed integer disjunctive model for transmission network expansion. IEEE Trans. Power Syst. **16**(3), 560–565 (2001)

M.L. Baughman, S.N. Siddiqi, J.W. Zarnikau, Integrating transmission into IRP part I: analytical approach. IEEE Trans. Power Syst. **10**(3), 1652–1659 (1995)

T. Bone, *The Disconnect Between Transmission Policies and Planning*. California Public Utilities Commission (2018)

E. Boyd, *Power Sector Modeling* (2016). www.energy.gov/ta

T. Clark, *Order No. 1000 at the Crossroads: Reflections on the Rule and Its Future* (2018). https://www.wbklaw.com/uploads/file/Articles-%20News/2018%20articles%20publications/WBK-%20TC-Order%201000%20whitepaper%20Final.pdf

S. De la Torre, A.J. Conejo, J. Contreras, Transmission expansion planning in electricity markets. IEEE Trans. Power Syst. **23**(1) (February 2008)

Y.P. Dusonchet, A. El-Abiad, Transmission planning using discrete dynamic optimizing. Syst. IEEE Trans. Power Appar. **PAS-92**(4), 1358–1371 (1973)

H. Faria, Jr., S. Binato, M.G. Resende, D.M. Falcão, Transmission network design by greedy randomized adaptive path relinking, IEEE Trans. Power Syst. **20**(1) (February 2005)

E.B. Fisher, R.P. O'Neill, M.C. Ferris, Optimal transmission switching. IEEE Trans. Power Syst. **23**(3), 1346–1355 (2008)

R.A. Gallego, A. Monticelli, Transmission system expansion planning by simulated annealing. IEEE Trans. Power Syst. **11**(1), 364–369 (1996)

R.A. Gallego, A. Monticelli, R. Romero, Transmission system expansion planning by an extended genetic algorithm. IEE Proc. Gener. Transm. Distrib. **145**(3), 329–335 (1998)

L.L. Garver, Transmission network estimation using linear programming. IEEE Trans. Power App. Syst. **PAS-89**(7), 1688–1697 (1970)

K.W. Hedman, R.P. O'Neill, E.B. Fisher, S.S. Oren, Optimal transmission switching—sensitivity analysis and extensions. IEEE Trans. Power Syst. **23**(3), 1469–1479 (2008)

K.W. Hedman, R.P. O'Neill, E.B. Fisher, S.S. Oren, Optimal transmission switching with contingency analysis. IEEE Trans. Power Syst. **24**(3), 1577–1586 (2009)

K.W. Hedman, M.C. Ferris, R.P. O'Neill, E.B. Fisher, S.S. Oren, Co-optimization of generation unit commitment and transmission switching with N-1 reliability. IEEE Trans. Power Syst. **25**(2), 1052–1063 (2010)

M.R. Hesamzadeh, J. Rosellón, S.A. Gabriel, I. Vogelsang, A simple regulatory incentive mechanism applied to electricity transmission pricing and investment. Energy Economics **75**, 423–439 (2018)

W. Hogan, *Transmission Cost Allocation* (Presentation to the Harvard Electricity Policy Group, 2010)

S. Huntoon, *Reliability Costs Fox-Henhouse Regulation* (RTO Insider, 2019)

P.L. Joskow, R. Schmalensee, *Markets for Power: An Analysis of Electrical Utility Deregulation* (The MIT Press, Cambridge, 1983)

A.K. Kazerooni, J. Mutale, Transmission network planning under security and environmental constraints. IEEE Trans. Power Syst. **25**(2), 1169–1178 (2010)

S.O. Kimbrough, F. Murphy, Y. Smeers, *Extending Cournot: When Does Insight Dissipate*?, Fox School of Business Research Paper No. 14-036 (October 15, 2014)

T.-O. Léautier, Regulation of an electric power transmission company. Energy J. **21**(4) (2000)

T.-O. Léautier, V. Thelen, Optimal expansion of the power transmission grid: why not? J. Regul. Econ. **36**, 127–153 (2009)

M. Loeb, W.A. Magat, A decentralized method for utility regulation. J. Law Econ. **22**(2), 399–404 (1979)

S.C. Littlechild, E.A. Ponzano, *Transmission Expansion in Argentina 5* The Regional Electricity Forum of Buenos Aires Province (2 December 2007). http://www.dspace.cam.ac.uk/bitstream/1810/195431/1/0762&EPRG0729.pdf

A. Liu, B.H. Hobbs, J. Ho, J.D. McCalley, V. Krishnan, M. Shahidehpour, Q. P. Zheng, *Co-optimization of Transmission and Other Supply Resources* (Department of Energy, National Energy Technology Laboratory, 2013)

E. Nicholson, M. Stone, D. Powers, *Building New Transmission Experience To-Date Does Not Support Expanding Solicitations*, prepared for: Ameren, Eversource Energy, ITC Holdings Corp., National Grid USA, and PSE&G (June 2019)

R.P. O'Neill, P.M. Sotkiewicz, B.F. Hobbs, M.H. Rothkopf, W.R. Stewart Jr., Efficient market-clearing prices in markets with nonconvexities. Eur. J. Oper. Res. **164**(1), 269–285 (2005a)

R.P. O'Neill, R. Baldick, U. Helman, M.H. Rothkopf, W. Stewart, Dispatchable transmission in RTO markets. IEEE Trans. Power Syst. **20**, 171–179 (2005b)

R.P O'Neill, E.A. Krall, K.W. Hedman, S.S. Oren. A model and approach to the challenge posed by optimal power systems planning. Math. Program. Ser. B **140**, 239–266 (2013)

J. Pfeifenberger, J. Chang, A. Sheilendranath, *Transmission Competition Under FERC Order No. 1000: What we Know About Cost Savings to Date* (October 25, 2018)

R. Romero, A. Monticelli, A hierarchical decomposition approach for transmission network expansion planning. IEEE Trans. Power Syst. **9**(1), 373–380 (1994)

E.E. Sauma, S.S. Oren, Economic criteria for planning transmission investment in restructured electricity markets. IEEE Trans. Power Syst. **22**(4) (2007)

R. Villasana, L.L. Garver, S.J. Salon, Transmission network planning using linear programming. IEEE Trans. Power Appar. Syst. **PAS-104**(2), 349–356 (1985)

I. Vogelsang, Can simple regulatory mechanisms realistically be used for electricity transmission investment? The case of H-R-G-V. Econ. Energy Environ. Policy **7**(1) (2018)

H.P Young, *Cost Allocation: Methods, Principles, Applications, North-Holland* (1985)

H.P. Young, *Equity: In Theory and Practice* (Princeton University Press, Princeton, 1995)

Transmission Planning and Co-optimization with Market-Based Generation and Storage Investment

Qingyu Xu and Benjamin F. Hobbs

1 Introduction

Optimal transmission expansion planning (TEP) is not simply the addition of lines to already congested corridors in order to lower fuel costs through more efficient dispatch of the existing generation fleet. This is because the amount and location of generation investment as well as its dispatch might shift to take advantage of changes in network capabilities, and these shifts will in general unfold over the multidecadal lifetime of the transmission assets. In sum, transmission investment will change not only operating costs of generation but also investment costs. Thus, a TEP planner should anticipate changes in generation plant siting, amounts, and mixes. The traditional approach of evaluating the economic benefits of transmission just by valuing the resulting savings in operating costs results in distorted estimates of the benefits of transmission reinforcements and potentially suboptimal grid expansion decisions (CAISO 2004; MISO 2010).

Transmission–generation expansion co-optimization tools are designed for this job: they help TEP planners to plan transmission in a proactive manner so that transmission planners are able to select the lines anticipating the market reactions of generation investors (Krishnan et al. 2015; Liu et al. 2013). Several generation–transmission co-optimization models have been published and are being tested by regional transmission agencies. Most are formulated as optimizations that minimize the total capital and operating cost of the joint transmission–generation system or as maximizing net market benefits (value of energy consumption minus those costs). The assumption of such models is that the underlying generation market is perfectly competitive with no major market failures (which is equivalent to net market benefits

Q. Xu · B. F. Hobbs (✉)
Whiting School of Engineering, Johns Hopkins University, 3400 North Charles Street, Baltimore 21218, Maryland, USA
e-mail: bhobbs@jhu.edu

M. R. Hesamzadeh et al. (eds.), *Transmission Network Investment in Liberalized Power Markets*, Lecture Notes in Energy 79,
https://doi.org/10.1007/978-3-030-47929-9_7

maximization for just generation) and that the transmission planner's objective is also to maximize net market benefits (van der Weijde and Hobbs 2012). Thus, the bi-level structure of decision making in the market (transmission acting as a "Stackelberg leader" with respect to generation followers) reduces to a convenient-to-solve single-level optimization. Other models, however, recognize that serious imperfections exist in the generation market (externalities, subsidies, market power, regulated prices) that mean that instead an explicitly bi-level optimization approach is called for. Such problems are inherently more difficult to solve, but recent progress has been made (Pozo et al. 2013; Tohidi et al. 2017).

In addition to market failures in generation markets, another challenge to TEP is the rise of new types of supply technologies, as well as storage and demand response. The challenges of a load growth together with renewables could be met with a greatly expanded grid, but storage and demand technologies hold the promise of lowering the cost of renewables integration and also being less costly in at least some cases than new transmission. A proactive TEP should therefore anticipate the response of investments in new technologies. This is the focus of this chapter; in particular, we expand least-cost types of co-optimization models to include storage as well as transmission and generation. With the cost of energy storage plummeting rapidly, consideration of storage might greatly affect TEP.

To anticipate energy storage sizing and siting, the classic generation–transmission co-optimization TEP model needs to be expanded. Here, we show an example of such an expanded TEP approach. The model selects the best set of lines while simulating the profit-maximizing reaction of competitive generation and storage investors in terms of the siting and sizing of new facilities. After parameterizing the model for the Western Electricity Coordinating Council (WECC, consisting of the western provides and states of Canada and the USA, respectively), we identify at what levels of battery investment costs it becomes economically valuable to consider storage expansion in TEP.

The rest of this chapter is organized as follows. In Sect. 2, we provide some background: the interactions of transmission and generation and the complications posed by storage; a historical view of co-optimization of transmission and generation expansion; finally, a procedure to calculate the economic value of considering storage expansion in TEP. In Sect. 3, we formulate a static (single year) co-optimization of transmission, generation, and storage expansion. In Sect. 4, we present a case study for the WECC regions. We conclude this chapter in Sect. 5.

2 Background

2.1 Interactions Among Transmission, Generation, and Storage

Generation and transmission expansions interact in complex ways. Fundamentally, they can be complements (investment in one increases the market value of investment in another) or substitutes (investment in one lowers the market value of the other). Transmission is valuable just because of its capability to deliver electricity from a cheap resource to the demand, avoiding turning on an expensive local generation; thus, transmission is a complement to the remote resource but a substitute for the local one. As specific examples, transmission and generation complement each other in cases such as mine-mouth coal power plants and wind farms that are distant from load centers: cheap power is only valuable when deliverable. The opposite can also be true: when local generation, such as gas turbines or rooftop solar panels, became cheap, it diminishes the value of new transmission into a load pocket, and so, generation and transmission become substitutes.

The rise of electricity storage, especially distributed storage in the form of batteries, is making this story more complicated. First, storage can both compete with and complement generation. Storage can compete with conventional generation, for instance in meeting peak loads. Regulators encourage this competition: Order No. 841 (FERC 2018) from the U.S. Federal Energy Regulatory Commission requires that independent system operators adjust their rules and market software so that storage can compete with the generation in the energy, ancillary service, and installed capacity markets. The fast ramping response of electric storage implies that storage and generation may compete fiercely in reliability markets as the cost of storage decreases. However, storage, because of its fundamental ability to shift supply from one time period to another, can be a complement to generation with less operational flexibility (e.g., base-loaded thermal plants) or intermittent availability (e.g., variable renewable energy, VRE). Indeed, pumped storage plants were often justified in the 1960s and 1970s because of this complementarity with nuclear plants which are most efficient when run flat out for all hours. Nowadays, however, the focus is on storage's complementarity with VRE; such storage will be essential to achieving the very high renewable penetrations that are the targets in some jurisdictions (e.g., 100% in Hawaii and California).

Storage also interacts with the transmission, but in a somewhat subtler way: they are both arbitragers of the energy, with the transmission arbitraging over space and storage doing so over time. They can both facilitate higher penetrations of VRE. A better interconnection can help in the following way: at a certain point in time, unexpected under-generation of VRE in one place can be made up by transmission delivering available production from another plant (e.g., another VRE) from hundreds of km away. This may, for instance, avoid starting up or ramping of local generation that is perhaps both costly and polluting. On the other hand, storage can also resolve local shortfalls by, in effect, delivering cheap output of a plant that was produced several hours or even days or months ago (e.g., from wind or hydro energy that would have otherwise been curtailed or "spilled").

Transmission and storage are not always competing. As a simple case, we can imagine a distant wind farm might be more economical because of a bundled storage facility, and hence, a transmission project also becomes valuable. On the other hand, this nearby storage could enable a transmission facility to be downsized and still deliver the same amount of VRE production.

Overall, the interactions among transmission, generation, and storage will strongly affect the economic value of transmission reinforcements. Hence, from the perspective of transmission planner, a planning model with the ability to capture the above interactions becomes valuable and informative. We shall next discuss co-optimization tools that have this capability.

2.2 Using Co-optimization to Support Transmission Expansion Planning

Co-optimization of transmission and generation planning is not a new topic. The mathematics problems describing siting generation and transmission together can be dated back at least to 1977 (Sawey and Zinn 1977). However, the meaning of co-optimization of transmission and generation expansion changed with time went by, and the major breaking point was the deregulation of the power sectors in Europe and the USA.

"*Co-optimization*" used to mean co-planning of just generation and transmission. When most of the power industry was still vertical integrated, generation planners and transmission planners were able to work together: generation expansion plans were first developed and handed to the transmission planners, transmission plan was then developed and may or may not be handed back to the generation planners for more iterations. In this iterative manner, the interaction between generation and transmission was at least partially accounted for by these vertically integrated monopolies.

The meaning of co-optimization has enriched since the deregulation of the power industry in Europe and the USA in the 1990s. In the newly established markets, the planning of transmission and generation expansions is separated and, respectively, performed by grid owners/transmission system operators (TSOs)/regional transmission organizations (RTOs) and generation companies. Without the full co-operation of the generation planners and, at the same time, lacking tools to anticipate how generation siting would respond to grid changes, many transmission planners have been forced to treat the locations and amounts of generation capacity as purely exogenous "boundary conditions": they would have to assume scenarios in which the generation siting is known and then plan the transmission expansion based on the scenarios. This is called "*reactive*" transmission expansion planning: transmission planners react to the generation expansion.

In contrast to "reactive" transmission expansion planning, "*proactive*" transmission expansion planning anticipates how generation investors will choose the sites, types, sizes, and timing of changes in their assets in reaction to the network plan and then choose the best set of transmission expansion projects. From the point of view of game theory, the game between transmission and generation is a bi-level or "Stackelberg" game. The transmission planner is a leader who optimizes subject to the anticipated reactions of a set of generation investors who are Nash players who do not anticipate how the grid plan would change in response to generation decisions. It is natural to place transmission in the role of a leader because transmission assets generally take much longer to plan and build than the natural gas-fired or renewable generating assets that constitute most or all of generation additions in North America and Europe today.

Transmission and generation co-optimization models can be seen as one of several types of "proactive" transmission expansion planning models if the following strong assumptions are made (Liu et al. 2013; Spyrou et al. 2017; Sauma and Oren 2006):

- The transmission expansion planner has an objective of maximizing market surplus (what the economists call "market efficiency" or "social welfare"). This is defined as the sum of surpluses accrued by all market parties, including profits earned by each resource and storage; transmission congestion surplus minus incremental grid costs; consumer surplus. If demand is perfectly inelastic (fixed), this objective is equivalent to minimizing the sum of resource, storage, and transmission costs.
- Short-run (spot) electricity markets, including energy, ancillary service, and capacity markets, are perfectly competitive. All suppliers are price takers and profit maximizers.
- Similarly, in the long run, generation expansion planners are siting optimally and competitively to maximize their profits, given the cost of transmission as reflected in locational marginal prices, which depend on the grid and all suppliers' decisions.

Of course, this basic proactive model simplifies reality but then does all models. These assumptions enable the bi-level game to be solved as a single optimization model since the TEP objective of maximizing market surplus is consistent with perfect competition on the lower level, which can be modeled by maximizing total market surplus as well. Relaxing any of those three assumptions will generate a new type of "proactive" transmission planning model that in general will have a difficult to solve bi-level structure in which the leader and follower objectives are not aligned. Although out of the scope of this chapter, readers that are interested in "proactive" transmission expansion models formulated explicitly as bi-level or multi-level games are referred to Pozo et al. (2013), Tohidi et al. (2017), Sauma and Oren (2006), Jenabi et al. (2013), Jin and Ryan (2014), Jin and Ryan (2014), Gonzalez-Romero et al. (2019).

Another way in which co-optimization models can be broadened is by including more types of market players, including consumers (i.e., demand response) and the storage. As mentioned before, in Feb. 2018, the FERC issued Order No. 841 to urge the US markets under its purview to modify their tariffs to make sure that electric storage can compete with the conventional generators in the energy, ancillary service, and capacity markets so that energy storage can participate fully in spot markets and are able to set prices.

With electricity storage coming into play, co-optimization models must now co-optimize (or anticipate) the siting and operation of storage. As a result, additional assumptions are needed, namely that storage owners are competitive. They therefore choose the timing, type, size, and location of storage facilities to maximize their profit subject to locational commodity prices that they assume they cannot alter. Reflecting the new FERC rules, practical co-optimization models usually assume that storage owners can either let the ISO dispatch their facilities optimally or, equivalently, they self-schedule with perfect foresight of the time-varying prices they will receive.

2.3 Quantify the Economic Value of Considering Storage Expansion in Transmission Expansion Planning

As battery costs continue to decline, batteries, fly wheels, compressed air, and other storage devices will more likely interact with and change the value of transmission and generation. Traditional vertically integrated utilities will likely adapt their generation and transmission planning methods to consider how possible investments in storage might change optimal investments in other assets. In restructured, vertically disintegrated markets, on the other hand, storage is another player whose operating and investment decisions will need to be anticipated by transmission planners in the proactive paradigm. If the effects of grid reinforcements on the siting, sizing, and timing of storage investment are disregarded in TEP, the result might be a different—and economically inferior—transmission plan. We now address the question: how can we quantify the value of considering storage in a proactive TEP? We propose and demonstrate a procedure for quantifying this value in the remainder of this chapter. The demonstration is for the western USA and Canada system (WECC) for the year 2034. Previous work (Spyrou et al. 2017) has quantified the value of anticipating how grid reinforcements affect generation expansion in TEP (i.e., the "value of generation-proactive TEP") for the eastern USA and Canada system. There, we show that iterating between (1) solving a TEP subject to a fixed generation build-out and (2) solving a generation expansion problem (GEP) subject to a fixed network can realize only part of the value of generation-proactive TEP.

In summary, the quantification of the value of considering storage in proactive TEP involves three steps: (1) plan with co-optimization of storage, generation, and transmission; (2) plan disregarding the possibility of storage installation and how it reacts to network expansions; and (3) evaluation of the latter, potentially flawed plan by modeling the "actual" reaction of storage and generation to that plan. The first step is the full co-optimization, where the transmission expansion is planned anticipating the reactions of both generation and storage installations. The results of this step are the optimal plan (a set of selected transmission projects) and a minimized system cost. In the second step, a transmission expansion plan is obtained from a "flawed" planning model, where the installation of storage is ignored and only generation is considered in a co-optimization framework. Finally, we evaluate this "flawed" plan by plugging it into the co-optimization model (fixing the network decision variables at their flawed values) and getting a new minimized cost for the generation and storage followers, which may involve installation of storage but at potentially different locations and in different amounts than the full co-optimization. The difference in the costs between steps 1 and 3 is the value of considering storage in transmission expansion planning. Because step 3 is more constrained than step 1, its cost will be no lower than the full co-optimized model and is potentially higher. We call this increase in cost the "*value of model enhancement for storage*" (VoMES). Another closely related term, "*value of storage*" (VoS), can also be defined as the objective function improvement if storage is allowed to be expanded in the system, i.e., the differences in the objective function values resulting from step 1 and 2. For example, the VoS under alternative incentive mechanisms for merchant transmission expansions is calculated for IEEE test-systems in Khastieva et al. (2019). These results show that the VoS is relatively small compared to system cost ($2 Million comparing to $442 Million) but can be more than three times higher than that amount if transmission expansion incentives are provided. The conceptual differences and relationship between VoMES and VoS will be discussed at the end of this section.

We now present the details of each step, including the TEP co-optimization models that we apply.

Step 1. Planning with Co-optimization (Benchmarking) Imagine we have a TEP tool which can select the best set of new transmission lines (T) by anticipating the construction of new generation (G), the installation of new storage (S), and the system operation (P) to minimize annualized system cost $C(T, G, S, P)$ (in $/yr) for some future scenario year. (Existing facilities are implicitly in the model as well.) All the decision variables are subject to the feasible region (F) which is defined by the physical operating constraints for the network as well as individual resources (e.g., Kirchhoff's laws, line and resource capacity limits, ramp limits, state-of-charge relationships, etc.) and policy constraints such as renewable portfolio standards or emissions limits. An abstract mathematical programming problem (MP1) can be shown as follows, whose detailed formulation can be found in the next section:

$$\begin{aligned} \text{Minimize}_{T,G,S,P} \quad & C(T, G, S, P) \\ \text{s.t.} \quad & (T, G, S, P) \in F \end{aligned}$$

If this is solved to optimality, it will return a solution of (T^*, G^*, S^*, P^*) and a system cost of $C(T^*, G^*, S^*, P^*)$. (Note that if demand is elastic, instead of minimizing cost, we would instead be maximizing net market surplus, recognizing the value of benefits associated with different levels of consumption as captured by the integrals of demand curves).

By definition, $C(T^*, G^*, S^*, P^*)$ is the lowest cost that the model can achieve and T^* is the optimal transmission plan provided by the model. In other words, any transmission plan other than T^* will lead to a system cost no lower than $C(T^*, G^*, S^*, P^*)$, and hence, that network configuration and the associated cost can be used as a benchmark.

Step 2. Planning without storage anticipation Imagine the planner chooses to ignore the storage installation in the TEP. Mathematically, it means forcing $S = 0$ in the formulation above (MP1). Thus, we are solving the following problem (MP2) instead:

$$\begin{aligned} \text{Minimize}_{T,G,P} \quad & C(T, G, 0, P) \\ \text{s.t.} \quad & (T, G, 0, P) \in F \end{aligned}$$

Let the solution of this TEP model be $(\hat{T}, \hat{G}, 0, \hat{P})$ and the associated system cost be $C(\hat{T}, \hat{G}, 0, \hat{P})$. $\hat{T}$, therefore, stands for the optimal transmission expansion plan that the planner can get if they ignore the possibility of installing storage.

Step 3. Plan Evaluation Imagine the transmission expansion plan from Step 2 is implemented. Mathematically, it means forcing $T = \hat{T}$ in MP1; equivalently, we are solving the following problem (MP3):

$$\begin{aligned} \text{Minimize}_{G,S,P} \quad & C(\hat{T}, G, S, P) \\ \text{s.t.} \quad & (\hat{T}, G, S, P) \in F \end{aligned}$$

Let $(\hat{T}, \overline{G}, \overline{S}, \overline{P})$ be the solution of MP3 and $C(\hat{T}, \overline{G}, \overline{S}, \overline{P})$ be the associated objective function. By definition, $C(\hat{T}, \overline{G}, \overline{S}, \overline{P})$ is no lower than $C(T^*, G^*, S^*, P^*)$ since the former is the system cost resulted from choosing a transmission plan $\hat{T}$ other than the optimal T^*. One can thus naturally conclude that the cost of ignoring storage installation leads to a different plan and a cost no lower than the optimal. And the difference between $C(\hat{T}, \overline{G}, \overline{S}, \overline{P})$ and $C(T^*, G^*, S^*, P^*)$ is the "*value of model enhancement to consider storage*" (VoMES) in TEP:

$$\text{VoMES} = C(\hat{T}, \bar{G}, \bar{S}, \bar{P}) - C(T^*, G^*, S^*, P^*).$$

In a sense, this is the value of "smart" planning that proactively anticipates how storage will be installed and used, versus a naïve plan that overlooks storage.

This value of smart planning is distinct from the overall "*value of storage*" VoS to the system, as in Khastieva et al. (2019), which is the cost improvement from a co-optimized plan that only includes transmission and generation to a plan that co-optimized storage as well. This is the reduction in cost from MP2 (no storage) to MP1 (all options):

$$\text{VoS} = C(\hat{T}, \hat{G}, 0, \hat{P}) - C(T^*, G^*, S^*, P^*)$$

Note that VoMES $\leq$ VoS because the cost of MP3 will necessarily be no higher than MP2's cost. This is because MP2 and MP3 have the same value of T, but MP3 is free to choose both G and S, while MP2 can only choose G since S is constrained to zero. Their relationship is shown in Fig. 1. One implication of this inequality is that the economic value that storage can potentially provide to the system can be offset by naively disregarding storage expansion and its response to transmission in TEP, in which case, the net benefit will be the remainder of (VoS–VoMES). Thus, the larger VoMES is (as proportion of VoS), the greater the loss of storage benefits will be if naïve rather than proactive transmission planning is undertaken; in other words, the benefits of storage to the system are more dependent on transmission expansion planning.

In this chapter, our focus is on the value of modeling to implement proactive TEP, so our major interest is in the calculation of VoMES to show what can be gained from proactive planning. But the calculation of VoS is also useful as it illustrates one of the many types of insights that can be obtained from applying TEP models. Readers should also bear in mind that the term of VoMES and VoS are not limited to the anticipated storage expansion, and we can easily extend such concepts to other aspects

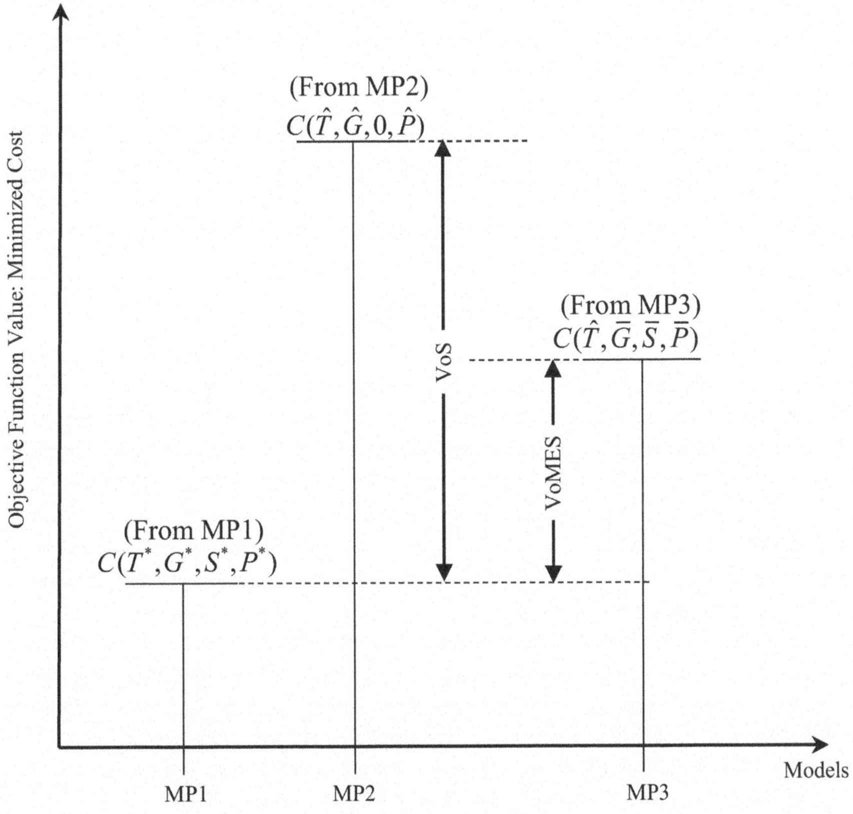

Fig. 1 Diagram of Value of Model Enhancement to consider Storage (VoMES) and the Value of Storage (VoS)

of electricity system. The value of enhancing a model with generation–transmission co-optimization is calculated in (Spyrou et al. 2017) (i.e., VoME of co-optimization), showing that co-optimization can double the net cost savings from transmission expansion, comparing to purely reactive TEP; iterative planning (alternating between transmission and generation capacity expansion models) can partially but not fully realize these benefits. For a review of enhancements that have been implemented in transmission expansion models, reader is referred to (Xu and Hobbs 2018).

3 Detailed Formulation

In this section, we show a formulation of the static transmission expansion problem co-optimized with generation and storage expansion. For a review of literature on co-optimization transmission and storage but omitting generation expansion, we refer the reader to (Khastieva et al. 2019; Qiu et al. 2017). Some general assumptions include the following.

- In general, TEP models need to consider both short- and long-run uncertainties since elsewhere we have shown that considering a range of long-run economic, regulatory, and technological scenarios in a two-stage stochastic programming framework can make a significant and economically important difference in transmission plans. However, for the sake of simplicity in this chapter, the consideration of uncertainty will be limited to short-term variability, namely load, wind, solar, and hydro conditions. For reviews of TEP models that consider long-term uncertainties, readers are referred to van der Weijde and Hobbs (2012), Ho et al. (2016), Munoz et al. (2014) and Park et al. (2018).
- The operating constraints and costs of this model include the linearized unit commitment formulation that was proposed in Kasina et al. (2013), in which start-up costs are included in the cost objective, while ramp rates, start-ups, and minimum output levels constrain generation levels. A more comprehensive version of this formulation with long-term planning and long-run uncertainties can be found in Xu et al. (2017). Meanwhile, classic unit commitment formulations that use binary variables to represent generator commitment status are given by Takriti et al. (1996) and Morales et al. (2013); such variables are difficult to include in long-term planning models due to the desire to avoid nonlinearities and impractically large MILP models, and so, transmission planning models tend to use simpler operating models.
- The network formulation is based upon a combination of a linearized DC load flow (DCOPF), which represents how Kirchhoff's Voltage Law induces parallel flows in the network, and disjunctive constraints that utilize the Big-M formulation (Winston et al. 2003). Only high voltage facilities are represented. For more advanced power flow modeling including transmission losses and reactive power, readers are referred to Zhang et al. (2013), Ozdemir et al. (2016).

- Renewable portfolio standards by state are represented, including rules allowing one state to use renewable energy credits generated in other states to meet renewable obligations. Carbon policy is represented by a tax on carbon emissions.

We begin by introducing notation, and then, the objective function and constraints of model MP1.

Sets

A Load areas, index a
ES Storage facilities, index esk
K Generators, index k
H Hours, index h
I Buses, index i
L Transmission lines, index l
RP Unidirectional renewable energy credit trading paths, index rp
ST States, index stt

Variables

$\mathrm{g}_{k,h}$ Power production by the generator (MW)
gcap_k Capacity of the generator (MW)
gexp_k Generation capacity expansion (MW)
$\mathrm{gr}_{k,h}$ Operating reserve provided by the generator (MW)
gret_k Generation retirement (MW)
$\mathrm{gsd}_{k,h}$ Minimum run capacity shut down at hour h (MW)
$\mathrm{gsu}_{k,h}$ Minimum run capacity started up at hour h (MW)
lav_l Transmission line availability (binary)
lexp_l Transmission expansion (binary)
$\mathrm{nl}_{i,h}$ Curtailed load on the bus i at the hour h (MW)
nr_{stt} Non-compliance with RPS policy (MWh)
rt_{rp} Renewable energy credit traded on the path rp (MWh)
sav_{esk} Storage availability (binary)
$\mathrm{sch}_{esk,h}$ Charging energy withdrawn from the network by storage (MW)
$\mathrm{sdc}_{esk,h}$ Injection of the storage into the network (MW)
sexp_{esk} Storage expansion (Binary)
$\mathrm{sl}_{esk,h}$ State of charge of storage at the beginning of the hour h
$\mathrm{sr}_{esk,h}$ Operating reserve provided by storage when injecting
$\mathrm{pf}_{l,h}$ Power flow on the transmission line (MW)
$\mathrm{pmin}_{k,h}$ Minimum run level of the generator (MW)
$\theta_{i,h}$ Voltage phase angle at bus i

Cost Parameters

CB Carbon tax ($/metric ton CO_2e)
CD_k Shutdown cost ($/MW)

CFG_k	Fixed operation and maintenance cost of the generator ($/MW-year)
CFS_{esk}	Fixed operation and maintenance cost of the storage ($/MW-year)
CG_k	Capital cost of generation expansion ($/MW-year)
CL_l	Capital cost of transmission expansion ($/year)
CS_{esk}	Capital cost of storage expansion ($/MWh-year)
CU_k	Start-up cost ($/MW)
$CVG_{k,h}$	Variable cost of generator injection without carbon cost ($/MWh)
SG_k	Salvage value of generator retirement ($/MW-year)

Constraint Parameters

ACP_{stt}	Alternative compliance penalty for RPS compliance ($/MWh)
B_l	Susceptance of line (p.u.)
$BAI_{i,a}$	Bus area incidence, 1 if bus *i* is located in the area *a*
BM_l	Big M for DCOPF disjunctive constraints
$BSI_{i,stt}$	Bus state incidence, 1 if bus *i* is located in state *stt*
$D_{i,h}$	Load on the bus *i* (MW)
ER_k	Carbon emission rate of generator *k* (Metric Ton CO_2e/MWh)
$GBI_{k,i}$	Generator bus incidence, 1 if generator *k* is located on bus *i*
$HAV_{k,h}$	Hourly resource availability (fraction of capacity)
HW_h	Number of hours that hour *h* is representing per year (hour/year)
$IGCAP_k$	Initial (existing) capacity of the generator (MW)
$ILAV_l$	Initial availability of the transmission line, 1 if available (binary)
$IRPS_{stt}$	In-state requirement for a particular state's RPS (fraction)
$ISAV_{esk}$	Existing availability of the storage, 1 if available to the grid (binary)
$LBI_{l,i}$	Line bus incidence, −1 if *i* is the from-bus of line *l*, 1 for the to-bus
MC_{esk}	Maximum charge capacity for the storage to expand (MW)
MD_{esk}	Maximum additional investment in storage discharge capacity (MW)
MDT_k	Minimum down time of the generator
MG_k	Maximum additional investment that can be added for a generator (MW)
ML_{esk}	Maximum energy capacity for the storage to expand (MWh)
MRT_k	Minimum (scheduled) retirement of the generator (MW)
MUT_k	Minimum up time of the generator
PBASE	The base power unit of the system (MW)
$QMIN_k$	Minimum run (fraction of maximum capacity)
$RE_{stt,k}$	RPS eligibility, 1 if generator *k* is tagged as renewable in state *stt*
$REX_{rp,stt}$	1 if trading path *rp* is from state *stt*
$RIM_{rp,stt}$	1 if trading path *rp* is to state *stt*
RM_a	Operating reserve margin of the area *a*
RPS_{stt}	Renewable portfolio standard of state *stt* (fraction of total annual MWh)
RR_k	Ramp-rate (fraction of started up capacity)
$SBI_{esk,i}$	Storage bus incidence, 1 if storage *esk* is located on the bus *i*
TM_l	Thermal limit of transmission line (MW)

VL	Value of loss load ($/MWh)
η_{esk}	Single-trip efficiency of the storage (%)

Objective Function

The co-optimization's objective is to minimize annualized system cost. The capital costs of generation, transmission, and storage are incorporated by annualizing those costs using capital recovery factors. Total system cost is as follows:

$$\begin{aligned}\text{Objective} = & \sum_k \left(\mathrm{CG}_k\,\mathrm{gexp}_k - \mathrm{SG}_k\,\mathrm{gret}_k\right) + \sum_k \mathrm{CFG}_k\,\mathrm{gcap}_k + \\ & \sum_{esk} \mathrm{CS}_{esk}\mathrm{ML}_{esk}\,\mathrm{sexp}_{esk} + \sum_{esk} \mathrm{CFS}_{esk}\mathrm{MD}_{esk}\,\mathrm{sav}_{esk} + \\ & \sum_{k,h} \mathrm{HW}_h \left(\mathrm{CVG}_{k,h}\,g_{k,h} + \frac{\mathrm{CU}_k}{\mathrm{QMIN}_k}\mathrm{gsu}_{k,h} + \frac{\mathrm{CD}_k}{\mathrm{QMIN}_k}\mathrm{gsd}_{k,h}\right) + \\ & \sum_l \mathrm{CL}_l\,\mathrm{lexp}_l + \\ & \sum_{i,h} \mathrm{HW}_h\,\mathrm{VL}\,\mathrm{nl}_{i,h} + \sum_{stt} \mathrm{ACP}_{stt}\,\mathrm{nr}_{stt} + \sum_{k,h} \mathrm{HW}_h\,\mathrm{CBER}_k\,g_{k,h}\end{aligned} \tag{1}$$

The objective function is composed of five components [lines 1–5 in (1)]. The first line is the build cost/salvage value of the generation as well as generation's fixed operation and maintenance costs. The second line is the build cost and fixed operation and maintenance cost of storage. The third line is the variable cost, start-up, and shutdown cost of generation. The fourth line is the build cost of transmission lines. And finally, the last line is the cost of curtailed load and two policy-related costs: the alternative compliance penalty of the RPS policy and the carbon tax.

Constraints–Investment

$$\mathrm{gcap}_k = \mathrm{IGCAP}_k + \mathrm{gexp}_k - \mathrm{gret}_k \quad \forall k \tag{2}$$

$$\mathrm{sav}_{esk} = \mathrm{ISAV}_{esk} + \mathrm{sexp}_{esk} \quad \forall esk \tag{3}$$

$$\mathrm{lav}_l = \mathrm{ILAV}_l + \mathrm{lexp}_l \quad \forall l \tag{4}$$

$$\mathrm{gret}_k - \mathrm{MRT}_k \geq 0 \quad \forall k \tag{5}$$

$$\mathrm{gcap}_k - \mathrm{MG}_k \leq 0 \quad \forall k \tag{6}$$

Constraints (2)–(6) establish the relationship between the investment decision and the availability of the generation capacity, storage facility, and the transmission line.

Constraints–Generation Operation

$$g_{k,h} + gr_{k,h} \leq HAV_{k,h}\, gcap_k \quad \forall k, h \tag{7}$$

$$g_{k,h} \geq pmin_{k,h} \quad \forall k, h \tag{8}$$

$$g_{k,h} + gr_{k,h} \leq \frac{pmin_{k,h}}{QMIN_k} \quad \forall k, h \tag{9}$$

$$pmin_{k,h} \leq QMIN_k\, gcap_k \quad \forall k, h \tag{10}$$

$$pmin_{k,h} - pmin_{k,h-1} = gsu_{k,h} - gsd_{k,h} \quad \forall k, h \tag{11}$$

$$\left(g_{k,h} + gr_{k,h} - pmin_{k,h}\right) - \left(g_{k,h-1} - pmin_{k,h-1}\right) \leq \frac{RR_k}{QMIN_k} pmin_{k,h-1} \quad \forall k, h \tag{12}$$

$$\left(g_{k,h} - pmin_{k,h}\right) - \left(g_{k,h-1} + gr_{k,h-1} - pmin_{k,h-1}\right) \geq -\frac{RR_k}{QMIN_k} pmin_{k,h-1} \quad \forall k, h \tag{13}$$

$$\left(g_{k,h-1} + gr_{k,h-1}\right) - gsd_{k,h} \leq \frac{pmin_{k,h}}{QMIN_k} \quad \forall k, h \tag{14}$$

$$\left(g_{k,h} + gr_{k,h}\right) - gsu_{k,h} \leq \frac{pmin_{k,h-1}}{QMIN_k} \quad \forall k, h \tag{15}$$

$$pmin_{k,h} \leq QMIN_k gcap_k - \sum_{(h-MDT_k \leq h' \leq h)} gsd_{k,h'} \quad \forall k, h \tag{16}$$

$$pmin_{k,h} \geq \sum_{(h-MUT_k \leq h' \leq h)} gsu_{k,h'} \quad \forall k, h \tag{17}$$

Constraints (7)–(17) constrain the operation of the generators. Constraint (7) is for the generators that are not subject to unit commitment constraints, i.e., (8)–(17); in particular, the wind, solar, hydro, and other intermittent resources are subject to hourly profiles, i.e., $HAV_{k,h}$ for those resources range between 0 and 1 depending on availability of the resource. Constraints (8)–(17) are the linearized version of the unit commitment, featured by the continuous variable $pmin_{k,h}$ with a unit of MW. The reader should notice that the linearized version of unit commitment enables the transmission planner to consider the limited generation flexibility in a large system with aggregated capacity without adding any binary variables, thus speed up the TEP model with generation to optimization. The explanation of the unit commitment constraints (8)–(17) is shown below.

Constraint (8) is the minimum run constraint, and constraint (9) is the maximum run constraint. Note if the minimum run started up is $pmin_{k,h}$, the maximum run started up is thus $\text{pmin}_{k,h}/\text{QMIN}_k$. Constraint (10) restricts that the maximum run can be started up cannot exceed the available capacity. Constraint (11) is the start-up–shutdown relation constraint.

Constraints (12) and (13) are the ramp-rate constraints: the generation above minimum run that needs to be ramped down/up in the next hour is subject to the ramp rate. In particular, constraints (12) and (13) are more conservative than a normal ramp-rate constraint: (12) is showing the ramp-up constraint is assuming the awarded operating reserve $\text{gr}_{k,h}$ will be activated in hour h; while (13) is assuming the awarded operating reserve $\text{gr}_{k,h-1}$ has been activated in hour $h-1$.

Constraint (14) is a type of shut-down-ready constraint: if at hour h, a part of the minimum run will be shut down, and the corresponding capacity must be operated at the minimum run (equals to $\text{gsd}_{k,h}$) in hour $h-1$. Furthermore, the remaining part of the generation in hour $h-1$ (i.e., $\text{g}_{k,h-1} + \text{gr}_{k,h-1} - \text{gsd}_{k,h}$) is subject to the remaining part of capacity $\text{pmin}_{k,h}/\text{QMIN}_k$. Similarly, in constraint (15), if some capacity is started up at h (i.e., $\text{gsu}_{k,h} > 0$), then the maximum electricity provided (excluding the newly started up capacity) by this generator is actually the maximum capacity at $h-1$.

Constraints (16) and (17) are the minimum uptime and downtime constraints. In particular, (16) is showing the minimum run at hour h cannot be higher than the total minimum run minus the minimum runs that are just shut down (i.e., $\text{gds}_{k,h'}$ where $h\text{-MDT}_k \le h' \le h$). The similar deduction can be made for constraint (17).

Constraints–Storage Operation

$$\text{sl}_{esk,h+1} = \text{sl}_{esk,h} + \eta_{esk}\text{sch}_{esk,h} - \frac{1}{\eta_{esk}}\text{sdc}_{esk,h} \quad \forall esk, h \tag{18}$$

$$\text{sl}_{esk,h} \le \text{ML}_{esk}\text{sav}_{esk} \quad \forall \text{esk}, h \tag{19}$$

$$\text{MD}_{esk}\text{sch}_{esk,h} + \text{MC}_{esk}\left(\text{sdc}_{esk} + \text{sr}_{esk,h}\right) \le \text{MD}_{esk}\text{MC}_{esk}\text{sav}_{esk} \quad \forall esk, h \tag{20}$$

$$\text{sl}_{esk} \ge \frac{1}{\eta_{esk}}\left(\text{sdc}_{esk,h} + 0.5\,\text{sr}_{esk,h}\right) \quad \forall esk, h \tag{21}$$

Constraints (18)–(21) are for storage operation simulation. Constraint (18) is tracking the state of charge of the storage, and (19) is the state-of-charge upper limit. Constraint (20) is a tight constraint for storage output upper limit and is particularly useful to mitigate the situation where charge and discharge are simultaneously non-zero, and discharge and charge capacities are different. If both are non-zero, (20) will make sure they limit each other since they shared the power capacity. If one of them is zero, (20) will become the capacity constraint of the other: for instance, if $\text{sch}_{esk,h} = 0$, (20) becomes $(\text{sdc}_{esk,h} + \text{sr}_{esk,h}) \le \text{MD}_{esk}\ \text{sav}_{esk}$. Constraint (21) guarantees that the state of charge is enough for generation and a half-hour activation of any operating reserve capacity that storage has provided to the market.

Constraints–Transmission

$$\sum_{k} \mathrm{GBI}_{k,i}\, g_{k,h} + \sum_{esk} \mathrm{SBI}_{esk,i}\left(\mathrm{sdc}_{esk,h} - \mathrm{sch}_{esk,h}\right) + \sum_{l} \mathrm{LBI}_{l,i}\, \mathrm{pf}_{l,h} + \mathrm{nl}_{i,h} - \mathrm{D}_{i,h} = 0 \quad \forall i, h \tag{22}$$

$$\left|\mathrm{pf}_{l,h}\right| \leq \mathrm{TM}_{l} \mathrm{lav}_{l} \quad \forall l, h \tag{23}$$

$$\left|\mathrm{pf}_{l,h} + \mathrm{PBASEB}_{l} \sum_{i} \mathrm{LBI}_{l,i} \theta_{i,h}\right| \leq \mathrm{BM}_{l}(1 - \mathrm{lav}_{l}) \quad \forall l, h \tag{24}$$

$$\left|\sum_{i} \mathrm{LBI}_{l,i} \theta_{i,h}\right| \leq \frac{\pi}{6} \quad \forall l, h \tag{25}$$

Constraints (22)–(25) are the network constraints. Constraint (22) is Kirchhoff's Current Law and (23) is the thermal limit constraint, which may also reflect security-based limits where such limits are tighter. Constraint (24) is Kirchhoff's Voltage Law and (25) limits the phase angle difference on the transmission line.

Constraints—Operating Reserve and RPS

$$\sum_{i} \mathrm{BAI}_{i,a}\left(\sum_{k} \mathrm{GBI}_{k,i}\, \mathrm{gr}_{k,h} + \sum_{esk} \mathrm{SBI}_{esk,i}\left(\mathrm{sr}_{esk,h} + \mathrm{sdc}_{esk,h}\right)\right) \geq \mathrm{RM}_{a} \sum_{i} \mathrm{BAI}_{i,a} \mathrm{D}_{i,h} \quad \forall a, h \tag{26}$$

$$\sum_{k,h} \mathrm{HW}_{h} \mathrm{RE}_{k,stt}\, g_{k,h} + \sum_{rp} \left(\mathrm{RIM}_{rp,stt} - \mathrm{REX}_{rp,stt}\right) \mathrm{rt}_{rp} + \mathrm{nr}_{stt} \geq \mathrm{RPS}_{stt} \sum_{i,h} \mathrm{HW}_{h} \mathrm{BSI}_{i,stt} \mathrm{D}_{i,h} \quad \forall stt \tag{27}$$

$$\sum_{k,h} \mathrm{HW}_{h} \mathrm{RE}_{k,stt}\, g_{k,h} - \sum_{rp} \mathrm{REX}_{rp,stt} \mathrm{rt}_{rp} + \mathrm{nr}_{stt} \geq \mathrm{IRPS}_{stt} \mathrm{RPS}_{stt} \sum_{i,h} \mathrm{HW}_{h} \mathrm{BSI}_{i,stt} \mathrm{D}_{i,h} \quad \forall stt \tag{28}$$

Constraint (26) is the operating reserve constraint, and constraints (27) and (28) are the RPS constraints. In particular, constraint (28), as the in-state RPS requirement constraint, shows the total local generation minus all the exported renewable credits has to be larger than the in-state RPS requirement.

4 Example

In this section, an example of co-optimization of transmission, generation, and storage is presented. This example is based on a 54-node network aggregated from the system of Western Electricity Coordinating Council (WECC) in the USA, and the planning target year is 2034. The network data are from the WECC 2026 Common Case (WECC 2026), and we plan for year 2034 based on the load, fuel cost, and policy data that are specified by WECC's Long-Term Planning Tool (WECC 2013). We shall use this example to demonstrate that anticipation of storage siting/sizing decisions can change the transmission expansion plan and this change to the plan can provide considerable economic benefits.

4.1 Test Case Description: 54-Node System for WECC

In this subsection, the test system, the 54-node system for WECC is summarized.

All 54 nodes are aggregated from the 2026 Common Case of WECC (2026). Each node stands for one or part of single Transmission Expansion Planning Policy Committee (TEPPC) subarea of WECC. When one TEPPC subarea is totally within one state, one node will be designated; when one TEPPC area has assets spanning several states, e.g., the Los Angeles Department of Water and Power (LADWP), several nodes will be designated, and one node will be defined for each state (see Fig. 2, where LADWP has nodes in States of California, Nevada and Utah).

There are 519 aggregated existing generators and 238 generator candidates in this network. These span 25 technologies, including different types of coal, gas, nuclear, hydro, wind, solar, geo, and biomass generation.

As for generation candidates, on each node, two types of generation can be invested without limit: Gas Combustion Turbine and Gas Combined Cycle. On the other hand, the renewables, i.e., wind, solar, bio, and geothermal, can only to be expanded at 53 candidate sites and will need new transmission lines to be interconnected with the existing grid. The 53 candidate sites (not the same as nodes) and their maximum installed capacity are identified in (Western Governors' Association and U.S. Dept. of Energy 2009). A system-wide view of the building cost and the expandable capacity is shown in Table 1.

There are two types of transmission lines: backbone reinforcements and renewable connections. Backbone reinforcement candidates, which are 39 in number, expand capacity on the arcs shown in Fig. 2. In addition, there are 53 renewable connection candidates, corresponding to the 53 renewable candidate sites. All of the transmission capacity expansion costs are calculated based on the length and the voltage level of the buses in the original network. The average line cost is 640 Million $/line, with a lifetime of 60 years. Assuming a 5%/year discount rate, the average annualized cost of transmission lines is about 34 million$/line-year.

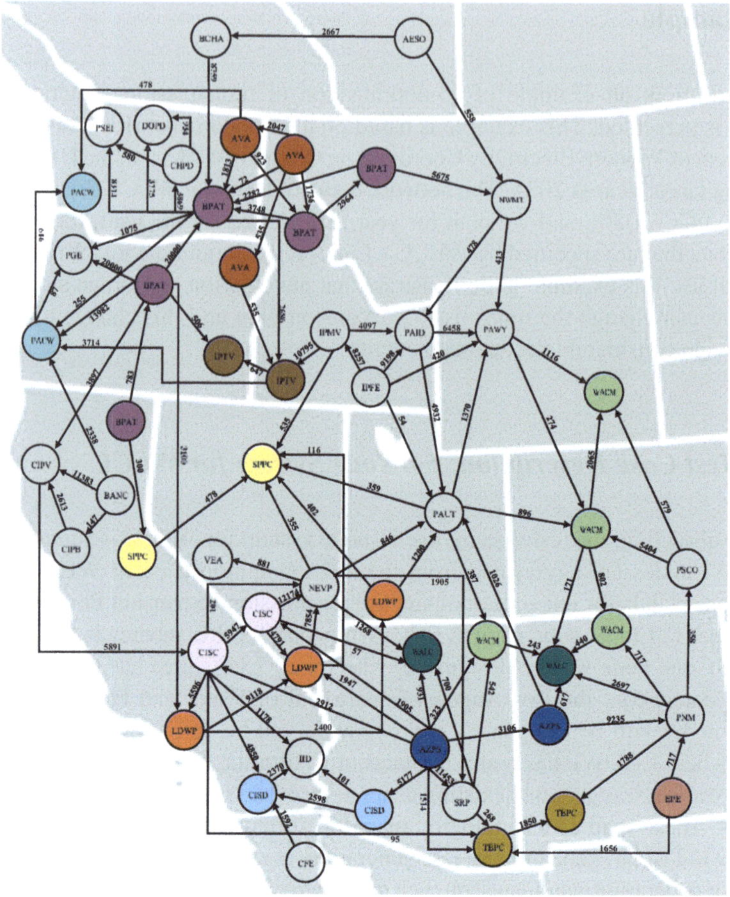

Fig. 2 Map of the test system. Colors represent different TEPPC subareas

The type of storage we consider is a battery electric storage system (BESS). We assume that a BESS will have 4-h of storage using Li-ion technology with round-trip efficiency of 92%. The build cost is assumed to be \$440/kWh in the year 2034 (i.e., \$1760/kW); with assumptions of 15-year lifetime and 5% discount rate, this corresponds to an annualized cost of \$42.5/kWh-year. Storage can be sited (1) at any of the 54 existing nodes in the system or (2) co-sited with the renewables at the 53 candidate renewable sites. Different siting locations will incur different fixed operation and maintenance costs (FOM cost), with the average being \$30/kW-year. More details on the cost assumptions can be found in WECC (2017). Storage is expandable up to a capacity of 1000 MW at each location.

There are four representative days that are selected, and each day is composed of 24 h. Thus, 96 h are simulated to represent the variability of load and renewable output conditions.

Table 1 System-wide Expansion Cost Assumptions for Generation in Year 2034

Gen. type	Fixed O&M ($/kW-year)	Overnight build cost ($/kW)	Lifetime (year)	Annualized build cost ($/kW-year)[a]	Potential capacity (MW)[b]	Capacity factor[c]
Biomass	120	4300	20	345.04	3272	–
Combined Cycle	10	1213	20	97.33	–	–
Combustion Turbine	9	825	20	66.20	–	–
Geothermal	120	5000	25	354.76	4719	–
Solar PV	20	1471	35	89.82	85144	26.0%
Onshore Wind	40	1355	20	108.72	95288	30.6%

[a]Assumes a 5% discount rate
[b]Summation over all candidate sites
[c]Weighted average over all candidate sites, weights are the potential capacity

We assume that future policies in the WECC region will incentivize significant increases in renewable generation. There are two types of environmental policies that are assumed to affect the system in the year 2034: Renewable Portfolio Standards (RPS) and Carbon Pricing. The RPS data for year 2034 are from the DSIRE [Database of State Incentives for Renewables and Efficiency, (N.C. Clean Energy Technology Center 2018)], and the demand data are from LTPT (WECC 2013) from WECC. RPS policies are implemented on the state-level, and we consider the fact that some states have in-state requirement. For examples, in 2034, California requires 60% of its demand to be supplied by renewables, and 90% of the renewables should come from within the State. Overall, in 2034, the WECC system requires 38% of its demand (1091 TWh/year) to be supplied by renewables; and for the USA part of the WECC, this requirement is 34% of a total energy demand of 854 TWh/year. The non-compliance penalty is assumed to be $100/MWh, which is imposed in the objective function if a given state's RPS is not met.

For carbon pricing policy, we assume a universal carbon tax will be implemented upon the WECC system (or equivalently, a carbon cap-and-trade system is implemented within WECC, and the carbon price reaches the assumed equilibrium level). The carbon tax varies among the different study cases we consider.

In the application of this chapter, we omit the voltage law constraint in the network representation in order to accelerate solution times. Our numerical experiments indicate that this assumption results in a minor overstatement of the network's transfer capability and results in only minor distortions in near-term transmission investments (Xu and Hobbs 2019). Thus, the power flow is a "pipe-and-bubbles" (transshipment) formulation. Furthermore, binary variables for both transmission and storage expansion are relaxed (i.e., are continuous in the range [0,1] rather than binary), again in the interest of faster computation times. In its use of continuous variables, the model resembles classical generation expansion planning models, which are formulated as linear programs. More realistic models can be used in an actual planning, but this model suffices for the purpose of this chapter which is to illustrate the use of co-optimization.

4.2 Questions to Be Answered and the Experimental Design

With the numerical results from the application of the above model and data, we shall answer the following questions:

- *Would the anticipation of the amount and siting of battery storage change the transmission expansion decisions and how? Will the electric storage incentivize more or less capacity expansion of transmission?* Less transmission indicates that, overall, batteries and transmission are substitutes; more would indicate that they are complements.
- *What is the economic value of enhancing the TEP model to include storage (VoMES)? And how will the VoMES change with the build cost of the storage?* Note that this is the not, per se, the benefit of storage itself, which is VoS, equal to the difference in cost between MP1 and the naïve model without any storage at all MP2. Rather, VoMES is the benefit of "smart TEP with storage," anticipating where storage will be sited and adjusting transmission decisions to take advantage of that; as explained at the end of Sect. 2, this is the difference between MP1 and MP3's objective function values.
- *Will the stringency of carbon prices impacting electricity markets change VoMES? I.e., if the carbon price is applied to the system, will anticipating the siting of storage be more or less valuable to the TEP?*
- *What are the sources of cost savings from proactive TEP? In particular, when there is a positive VoMES, were the cost savings from investment in transmission or generation, or from reduced fuel or carbon costs?* Ignoring the storage in transmission expansion planning will change the transmission expansion plan and may consequently incentivize investors to make suboptimal siting and the operating decisions—which of those will be distorted more? It is also conceivable that transmission costs will also increase; perhaps disregarding the possibility of storage in model MP2 will result in overbuilding of transmission versus that optimal TEP from model MP1, which might find that transmission and storage substitutes. That would indicate that, overall, transmission and storage are substitutes. On the other hand, reduced investment in T in MP2 (no storage S) would indicate that T and S are instead complements.

We design the experiments as shown in Table 2 to answer the questions above.

Table 2 Experimental design for value of storage in TEP: Sets of model runs

Set ID	Set name	Planning model description
MP1	TEP with storage and generation expansion	10 levels of build cost of storage (from 100% of base level \$42.5/kWh-year to 10% of base level); 10 levels of WECC-wide carbon tax from \$0 to \$90/t. There are a total of $10 \times 10 = 100$ runs
MP2	TEP with generation expansion	10 levels of WECC-wide carbon tax from \$0 to \$90/t. There are 10 runs
MP3	Storage and generation expansion	Same as Set MP1; except that transmission expansion plan is fixed at the levels selected in MP2 with the same carbon tax. There are $10 \times 10 = 100$ runs

4.3 *The Impact of Storage on Transmission Expansion Plans*

In this section, we show how the storage expansion would affect the transmission expansion plan. Several conclusions can be drawn from the detailed results below:

(1) Anticipation of storage siting/sizing will change the transmission expansion plan. An example is given in Fig. 3, where cheaper storage results in more line construction in some places (substitution relationship) and less in others (complementary relationship);
(2) The greater the level of carbon tax that is applied to the system, the more impact the storage expansion anticipation will change the transmission expansion plan;
(3) Storage expansion anticipation can both encourage and discourage transmission expansion, with complement effects dominating under some assumptions and substitution effects in other; and finally,
(4) The way that the transmission expansion plan changes differs between types of transmission candidates, i.e., backbone reinforcement and renewable interconnectors. While the interactions between the backbone reinforcement and storage expansion are mixed and location dependent, the interaction between the renewable interconnectors and the storage expansion is more clear and is larger in magnitude: (a) while carbon cost is low, storage substitutes for renewable interconnectors, while (b) when carbon cost is high, then as the BESS cost is decreased, storage first substitutes for renewable interconnectors and then complements them.

Now, we shall examine the numerical result more closely.

Figure 4 shows the difference between MP1 and MP2's investment in the backbone reinforcements (on inter-regional lines) in 33 (out of 110) study cases: carbon tax = \$0, 60, 80/t CO_2e, and battery cost ranges from \$42.5 to \$4.25/kWh-year. The capacity of all new backbone lines, in MW, is added up to create this index. The figure shows that in cases where carbon tax = \$0/t, anticipating storage expansion does not change the total backbone reinforcements from the "No BESS" case. The

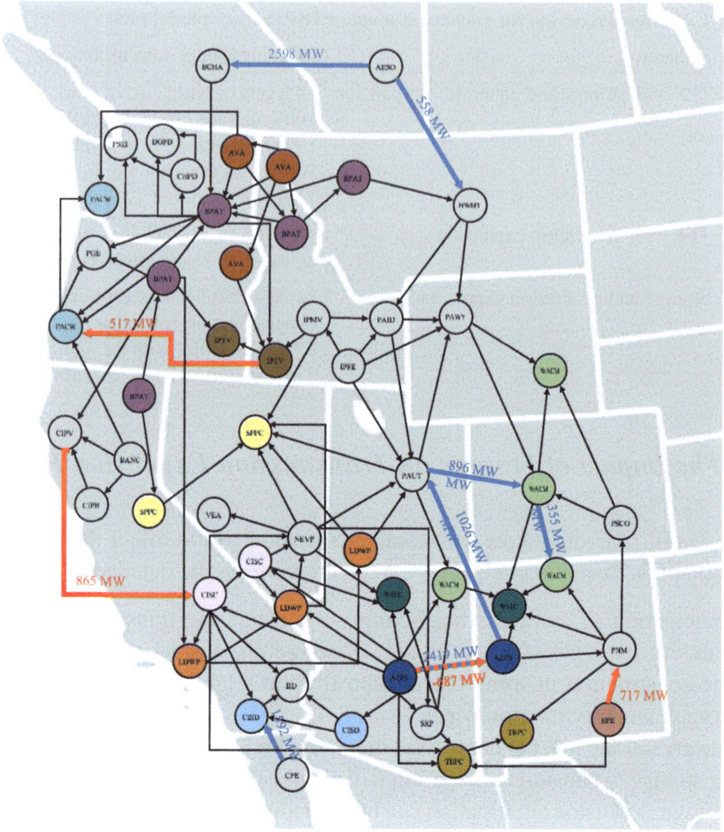

Fig. 3 Map of Backbone Reinforcement Expansion: Comparison between battery costs of 100% of the base case level ($42.5/kWh-year) and 10% of that level. Blue lines represent the expansion plan at a battery cost level of 100%, and solid red lines are additional lines included in the expansion plan when the battery cost level becomes 10%. Note the additional lines expanded between Idaho and Oregon, Northern and Southern California, and within Southern New Mexico when the battery cost is decreased; meanwhile, one line between Arizona and New Mexico is canceled (dashed line). Carbon Tax is $80/t CO_2e

locations of additions do not change either. On the other hand, the results show some impact when carbon price is high, and the battery cost is lower, in particular, when carbon price is set to $80/t CO_2e, considering storage expansion can cause both the addition and the cancelation of lines, depending on the cost of batteries. So, whether backbone lines and storage or complements depend on battery cost assumptions, and surprisingly, this effect is nonmonotonic. Under the highest carbon cost, the magnitude of the effect does not increase uniformly as battery cost falls, and the direction of the effect changes twice as that cost is adjusted.

We now turn to locational effects. Figure 5 is a zoom-in of the case of carbon tax = $80/t CO_2e in Fig. 4. When the 4-h battery cost dropped from 40 to 30%

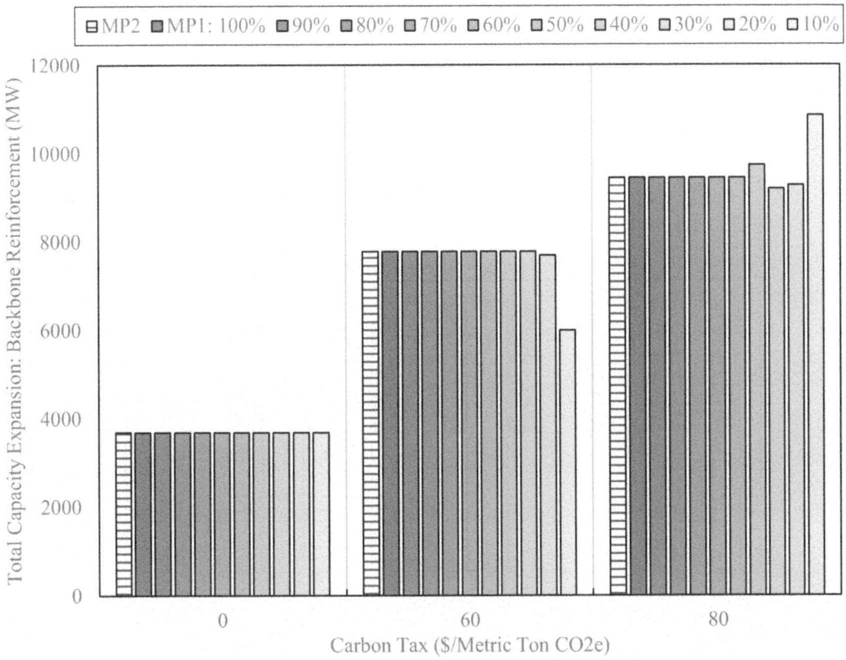

Fig. 4 Transmission capacity expansion (backbone reinforcements only) by proactive TEP models MP1 with different BESS costs compared to the result of the TEP model with "No BESS" MP2 (Energy Storage Cost at 100% = $42.5/kWh-year)

(corresponding to $16.98 and $12.74/kWh-year, respectively), one line from Arizona to New Mexico is canceled; while the battery cost goes lower, several line capacities are added to the system, encouraged by the storage expansion. The locations of those additions are scattered throughout the west, some near load centers (California) and others closer to renewable solar resources (New Mexico). This is essentially showing that the storage system can both substitute (in cases where lines are canceled because of lower storage cost) and complement (in cases where lines are built because of lower storage cost) the transmission expansion.

When we turn from backbone line expansion to renewable interconnections, the story goes in a similar direction but with a much larger magnitude. Renewable interconnectors are the lines necessary to deliver new renewable developments to market. The expanded capacity of those interconnectors is much higher than the backbones. For instance, backbone reinforcement expansion ranges from 3.7 to 11 GW, while for renewable interconnectors, the range of additions is 31–86 GW. This much higher expansion of interconnectors reflects the impetus toward renewable development throughout the west resulting from our assumed renewable and carbon policies as well as declining costs of renewables. Figure 6 shows that anticipation of the storage expansion can both discourage or encourage interconnector expansion. We highlight that in both cases with carbon tax = $60 and $80/t CO_2e, lower battery costs will first

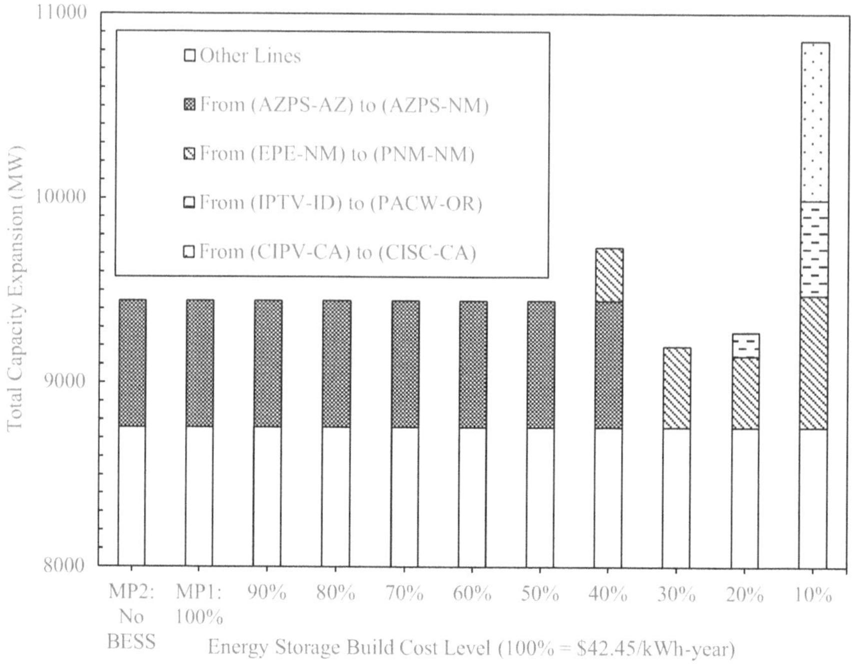

Fig. 5 Transmission capacity expansion of backbone reinforcements selected by models with carbon tax = $80/t CO_2e in the year 2034

slightly complement the renewable interconnector expansion (expanded capacity is slightly higher when battery costs go lower), then substitute for expansion (expanded capacity is lower with battery cost goes lower), and then, reverse again, returning to a complementary effect.

We can intuitively understand how the storage can substitute for interconnector expansion: you either transport the excessive energy out for consumption, i.e., transmission expansion, or save it for later, i.e., storage expansion, and the model (and assumedly the market) will choose the most economical approach. Meanwhile, in cases where the storage expansion encourages renewable interconnectors, the reason is basically that the cheaper storage makes some originally uneconomical intermittent power become economical and worthwhile to be connected. An example is solar in New Mexico that is only available but very strong in the middle of the day; it is not developed at all in high battery cost cases, but at some levels of battery costs, we see expansion of that renewable source. In one case where carbon price is at $80/t CO_2e and battery cost is at 10% of the base level, a 1000 MW BESS is co-sited with a 1575 MW Solar PV facility at a renewable candidate site at Southwestern New Mexico, and a transmission line with 850 MW capacity connects both of them to a main grid node at El Paso Electric (EPE) at New Mexico; however, none of these lines are invested in when battery cost is above 20% of base level.

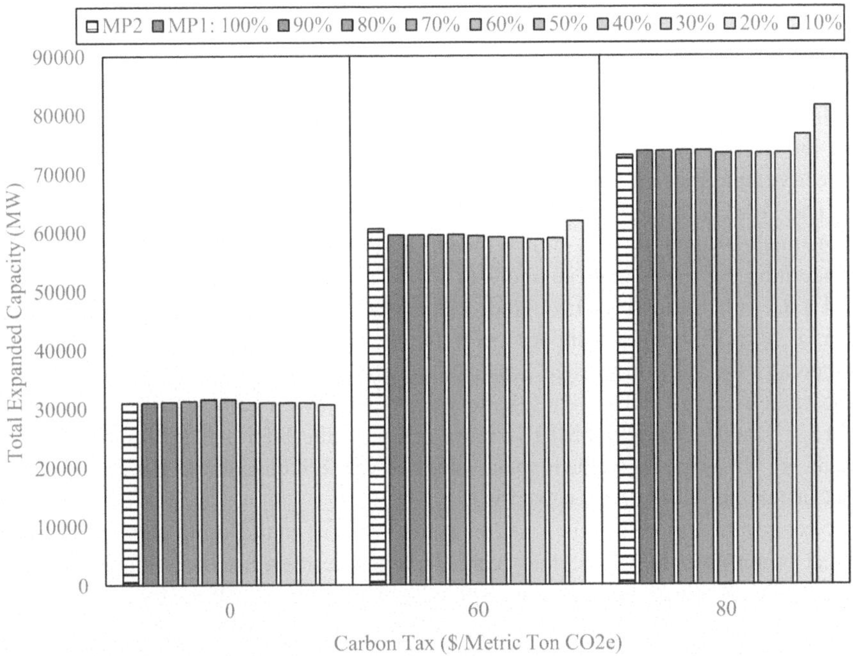

Fig. 6 Transmission capacity expansion of renewable interconnectors by proactive TEP models MP1 with different BESS costs compared to the result of the TEP model with "No BESS" MP2 (Energy Storage Cost at 100% = 42.5/kWh-year)

Overall, we see from the results that anticipation of storage expansion will change the transmission expansion plan from our TEP model, sometimes encouraging it, and at other times the opposite. How much does this anticipation, with the resulted expansion change, benefit us? Or equivalently, if we transmission planners ignore storage siting/sizing while making the plan, what is the cost we will bear? As was explained in Sect. 2, this benefit/cost is called VoMES, and the value of TEP model enhancement to proactively anticipate storage will be discussed next.

4.4 *Value of Considering Storage in Co-optimized Transmission Expansion Planning*

In this subsection, we calculate the value of storage in transmission expansion planning VoMES. As a reminder, we first plan transmission expansion T anticipating both generation G and storage S investments (MP1); second, we plan the transmission expansion without considering storage (MP2, having only T and G as variables); finally, we plug the resulting naïve plan from MP2 into a co-optimization model that includes storage expansion to simulate the reaction from the market to the naïve transmission plan (MP3, optimizing S and G, but freezing T at MP2's levels). The intent of VoMES is to simulate the efficiency loss resulting from situation that transmission expansion planner naively ignores the possibility of storage investment, as well as the reaction of storage siting and operation to transmission reinforcements, but the storage investors still have the chance to react. The difference between the objective function values of MP1 and MP3 is this index.

The VoMES in TEP in all 100 test cases are shown in Fig. 7, and the amount of investment for new lines is shown in Fig. 8. Two basic observations can be made concerning the trends in these figures.

First, with the carbon tax fixed at a certain level, VoMES is monotonically increasing as the battery cost goes lower. In other words, the lower the battery cost is, the greater the value of storage expansion anticipation is the transmission planners. The value is zero for the highest battery costs and lowest carbon costs, because no storage is added by model MP1 in those cases, so the MP1 and MP3 solutions are identical. Unsurprisingly, the highest values of VoMES are associated with solutions that install the most battery capacity.

Second, the carbon tax is a factor in the value of anticipating storage, but the effect is not monotonic. In other words, a higher carbon tax does not necessarily make VoMES higher. For example, when the battery cost is half the base level (50% case), as the carbon tax goes higher, the VoMES will first go down then up.

To help interpret the magnitude of VoMES, first, we compare it to the incremental transmission investments. Their ratio gives an indication of the relative importance of incorporating the proactive/anticipative perspective in planning. Figure 8 shows the transmission expansion cost in all 100 MP1 test cases as well as the 10 MP2 cases that is without the storage siting. Sixty eight out of 100 MP1 test cases have lower transmission expansion investment costs than the corresponding the "No BESS" case, implying that anticipating storage results in less transmission investment (substitution effect). In the remaining 32 cases, proactive planning including storage results in more transmission (complementary effect). The ratios of VoMES to the MP1 transmission investments are shown in Fig. 9. This shows that the value of proactive planning that recognizes storage is a significant fraction of total transmission investment under the higher carbon cost assumptions and lower battery costs, which are the runs that have the most battery investment.

Although how carbon policy will affect the transmission is largely out of the scope of this chapter, Fig. 8 also shows that carbon policy has more impact on the transmission expansion than the storage expansion, the major topic here.

The overall value of storage to the system (VoS) results is shown in Fig. 10. As pointed out in the Sect. 2, the larger VoMES is (as a proportion of VoS), the stronger the impact that naïve transmission expansion decisions (which disregard storage reactions) will have upon the final realization of the economic value of storage. Among all the test cases, VoMES is about 0–27% of the VoS, and the average is about 14%. Thus, anticipating how storage siting and amounts will react to grid expansion can significantly enhance the value of storage.

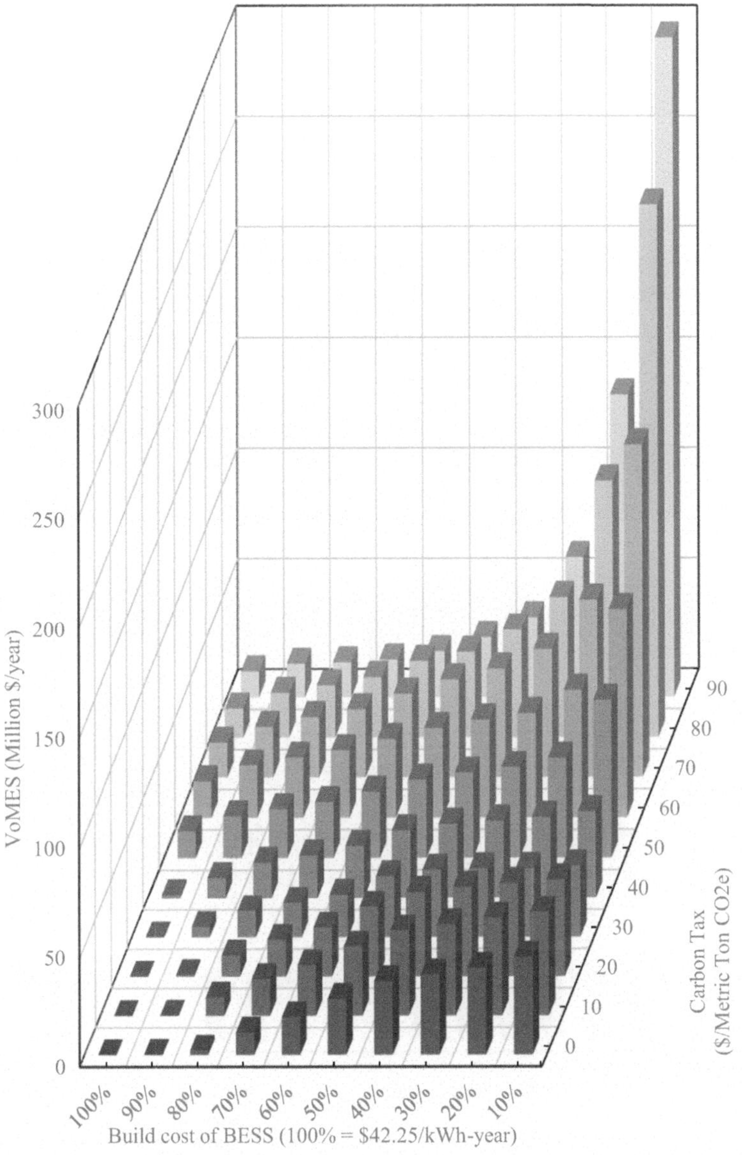

Fig. 7 VoMES in TEP in different test cases

4.5 Sources of VoMES in Transmission Planning

We have seen that anticipating the sizing/siting of the storage will change the transmission expansion and this change will provide an economic benefit (VoMES in TEP)

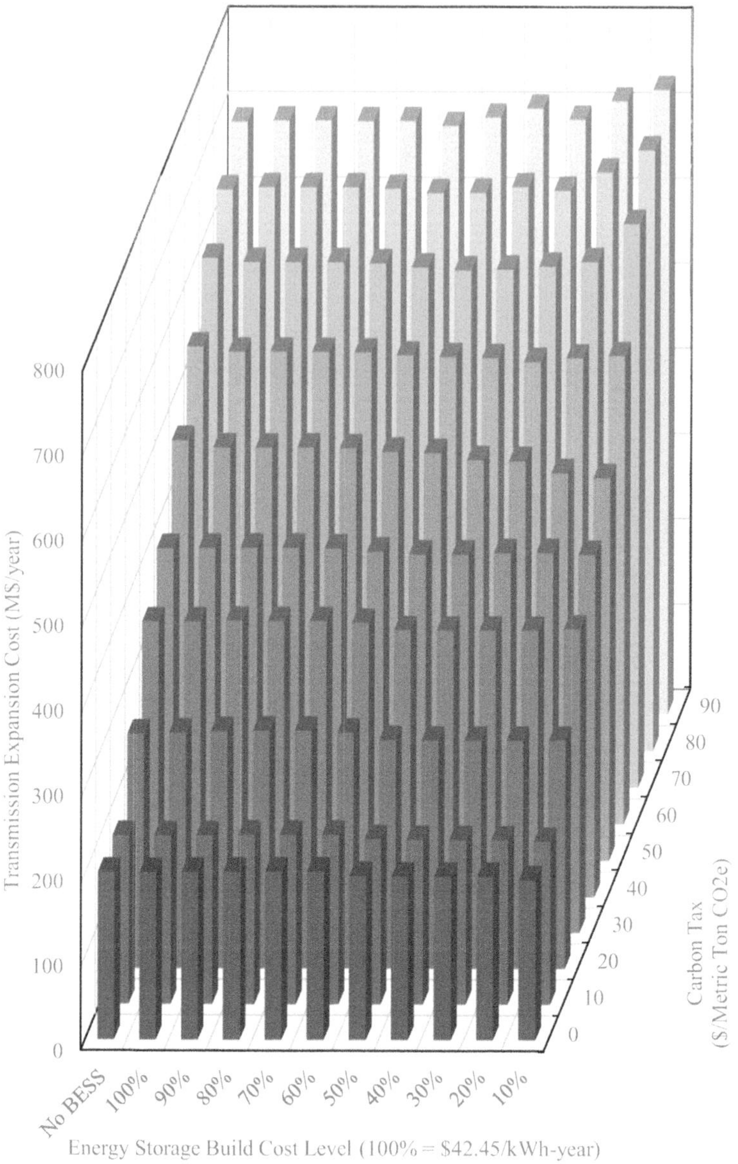

Fig. 8 Backbone and Renewable Interconnection Transmission Investment Cost in TEP in different test cases

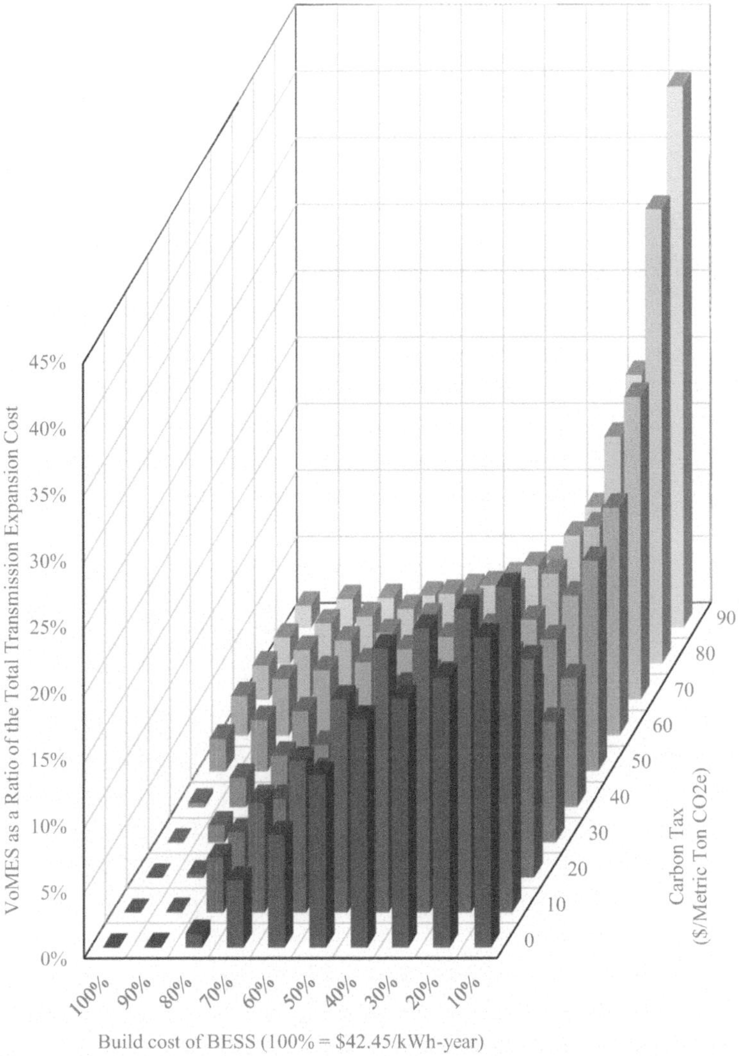

Fig. 9 VoMES as a Ratio of total Transmission Expansion Cost

to transmission expansion planners. To understand why, it is important to examine the sources of the VoMES, in terms of whether it is reduced investment (and if so, of what type) or reduced operating costs. Is VoMES positive because given the changed transmission plan, the market will react with different generation/storage expansion, or are those investments relatively unchanged and it is transmission investments that shift? Is most of VoMES comprised of fuel and carbon cost savings, or do capital cost savings contribution a large portion? We shall see the source of VoMES in the figures as follows.

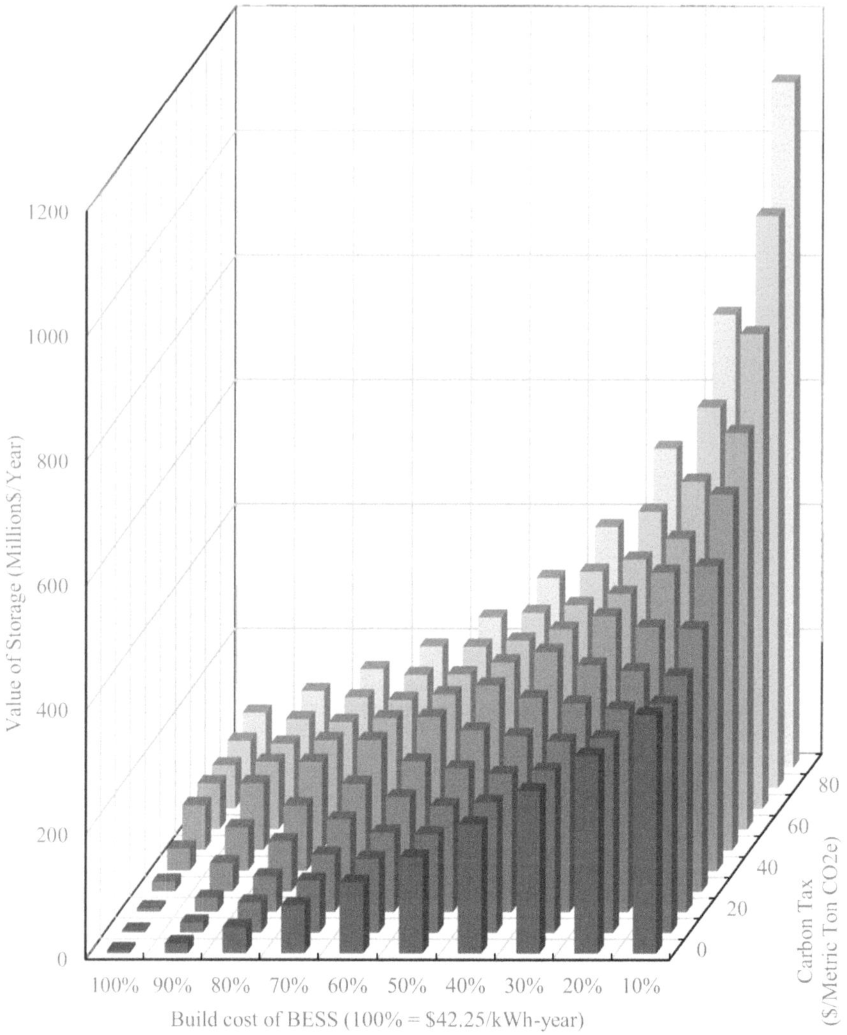

Fig. 10 VoS in TEP in different test cases

Figures 11, 12, and 13 show the components of VoMES for 60 different test cases (one figure per carbon price = \$0, 60, 80/t CO_2e, and within each figure, BESS costs from 100% level to 10%). As a reminder, VoMES is calculated by taking the difference between two objective functions: (1) the objective of MP1, i.e., TEP with generation–storage anticipation and (2) the objective of MP3, i.e., generation/storage expansion simulation with transmission expansion fixed from the "No BESS" case (MP2). Here, we now consider the differences in individual sets of objective function terms, shown in Eq. (1) in Sect. 3. The five components we break out are the separate investments in transmission, generation, and storage; fuel and variable O&M costs

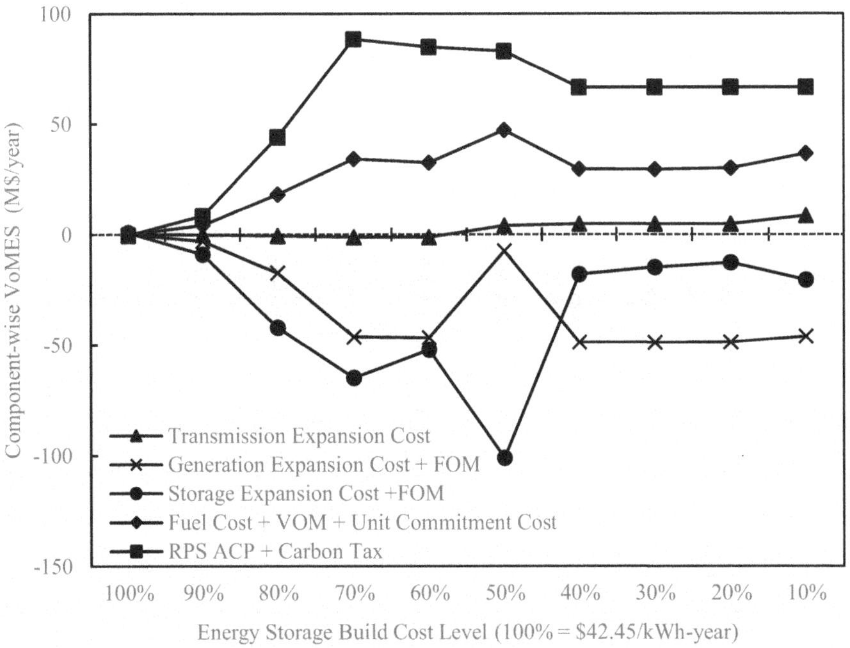

Fig. 11 Component-wise VoMES in TEP, carbon tax = $0/t

of generation (excluding carbon costs); and environmental terms, namely the carbon tax and any penalties ("ACP") associated with non-compliance with the state-level renewable portfolio standards.

All three figures show the same pattern:

(1) The proactive transmission plan (MP1, which anticipates storage in TEP) is introducing more generation and storage expansion than the naïve plans (MP2, without storage anticipation), and thus, the VoMES components associated with generation and storage investments are negative. Thus, by proactively planning, transmission planners also encourage investment in generation and storage.
(2) VoMES arises mostly from savings in operating costs and policy compliance: the additional *G* and *S* investment just discussed more than pays for itself in terms of lower fuel costs, variable operation and maintenance costs, start-up and shutdown costs, carbon taxes and the RPS alternative compliance penalty.
(3) Consistent with the changes in transmission expansion cost discussed in Sect. 4.3, most scenarios have slightly more transmission investment but about a third have less investment. However, the changes in transmission investment itself are not a significant portion of VoMES.

Interestingly, these results imply that although the total amount of transmission investment does not change greatly, there is a magnification effect in which

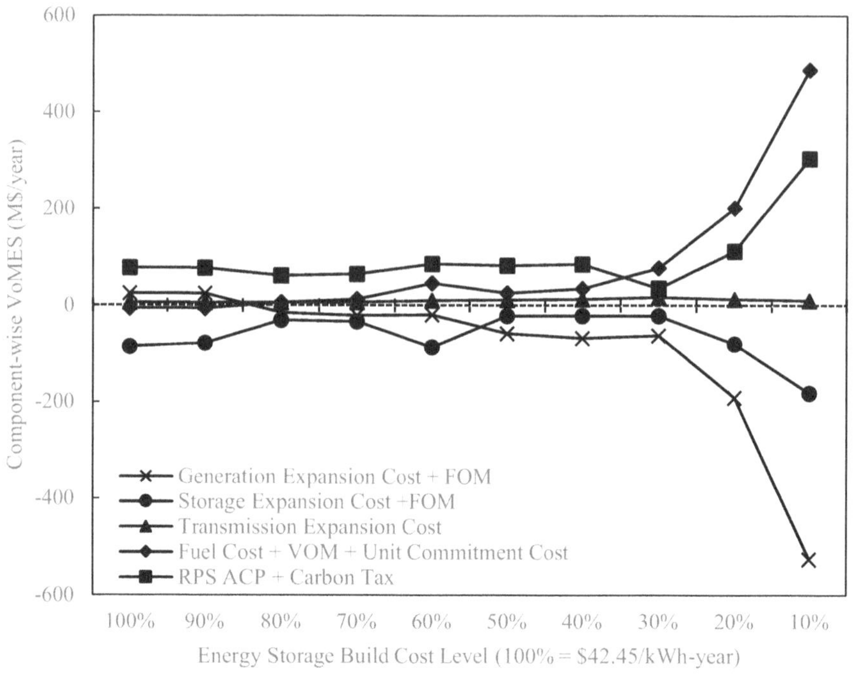

Fig. 12 Component-wise VoMES in TEP, carbon tax = $60/t

the changes that do occur in amount and location induce much larger changes in generation and storage investment.

We see an example of this impact in Fig. 14. There, the generation expansion and storage expansion gave different transmission plans. (Only wind and solar are shown in the figure because other generation expansions are minor.) MP1 is showing the optimal expansions, and while MP3 is the reaction of the market if instead, the naïve transmission plan is implemented. The results first show that in both models MP1 and MP3, solar is more impacted than wind by battery installations spurred by low battery prices. Second, they show that the effect of naïve TEP is correspondingly greater on solar investments than wind investments. Proactive TEP that anticipates storage will facilitate a roughly doubling of the amount of storage installation under low battery prices and up to a 30% increase in solar installations. There are much smaller increases in wind capacity. The reason is that solar is only available during the day, and the storage is potentially more valuable to it than the wind resource, which is distributed more evenly over all 24 h. Thus, ignoring storage expansion in TEP will undervalue the combination of solar and storage, resulting in less transmission being built for solar and, ultimately, less solar development since the ability to convey remote inexpensive solar to markets is reduced.

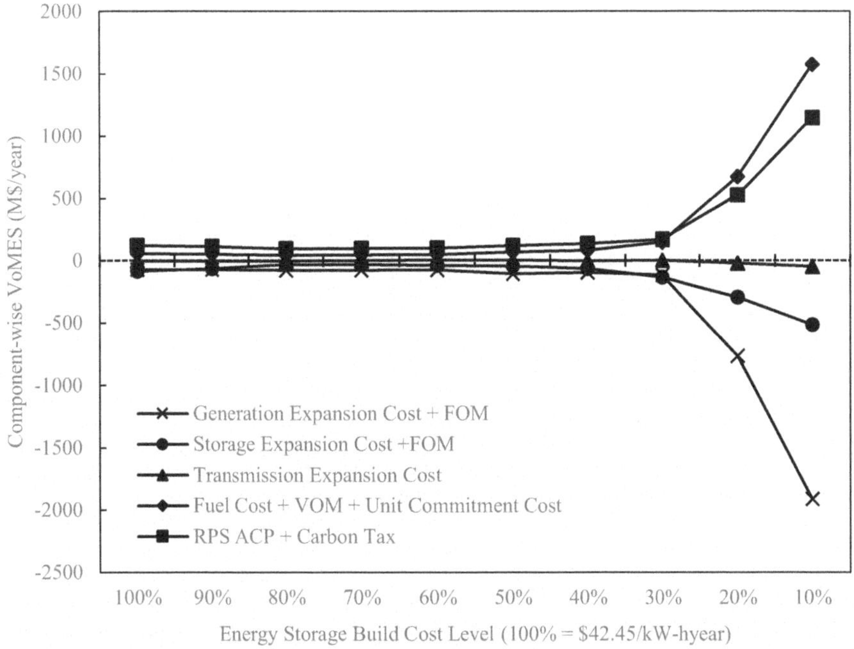

Fig. 13 Component-wise VoMES in TEP, carbon tax = $80/t

5 Conclusion

With renewable penetration increasing in many power systems, the need for transmission to bring remote renewables to market is growing, as is the need for storage. Because of the ten year or longer lead times for grid reinforcements, this transmission should be planned in a proactive manner, anticipating how generation and storage siting, amounts, types, and timing will be affected (Krishnan et al. 2015; Liu et al. 2013; Spyrou et al. 2017; Sauma and Oren 2006). Will the best plans for integrating renewables include large amounts of transmission, large amounts of storage, neither, or both? It remains to be seen. Whatever the answer is, a transmission expansion planning tool with generation and storage co-optimization will decrease the cost of renewable integration relative to naïve planning that does not anticipate how supply and storage investors will react to changes in the grid.

This chapter presents and applies a proactive transmission expansion planning model with generation–storage co-optimization, building on our previous work on transmission–generation co-optimization (Ho et al. 2016). After applying this model to the test case, we show examples to calculate the economic value of model enhancements to proactively consider storage expansion (VoMES) in TEP.

The results show that considering storage expansion in TEP will change the transmission plan by helping to identify and correct: (1) overbuilt line capacities that can be avoided by building storage, primarily near renewable energy generation locations

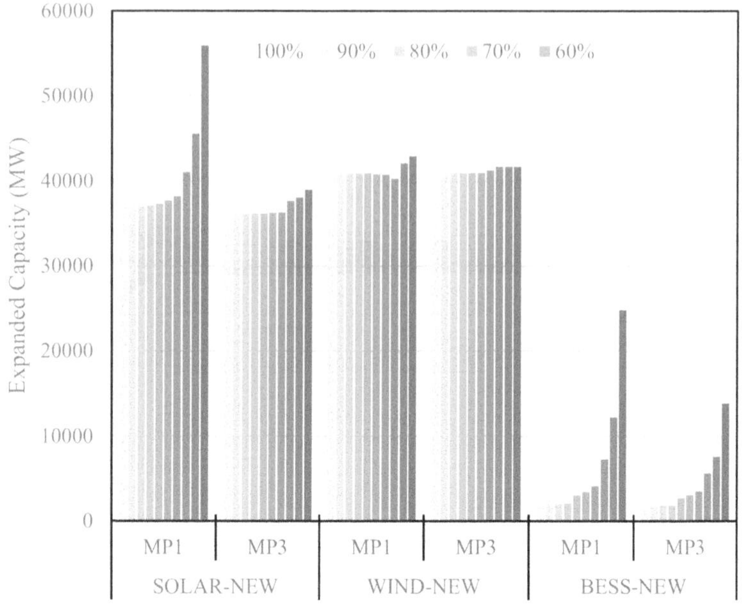

Fig. 14 Solar, Wind and Storage Expansion given transmission plans from different TEPs, MP1 is the TEP model with storage siting anticipation, MP3 is the model reoptimizing *G* and *S* given the transmission plan from model MP2 (TEP without storage anticipation). (Carbon Price = \$80/t CO2e, Battery cost at 100% level = \$42.45/kWh-year)

and (2) underbuilt line capacities that convey renewable resources that turn out to be economic only when accompanied by storage. In other words, the result shows that the storage can both complement and substitute for transmission expansion.

The VoMES in our example is primarily the net of two cost changes: the incremental investment for larger amounts of generation and storage expansion in a fully proactive TEP model and the savings that the increased investment makes possible in operating costs, such as fuel and carbon costs. Both occur because of improved transmission planning resulting from co-optimization with storage. On the other hand, a naïve transmission plan, which is the result of a planning process that disregards potential storage expansion, can discourage investment in solar generation and storage expansion.

As shown in the example, application to western USA and Canada, as storage costs are reduced in year 2034, the VoMES in TEP increases. This highlights the needs for transmission planner to consider storage expansion in the planning process. However, this VoMES is sensitive to the policies that are affecting the power system: in our case, the carbon price will affect the VoMES in TEP significantly.

To conclude, improved TEP models have value if they result in system plans with lower costs. This chapter has shown how this value can be quantified for one particular improvement, the incorporation of storage. Elsewhere, we have quantified the value of enhancing transmission models to include just generation co-optimization

(Spyrou et al. 2017) and the value of recognizing long-run uncertainties in regulatory, economic, and technology conditions. In several cases, these values are comparable in magnitude to the size of the transmission investments themselves.

References

CAISO, *Transmission Economic Assessment Methodology*. California Independent System Operator, Folsom, CA (2004). Available: www.caiso.com/Documents/TransmissionEconomicAssessmentMethodology.pdf

N.C. Clean Energy Technology Center. (2018, 04/30/2018). *Database of State Incentives for Renewables & Efficiency (DSIRE)*. Available: www.dsireusa.org/

FERC, "Order No. 841," Feburary 25, 2018

I.-C. Gonzalez-Romero, S. Wogrin, T. Gomez, Proactive transmission expansion planning with storage considerations. Energy Strateg. Rev. **24**, 154–165 (2019)

J. Ho, B.F. Hobbs, P. Donohoo-Vallett, Q. Xu, S. Kasina, S.W. Park, Y. Ouyang, *Planning Transmission for Uncertainty: Applications and Lessons for the Western Interconnection*. (The Western Electricity Coordinating Council, 2016)

M. Jenabi, S.M.T. Fatemi Ghomi, Y. Smeers, Bi-level game approaches for coordination of generation and transmission expansion planning within a market environment. IEEE Trans. Power Syst. **28**(3), 2639–2650 (2013)

S. Jin, S.M. Ryan, A tri-level model of centralized transmission and decentralized generation expansion planning for an electricity market—Part I. IEEE Trans. Power Syst. **29**(1), 132–141 (2014a)

S. Jin, S.M. Ryan, A tri-level model of centralized transmission and decentralized generation expansion planning for an electricity market—Part II. IEEE Trans. Power Syst. **29**(1), 142–148 (2014b)

S. Kasina, S. Wogrin, B.F. Hobbs, Approximations to unit commitment in planning models, in *INFORMS Annual Meeting*, Minneapolis, MN, 2013

D. Khastieva, M.R. Hesamzadeh, I. Vogelsang, J. Rosellón, M. Amelin, Value of energy storage for transmission investments. Energy Strateg. Rev. **24**, 94–110 (2019)

V. Krishnan, J. Ho, B.F. Hobbs, A.L. Liu, J.D. McCalley, M. Shahidehpour, Q.P. Zheng, Co-optimization of electricity transmission and generation resources for planning and policy analysis: Review of concepts and modeling approaches. Energy Syst. **7**(2), 297–332 (2015)

A.L. Liu, B.F. Hobbs, J. Ho, J. McCalley, V. Krishnan, M. Shahidehpour, Q. Zheng, *White Paper: Co-Optimization of Transmission and Other Supply Resources*. EISPC (2013)

MISO, *Regional Generation Outlet Study*. MISO, Carmel, IN (2010). Available: www.misoenergy.org/Library/Repository/Study/RGOS/Regional%20Generation%20Outlet%20Study.pdf

E. Morales, J.M. Latorre, A. Ramos, Tight and compact Milp formulation for the thermal unit commitment problem. IEEE Trans. Power Syst. **28**(4), 4897–4908 (2013)

F.D. Munoz, B.F. Hobbs, J.L. Ho, S. Kasina, An engineering-economic approach to transmission planning under market and regulatory uncertainties: WECC case study. IEEE Trans. Power Syst. **29**(1), 307–317 (2014)

O. Ozdemir, F.D. Munoz, J.L. Ho, B.F. Hobbs, Economic analysis of transmission expansion planning with price-responsive demand and quadratic losses by successive Lp. IEEE Trans. Power Syst. **31**(2), 1096–1107 (2016)

S. Park, Q. Xu, B.F. Hobbs, Comparing scenario reduction methods for stochastic transmission planning. IET Gener. Transm. Distrib. **13**(7), 1005–1013 (2018). https://doi.org/10.1049/iet-gtd.2018.6362

D. Pozo, E.E. Sauma, J. Contreras, A three-level static milp model for generation and transmission expansion planning. IEEE Trans. Power Syst. **28**(1), 202–210 (2013)

T. Qiu, B. Xu, Y. Wang, Y. Dvorkin, D.S. Kirschen, Stochastic multistage coplanning of transmission expansion and energy storage. IEEE Trans. Power Syst. **32**(1), 643–651 (2017)
E.E. Sauma, S.S. Oren, Proactive planning and valuation of transmission investments in restructured electricity markets. J. Regul. Econ. **30**(3), 261–290 (2006)
R.M. Sawey, C.D. Zinn, A mathematical model for long range expansion planning of generation and transmission in electric utility systems. IEEE Trans. Power Appar. Syst. **96**(2), 657–666 (1977)
E. Spyrou, J.L. Ho, B.F. Hobbs, R.M. Johnson, J.D. McCalley, What are the benefits of co-optimizing transmission and generation investment? Eastern interconnection case study. IEEE Trans. Power Syst. **32**(6), 4265–4277 (2017)
S. Takriti, J.R. Birge, E. Long, A stochastic model for the unit commitment problem. IEEE Trans. Power Syst. **11**(3), 1497–1508 (1996)
Y. Tohidi, L. Olmos, M. Rivier, M.R. Hesamzadeh, Coordination of generation and transmission development through generation transmission charges—a Game theoretical approach. IEEE Trans. Power Syst. **32**(2), 1103–1114 (2017)
A.H. van der Weijde, B.F. Hobbs, The economics of planning electricity transmission to accommodate renewables: using two-stage optimisation to evaluate flexibility and the cost of disregarding uncertainty. Energy Econ. **34**(6), 2089–2101 (2012)
WECC, *WECC Long-Term Planning Scenario Report*. Salt Lake City, UT (2013). Available: www.wecc.biz/Administrative/WECC%202032%20Scenario%20Book%20Supporting%20Docs.pdf
WECC, 2026 Common Case Version 2.0, Salt Lake City, UT (2017a). Available: www.wecc.biz/Reliability/WECC%202026%20Common%20Case%20Version%202.0%20Release%20PackagePublic.zip
WECC and Energy and Environmental Economics, Review of Capital Costs for Generation Technologies, Salt Lake City, UT (2017b). Available: https://www.wecc.biz/Reliability/E3_WECC_CapitalCosts_FINAL.pdf
Western Governors' Association and U.S. Dept. of Energy, Western Renewable Energy Zones—Phase 1 Report (2009)
W.L. Winston, M. Venkataramanan, J.B. Goldberg, *Introduction to Mathematical Programming* (Thomson/Brooks/Cole Duxbury, Pacific Grove, CA, 2003)
Q. Xu, B.F. Hobbs, JHSMINE Formulation Version 1 (2017). Available: https://hobbsgroup.johnshopkins.edu/docs/papers/JHSMINE_Formulation.pdf
Q. Xu, B.F. Hobbs, Value of model enhancements: quantifying the benefit of improved transmission planning models. IET Gener. Transm. Distrib. **13**(13), 2836–2845 (2019)
H. Zhang, G.T. Heydt, V. Vittal, J. Quintero, An improved network model for transmission expansion planning considering reactive power and network losses. IEEE Trans. Power Syst. **28**(3), 3471–3479 (2013)

Qingyu Xu received the Ph.D. degree in environmental systems and economics from Johns Hopkins University, Baltimore, MD.

Benjamin F. Hobbs received the Ph.D. degree in environmental systems engineering from Cornell University, Ithaca, N.Y. He is the Theodore K. and Kay W. Schad Professor of Environmental Management in the Department of Environmental Health and Engineering and the Department of Applied Mathematics and Statistics, The Johns Hopkins University, Baltimore, MD. He also directs that university's Environment, Energy, Sustainability & Health Institute. Dr. Hobbs chairs the California ISO Market Surveillance Committee. He is a Fellow of INFORMS and IEEE.

A Parametric Programming Approach to Bilevel Merchant Electricity Transmission Investment Problems

Henrik C. Bylling, Trine K. Boomsma, and Steven A. Gabriel

Abstract Nowadays, electricity transmission investments are made in a liberalized market environment, in which the transmission system operator, the market, producers, and investors have different objectives. The transmission expansion problem can account for this by bilevel programming, with an investor making expansion decisions in an upper level while anticipating the result of a lower-level market-clearing. In this work, we formulate a stochastic transmission expansion problem of a merchant investor collecting congestion rents determined by the differences between nodal market prices. The bilevel program can be recast as a mathematical program with equilibirium constraints (MPEC), but does not allow for linearization and reformulation by mixed-integer linear programming. Instead, we apply a parametric programming approach that facilitates decomposition with respect to both time periods and scenarios. A numerical study illustrates its ability to solve the problem, even though standard solvers for non-linear MPECs fail.

1 Introduction

Transmission expansion in power markets may involve many players with different objectives. For instance, a system operator aims to improve the functioning of the power system, for example, through social welfare maximization or with respect to reliability of the network. Generation companies assess the effects of transmission expansions on their profits, since changes to the network topology involve changes to supply and demand. In this chapter, we take the perspective of a merchant transmission investor, i.e., a company that installs new transmission lines in order to profit

H. C. Bylling (✉) · T. K. Boomsma
Department of Mathematical Sciences, University of Copenhagen, Copenhagen, Denmark
e-mail: bylling@math.ku.dk

S. A. Gabriel
Department of Mechanical Engineering, University of Maryland, College Park, MD, USA

Department of Industrial Economics and Technology Management,
Norwegian University of Science and Technology, Trondheim, Norway

M. R. Hesamzadeh et al. (eds.), *Transmission Network Investment in Liberalized Power Markets*, Lecture Notes in Energy 79,
https://doi.org/10.1007/978-3-030-47929-9_8

from their use. We assume a power network with nodal prices, also known as locational marginal prices (LMPs) (Sorokin et al. 2012). As a producer sells its power in the node it is located and at the local LMP, flow of power to another node with a different LMP may involve profits to the owner of the line. The profits from using both existing and newly installed transmission lines consist of congestion rents defined by differences in nodal prices (Sorokin et al. 2012). In many cases, a transmission system operator (TSO) owns and operates the network and the profits translate into financial transmission rights (FTRs) which are sold in a secondary market or in an auction. The merchant-investor perspective to transmission investments is based on profits from long-term financial transmission right (LTFTR) offsetting the investment costs (Rosellón and Kristiansen 2013). Examples of this power market setup can be found in PJM, New York, California and New England (Kristiansen 2004).

As transmission expansions change the network topology, supply and demand is affected and the market adopts new LMPs. In particular, the installation of new transmission lines can connect producers to new nodes, in which the merit order and therefore the local LMP changes. To model this feedback mechanism between investment decisions and LMPs and the different objectives of the merchant and the market, we use bilevel programming. A bilevel programming problem (BPP) consists of an upper level and a lower level, often illustrated by the leader–follower paradigm (or Stackelberg game), in which a leader makes an upper-level decision while accounting for the reaction of one or more followers in the lower level. We consider the merchant investor as a leader making upper-level investment decisions while anticipating lower-level market-clearing. Our problem of long-term transmission expansions is static, but accounts for short-term dynamics of the power system, including market-clearing. Moreover, by including demand uncertainty, our problem becomes a two-stage stochastic program with recourse, with the first and second stages being the upper level and lower level, respectively.

A popular approach to solve a BPP is based on replacing the lower-level problem by its Karush-Kuhn-Tucker (KKT) optimality conditions, assuming these are sufficient (Dempe et al. 2015). The resulting problem is a mathematical program with equilibrium constraints (MPEC), for which solution approaches include reformulation by mixed-integer programming (MIP), non-linear methods or heuristics. In case the BPP has linear constraints and objectives, a widely applied method is linearization and reformulation by mixed-integer linear programming (MILP). In our case, the lower-level problem of the BPP is a linear program, meaning that the KKT-conditions are necessary and sufficient for optimality. Also, the upper-level problem has linear constraints. However, the upper-level objective function involves bilinear congestion rents, determined by products of LMPs (lower-level dual variables) and line flows (lower-level primal variables). These bilinear terms make the resulting MPEC non-linear and non-convex, and thus, difficult to solve to global optimality.

As an alternative to MILP and MPEC methods, we apply a solution approach for the merchant investor BPP that is based on parametric programming (Bylling et al. 2020). This method can solve a BPP with bilinear objective terms to global optimality. Furthermore, it facilitates decomposition with respect to both time periods and scenarios. In a numerical study, we illustrate its ability to solve the problem, even though standard solvers for non-linear MPECs fail.

The main objectives of this chapter are:

- To formulate a bilevel programming problem for transmission expansion of a merchant investor.
- To illustrate the application of parametric programming and its advantages for the transmission expansion problem.
- To obtain numerical results for a case study of electricity investments in transmission lines.

The rest of this chapter is organized as follows: Sect. 2 provides an overview of the existing literature and positions this chapter within recent research. In Sect. 3, we present the bilevel programming problem of transmission expansion and Sect. 4 describes the parametric programming approach. Numerical results are provided in Sect. 5 and Sect. 6 concludes.

2 Literature Review

The existing literature includes a number of transmission expansion problems formulated as BPPs. For instance, Conejo et al. (2016) present a bilevel transmission and generation expansion problem with market-clearing in the lower level and profit maximization at the upper level. Similarly, Garcés et al. (2009) propose a bilevel problem of a transmission planner who minimizes network expansion costs in the upper level subject to market-clearing at the lower level. Baringo and Conejo (2012) likewise, consider a joint generation and transmission expansion, but with the objective to minimize consumer payments when installing wind power units and the required network reinforcements. These references reformulate the bilevel problem to a mixed-integer linear program (MILP) via the KKT-conditions. In fact, although the upper-level objective function by Conejo et al. (2016) and Baringo and Conejo (2012) is bilinear, it can be linearized using the KKT-conditions and strong duality of the lower-level problem.

The perspective of a merchant transmission investor is proposed by Joskow and Tirole (2005). This view is taken by Maurovich-Horvat et al. (2014), who formulate a stochastic bilevel problem and compare transmission investments of a merchant investor and a TSO. Buijs and Belmans (2012) likewise present a bilevel transmission expansion problem and analyze different upper-level objectives, including the merchant's. Rosellón and Kristiansen (2013) investigate a merchant mechanism to transmission expansion, using LTFTR as incentive to construct new lines. The resulting problem becomes an MPEC, which is solved via its KKT-conditions. Since the MPEC is non-convex, the KKT-conditions may not be sufficient for optimality, and thus, the solution may not be globally optimal.

We continue to consider a merchant perspective on transmission expansion. Unfortunately, to the best of our knowledge, the structure of our problem does not allow for linearization and reformulation by MILP. For example, our problem fails to satisfy the sufficient conditions for linearization by Bylling et al. (2020). Also, the above solution methods may not solve bilevel transmission expansion problems with a bilinear objective to global optimality. In contrast, we apply a new solution method that guarantees global optimality.

For other solution methods to BPPs, we refer to the reviews by Dempe et al. (2015) and Colson et al. (2007). For a review of transmission expansion problems in general, we refer to Hemmati et al. (2013).

3 The Bilevel Transmission Expansion Problem

This section presents the bilevel programming problem of merchant electricity transmission investments. Our problem consists of two levels: a lower-level market-clearing problem and an upper-level investment problem. A nomenclature is provided in Table 7 in the Appendix.

In the lower-level market-clearing problem, we assume a perfectly competitive power market, such that producers offer generation at their marginal production cost. By further assuming inelastic demand, market-clearing can be formulated as a linear cost minimization problem; cf. Gabriel et al. (2013). In our setup, market-clearing accounts for network flow, which is modeled using a DC load flow representation. To capture short-term dynamics, we consider a number of time periods, e.g., hours, for which the power market clears. To represent demand uncertainty, we assume a discrete distribution with a finite number of scenarios. For fixed upper-level decisions, the lower-level problem decomposes into a number of subproblems, one for each time period and each scenario. The lower-level subproblem of time period t and scenario s is the following:

$$\min_{\mathbf{y}_{ts},\mathbf{p}_{ts},\boldsymbol{\theta}_{ts}} \quad \sum_{g\in\mathcal{G}} c_g y_{gts} \tag{1a}$$

$$\text{s.t.} \quad \sum_{g\in\mathcal{G}(i)} y_{gts} - \sum_{j\in\mathcal{I}(i)} p_{ijts} = d_{its} \ : \lambda_{its}, \quad i \in \mathcal{I} \tag{1b}$$

$$0 \leq y_{gts} \leq y_g^{max} \ : \mu_{gts}^{y}, \quad g \in \mathcal{G} \tag{1c}$$

$$p_{ijts} = B_{ij}(\theta_{its} - \theta_{jts}) \ : \mu_{ijts}^{p}, \quad (i, j) \notin \mathcal{J} \tag{1d}$$

$$p_{ijts} = x_{ij} B_{ij}(\theta_{its} - \theta_{jts}) \ : \mu_{ijts}^{p,\mathcal{J}}, \quad (i, j) \in \mathcal{J} \tag{1e}$$

$$-F_{ij}^{max} \leq p_{ijts} \leq F_{ij}^{max} \ : \mu_{ijts}^{F,\min}, \mu_{ijts}^{F,max}, \quad (i, j) \notin \mathcal{J} \tag{1f}$$

$$-\mathcal{F}_{ij} \leq p_{ijts} \leq \mathcal{F}_{ij} \ : \mu_{ijts}^{\mathcal{F},\min}, \mu_{ijts}^{\mathcal{F},max}, \quad (i, j) \in \mathcal{J} \tag{1g}$$

$$-\pi \leq \theta_{its} \leq \pi \ : \mu_{its}^{\theta,\min}, \mu_{its}^{\theta,max}, \quad i \in \mathcal{I} \tag{1h}$$

$$\theta_{its} = 0 \ : \mu_{ts}^{\theta,ref}, \quad i = ref \tag{1i}$$

where $\mathbf{y}_{ts} = \{y_{gts}\}_{g\in\mathcal{G}}$, $\mathbf{p}_{ts} = \{p_{ijts}\}_{i,j\in\mathcal{I}}$ and $\boldsymbol{\theta}_{ts} = \{\theta_{its}\}_{i\in\mathcal{I}}$. The objective function minimizes production costs, while the constraints (1b) balance demand and supply at each node. The constraints (1c) limit power generation by existing capacity for each generating unit. Similarly, the constraints (1d) and (1e) define the power flow on existing and candidate lines, respectively, and (1d) and (1e) limit flow by existing and potential capacity. Potential capacity depends on whether a candidate line has been installed ($x_{ij} = 1$) or not ($x_{ij} = 0$), which is an upper-level decision fixed in the lower-level problem. Finally, the constraints (1h) restrict the voltage angle at each node and (1i) define the voltage angle for some reference node of the network to be zero.

In the upper-level investment problem, the merchant maximizes profits, i.e., congestion rents less investment costs, subject to a total budget. The upper-level problem is as follows:

$$\max_{\mathbf{x},\boldsymbol{\mathcal{F}},\mathbf{p},\boldsymbol{\lambda}} \sum_{t\in\mathcal{T}} \rho_t \sum_{s\in\mathcal{S}} \phi_s \sum_{i,j\in\mathcal{I}:i<j} p_{ijts}\left(\lambda_{its} - \lambda_{jts}\right) - \sum_{(i,j)\in\mathcal{J}} \left(K_{ij}x_{ij} + k_{ij}\mathcal{F}_{ij}\right) \tag{2a}$$

$$\text{s.t.} \quad 0 \le \sum_{(i,j)\in\mathcal{J}} \left(K_{ij}x_{ij} + k_{ij}\mathcal{F}_{ij}\right) \le K^{max} \tag{2b}$$

$$0 \le \mathcal{F}_{ij} \le x_{ij}\mathcal{F}_{ij}^{max}, \quad i, j \in \mathcal{J} \tag{2c}$$

$$x_{ij} \in \{0, 1\}, \quad i, j \in \mathcal{J} \tag{2d}$$

$$\mathbf{p}_{ts} \text{ is a primal optimal solution to (1)}, \quad t \in \mathcal{T}, s \in \mathcal{S} \tag{2e}$$

$$\boldsymbol{\lambda}_{ts} \text{ is a dual optimal solution to (1)}, \quad t \in \mathcal{T}, s \in \mathcal{S} \tag{2f}$$

where $\mathbf{x} = \{x_{ij}\}_{i,j\in\mathcal{I}}$, $\boldsymbol{\mathcal{F}} = \{\mathcal{F}_{ij}\}_{i,j\in\mathcal{I}}$, $\mathbf{p} = \{p_{ijts}\}_{i,j\in\mathcal{I},t\in\mathcal{T},s\in\mathcal{S}}$ and $\boldsymbol{\lambda} = \{\lambda_{ts}\}_{t\in\mathcal{T},s\in\mathcal{S}}$. The objective function maximizes profits from installation of new lines. Profits consists of accumulated hourly congestion rents determined by the differences between nodal market prices and less fixed and variable investment costs. Constraints (2b) ensure compliance with the investment budget and the constraints (2c) limit the maximum capacity installed at each line.

4 The Parametric Programming Method

By replacing the lower-level problem of the BPP by its Karush-Kuhn-Tucker (KKT) optimality conditions, the resulting problem is a mathematical program with equilibrium constraints (MPEC). The bilinear term of the upper-level objective function makes the objective function of the MPEC non-linear. To the best of our knowledge, it is not possible to linearize this bilinear term and the problem can only be solved to local optimality by non-linear methods.

Instead, we propose a solution approach for the BPP based on parametric programming. The approach applies to a linearly constrained BPP with continuous variables at both levels, and thus, does not directly apply to the transmission expansion problem with binary variables in the upper level. For a limited number of candidate lines, however, the number of binary solutions is moderate (for $|\mathcal{J}|$ candidate lines, the number of solutions is $2^{|\mathcal{J}|}$). We therefore use the parametric programming approach in combination with complete enumeration of the of binary solutions. Our method has the advantage that it solves the bilevel problem with bilinear objective to global optimality.

In Sect. 4.1 we present the parametric programming method for a BPP with only continuous variables and in Sect. 4.2, we briefly explain the enumeration of binary solutions.

4.1 Continuous Upper Level

In this section, we fix the binary decisions $\mathbf{x} \in \{0, 1\}^{|\mathcal{J}|}$ to install candidate lines or not and consider only the continuous line capacities $\boldsymbol{\mathcal{F}} \in \mathbb{R}^{|\mathcal{J}|}$ as upper-level decision variables.

We define the upper-level feasibility set $S \subseteq \mathbb{R}^{|\mathcal{J}|}$ as the set of upper-level solutions that satisfy the upper-level constraints (2b) and (2c) and that render the lower-level problem (1) feasible.

The idea behind the parametric programming method is to parameterize the lower-level primal and dual optimal solutions by the upper-level feasible solutions, i.e.,

$$\mathbf{p}(\boldsymbol{\mathcal{F}}) \quad \text{and} \quad \boldsymbol{\lambda}(\boldsymbol{\mathcal{F}}), \quad \boldsymbol{\mathcal{F}} \in S, \tag{3}$$

such the upper-level objective function can be expressed in terms of upper-level variables only.

To inspect the optimal solutions to the lower-level problem, let the upper-level solution $\boldsymbol{\mathcal{F}} \in S$ be fixed and let B be a basis for the lower-level linear programming problem, i.e., a set of linearly independent columns of the constraint matrix. We consider the corresponding basic solution to the lower-level problem, i.e., for which the variables corresponding to columns of the basis are called basic variables and the remaining variables are called non-basic and equal zero.

The following definition stems from parametric programming (Gal 1995).

Definition 1 The *critical region* $\Lambda_B \subseteq S$ corresponding to the basis B is the set of upper-level feasible solutions for which the corresponding basic solution is optimal in the lower-level problem.

It can be shown that a critical region is a polyhedron; (cf. Gal 1995).

On each critical region, we can characterize the upper-level objective function in terms of upper-level variables only. This result follows from Bylling et al. (2020).

Proposition 1 Let Λ_B be the critical region corresponding to the basis B. Then, the bilinear term $p_{ijts}(\boldsymbol{\mathcal{F}})(\lambda_{its}(\boldsymbol{\mathcal{F}}) - \lambda_{jts}(\boldsymbol{\mathcal{F}}))$ is an affine function of $\boldsymbol{\mathcal{F}}$ on the interior of Λ_B and for all i, j, t, s.

In other words, the upper-level objective function is a piece-wise linear (but not necessarily continuous) function which is affine on each critical region. It is easy to determine the gradient of the affine functions, see Bylling et al. (2020). With the gradient and a function value of the upper-level objective function for each critical region, we can obtain an explicit expression for the upper-level objective function. Furthermore, with an affine objective function and a polyhedral feasibility set, the restriction of the BPP to a single critical region is a linear programming problem. We use this to solve the BPP.

Our strategy is to find a cover of the upper-level feasibility set by critical regions, i.e., a set of bases $\mathcal{B}$ such that

$$S = \bigcup_{B \in \mathcal{B}} \Lambda_B, \tag{4}$$

to solve the restricted problems for all critical regions in the cover and finally obtain a global optimal solution by simply comparing candidate solutions.

To find a cover of S by critical regions, we define neighboring critical regions as follows, cf. Gal (1995).

Definition 2 Two critical regions, Λ_1, Λ_2, are *neighbors* if the following holds for their corresponding bases B_1, B_2:

1. There exists an $\mathcal{F} \in S$ for which B_1 and B_2 are both optimal bases to (1).
2. It is possible to pass from B_1 to B_2 in one iteration of the dual simplex method.

By Gal (1995), the union of all neighboring critical regions forms a cover of S. Thus, it is unnecessary to consider all possible bases of the lower-level problem. Neighboring critical regions are obtained by the following algorithm by Gal (1995), based on dual simplex.

Algorithm 1 Parametric programming algorithm

Step 0 (initialization) Set $h := 0$. Given an initial upper-level solution, solve the lower-level problem (1). Store an optimal basis B_0 and set $\mathcal{B} := \{B_0\}$.
Step 1 (iteration h) If $\mathcal{B} = \emptyset$, then stop. Otherwise, set $h := h + 1$, select $B_h \in \mathcal{B}$ and set $\mathcal{B} := \mathcal{B} \setminus \{B_h\}$.
Step 2 (determine leaving variable) Let $B := B_h$. Select a basic variable that has not yet been inspected and determine if a neighbor exists. If not, return to Step 2. If all basic variables have been inspected, return to Step 1.
Step 3 (determine entering variable) Carry out an iteration of the dual simplex method with the basic variable as the leaving variable. Store a neighboring basis B_j and set $\mathcal{B} := \mathcal{B} \cup \{B_j\}$. Return to Step 1.

For further details on the parametric programming approach, see Bylling et al. (2020).

4.1.1 Decomposition

For fixed upper-level decisions, the lower-level problem of the BPP decomposes into a number of subproblems, one for each time period and each scenario. We refer to the BPP with one time period and one scenario as a BPP subproblem. We process the subproblems individually, which allows for parallel computations and is likely to provide computational advantages.

By processing a BPP subproblem, we obtain neighboring critical regions for one time period and scenario. By processing all subproblems, the union of all critical regions forms a cover of S. Observe that an optimal solution to the restricted BPP can be found at a vertex of the critical region. Unfortunately, an optimal solution to the BPP may not be found among the optimal solutions to the restricted BPP subproblems. However, the vertices of the critical region must be found among the vertices of the critical regions obtained for one time period and scenario at a time. Thus, to find an optimal solution to the BPP, we enumerate and evaluate all vertices of the critical regions of the BPP subproblems. This provides us with a global optimal solution. For vertex enumeration, we use the procedure of Avis and Fukuda (1996).

The solution algorithm is as follows:

Algorithm 2 Decomposition

Step 1 (parametric programming) Apply the parametric programming Algorithm 1 to the BPP subproblems.
Step 2 (vertex enumeration) Use vertex enumeration for each of the critical regions obtained in Step 1.
Step 3 (comparison) Collect all solutions from Step 2 and evaluate their upper-level objective function values.

As an alternative to Algorithm 2, we also propose a heuristic that omits the computationally costly vertex enumeration. In Step 2, we obtain optimal solutions to the restricted BPP subproblems.

The heuristic can be summarized as:

Algorithm 3 Heuristic

Step 1 (parametric programming) Apply the parametric programming Algorithm 1 to the BPP subproblems.
Step 2 (restricted optimization) Solve the BPP subproblems restricted to each of the corresponding critical regions obtained in Step 1.
Step 3 (comparison) Collect all vertices from Step 2 and evaluate their upper-level objective function values.

4.2 Binary Upper Level

This section outlines the combination of the parametric programming approach and complete enumeration. We simply iterate through the upper-level, binary solutions, i.e., all potential configurations of the network. For fixed binary decisions to install candidate lines or not, $\mathbf{x} \in \{0, 1\}^{|\mathcal{J}|}$, we apply parametric programming.

The procedure is as follows:

Algorithm 4 Enumeration

Step 1 (enumeration) Enumerate all binary solutions, $\mathbf{x}$.
Step 2 (parametric programming) For each solution, solve the BPP using Algorithm 1, Algorithm 2 or the Heuristic 3.
Step 3 (comparison) Collect all solutions from Step 2 and their upper-level objective function values.

4.3 Non-linear Programming

As benchmarks, we also implement a non-linear MPEC formulation and a mixed-integer non-linear programming (MINLP) formulation of the problem. These can be solved using standard software, with the upper-level variables $\mathbf{x}$ defined as binary. Since the MPEC and MINLP are non-convex, however, we can only obtain local optimality.

The MPEC formulation is derived by replacing the lower-level problem, (1), by the necessary and sufficient KKT-conditions. This formulation is:

$$
\begin{aligned}
\max\quad & (2a) && (5a)\\
\text{s.t.}\quad & (2b) - (2d) && (5b)\\
& (1b) - (1i) && (5c)\\
& c^g - \lambda_{its} + \mu^y_{gts} \geq 0 \quad \forall g, t, s && (5d)\\
& \lambda_{its} - \mu^p_{ijts} - \mu^{F,\min}_{ijts} + \mu^{F,\max}_{ijts} = 0 \quad \forall t, s, (i, j) \notin \mathcal{J} && (5e)\\
& \lambda_{its} - \mu^p_{ijts} - \mu^{\mathcal{F},\min}_{ijts} + \mu^{\mathcal{F},\max}_{ijts} = 0 \quad \forall t, s, (i, j) \in \mathcal{J} && (5f)\\
& -\mu^{\theta,\min}_{its} + \mu^{\theta,\max}_{its} = 0 \quad \forall t, s, i \neq \text{ref.} && (5g)\\
& -\mu^{\theta,\min}_{its} + \mu^{\theta,\max}_{its} + \mu^{\theta,\text{ref}}_{ts} = 0 \quad \forall t, s, i = \text{ref.} && (5h)\\
& y_{gts}\mu^y_{gts} = 0 \quad \forall g && (5i)\\
& (p_{ijts} + F^{\max}_{ij})\mu^{F,\min}_{ijts} = 0 \quad \forall t, s, (i, j) \notin \mathcal{J} && (5j)\\
& (F^{\max}_{ij} - p_{ijts})\mu^{F,\max}_{ijts} = 0 \quad \forall t, s, (i, j) \notin \mathcal{J} && (5k)\\
& (p_{ijts} + \mathcal{F}_{ij})\mu^{\mathcal{F},\min}_{ijts} = 0 \quad \forall t, s, (i, j) \in \mathcal{J} && (5l)\\
& (\mathcal{F}_{ij} - p_{ijts})\mu^{\mathcal{F},\max}_{ijts} = 0 \quad \forall t, s, (i, j) \in \mathcal{J} && (5m)\\
& (\theta_{its} + \pi)\mu^{\theta,\min}_{its} = 0 \quad \forall i, t, s && (5n)\\
& (\pi - \theta_{its})\mu^{\theta,\max}_{its} = 0 \quad \forall i, t, s && (5o)\\
& \mu^y_{gts}, \mu^{F,\min}_{ijts}, \mu^{F,\max}_{ijts}, \mu^{\mathcal{F},\min}_{ijts}, \mu^{\mathcal{F},\max}_{ijts}, \mu^{\theta,\min}_{its}, \mu^{\theta,\max}_{its} \geq 0. && (5p)
\end{aligned}
$$

A challenge for standard solvers is that all feasible points of the MPEC are non-regular, i.e. the gradients of the binding constraints are linearly dependent. Most non-linear optimization solvers even fail to obtain a locally optimal solution. A way to overcome the non-regularity is by the regularization approach of Scholtes (2001) and Ralph and Wright (2004). Using this approach, the equality constraints of complementary slackness are replaced by inequalities and the infeasibility gap is iteratively reduced. With inequality constraints, the MPEC is regular.

Alternatively, the complementary slackness constraints can be linearized using disjunctive constraints. Disjunctive constraints introduces a binary variable for each complementary slackness constraint, i.e., the constraints (5i)–(5o), and a large constant. The binary variable ensures that the two factors of the product cannot both be non-zero. The constant, usually denoted by M, has to be sufficiently large not to cut off any feasible solutions. At the same time, it must be sufficiently small not to create computational difficulties, see (Pineda et al. 2017) for more details. The resulting problem is a mixed-integer problem but remains non-linear due to the bilinear term in the objective function, i.e., is a MINLP. Usually, such problems can only be solved to local optimality.

Since the above are standard methods, we use them as benchmarks for the parametric programming methods. To the best of our knowledge, no other existing methods can solve this problem to global optimality.

5 Numerical Results

We present a case study of transmission expansion in the Nordic region, with 4 nodes representing Norway, Sweden, and the two Danish pricing regions: DK1 as Western Denmark and DK2 as Eastern Denmark; cf. Nord Pool AS (2017).

5.1 Data

We assume that three DC cables are already in place: One connecting the two Danish price regions, one connecting the Eastern Danish pricing region, and Sweden and one connecting Sweden and Norway. The existing cables each have a capacity of 1.000 MW. Three additional DC cables can be installed, providing connections where not already. These are the cables $(N, DK1)$, $(N, DK2)$ and $(SE, DK1)$, see Fig. 1. The topology of the network is not as the current one, but is chosen for the purpose of illustration. Variable investment costs of each candidate line are assumed to be 20.000 DKK/MW, whereas we disregard fixed investment costs. We likewise disregard the budget and limitations for installed capacities of candidate transmission lines.

Hourly demand data at each node is available from Nord Pool AS (2017) and we select the year 2015. This data is clustered into a number of representative hours

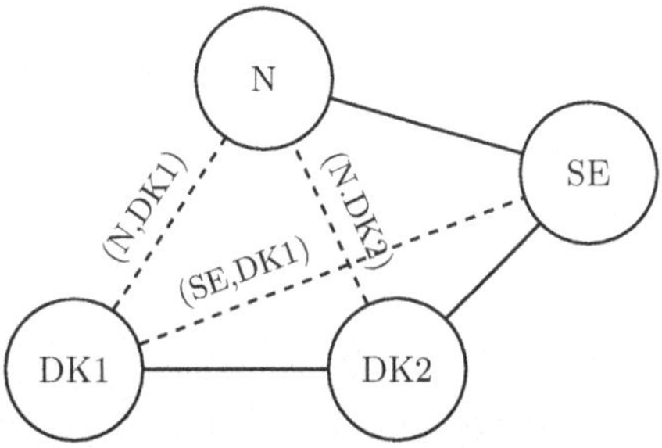

Fig. 1 Network topology. Solid lines represent existing lines, dashed lines represent candidate lines

Table 1 Generation capacities and production costs

	DK1	DK2	SE	N
Centralized cap. (MW)	1.800	2.400	13.800	15.000
Decentralized cap. (MW)	1.200	1.600	10.000	9.200
Centralized cost (DKK/MWh)	500	450	400	300
Decentralized cost (DKK/MWh)	760	760	700	700
Average demand (MWh)	2299	1526	15275	14369

using k-means clustering (Hartigan and Wong 1979). We obtain results for different numbers of representative hours. For simplicity, we disregard demand uncertainty.

Generation capacities and costs for DK1 and DK2 are obtained from Energianalyse (2014) that divides generation into centralized and decentralized units. Generation capacities are adjusted to the Norwegian and Swedish nodes by considering historical production data. As opposed to Denmark, Norway and Sweden have considerable amounts of hydropower, which is reflected in the lower production costs of the centralized plants. The generation capacities and production costs are shown in Table 1.

5.2 *Implementation*

The parametric programming approach to decomposition and the heuristic has been implemented in R using the interfaces by Berkelaar (2015) to solve LPs and Robere (2015) for vertex enumeration. The software is open source and free. The MPEC and MINLP have been implemented in GAMS (2017) and solved using the DICOPT solver. All problems are solved on an HP ProLiant server with 4 AMD 2.50 GHz CPUs and with a total of 64 cores and 256 GB of RAM.

5.3 *Optimal Investments*

For the most detailed case with 1000 representative hours, an optimal solution is given in Table 2.

As the table shows, investment are made in all candidate lines with maximum capacity on $(N, DK1)$ and $(N, DK2)$. We use this solution as a benchmark.

The investments in all candidate lines are justified by total congestion rents offsetting the investment costs. In fact, the transmission of power and differences in nodal prices generate significant revenues for the merchant investor. We explain this as follows.

Since the costs of centralized generation are significantly lower than those of decentralized generation, demand is satisfied by central production unless generation capacity is binding. As the production costs of Norway and Sweden are lower than those of the Danish nodes, demand of all nodes is satisfied by central production in Norway and Sweden, using both existing and newly constructed transmission lines. Thus, power is transmitted from Norway and Sweden to Denmark unless transmission capacities are binding, i.e., congestion occurs. As a result, the nodal prices are determined by the marginal costs of centralized Norwegian and Swedish generation in many of the representative hours. When congestion occurs, however, market prices of the Danish nodes are higher than for Norway and Sweden.

Average nodal prices are given in Table 3. As expected, average prices are higher for the Danish nodes than for the Norwegian and Swedish nodes, but the same for both of the Danish nodes.

In Table 4, we list the number of hours (out of the 1000 representative hours) for which the transmission lines are congested. Furthermore, the direction of the power flow is indicated by the number of hours with positive and negative flow. We note that power always flows into the Danish nodes from the (N,DK1), (SE,DK1) and

Table 2 Investment decisions and capacities in candidate lines

(i, j)	(N,DK1)	(N,DK2)	(SE,DK1)
x_{ij}	1	1	1
$\mathcal{F}_{ij}$	1000	1000	16

Table 3 Average prices at the four nodes in DKK/MWh

DK1	DK2	SE	N
491	491	451	357

Table 4 Number of hours (out of 1000) with congested lines, positive flow and negative flow for each transmission line

	(N,SE)	(N,DK1)	(N,DK2)	(SE,DK1)	(SE,DK2)	(DK1,DK2)
Congested lines	947	980	997	1000	812	0
Positive flow	981	990	1000	1000	1000	90
Negative flow	19	10	0	0	0	907

(SE,DK2) lines, clearly confirming the relatively low-cost generation from Norway and Sweden supplied to the Danish market. The (N,SE) and (N,DK1) lines mainly have flow from Norway to DK1 and Sweden (in 981 and 990 out of 1000 hours, respectively). All but the (DK1,DK2) line have power flow during all hours and the (DK1,DK2) line only has 3 out of 1000 hours without flow of power. Thus, the markets exploits the network at all times.

As can be seen, the line connecting Sweden and DK1 is always congested, meaning the merchant investor collects congestion rents in all hours. Also, the transmission lines connecting Norway and Sweden, Norway and DK1, Norway and DK2, and Sweden and DK2 are almost always congested (between 812 and 997 hours out of the 1000 representative hours). The only line that is never congested is the one connecting the two Danish regions, DK1 and DK2.

5.4 *Comparison of Solution Methods*

We apply two solution methods based on parametric programming: The parametric programming approach to decomposition (Decomp.) that guarantees global optimality and the parametric programming heuristic (Heuristic). We compare with the three non-linear programming methods: A standard MPEC solver, a regularization approach (reg. MPEC), and a reformulation by disjunctive constraints (MINLP). We solve the BPP with all these methods, varying the number of representative hours by 10 from 10 to 100 and by 100 from 100 to 1000, the result of which is a total of 19 problem instances of increasing size.

The standard MPEC solver returned local infeasibility for all instances, and thus, we do not report further results of using this solution method. The MINLP method likewise did not provide any results, with the solver reporting that the search stopped as the objective function of the NLP subproblems started to deteriorate. While the regularization approach returned local optimal solutions for all 19 instances, all these solutions had $x_{ij} = 0$ and $\mathcal{F}_{ij} = 0$ for all $(i, j) \in \mathcal{J}$, i.e., no investments were made. This results in an optimality gap of 99% and is of no practical use.

Table 5 Investment decisions from the two solution methods, the decomposition approach and the heuristic

No. of rep. days	Decomp. (N,DK1)	(N,DK2)	(SE,DK1)	Heuristic (N,DK1)	(N,DK2)	(SE,DK1)
10	1000	1000	275	1000	1000	275
20	1000	1000	126	1000	1000	126
30	1000	1000	33	1000	1000	33
40	1000	1000	128	1000	1000	128
50	1000	994	0	994	1000	0
60	1000	1000	77	1000	1000	77
70	1000	1000	0	1000	1000	0
80	1000	1000	0	1000	1000	0
90	1000	1000	51	1000	1000	51
100	1000	1000	11	1000	1000	11
200	1000	1000	38	1000	1000	38
300	1000	1000	38	1000	1000	62
400	1000	1000	0	1000	1000	0
500	1000	1000	0	1000	1000	4
600	1000	1000	9	769	1000	0
700	1000	1000	10	913	1000	0
800	1000	1000	11	1000	1000	4
900	1000	1000	16	993	1000	0
1000	1000	1000	16	901	1000	0

All numbers in MW

To compare the solutions of the decomposition approach and the heuristic, we report the investment capacities of the three candidate lines in Table 5. We see that the two solution methods agree in 14 out of 19 cases, as also indicated by the zero optimality gap in Table 6. For both methods, investments are made in lines (N,DK1) and (N,DK2) at maximum capacity in all but one instance (50 representative days). The investment in line (SE,DK1) is of a smaller capacity, although in many instances (14 and 11 out of 19 for the decomposition approach and the heuristic, respectively), some investment is profitable. In fact, a small capacity is enough to create congestion and generate some revenue. For the larger instances, however, the heuristic fails to capture small investments, which results in a significant optimality gap. In particular, for 600–1000 representative days, the exact approach suggests investment in line (SE,DK1), whereas in four out of five instances, the heuristic does not.

Table 6 provides the solution times of the exact parametric programming approach and the heuristic as well as their differences in objective function values, i.e., optimality gaps. For a number of representative days higher than 500, the optimality gaps produced by the heuristic varies from 0.1% to 10.1%. When the number of representative days is 500 or lower, the heuristic obtains an optimal solution. For instances

Table 6 Solution times and optimality gaps for the two solution methods, the decomposition approach and the heuristic

Number of rep. days	Sol. time, decomp. (s)	Sol. time, heuristic (s)	Optimality gap (%)
10	150.7	2.2	0
20	134.7	3.6	0
30	177.6	6	0
40	201.1	9.1	0
50	363.8	15.5	0
60	373.3	16.3	0
70	378.9	21.7	0
80	421.8	28.2	0
90	537.1	34	0
100	838.3	40.7	0
200	1013.9	143	0
300	1182.5	305.5	0.3
400	2308.9	528.2	0
500	3391	818.7	0.1
600	6764.8	1146.4	1.7
700	9142.5	1587.8	7.8
800	10222	2567	3.3
900	13543	3215.5	10.1
1000	16733.8	4238.5	7

with 100 representative days or lower, the heuristic obtains an optimal solution 15–70 times as fast as the decomposition approach. For problems with 200 representative days or more, the heuristic maintains lower solution times for almost all instances but with a factor between 4 and 7. While the heuristic provides no guarantees of optimality, our case study suggests that for small to moderate sized bilevel problems, it works very well. Furthermore, it solves even large problems relatively fast and provides solutions within a 10% optimality gap. Its main disadvantage is that the solutions may be structurally different from the optimal, and thus, this method may be better suited for cost assessments than for investment planning.

6 Conclusion

This chapter adopts a merchant investor perspective on transmission expansion. Investment is incentivized by the merchant collecting congestion rents on installed transmission lines. We formulate a bilevel programming problem in which investment decisions are made in an upper level and in anticipation of lower-level market-clearing. With the inclusion of congestion rents, the formulation involves a bilinear

Table 7 Nomenclature

Sets	
$\mathcal{T}$	Set of time periods
$\mathcal{S}$	Set of scenarios
$\mathcal{I}$	Set of network nodes
$\mathcal{I}(i)$	Set of nodes connected to node i
$\mathcal{J}$	Set of candidate lines
$\mathcal{G}$	Set of all production units
$\mathcal{G}(i)$	Set of production units at node i
Parameters	
ρ_t	Duration of time period t (p.u.)
ϕ_s	Probability of scenario s (p.u.)
k_{ij}	Linear investment cost for candidate line between nodes i and j (€/MW)
K_{ij}	Fixed investment cost for candidate line between nodes i and j (€)
$K^{\max}$	Investment budget (€)
$\mathcal{F}_{ij}^{\max}$	Maximum capacity available for candidate transmission line between nodes i and j (MW)
$F_{ij}^{\max}$	installed capacity of existing transmission line between nodes i and j (MW)
c_g	Linear production cost for unit g (€/MWh)
$y_g^{\max}$	Maximum production for generation unit g (MW)
d_{its}	Demand in node i, at time t and in scenario s (MW)
Variables	
x_{ij}	Binary investment decision for candidate line between nodes i and j
$\mathcal{F}_{ij}$	Installed capacity of candidate transmission line between nodes i and j (MW)
p_{ijts}	Power flow between node i and j, at time t and in scenario s (MWh)
θ_{its}	Voltage angle at node i, at time t and in scenario s
y_{gts}	Production of unit g, at time t and in scenario s
λ_{its}	Shadow price/dual variable of the balancing constraint at node i, at time t and in scenario s (€/MWh)
μ_{gts}^{y}	Dual variable of the capacity constraint for unit g, at time t and in scenario s
$\mu_{ijts}^{p}, \mu_{ijts}^{p,\mathcal{J}}$	Dual variable of the flow constraint between nodes i and j, at time t and in scenario s
$\mu_{ijts}^{F,\min}, \mu_{ijts}^{F,\max},$	Dual variable of the capacity constraint between nodes i and j, at time t and inscenario s
$\mu_{ijts}^{\mathcal{F},\min}, \mu_{ijts}^{\mathcal{F},\max}$	Dual variable of the capacity constraint between nodes i and at time t and in scenario s j
$\mu_{its}^{\theta,\min}, \mu_{its}^{\theta,\max}, \mu_{ts}^{\theta,\mathrm{ref}}$	Dual variable of the voltage angle constraint of node i, at time t and in scenario s

Note Bold-face indicates a vector of variables, e.g., $\mathbf{x} = \{x_{ij}\}_{i,j}$ and $\boldsymbol{\lambda} = \{\lambda_{its}\}_{i,t,s}$

revenue term in the upper-level objective function. This makes the problem difficult to solve to global optimality by standard approaches, such as MPEC or MILP reformulations.

Instead, we apply an exact algorithm based on parametric programming that solves the bilinear bilevel programming problem to global optimality. Furthermore, it allows for decomposition of the lower-level problem and thereby has potential to provide computational advantages. We also present a faster, but heuristic version of the algorithm.

We illustrate the problem and the solution methods on a case study of transmission investment in the Nordic region. The numerical results indicate that is it profitable to be a merchant investor in an electricity network. The parametric programming approach is able to solve problem instances with up to 1000 representative days within 4.5 hours while the heuristic terminates in 1.2 hours and with an optimality gap of up to 10%. For small and moderately sized instances the heuristic found the optimal solution in 14 out of 19 cases with significantly lower solution times than the parametric approach. For large instances, however, the structure of the solutions produced by the heuristic often differ from the optimal.

Acknowledgements T. K. Boomsma acknowledges support from the project AHEAD, Analyses of Hourly Electricity Demand, funded by Energiteknologisk Udviklings- og Demonstrationsprogram (EUDP) under the Danish Energy Agency.

Nomenclature

See Table 7.

References

D. Avis, K. Fukuda, Reverse search for enumeration. Discr. Appl. Math. **65**(1–3), 21–46 (1996)

L. Baringo, A.J. Conejo, Transmission and wind power investment. IEEE Trans. Power Syst. **27**(2), 885–893 (2012)

M. Berkelaar, *lpSolve: Interface to 'Lp_solve' v. 5.5 to Solve Linear/Integer Programs* (2015)

P. Buijs, R. Belmans, Transmission investments in a multilateral context. IEEE Trans. Power Syst. **27**(1), 475–483 (2012)

H.C. Bylling, S.A. Gabriel, T.K. Boomsma, A *Parametric programming Approach to Bilevel Optimization with lower-level Variables in the upper Level*, J. Oper. Res. Soc. **71**(5), 846–865, (2020)

B. Colson, P. Marcotte, G. Savard, An overview of bilevel optimization. Ann. Oper. Res. **153**(1), 235–256 (2007)

A.J. Conejo, L. Baringo, S.J. Kazempour, A.S. Siddiqui, *Investment in Electricity Generation and Transmission* (Springer, Berlin, 2016)

S. Dempe, V. Kalashnikov, G.A. Pérez-Valdés, N. Kalashnykova, *Bilevel Programming Problems* (Springer, Berlin, 2015)

Ea Energianalyse, Elproduktionsomkostninger. Energistyrelsen **1**(1), 1–42 (2014)

S.A. Gabriel, A.J. Conejo, J.D. Fuller, B.F. Hobbs, C. Ruiz, *Complementarity Modeling in Energy Markets* (Springer, Berlin, 2013)

T. Gal, *Postoptimal Analyses, Parametric Programming, and Related Topics* (Walter de Gruyter, New York, 1995)

GAMS, *General Algebraic Modeling System* (2017)

L.P. Garcés, A.J. Conejo, R. García-Bertrand, R. Romero, A bilevel approach to transmission expansion planning within a market environment. IEEE Trans. Power Syst. **24**(3), 1513–1522 (2009)

J.A. Hartigan, M.A. Wong, Algorithm AS 136: a K-means clustering algorithm. J. R. Statistical Soc. Ser. C (Appl. Statistics) **28**(1), 100–108

R. Hemmati, R.A. Hooshmand, A. Khodabakhshian, Comprehensive review of generation and transmission expansion planning. IET Gener. Transmiss. Distrib. **7**(April), 955–964 (2013)

P. Joskow, J. Tirole, Merchant transmission investment. J. Ind. Econ. **53**(2), 233–264 (2005)

T. Kristiansen, Markets for financial transmission rights. Energy Stud. Rev. 35 (1979)

L. Maurovich-Horvat, T.K. Boomsma, S.-E.S.-E.E. Fleten, A.S. Siddiqui, Transmission and wind investment in a deregulated electricity industry. IEEE Trans. Power Syst. **99**, 1–7 (2014)

Nord Pool AS. Nord pool (2017)

S. Pineda, H.C. Bylling, J.M. Morales, Efficiently solving linear bilevel programming problems using off-the-shelf optimization software. Optimization and Engineering (2017)

D. Ralph, S.J. Wright, Some properties of regularization and penalization schemes for MPECs. Optimi. Methods Softw. **19**(5), 527–556 (2004)

R. Robere, *vertexenum: Vertex Enumeration of Polytopes* (2015)

J. Rosellón, T. Kristiansen. *Financial Transmission Rights Analysis, Experiences and Prospects.* Lecture Notes in Energy (Springer, Berlin, 2013)

S. Scholtes, Convergence properties of a regularization scheme for mathematical programs with complementarity constraints. SIAM J. Optim. **11**(4), 918–936 (2001)

A. Sorokin, S. Rebennak, P.M. Pardalos, N.A. Iliadis, *Handbook of networks on power systems* (Springer, Berlin, 2012)

Merchant Investment in Transmission Network

Market Versus Planning Approaches to Transmission and Distribution Investment

Frank A. Felder

1 Introduction

What role should markets play in electricity transmission and distribution? At the inception of the industry, the entire system—generation, transmission, and distribution—was competitive, and multiple electric companies competed to supply and distribute electricity. Electricity was also competing with well-established alternative forms of lighting and energy. In the USA, once economic regulation of electric utilities took hold at the state level at the start of the 1900s, the answer to this question was turned on its head (Kiessling 2008). The entire system was operated, planned and owned by the utility. The motivation for economic regulation was based upon vertical and horizontal economies of scale and scope of building, maintaining, and operating a capital-intensive system that requires instantaneous real-time integration of generation and consumption.

Economic regulation of the transmission and distribution sectors continued when non-utility generation (NUGs) was introduced in 1978 in the USA with the passage of the Public Utility Regulatory Policy Act. This federal law mandated utilities to purchase power from cogeneration facilities (i.e., facilities more efficient than large-scale generation stations due to the combined production of heat and electricity) and power plants using renewable fuels at the utilities' avoided costs (Joskow 1997). The rationale for permitting NUGs was to reduce dependence on foreign oil supplies. The policy of regulating transmission and distribution assets also continued with the introduction of wholesale US electricity markets in the early 1990s, which allowed open access to the transmission sector but not the distribution sector. Nonetheless,

F. A. Felder (✉)
Edward J. Bloustein School of Planning and Public Policy, Rutgers University, 33 Livingston Ave, New Brunswick, NJ 08901, USA
e-mail: ffelder@ejb.rutgers.edu

M. R. Hesamzadeh et al. (eds.), *Transmission Network Investment in Liberalized Power Markets*, Lecture Notes in Energy 79,
https://doi.org/10.1007/978-3-030-47929-9_9

the utility's regulated monopoly on owning, maintaining, and operating transmission and distribution was preserved.

In the early 2000s, motivated in part with the establishment of wholesale electricity markets based upon locational marginal prices (LMPs) in parts of the USA, the question of merchant transmission arose in general and as part of the discussion of financial transmission rights (FTRs). FTRs, which have different names in different parts of the USA,[1] allow buyers and sellers of power to hedge congestion risk and allow market participants to speculate on those risks. Thus, a merchant transmission developer may want to invest in a transmission facility based upon the projected revenues that it would receive from the additional FTRs that the project would create.

The development of merchant transmission has not been widespread in the world, although some merchant transmission projects were successful in Australia, New Zealand, and the USA It has played a minor role in the overall transmission investment in regions that have wholesale electricity markets. Valid concerns, which are discussed further below, have been raised regarding the viability of merchant transmission and whether it would result in optimal transmission expansion (Joskow and Tirole 2005). Nonetheless, as a matter of policy but not widespread practice, policies allowing merchant transmission are a breach of the utilities' transmission monopoly.

The question of markets for distribution is now on the policy agenda. The push to decarbonize the electric power system along with advances in distributed resources and communication and computational technologies has extended this breach to the distribution system (Burger et al. 2018). For instance, New York State has embarked on an ambitious project to decentralize the distribution system and create competitive entry for the provision of distribution services (Felder and Athawale 2016).

In addition, some states are advancing microgrid and community solar policies that also chip away at the utilities' distribution monopoly. For example, in New Jersey, offsite distribution of electricity must be done by the electric public utility that has the franchise rights for that municipality (NJ BPU 2016). This legal requirement prevents microgrids owned by retail customers to connect across streets, which hampers microgrid development. On the other hand, by protecting the utility's franchise, it lowers the utility's cost of capital (Hausman and Neufeld 2002).

What explains these different answers to the question of what should be the extent, if any, of markets for electricity transmission and distribution? What part is due to changes in technology versus changes in regulatory objectives? Why is there a push to expand the role of markets and competition with respect to transmission and distribution while at the same time, regulators are pursuing policies that undercut wholesale electricity markets? Stepping back, when analyzing the question of transmission and distribution planning versus markets, what are the fundamental elements that need to be considered?

This chapter identifies three such elements: the physics of the electric power system, the wholesale and retail market structure, and policymakers' objectives and

[1]Other names for concepts similar to FTRs are congestion revenue rights (CRR) and transmission congestion contracts (TCC) (Hogan 2003).

opportunities for strategic behavior. Although the examples presented are based upon the US context, the discussion and findings are also relevant for other countries.

Important work has and is being done on how to model and therefore improve efficiency outcomes by aligning wholesale electricity markets that may contain market power with transmission expansion policies that may permit both market-based and traditional utility-based investments either performed by a utility that also administers the wholesale market (i.e., a Transco) or a separate entity (i.e., a Regional Transmission Operator/Indpendent System Operator (RTO/ISO). Some of this work is referred to and briefly discussed in this chapter, including referencing other chapters in this book.

The thrust of this chapter, however, is that attempts to improve the alignment of wholesale electricity markets and transmission policy should consider issues beyond transmission. These broader issues are (1) the practical policy decisions that arise from the liberalization of the power sector; (2) the potential expansion of liberalization to the distribution portion of the power system; (3) the multiple policy objectives that policymakers are claiming to be pursuing beyond economic efficiency; and (4) the political motivations and associated strategic behavior that is behind this expansion.

2 The Dominant Engineering and Economic Framework for Transmission and Distribution

The transmission and distribution of electricity have several well-known physical properties that distinguish them from the transportation of other goods (Hogan 1992; Joskow 1997; Kiesling 2008). The amount of power injections and withdrawals must (almost) be equal at every instance in time to maintain system frequency within the required bandwidth. If an imbalance between generation and load occurs for too long of a time in too large of an amount, the flows of power over larger regions of the grid may become unstable, resulting in large-scale blackouts. Although the costs of various storage technologies are declining, it is still too expensive to store large amounts of electricity. Input fuels are stored at or near power plants—for example, water in a reservoir at a pumped hydro facility or a coal pile at a power plant—so that, they can be quickly converted into electricity as needed.

Even when the supply and demand of electricity are in balance, the locations and amounts of the injections and withdrawals affect the ability to generate and consume electricity at every other location on the network, which makes establishing physical property rights on the network challenging. More precisely, the injection or withdrawal of power at one point in the network affects—sometimes in positive and other times in negative ways—the injection and withdrawal of power at potentially every other point in the network. These network effects are complex and non-trivial. Moreover, electricity cannot be practically directed or switched from generators to loads and instead flows in parallel paths and multiple loops. These network externalities can be addressed, with varying degrees of success, depending on the type of congestion pricing system that an electricity market uses and the availability of engineering control options to manage operations.

Thus, the terms "transmission system" and "distribution systems" are misnomers. Transmission and distribution networks, along with generation and demand, form a single, integrated system whose reliability and economic performance depend on one another. Due to vertical economies of scale and scope, the operation and planning of the power system must be coordinated across all of these subsystems (Hughes 1993; Joskow and Schmalensee 1988).[2] With the introduction of electricity markets, generation and retail supply are scheduled and procured via markets, but the transmission and distribution of electricity are still managed via planning, perhaps complemented with active markets for transmission. Thus, the introduction of wholesale and retail electricity markets into the electric power sector is a distinct organizational strategy and approach from transmission and distribution planning, yet these two different strategies are currently implemented in the same system.

Mixing planning and markets is a challenge (Joskow and Tirole 2005). One tangible illustration of this potentially potent combination involves reactive power, a component of alternating current (AC) electricity. AC electricity has both real and reactive components that must be supplied and consumed almost simultaneously. Generation, transmission and distribution, and load interact in complicated and non-linear ways to supply and consume reactive power. In practice, in the USA, electricity markets supply real power, whereas the provision of reactive power is done through non-market means. Reactive power problems were found to be a contributing factor to the large-scale 2003 US blackout perhaps exacerbated by the fact that real power was procured via markets and reactive power was procured via a planning model (Liscouski and Elliot 2004; Taylor 2006). Reactive power can be supplied, in varying degrees, in the dispatch of generation, the operation of transmission and distribution systems, and the investment in transmission and distribution assets. Reactive power is even more of an issue for the distribution system than the transmission system, further complicating the problem of determining the appropriate mix of distribution planning and markets for distributed resources.

The dominant economic and regulatory framework in the electricity sector is well-known and has been discussed extensively (e.g., Joskow 1997; Kiesling 2008). The combination of power system physics and natural monopoly in the generation and transmission and distribution of electricity with its downward-sloping, long-run

---

[2]In the context of electricity, the term "system" also has a broader meaning beyond a technical description of the physical components and including institutional arrangements and values. "Electric power systems embody the physical, intellectual, and symbolic resources of the society that constructs them. Therefore, in explaining changes in the configuration of power systems, the historian must examine the changing resources and aspirations of organizations, groups, and individuals. Electric power systems made in different societies–as well as in different times–involve certain basic technical components and connections but variations in the basic essentials often reveal variations in resources, traditions, political arrangements, and economic practices from one society to another and from one time to another. In a sense, electric power systems, like so much other technology, are both causes and effects of social change" (Hughes 1993, p. 2).

average costs and subadditivity calls for the creation of a regulated utility with a monopoly on the provision of electricity within its service territory (Kiesling 2008). In the USA, with its split jurisdiction, federal and state regulators set tariffs, terms and conditions of service, entry restrictions, service quality, and other aspects of utility behavior (Kahn 1988; Joskow 1997). As part of this regulatory compact, utilities have an obligation to serve all customers within their service territory (Rossi 2000).

Rent-seeking behavior by interest groups combined with the normative case for restructuring based upon widely noted inefficiencies and limitations of the regulatory framework pushed, both internationally and in the USA, for introducing competition into the power sector (Borenstein and Bushnell 2014). These efficiency concerns range from poor performance of power plants, low labor productivity, inefficient investments, slow adoption of new technologies, to the regulatory capture by industry (Stigler 1971; Joskow 1997; Kiesling 2008). It is within this context that liberalization was introduced into the industry.

3 The Market's Nose Under the Planning Tent

This section outlines a conceptual, non-historical approach to introduce wholesale and retail supply electricity markets, starting with the least intrusive situation in which there is only a market for wholesale generation and works through the implications and possible adjustments to transmission and distribution planning that may be required in response. The claim is that even the least intrusive introduction of competition in generation affects transmission planning requiring changes to the policies surrounding transmission planning.

Wholesale markets are a different means to the stated efficiency ends, at least in theory, of the regulated system that they replaced. Prior to industry restructuring and liberalization, the electric power industry, at least on paper, was supposed to satisfy electricity demand now and into the foreseeable future, ensuring reliability at the least cost. In fact, the intended goal of regulation can be viewed as achieving the same outcome that would have occurred if the industry had both a physics and economic structure conducive to workably competitive markets.[3] This objective was reflected in the five-minute dispatch, daily generation unit commitment, and decadal generation and transmission planning. Historically, the emphasis on achieving low-cost predominated, although there were notable exceptions (Hughes 1993).[4] With the push for a market-based framework for the distribution sector in a multiple objective context, not only are the means being replaced but also the end objectives. Of

[3]See Clark (1940) for an early discussion of workable competition.

[4]Viewed from an economic perspective, the intended goal is to maximize social welfare, which reduces to a cost minimization problem if demand is assumed to be inflexible due to regulatory policies that do not result in retail customers paying time varying prices. See chapters "An introduction to transmission network investment in the new market regime" and "Regulated expansion of the power transmission grid" as well as Biggar and Hesamzadeh (2014).

course, the relative merits of liberalization and its extent compared to the traditionally regulated, integrated utility depends on the assessment of the tradeoff between imperfect regulation and imperfect competition (Joskow and Schmalensee 1988). These tradeoffs involve reducing horizontal and vertical economies of scale and scope for increased efficiency, subject to market power concerns, of generation and retail supply (Michaels 2004).

3.1 Competitive Supply of Power Purchase Agreements for Generation

A good place to start understanding the interaction and implication of electricity markets with transmission and distribution planning is by analyzing the most limited type of competition to generate electricity. A utility that owns generation and transmission assets could procure additional generation (as well as transmission) via a competitive procurement process in which developers of power plants including renewables compete to provide electricity supply via a long-term power purchase agreement. In this example, there are no organized wholesale electricity markets, either spot or forward markets.

Unless the location and type of power plant is specified beforehand as a part of the procurement process, the utility's transmission and distribution planning would have to be adjusted to account for potentially different proposed locations (Tohidi et al. 2016). Given the physical characteristics of parallel or loop flow on the grid combined with constraints on the system, the electrical location of generating plants can matter tremendously. It can affect the interconnection cost of generation, the need for and costs of additional transmission upgrades, and the potential operation and dispatch of existing generation units.

This immediately raises the question of who pays for the associated interconnection costs. If the utility pays, i.e., the interconnection costs are not part of the generation developer's offer, then the developer does not have an incentive to locate its proposed project where the interconnection costs are low. If the developer pays for the interconnection costs, then the utility, using its transmission and distribution planning process, would need to provide the developer with those costs as part of the procurement process so that the developer can account for them in its offer.

Putting aside the obvious fact that the utility, and not the developer, is in the best position to estimate the interconnection costs, the developer should not be required to do so because it requires modeling the entire system over time to assess how the developer's interconnection will affect the power system.[5] In other words, the developer needs to, in effect, determine a transmission plan for the entire system in order to identify the size and therefore costs of its interconnection of its proposed facility because of the network externalities of the power system and the fact that

[5] In practice the utility may not be willing to share the necessary data with the generation developer due to the sensitive nature of the data.

transmission assets have economies of scale. It also may be the case that for reasons of reliability and economics, a larger interconnection investment should be made at the time of the developer's investment. This may be necessary given likely or even possible future generation development not only at or near the developer's proposed location, but elsewhere on the grid. The value of having additional transmission capacity available (which may not be that costly due to economies of scale), even with a low probability of that additional capacity being utilized, may be sufficient to warrant a larger investment.

Note that via the procurement process, the utility could require that the developer only pays for the interconnection costs associated with its project, but instead has to install a larger interconnection to accommodate future needs that the utility pays for and perhaps recovers from generation developers in the future. Even if the generation interconnection is appropriately sized, there may be surplus interconnection transmission capacity available for future use by others. In such cases, how to define, value, and transfer these rights and reliabilities requires a market or market-like mechanism.

3.2 Generation Development and Transmission and Distribution Interconnections

The situation becomes more complicated when investment in generation is via an open market that is not restricted by the utility initiated procurement process (Hesamzadeh et al. 2011; Motamedi et al. 2010). In this market design, multiple developers invest in new generation based upon their market expectations. For any proposed generation project, there must be an interconnection process to determine the size and cost of the interconnection. Moreover, since there are multiple developers competing for projects over different locations, the interconnection process must have some policy that determines the order in which projects are analyzed and which other projects, if any, should be assumed to be developed along with their associated interconnection investments. Crafting an interconnection policy that is practical, i.e., analytically tractable, timely, and low cost, supports a reliable and efficient transmission and distribution plan, and simultaneously limiting costly strategic behavior by developers is not an easy endeavor.

One possible rule to analyze the proposed generation projects could be on a first-come-first-serve basis while assuming that all prior projects to the current project being analyzed will be developed. This rule creates both a practical problem and a path for possible strategic behavior by developers. The practical problem is that the transmission entity may be evaluating a long list of projects. If the interconnection facilities of prior projects are assumed to be built in the analysis of subsequent projects in the queue, the results of many analyses may not be relevant, particularly if there is a strong interaction or dependency among projects because many proposed projects do not make it to fruition.

Strategic behavior can also occur with this rule. The most straightforward type of such behavior is for a developer to propose a long list of projects—even ones that are not likely to be developed—in order to reserve its place in line. This would congest the queue, so to speak, to delay competitors' projects that adversely affect their existing or planned projects. Likewise, a developer may propose projects at strategic locations to raise the interconnection costs of its competitors' projects.

Of course, there are other rules that could be added to reduce both the practical problem and strategic behavior just described. Developers could be charged for the interconnection studies, so there is some disincentive for them to propose phantom projects. The transmission entity could remove proposed projects from the queue if the project has not completed certain milestones within a specified time. In addition, developers could be required to post-increasing deposit amounts at key milestones in the interconnection process or otherwise lose their place in the queue. Nonetheless, how to strike an optimal balance from a social welfare perspective that reduces strategic behavior without stifling legitimate generation development is unclear because there is no unique method where the pros obviously outweigh the cons.

The above analysis assumes that there is only one level of interconnection for a project. What if the developer wants a deeper level of interconnection and network integration so that the electricity from its project obtains a higher energy price in the wholesale market or access to a more lucrative capacity market (Joskow 2006)?[6] Whether deeper levels of interconnection should be offered, how many of them, and how those variations are accounted for in the order in which they are performed are important questions. For example, should Developer A's deeper interconnection study be conducted before or after Developer B's standard interconnection study, if Developer A is before Developer B in the queue?

Similar to the utility-procurement case, the option of deeper interconnection accentuates the need to consider developing property rights associated with transmission investments made by developers. A particular interconnection investment may provide excess transmission capacity beyond what the developer needs due to the fact that transmission facilities come in discrete sizes or a developer may wish to invest in overcapacity either to use for subsequent expansion at its proposed site or for resale to other developers. It may be more cost effective for the developer to invest in overcapacity, which would only occur if the developer then could profit from such an investment, than for the transmission utility to make a separate transmission upgrade or expansion. Such an investment by a developer is speculative because market and system conditions may change. The questions of whether developers should be able to own transmission rights, if they own these rights must they use them or lose them

---

[6]Capacity markets are designed to ensure that there is sufficient generation to satisfy demand accounting for the random failure of generation units. See Jaffe and Felder (1996).

by a certain time to limit strategic behavior, and are those rights affected or not by future generation and transmission investments must all be answered (Pollitt 2008).[7]

The above analysis applies to the distribution subsystem as well. Many jurisdictions in the USA and elsewhere have policies that are promoting large investments in distributed generation, including solar. Although much of this distributed generation is behind the meter with little or no net injection into the distribution system, there are also many large distribution projects, such as large solar farms, that are connected to the distribution system, some of which will result in the need for additional distribution investments. These situations raise the analogous issues discussed previously with respect to transmission.

3.3 Markets for Transmission Investment

In general, the transmission investment problem can be addressed in one of four ways: planning with rate-of-return regulation, performance-based regulation (PBR), merchant investment based upon financial transmission rights (Stoft 2006), or a blend of these approaches (Hesamzadeh et al. 2018; Chap. 12). Of the four, the first two envision a small (if any) role for merchant transmission. The case for planning over PBR depends on one's assessment of the tradeoffs between rate-of-return and performance-based regulation (Stoft 2006; Joskow 2008). In addition, the comparison of planning vs. PBR also overlaps with the institutional design of wholesale markets. The planning approach does not preclude having Regional Transmission Organizations/Independent System Operators (RTOs/ISOs) administering wholesale electricity markets. In contrast, the PBR approach centers on "a single independent transmission company that spans a large geographic area, and integrates system dispatch, congestion management, network maintenance, and investment under PBR regulation" (Joskow 2008).[8] Under these first two approaches, there is a role for competitive procurement of transmission (chapter "Market Versus Planning Approaches to Transmission and Distribution Investment"). For example, once a particular transmission project is determined to be needed, either by the utility's planning process or transmission customers, then that project is put out to competitive bid (Littlechild and Skerk 2008).[9]

[7]The Federal Energy Regulatory Commission issued Order No. 845 Reform of Generator Interconnection Procedures and Agreements on April 19, 2018, which includes energy storage as part of the definition of generating facility.

[8]It is conceivable to use PBR to incentivize transmission utilities to build transmission within an RTO/ISO structure, but since the transmission companies would not be directly responsible for dispatch and congestion management but responsible for maintenance, the PBR's advantage of having a single company tradeoff operating costs and capital expenditures would be lost. Note that RTOs/ISOs make the final decision regarding transmission maintenance not transmission companies.

[9]See also FERC Order No. 1000 regarding limiting transmission utilities' first right of first refusal.

Both the planning and PBR approaches could permit non-transmission alternatives—such as demand response, generation, or energy storage—to be procured to satisfy part or all of the transmission need. The advantage of including non-transmission solutions is that they may be less expensive than transmission ones or the only options that are feasible due to space limitations. The concern, however, is that this option may undercut the market since the recovery of these investments would be through the regulatory process, presumably at a lower cost of capital given the allocation of risk to ratepayers instead of investors.

If merchant transmission is viewed as both a viable and desirable feature of wholesale electricity markets, then one important strand of this debate is whether wholesale markets with FTRs will fund efficient levels of market-based transmission (chapters "Competition for Electric Transmission Projects in the USA: FERC Order 1000" and "Merchant Transmission Investment Using Generalised Financial Transmission Rights"). The developer of merchant transmission does not recover its transmission investment through captive ratepayers but instead attempts to recover its costs through market pricing, such as differences in LMPs as reflected in FTRs and, if applicable, in generation capacity rights associated with capacity markets. Another approach to funding merchant transmission is that a large load serving entity in a load pocket funds the merchant transmission line from both the FTRs and reductions in its energy and capacity procurement costs.

Not surprisingly, merchant transmission developers want to build additional transmission between high-cost and low-cost regions to use the large difference in locational prices (or even zonal prices if the underlying energy market is zonal based) to fund their investment. Conceptually, the controllable switching of transmission lines could also increase FTRs (along with the concern that such capability could be used for market manipulation purposes). As the size of the transmission capacity between the two regions increases the difference in prices decreases, which reduces the revenues from FTRs that the transmission developer will earn to fund its project. If, in the US context, the two regions are two different states, then the state with the lower electricity cost may use its siting authority to hinder the development of additional transmission interconnection. This dynamic is the intuition behind the assessment that merchant transmission markets will undersupply the optimal level of transmission needed for economic efficiency compared to a regulatory model or one with a regulatory backstop (Hogan 2003).

This discussion is not about whether transmission markets would invest in the necessary transmission investments for reliability purposes. They would not because there does not currently exist a market mechanism and associated pricing related to the power system satisfying the security component of system reliability or transmission adequacy (except perhaps as part of a capacity market that has a deliverability requirement). This division of a transmission investment's benefits into either reliability or economic is simplistic, since for any given transmission investment, it is likely to have a combination of reliability (i.e., adequacy and security) impacts as well as economic ones. The reliability benefit of transmission lines can also change with the introduction of intermittent sources of renewable generation (Lamadrid et al. 2016; Chap. 14). Moreover, if a transmission project has this dual reliability and economic benefit, it provides a further reason to believe that merchant transmission alone would underinvest in transmission since the reliability benefits are not captured in FTRs.

Instead, the discussion revolves around optimal investment in transmission for economic efficiency reasons, assuming that any reliability requirements are satisfied (Joskow and Tirole 2005). Even advocates for transmission markets acknowledge that some type of regulatory backstop that results in additional transmission investment beyond that required to meet reliability-security standards is necessary (Hogan 2003). The discussion is therefore centered on if and how to structure markets for transmission along with the appropriate policies needed for the regulatory backstop.

A complete analysis of the above requires specifying the incentives and therefore the conditions under which a regulated transmission investment occurs (chapters "Transmission Planning and Operation in the Wholesale Market Regime" and "A Parametric Programming Approach to Bilevel Merchant Electricity Transmission Investment"). This depends on both the regulatory structure, i.e., cost-of-service or incentive based, and how that regulatory structure works in practice. In short, analysis needs to account for both market and regulatory imperfections and not compare imperfect competition to perfect regulation (or vice versa).[10] Other factors also suggest that relying solely on merchant transmission will not result in efficient amounts of investments such as market power in wholesale markets, lumpiness in transmission projects, stochastic characteristics of power systems, property right definitions, and potential strategic behavior by merchant transmission owners (Joskow and Tirole 2005). Joskow and Tirole argue that the merchant transmission model assumes that prices in the two markets that are being interconnected reflect the marginal cost of production and the marginal willingness to pay, which may not be the case for the list of reasons they discuss.

Merchant transmission investment is not just competing with other transmission investments (merchant or regulated) but also with grid-based generation, distributed generation, energy storage, and demand response. The effects of these types of investments on the grid can be non-trivial and challenging to analyze due to the systems' complex physics and because some of these options are connected at the transmission level whereas others are connected at the distribution level. For instance, if there are no retail electricity markets, then the local transmission and distribution utility retains an obligation to serve likely using its fleet of regulated generation units. If there are retail electricity markets, they may result in different types of generation investments than those under economic regulation. In addition, how retail electricity customers are served if they decide not to choose a third-party energy supplier can also affect generation expansion and therefore transmission expansion decision.

Furthermore, to evaluate the efficiency of transmission investments, how costs are recovered via the regulatory backstop must be considered (Rotger and Felder 2001). If the backstop transmission costs are socialized, that is recovered from transmission customers with a broad region and not just from the beneficiaries of the project, then

[10] "At present, there is no first-best solution available at either extreme to guarantee perfect economic efficiency in transmission investments" (Hogan 2003).

that may undercut investment market investment, whether generation, merchant or demand response.

To illustrate how transmission cost recovery can affect market decisions, consider a merchant developer's options. One option is to invest in a transmission asset and own all the FTRs, that is taking all of the risk associated with the differences in LMPs, which may be substantial. Another is to diversify some of its risk by selling the FTRs to the load that is located with the low-cost region or load pocket. If the regulatory backstopped transmission costs are socialized and not entirely borne by the load in the load pocket, then that load may not want to purchase merchant-based FTRs. The reason is that the costs borne by the load under the socialized backstop mechanism may be less expensive to the load (Rotger and Felder 2001). It is possible that the regulatory backstop investment is more expensive from a societal perspective than either a corresponding market-based generation or transmission investment. As discussed in Sect. 4 below, regulators behave strategically. In the US federal system, which has a dual jurisdiction with respect to federal transmission expansion policy and state siting of transmission lines, regulators in the low-cost region (e.g., State A) may not want new transmission investment to the high-cost region (State B) due to concerns that electricity prices will rise in the low-cost region.

Another factor to consider is what should be the specific decision rule that is used to determine when to trigger the regulatory backstop. A reasonable rule is whether the present value of the reduction in energy costs on a discounted cashflow basis exceeds the cost of the backstop transmission investment using the cost of capital of the transmission utility. The market risk that the merchant transmission investor faces is different from that of the transmission utility, i.e., the two costs of capital are not identical. If the merchant transmission developer's cost of capital corresponds to the appropriate market risk and the transmission utility cost of capital is less due to ratepayers backstopping utility losses, then again, the merchant transmission investment is at a disadvantage with a corresponding reduction in social welfare. Of course, it is plausible that the regulatory structure or its application is such that the merchant transmission investment is unduly favored compared to the backstop project.

Another version is that a particular need that has not been resolved through market investments in generation, transmission, energy storage, or demand side management is addressed through a competitive procurement process. The selected solution may not necessarily be a transmission project (Rotger and Felder 2001). In this model, the planning process is a regulatory backstop to the market, which includes merchant transmission. By allowing non-transmission solutions to participate as part of the regulatory backstop process, there may be a non-transmission solution that was not supported by the market but that is nonetheless less expensive than a transmission solution. On the other hand, allowing this option may further undercut the market by creating a slippery slope that would lead to integrated resource planning with the possible end state of all investments being made outside the market (Hogan 2003).

3.4 *Markets for Distribution Investment*

One of the major texts analyzing electricity markets assumed, along with many other analysts, that the electric distribution function would be performed by a franchised monopoly (Joskow and Schmalensee 1988). The emergence of distributed resources is challenging this assumption for three major reasons. First is the reduction and its expected continuation in the costs of distributed resources such as solar and energy storage, which is attractive to many policymakers as a means of reducing greenhouse gas emissions and increasing renewable energy. Another is the general decrease in costs of smart grid technologies, particularly on the distribution subsystem including the reduction in communication and control technologies (Kiesling 2008). Third is at least in the USA, many states are linking climate change mitigation policies with economic policies to offset the additional costs of non-traditional generation technologies over fossil-fuel technologies. States do so by incentivizing distributed resources to be located within the state to obtain in-state economic benefits, as opposed to large-scale generation, which could be located out-of-state along with the associated economic impacts.

The potential unwinding of the distribution franchise affects the rest of the power system due to the underlying physics. Thus, just as transmission planning must account for and be integrated with wholesale electricity markets, distribution policy should also be aligned and integrated with how the rest of the system works (chapter "The Impact of Transmission Development on a 100% Renewable Electricity Supply—A Spatial Case Study on the German Power System"). In the USA, states are the sole regulators of distribution whereas the Federal government, outside most of Texas, Alaska and Hawaii, regulates wholesale generation and transmission. Given cost reductions in distributed resources, state regulators unsatisfied with the outcome of wholesale electricity markets can extend their policy reach via the distribution system. In particular, one view among some states is that these markets have not quickly enough responded to environmental and economic challenges that regulators are interested in addressing.[11] This gives states some maneuvering room to adopt policies that are inconsistent, if not contradictory, to federal policy as well as the ability to behave strategically by intentionally adopting policies that advance state interests over federal ones.[12]

The introduction of distribution markets opens up additional sets of concerns beyond the issues with transmission markets. Regarding distribution markets, regulators are becoming more adamant about pursuing multiple objectives compared to wholesale electricity markets. In the case of wholesale markets, the stated goal was

[11] For example, many Northeastern States are pursuing ambitious clean energy targets in addition to implementing carbon pricing on electricity via the Regional Greenhouse Gas Initiative, a carbon dioxide emission cap-and-trade program.

[12] It is not a necessary condition that the wholesale regulator be different from the state regulator for states to undercut wholesale policy via retail policy. Whether single state electricity markets—such as most of Texas, much of California, and New York—are less susceptible to this type of strategic regulatory behavior is an open empirical question.

either reducing costs or improving efficiency. Except for economists, many policy-makers and analysts interpreted these two goals as synonymous.[13] Of course, the stated goal does not have to be the actual goal, and wholesale markets would not have been introduced if there were not a sufficient politically powerful coalition that supported it (Joskow 1997). Regarding the formation of distribution markets, the regulators have a much longer list of goals. For example, the New York policy initiative—Reforming the Vision (REV)—includes nine objectives:[14]

1. Make energy more affordable for New Yorkers
2. Cut greenhouse gas emissions 80% by 2050
3. Empower New Yorkers to make more informed energy choices
4. Improve New York's existing energy infrastructure
5. Create new jobs and business opportunities
6. Protect New York's natural resources
7. Build a more resilient energy system
8. Support cleaner transportation
9. Help clean energy innovation grow.

An important fact used to support the case for introducing generation competition was that the economies of scale of generation, both in absolute terms and relative to the size of a utility's load, were decreasing, making competition for supply economically viable. For distribution markets, the corresponding argument is that new distributed technologies—generation and storage—can partially, if not entirely, replace transmission and perhaps even distribution. Unlike the situation with newer generation technologies that had lower economies of scale and costs than established ones, distributed resources may be more expensive than the distribution components that they would need to replace.

Moreover, the role of transmission markets as part of the creation of wholesale markets still relied upon a regulated transmission (and distribution) system. Even the most pro-transmission market policies pale in comparison to what is being proposed and pursued on the distribution subsystem. Since the distribution subsystem, in general, is much more radial then then transmission subsystem, there may be more opportunities for the exercise of market power at the distribution level versus the transmission level. The radial nature means that distribution systems are typically voltage-constrained, and many distributed energy resources have the ability to control their active and reactive power inputs and terminal voltages, which provides a means for exercising market power (Roveto and Dvorkin 2019). In the case of some of the initiatives for distribution markets, the push is to fundamentally change the role of the electric distribution company as opposed to modifying its role to account for the introduction of distributed generation.

---

[13] Obviously, improving efficiency of the electric power system may result in prices increasing, for example, if prices are subsidized or if they do not include negative externalities, such as the social cost of air emissions.

[14] See New York Reforming the Energy Vision website, https://static1.squarespace.com/static/576aad8437c5810820465107/t/5aec725baa4a99171e5890d4/1525445212467/REV-fm-fs-1-v8.pdf, assessed November 26, 2018.

4 What Policymakers Are Really Trying to Achieve: Regulatory Competition and Strategic Behavior in the Context of Transmission and Distribution Investment

One interpretation of the pursuit of distribution markets is that regulators, based upon their experience with wholesale markets, find imperfect markets preferable to imperfect regulation if the markets will increase economic efficiency, which was a major stated motivation for the adoption of wholesale electricity markets. This is supported, at the federal level, by passing the Energy Policy Act of 1992, and the subsequent adoption by many states of similar legislation to implement retail competition in electricity (and natural gas). Moreover, many states have adopted other market mechanisms for renewable energy (renewable portfolio standards), carbon emissions (cap-and-trade), and even for energy efficiency in some cases (Transue and Felder 2010).

Some states have pursued policies that subsidize fossil-fuel generation that some analysts believe were designed to suppress wholesale electricity prices.[15] In addition, many have argued that renewable portfolio standards, energy efficiency, and nuclear subsidies also suppress prices (Felder 2011). These policies, combined with lower wholesale natural gas prices, have a compounding effect, requiring other subsidies (e.g., nuclear), which reinforce this dampening trend. During the introduction of wholesale electricity markets, research has indicated that states with much higher than average electricity prices (e.g., California and the Northeast) pushed for liberalization and restructuring (White 1996; Joskow 1997). State policymakers, not surprisingly, want low electricity (and other forms of energy) prices.

Another motivation for policymakers is to take credit now for the benefits of policies whose costs are pushed into the future. The policy of renewable portfolio standards is a good example. When states adopted these policies, the renewable requirements started low and were increased over time. Granted, expected costs of renewable were expected to decline over time, but a lot of the costs will be incurred in the future by a future legislator or regulator rather than the current ones. For instance, New Jersey has recently released a draft Energy Master Plan that extends to the year 2050, well beyond the career trajectory of the current governor.

Of course, policymakers may have multiple objectives that result in particular strategic behavior as well as face various real and perceived political constraints that may force them to adopt policies that are not their first choice. For instance, given political pressure and sensitivities, policymakers may reject internalizing the social costs of environmental impacts, in particular greenhouse gases, as politically viable even if they believe such a policy is preferable to other alternatives (Vollebergh et al. 1997). This rejection, however, may be influenced by political ideology (Baranzini et al. 2017).

[15] See Supreme Court (2015) and Johnson (2018). The federal government is also considering similar subsidies for coal and nuclear power. See U.S. Department of Energy (2017).

With motivations besides economic efficiency, regulators may have found another strategic means—restructuring the distribution system—to achieve their political, economic and environmental objectives. Thus, an analysis of the tradeoffs between markets and planning for transmission and distribution requires understanding the actual, not just stated, motivations of both federal and state policymakers as well as the various policies and regulatory mechanisms they have at hand to pursue their actual objectives.

5 Conclusions and Questions for Further Research

Stepping back, three fundamental questions confront the electric power sector. First, what is its purpose? Historically, it has been to provide low-cost power that meets high levels of reliability. This objective is being expanded to include many others as exemplified by New York's REV.

Second, what is the technological strategy that the industry should use? Again, the strategy has been based upon economies of scale and scope, which means largescale, centralized power plants that transmit power long distances to distribution systems and their customers. With advances in distributed generation, storage and control and communication technologies, many analysts and advocates are claiming that a distributed strategy is preferable across many, if not all, of the growing list of policy objectives (Kiesling 2008; Bauknecht 2011). If and how the power sector shifts to a more decentralized and distributed architecture, market, and institutional setup will affect generation, transmission and distribution. Some have even speculated that in the not too distant future, microgrids will be the organizing unit that may not be connected to the public grid during routine operations, if at all.

Third, to achieve the stated objective(s) for a given technological strategy, what combination of economic regulation and markets should be used? Distributed generation is both a complement of and a substitute for transmission and distribution.[16] It needs a distribution system to sell its excess power but can also reduce the need for distribution facilities. Some recent scholarship is directed at reconciling liberalization and sustainability, focusing on transformation rather than particular innovations, finding that governance not prices are in need of reform, and emphasizing the importance of distributed generation (Bauknecht 2011). The claim in this strand of the literature is that the electric power sector is locked into a suboptimal structure that requires fundamental changes in governance, not just internalizing the cost of greenhouse gases and other environmental externalities, in order to arrive at an optimal structure. State markets for the retail supply of electricity also play a role by enabling "consumers to communicate information about their preferences through the value chain and into generation and fuel choice decisions," which may open up new institutional arrangements and decentralization in the power sector (Kiessling 2008). Thus,

[16]This is the analog of transmission being both a complement to and substitute for centralized generation (Stoft 2006).

it is not clear whether internalizing environmental externalities in the electric sector will be sufficient to transit the system to the appropriate technological strategy or if more fundamental institutional and governance reforms are necessary.

The question of markets versus planning approaches to the transmission and distribution sectors depends on and is embedded in these three questions. Unless the objectives are stipulated, the technological strategy articulated, and the combination of markets and planning for other parts of the system set, the relative merits of transmission and distribution markets and planning cannot be assessed. The ability for market participants and policymakers to behave strategically must also be evaluated, particularly in the US context in which political authority to regulate the industry is split between the federal government and states.

Acknowledgements The author would like to thank Yury Dvorkin, Albert J. Lamadrid, and the editors for their comments. The standard caveat applies.

References

A. Baranzini, et al., Carbon pricing in climate policy: seven reasons, complementary instruments, and political economy considerations. Wiley Interdiscip. Rev.: Clim. Change **8.4**, e462 (2017)

D. Bauknecht, *Transforming the Grid: electricity system governance and network integration of distributed generation* (University of Sussex, Diss, 2011)

D.R. Biggar, M.R. Hesamzadeh. *The Economics of Electricity Markets*. (Wiley, New York, 2014)

S. Borenstein, J. Bushnell, The US electricity industry after 20 years of restructuring. Annu. Rev. Econ. **7**(1), 437–463 (2015)

S.P. Burger, J.D. Jenkins, C. Batlle, J. Pérez-Arriga, *Restructuring Revisited: Competition and Coordination in Electricity Distribution Systems* (CEEPR WP 2018-007, 2018)

J.M. Clark, Toward a concept of workable competition. Am. Econ. Rev. 241–256 (1940)

F.A. Felder, Examining electricity price suppression due to renewable resources and other grid investments. Electr. J. **24**(4), 34–46 (2011)

F.A. Felder, R. Athawale, Optimizing New York's reforming the energy vision. Utilities Policy **41**, 160–162 (2016)

W.J. Hausman, J.L. Neufeld, The market for capital and the origins of state regulation of electric utilities in the United States. J. Econ. Hist. **62**(4), 1050–1073 (2002)

M.R. Hesamzadeh, et al., Transmission augmentation with mathematical modeling of market power and strategic generation expansion—Part I. IEEE Trans. Power Syst. **26**(4), 2040–2048 (2011)

M.R. Hesamzadeh, et al., A simple regulatory incentive mechanism applied to electricity transmission pricing and investment. Energy Econ. **75**, 423–439 (2018)

W.W. Hogan, Contract networks for electric power transmission. J. Regul. Econ. **4**(3), 211–242 (1992)

W.W. Hogan, Transmission market design (2003)

T.P. Hughes, *Networks of Power: Electrification in Western Society, 1880–1930* (JHU Press, 1993)

A.B. Jaffe, F.A. Felder, Should electricity markets have a capacity requirement? If so, how should it be priced? Electr. J. **9**(10), 52–60 (1996)

T. Johnson, *Federal Move Could Undermine Murphy Clean Energy Policies* (New Jersey Spotlight, 2018)

P.L. Joskow, R. Schmalensee, *Markets for Power: An Analysis of Electrical Utility Deregulation* (MIT Press Books, 1988)

P.L. Joskow, Restructuring, competition and regulatory reform in the US electricity sector. J. Econ. Perspect. **11**(3), 119–138 (1997)
P. Joskow, Patterns of transmission investments. Compet. Electr. Markets Sustain. **131** (2006)
P.L. Joskow, *Lessons Learned from the Electricity Market Liberalization* (Center for Energy and Environmental Policy Research, Massachusetts Institute of Technology, Cambridge, 2008)
P. Joskow, J. Tirole, Merchant transmission investment. J. Ind. Econ. **53**(2), 233–264 (2005)
A.E. Kahn. *The Economics of Regulation: Principles and Institutions*, Vol. 1 (MIT Press, Cambridge, 1988)
L.L. Kiesling, *Deregulation, Innovation and Market Liberalization: Electricity Regulation in a Continually Evolving Environment* (Routledge, Abingdon, 2008)
A.J. Lamadrid, S. Maneevitjit, T.D. Mount, The economic value of transmission lines and the implications for planning models. Energy Econ. **57**, 1–15 (2016)
S.C. Littlechild, C.J. Skerk, Transmission expansion in Argentina 4: a review of performance. Energy Econ. **30**(4), 1462–1490 (2008)
B. Liscouski, W. Elliot, *Final Report on the August 14, 2003 Blackout in the United States and Canada: Causes and Recommendations*. A report to US Department of Energy 40.4 (2004), p. 86
R.J. Michaels, Vertical integration and the restructuring of the US electricity industry (2004)
A. Motamedi, et al., A transmission planning framework considering future generation expansions in electricity markets. IEEE Trans. Power Syst. **25**(4), 1987–1995 (2010)
New Jersey Board of Public Utilities, *Draft 2019 New Jersey Energy Master Plan: Policy Vision to 2050* (2019)
New Jersey Board of Public Utilities, *Microgrid Report* (2016)
M. Pollitt, The arguments for and against ownership unbundling of energy transmission networks. Energy Policy **36**(2), 704–713 (2008)
J. Rossi, Universal service in competitive retail electric power markets: whither the duty to serve. Energy LJ **21**, 27 (2000)
J.A. Rotger, F.A. Felder, Reconciling market-based transmission and transmission planning. Electr. J. **14**(9), 31–43 (2001)
M. Roveto, Y. Dvorkin, *Market Power in Electric Power Distribution Systems*. Submitted to the 2019 North American Power Symposium (NAPS), Wichita, Kansas, (2019), pp. 1–6
G.J. Stigler, The theory of economic regulation. The Bell J. Econ. Manag. Sci. 3–21 (1971)
S. Stoft, Problems of transmission investment in a deregulated power market. Compet. Electr. Markets Sustainab. 87–130 (2006)
Supreme Court of the United States. *On Writs of Certiorari: Brief of Robert L. Borlick, Joseph Bowring, James Bushness, and 19 Other Leading Economists as Amici Curiae in Support of Respondents. Nos. 14-840, 14-841* (Washington, DC, 2015)
C.W. Taylor, In my view-Reactive power today best practices to prevent blackouts. IEEE Power Energ. Mag. **4**(5), 104–102 (2006)
Y. Tohidi, M.R. Hesamzadeh, F. Regairaz, Sequential coordination of transmission expansion planning with strategic generation investments. IEEE Trans. Power Syst. **32**(4), 2521–2534 (2016)
M. Transue, F.A. Felder, Comparison of energy efficiency incentive programs: rebates and white certificates. Utilities Policy **18**(2), 103–111 (2010)
U.S. Department of Energy, *Staff Report to the Secretary on Electricity Markets and Reliability* (2017)
H.R.J. Vollebergh, J.L. De Vries, P.R. Koutstaal, Hybrid carbon incentive mechanisms and political acceptability. Environ. Resour. Econ. **9**(1), 43–63 (1997)
M. White, *Power Struggles: Explaining Deregulatory Reforms in Electricity Markets* (Brookings Papers on Economic Activity: Microeconomics, 1996), pp. 201–50

Competition for Electric Transmission Projects in the USA: FERC Order 1000

Paul L. Joskow

1 Introduction and Background

The electric power sector restructuring and competition reforms that have been realized over the last 25+ years in the USA, Canada, Europe, Latin America, Australia, and elsewhere were built upon a number of institutional assumptions.[1] First, vertically integrated geographical monopolies subject to government regulation or public ownership, embodied a number of inefficiencies. The most serious inefficiencies were associated with the construction, operation, maintenance, and retirement of generating facilities. Other inefficiencies arose from the limited array of price and service options available to end-use (retail) customers and inefficient retail rate designs mandated by many regulators.

[1]The restructuring model that I will discuss has not been adopted everywhere in these countries and regions. For example, in the USA, most of the South and West (except California and Texas) continue to rely on regulated vertically integrated firms, though they rely more on PPAs with independent power producers selected through a competitive process, rather than ownership, than was once the case.

Elizabeth and James Killian Professor of Economics, MIT and Research Associate, National Bureau of Economic Research. The views expressed here are my own and do not reflect the views of MIT, the National Bureau of Economic Research or any other entities with which I an affiliated. MIT provided support. A list of my affiliations can be found at http://economics.mit.edu/files/15081. I note in particular that I am on the board of directors of Exelon Corporation which has an interest in the issues discussed here. I want to thank Hannes Pfeifenberger and his team at the Brattle Group, Craig Glazer and Suzanne Glatz at PJM, and Mike Kormos formerly at PJM and now at Exelon for assistance with understanding the implementation of Order 1000 and with the data utilized here. I am grateful to Ingo Vogelsang, Mohammad Hesamzadeh and Stephen Littlechild for helpful comments on an earlier version of this paper.

P. L. Joskow (✉)
MIT, Cambridge, MA, USA
e-mail: pjoskow@mit.edu

M. R. Hesamzadeh et al. (eds.), *Transmission Network Investment in Liberalized Power Markets*, Lecture Notes in Energy 79,
https://doi.org/10.1007/978-3-030-47929-9_10

Second, reformers accepted the view that vertical integration between the generation, distribution, transmission, and retail supply segments was not necessary to supply electricity economically and reliably. The generation segment was viewed as being potentially competitive. So too, the retail supply segment. Accordingly, restructuring involved the separation (structural or functional unbundling) of generation from transmission and distribution, de-concentration of the generation segment through divestiture in some cases, as well as the unbundling of the retail supply segment (financial agreements to support the supply of electricity to end-use customers) from the physical distribution (wires) segment. Competition and market incentives would replace central planning and regulation in the generation segment and the retail supply segment. The restructuring reforms focused on developing and implementing wholesale market designs for energy and ancillary services that would support efficient scheduling and dispatch of generation to balance supply and demand continuously and efficiently, efficient spot market price formation, and efficient entry and exit decisions by owners of generating plants. The retail supply segment was then supported by a competitive wholesale market, unbundling of retail supply, and more recently, the diffusion of advanced metering infrastructure.

Third, it was generally assumed that the physical operation, ownership of, and investment in distribution and transmission networks would continue to be treated as geographic monopolies subject to cost-of-service regulation, perhaps enhanced with the application of incentive regulation reforms (Joskow 2014). Finally, policymakers concluded that the operation of and planning for a regional transmission network could be separated from ownership, investment, and maintenance of the physical transmission facilities that compose the transmission network. In restructured regions, an independent system operator (ISO) was needed both to perform these functions and to manage wholesale markets for energy and ancillary services in coordination with the reliable operation of the transmission network and the management and pricing of transmission congestion. It was expected that the ISOs would be responsible as well for transmission planning and the assignment of investment responsibility to incumbent transmission owners for designated regional facilities. The ISO itself would own no generation, transmission, or distribution assets but rather would act like the conductor of a symphony, making sure that the "decentralized" instruments all played their parts in a way that led to a high-quality symphonic result. This vision has been realized in several regions of the USA but certainly not everywhere.

As competitive wholesale market designs and the retail supply industry matured, interest turned to the question of whether decentralized investment in new transmission facilities might be feasible and economical as a complement to or complete substitute for centralized transmission planning, ownership, regulation of the charges for transmission service. Merchant transmission investment models that anticipated the free entry of developers to own and operate transmission facilities in return for payments reflecting time varying differences in locational prices mediated through the sale of tradeable financial or physical transmission (property) rights, first appeared in the academic literature in the 1990s as restructuring activity gained steam in the

USA. (Hogan 1992; Bushnell and Stoft 1996; Chao and Peck 1996). These transmission rights and any revenue associated with their sale would accrue to the owner of the transmission assets and would be the sole source of revenue to compensate the owners for their capital and operating costs. Thus, merchant transmission owners would take on all risks associated with uncertainty over the future value of transmission rights, capital and operating costs, and the profitability of the investment. Traditional centralized transmission planning, investment in designated facilities by the incumbent transmission owners, and the reliance on traditional cost-of-service regulation, is replaced with decentralized investment decisions driven by market incentives—locational price differences. The properties of this classical merchant model are indeed very elegant. In 2005, Jean Tirole and I wrote:

> Under a stringent set of assumptions, the merchant investment model has a remarkable set of attributes that appears to solve the natural monopoly problem and the associated [exclusive] need for regulated electric transmission companies. (Joskow and Tirole 2005, p. 233)

In short, if the merchant model could be relied on in practice "... it would lead to the remarkable conclusion that both the generation of electricity and the transmission of electricity could be largely deregulated." (Joskow and Tirole 2005, p. 235)

However, after examining this model under a large set of what we viewed as more realistic assumptions about the attributes of transmission investments, transmission network operating practices, imperfections in wholesale energy markets, potential strategic behavior by transmission owners of various kinds, and other considerations, we concluded that:

> Unfortunately, these assumptions do not reflect the attributes of the transmission investment opportunities that are likely to be most conducive to merchant investment, the stochastic properties of real transmission networks or widely documented imperfections in wholesale electricity markets. (Joskow and Tirole 2005, p. 235)

> Clearly, policymakers cannot proceed under the assumption that they can avoid dealing with the difficult issues associated with stimulating efficient investment in electric transmission network simply by adopting the merchant transmission model. The merchant model ignores too many important attributes of transmission network and the behavior of transmission owners and system operators. (Joskow and Tirole 2005, p. 262)

Policymakers too have been less than enthusiastic about adopting the classical merchant transmission model, despite the widespread reliance on locational marginal pricing (LMP), at the nodal (primarily in the USA) or Zonal levels (elsewhere), and the creation of tradeable financial or physical transmission rights assigned initially to transmission owners for sale. The only transmission projects of which I am aware that were developed entirely based on the classical merchant model are the MurrayLink and DirectLink projects in South Australia, and the Montana-Alberta tie-line in the USA, though the Montana-Alberta line benefited from government financing managed through the Western Area Power Administration.[2] Both of the Australian

[2]https://m-m.net/montana-alberta-tie-line/; https://ieeexplore.ieee.org/document/5589741; https://www.fool.com/investing/general/2013/11/18/montana-alberta-tie-line-merchant-transmission-inv.aspx.

projects[3] ran into financial difficulties after they were completed and were ultimately converted to regulated projects.[4] The Montana-Alberta line continues to operate though little is known about its behavior and performance.[5] There may be other merchant transmission links of this kind that rely on nodal price differences for revenues that I am not aware of, but it is clear that the classic merchant model has hardly taken the world by storm.[6]

I do not want to revisit or expand on the analysis in our 2005 paper. The theoretical research on the classic merchant model was elegant and useful. For good reasons, it just is not being applied widely in practice. However, this does not mean that competition cannot be introduced into the development of new transmission projects. An alternative competitive transmission model is available that is more compatible with the technical and institutional attributes of transmission networks in the US ISO regions, regional planning processes and with the regulatory institutions that govern compensation for transmission owners and the design of tariffs for transmission customers. I will call this model the "competitive transmission procurement model" as it is based on competitive bidding for ISO-designated transmission projects or identified reliability violations or opportunities economically to reduce congestion costs.

In the US context, under this model, the construction and operating costs incurred by the winning bidder are still recovered through traditional cost-of-service regulation, perhaps after performance incentive provisions as well as regional cost allocation and transmission tariff design protocols are applied. Accordingly, this model involves ex ante competition for transmission projects that are regulated ex post. The

[3] Basslink is a third "merchant" link between Tasmania and Victoria that was completed in 2006. The link was originally developed by a subsidiary of National Grid PLC following a competitive process run by Hydro-Tasmania. The rights to the link are contracted to Hydro-Tasmania, the government owned electricity company in Tasmania, under a long-term contract with the owner of the link. Pursuant to the long-term contract Hydro-Tasmania pays an annual facility fee to have exclusive access to the link. While Hydro-Tasmania developed the link in anticipation of profitable energy price arbitrage between Tasmania and the South Australian market, the annual facility fee payed to the owner of the link is based on the project's cost.

[4] https://www.aer.gov.au/networks-pipelines/determinations-access-arrangements/directlink-determination-2006-15; https://www.aer.gov.au/networks-pipelines/determinations-access-arrangements/murraylink-determination-2018-23/proposal. Littlechild (2012) argues that the merchant projects in Australia did not exhibit the kinds of market imperfections discussed by Joskow and Tirole (2005). However, the two Australian merchant projects discussed did not succeed financially and were converted to regulated projects. I do not disagree with the observations about regulatory imperfections and that the potential imperfections of merchant projects should be assessed in the context of the imperfections of regulated projects. Joskow and Tirole (2005) also consider classical merchant projects financed based on the differences between nodal prices to have a much narrower definition than the broader category of projects based on "private initiative," including the kinds of competitive projects selected pursuant to Order 1000.

[5] The owner of the line conducts auctions for transmission right. The line also appears to experience frequent outages. http://www.oatioasis.com/matl/.

[6] I will discuss briefly below a couple of additional US projects that have components that follow the classical merchant model.

terms and conditions that the winning bidder agrees to then become a sort of (long-term) regulatory contract that defines how compensation for these projects will be determined over time by regulators. The "regulatory contract" here is defined by the selection criteria used by the ISO, performance commitments made by the winning bidder and Federal Energy Regulatory Commission (FERC) cost-of-service recovery rules. For example, the winning bidder could make a firm commitment for construction and maintenance costs of a proposed project, or cap cost recovery (revenues) for some period of time, or propose a sharing mechanisms if costs are lower or higher than specified, or simply agree to be compensated based on traditional cost-of-service regulatory principles (prudent capital costs and reasonable operating costs) without any special performance incentives—the default—once they are selected through competitive procurement. I will discuss in more detail how transmission owners are compensated through FERC cost-of-service regulation further below.

Variants of the competitive procurement model have been used in some countries outside North America for years (Mountain and Carstairs 2018). The model began to evolve slowly and idiosyncratically in the USA and in Canada in the early twenty-first century. More recently, the application of this model in the USA has now been stimulated by FERC Order 1000 issued in 2011 with compliance dates starting in 2014.[7,8]

The paper proceeds as follows. The first section discusses the institutional context in which electric transmission planning, investment, and revenue/cost recovery takes place in the US ISO regions. The next section discusses why competitive procurement can improve the information about costs available to regulators and supplement any performance incentives that are part of standard regulatory protocols. The fourth section discusses FERC Order 1000. The fifth section briefly discusses pre-Order 1000 experience with competitive procurement in the USA and other countries.

---

[7]https://www.ferc.gov/industries/electric/indus-act/trans-plan.asp.

[8]There is another competitive transmission model that has never drawn much interest in the USA. This is the "natural gas pipeline" model. In the USA, new interstate natural gas pipelines and expansions of existing pipelines are developed in the following way. An interstate pipeline developer sees a market opportunity to transport additional natural gas from point A to point B. The developer then proposes a pipeline project to increase gas transportation capacity from point A to point B and seeks long-term contractual commitments from shippers to ship their gas from one point to another served by the proposed pipeline. The bids are solicited through a competitive procurement process (open season). If enough shippers ("anchor shippers") agree to make long-term commitments to the project at negotiated prices and price adjustment provisions subject to regulated price caps determined by FERC using traditional cost-of-service principles to make the project financially viable, then the developer may proceed. Pipeline capacity that is not sold under long-term contracts can be marketed under short-term contracts by the owner at market prices. There is no collective regional planning or approval process for natural gas pipelines as there is for electric transmission facilities. This model has not attracted much interest in the electric power sector in the USA, though there are a few project developers who have attempted or are attempting to develop projects using this model. See for example, "1000-MW New York power transmission project signs up anchor customer," *Megawatt Daily*, November 27, 2018, p. 4. In the US electric power sector, this model seems to me to be best suited for DC lines where the "point A to point B" model better reflects the attributes of electric transmission facilities and the transmission networks of which they are components.

The fifth and sixth sections discuss experience with the implementation of Order 1000 to date, including a discussion of the attributes of proposals submitted and evaluation metrics used in all of the competitive procurements undertaken in the USA between 2013 and 2018. The final substantive section discusses lessons learned and opportunities to enhance the value of competitive procurement for transmission in the U.SA. There is a short concluding section.

2 Transmission Planning, Project Selection, and Economic Regulation in the USA

Transmission facilities in an ISO region[9] are owned by multiple transmission utilities, primarily incumbents which have been in operation for a century or more. Transmission owners typically also own proximate regulated distribution utilities and may own generating facilities in their region. Transmission owners' facilities typically cover a geographic "footprint" around their distribution utilities and connecting generating facilities which they owned prior to restructuring or continue to own. There are also jointly owned transmission facilities and a few independent transmission utilities that do not own generating or distribution facilities in the region.

The ISO is responsible for managing energy and ancillary service markets, capacity markets where they exist, and the integrated operation of the transmission network in conjunction with these markets and various reliability criteria. The ISO does not own any generating, transmission, or distribution facilities. The ISO is the region's transmission operator, operating transmission facilities owned by others.

FERC Order 888 requires that transmission must be at least be "functionally" unbundled from generation and distribution and that all transmitting entities must file Open Access Transmission Tariffs (OATT). All ISOs have filed Open Access Transmission Tariffs (OATT) that have been approved by FERC. The OATT for each ISO includes rules governing the scope and definition of transmission services, ancillary services, congestion management, congestion prices, congestion revenue rights, transmission cost allocation rules,[10] transmission planning procedures, regional transmission tariffs, and anything else related to the operation of, investment in, interconnection of, use of, and planning for the transmission facilities under the ISO's umbrella. The OATT for the New England ISO, for example, is 500 pages in length, including schedules and appendices; there are 45 pages of text just for cost allocation rules for incumbent and non-incumbent transmission owners. PJM's OATT is 3500 pages and the companion PJM Operating Agreement is over 600 pages long.[11]

ISOs also manage a regional transmission planning process and coordinate their transmission planning with local utility transmission planning and investment in their

[9]https://www.ferc.gov/industries/electric/indus-act/rto.asp.

[10]FERC Order 1000 requires that regional cost allocation follow a "beneficiary pays" principle.

[11]See FERC orders 888, 889, 890, 1000, https://www.ferc.gov.

region. The nature of the interaction between regional planning and local planning varies from ISO to ISO. Pursuant to FERC Order 890, both regional and local transmission planning processes are supposed to be open to all stakeholders, including non-incumbent transmission developers. The regional planning process specifies transmission expansion needs and designates which transmission developer/owner is responsible for building the facilities. Prior to Order 1000, incumbent transmission owners had a right of first refusal to build, own, and operate the designated transmission facilities. Order 1000 ended any (federal) right of first refusal, so non-incumbents may seek to develop projects selected by the ISO through the planning process.

Prior to Order 1000, there was nothing in principle that kept an independent transmission developer from proposing to an ISO to build a project and to recover its costs from the revenues it anticipates receiving from the sales of congestion revenue rights alone. That is a classical merchant transmission project. Except for a few special cases noted below, the planning, regulatory, and economic environments have made classical merchant investments unattractive.

ISOs are not economic regulators; they do not determine the rules for compensating transmission owners for the use of its facilities or the tariffs that specify how different categories of transmission customers pay for the use of their facilities. This type of economic regulation is the responsibility of FERC except for Texas which is subject only to state regulatory jurisdiction.[12] FERC also regulates the ISOs. States retain the authority to review the siting of proposed transmission projects, including environmental impacts and the "need" for the facilities. State certification is required in order for at least a major transmission project to proceed. Certification procedures and requirements vary from state to state.

Transmission owners are compensated through the FERC regulatory process using fairly traditional rate of return/cost-of-service procedures. A transmission owner in an ISO (whether an incumbent with distribution and perhaps generating facilities or an independent) must file with FERC the information necessary to form a transmission "revenue requirement." The revenue requirement starts with the specification of a rate base. The rate base is equal to the depreciated original cost of the transmission owner's facilities. The capital cost-related portion of the revenue requirement is then the annual depreciation on the rate base plus carrying charges on the rate base. The carrying charges are derived by the regulator specifying an allowed rate of return which is multiplied by the rate base. To these two elements of capital cost recovery are added each transmission owner's operating and maintenance costs, fees, and adjustments for taxes. This yields the transmission owner's initial revenue requirement. However, since there are multiple transmission owners within the region and shared regional cost responsibility for some projects, some portion of the initial transmission revenue requirement may be allocated to or from other transmission owners. FERC relies on a formula rate approach that adjusts the revenue requirement annually for changes in capital, operating and maintenance costs, allowed rate of return, etc.

[12]Aside from implementing cost allocation and regional transmission tariff provisions in OATT.

The precise cost allocation and tariff structures that determine how transmission owners actually collect their revenue requirement vary from ISO to ISO. The procedures are fairly complicated in practice, though the basic principles are similar. For example, in the CAISO, the transmission owner's total revenue requirement is divided between a Regional Revenue Requirement (R-TRR) for high voltage facilities and a Local Revenue Requirement (L-TRR) for lower voltage facilities that the transmission owner relies on to serve distribution customers in its area. The CAISO has a tariff to charge transmission customers, primary distribution utilities and exporters, a postage stamp rate based on their load ratios for use of the regional high voltage network. The revenues are then remitted to the transmission owners based on their Regional Revenue Requirements. The transmission owners each have their own FERC approved tariff for recovery of their Local Revenue Requirement from distributors designated to be in their transmission footprints.

I recognize that this cost allocation and cost recovery process is fairly complicated. It is depicted graphically in Fig. 1. To complicate matters further, there are a variety of grandfathered transmission contracts and transmission cost allocation agreements that have been carried over from the pre-restructuring/ISO period.

This compensation, cost allocation, and revenue collection process was clearly developed initially with incumbent transmission owners whose facilities primarily serve their local distribution companies in mind. How does it work for a non-incumbent independent transmission developer/owner no distribution assets in the

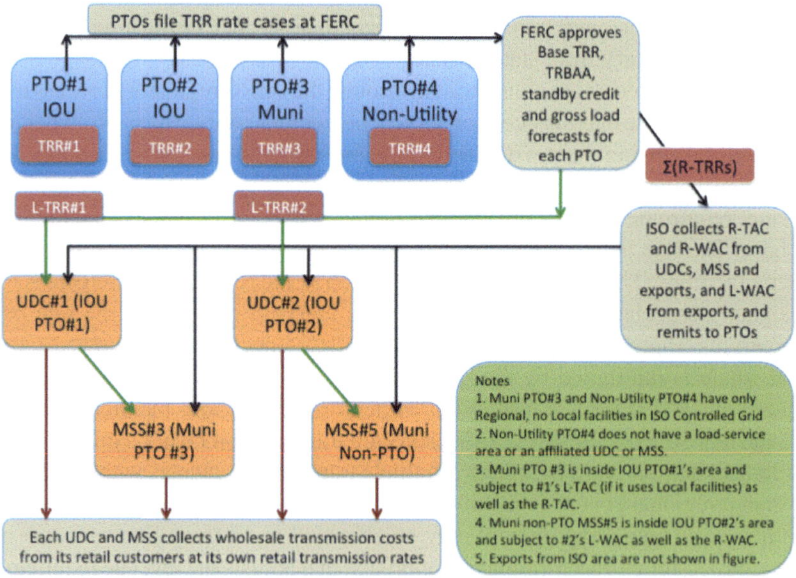

Fig. 1 Transmission Cost Allocation in the CAISO
Source "How Transmission Cost Recovery Through the Transmission Access Charge Works Today," CAISO, April 12, 2017

area? The process would be similar. The transmission owner would still file with FERC for a transmission revenue requirement which would include any incentive arrangements it has agreed to through a competitive procurement process and would be adjusted by formula annually. The ISO would then apply its cost allocation rules to the transmission owner's facilities to determine how the costs would be recovered through a regional ISO tariff, the tariff of a transmission owner with local facilities, or directly from transmission customers. The revenues are then remitted back to the independent transmission owner.

As noted, FERC is the economic regulator for high voltage transmission facilities operated by the ISO. In principle, it applies prudent investment and reasonable cost standards to the capital and operating costs presented to it by a transmission owner. Costs that are determined to be imprudent or unreasonable can be disallowed and excluded from the revenue requirement. However, at best, such exclusions are rare. Nor does FERC apply conventional incentive/performance-based regulation mechanisms (Joskow 2014; Vogelsang 2001, 2018). For all intents and purposes, the FERC regulatory process is a model of cost pass-through regulation with little scrutiny of costs. FERC does offer financial incentives for transmission investments meeting several specified goals, but these are different from traditional incentive/performance-based regulatory mechanisms.

According to FERC, the following types of incentives are available:

> The Energy Policy Act of 2005 directed the Commission to develop incentive-based rate treatments for transmission of electric energy in interstate commerce, adding a new section 219 to the Federal Power Act. The rule implemented this new statutory directive through the following incentive-based rate treatments:
>
> a. Incentive rates of return on equity for new investment by public utilities (both traditional utilities and stand-alone transmission companies, or transcos);
> b. Full recovery of prudently incurred construction work in progress;
> c. Full recovery of prudently incurred pre-operations costs;
> d. Full recovery of prudently incurred costs of abandoned facilities;
> e. Use of hypothetical capital structures;
> f. Accumulated deferred income taxes for transcos;
> g. Adjustments to book value for transco sales/purchases;
> h. Accelerated depreciation;
> i. Deferred cost recovery for utilities with retail rate freezes; and
> j. A higher rate of return on equity for utilities that join and/or continue to be members of transmission organizations, such as (but not limited to) regional transmission organizations and independent system operators.
>
> All rates approved under the rules are subject to Federal Power Act rate filing standards. The rule allows utilities on a case-by-case basis to select and justify the package of incentives needed to support new investment. Additionally, the rule provides expedited procedures for the approval of incentives to provide utilities greater regulatory certainty and facilitate the financing of projects. The rule became effective on September 29, 2006.[13]

[13]https://www.ferc.gov/industries/electric/indus-act/trans-invest.asp.

Aside perhaps from removing the disincentive a regulated utility might have to continue to operate a project which should be abandoned so as to avoid having the undepreciated costs stranded, most of these incentives are either incentives to invest in transmission generally or to join an RTO/ISO or to encourage the creation of independent transmission companies. They are not the kind of cost control and operating performance incentives that would normally be an important part of a performance-based incentive regulation tool kit. Rather, the incentive scheme is basically cost-of-service regulation with higher returns to take certain actions that advance FERC policies—especially encouraging investing in transmission by offering higher returns than the standard pro-forma rate of return and higher cash flows by doing so and encouraging independent transmission companies. FERC has been uninterested in applying more traditional incentive regulation mechanisms.

3 FERC Order 1000

FERC Order 1000 was issued on July 21, 2011 and became effective on October 11, 2011. This Order (or technically "Rule") establishes revised obligations regarding "Transmission Planning and Cost Allocation by Transmission Owning and Operating Utilities." The Rule builds on earlier transmission reforms, especially Order 890, related to open transmission planning processes and cost allocation methods.

Regarding transmission planning reforms, the new Rule specifies that[14]:

a. Each public utility transmission provider must participate in a regional transmission planning process that satisfies the transmission planning principles of Order No. 890 and produces a regional transmission plan.
b. Local and regional transmission planning processes must consider transmission needs driven by public policy requirements established by state or federal laws or regulations. Each public utility transmission provider must establish procedures to identify transmission needs driven by public policy requirements and evaluate proposed solutions to those transmission needs.
c. Public utility transmission providers in each pair of neighboring transmission planning regions must coordinate to determine if there are more efficient or cost-effective solutions to their mutual transmission needs.
d. The Rule specifies 15 regional transmission planning areas, with each RTO/ISO a separate planning area. Transmission utilities that are not members of an RTO/ISO were assigned to a regional planning area as well.

Regarding cost allocation reforms, the rule provides that:

a. Each public utility transmission provider must participate in a regional transmission planning process that has a regional cost allocation method for new

[14]This description is taken directly from https://www.ferc.gov/industries/electric/indus-act/trans-plan.asp?csrt=16071104753993213338, which also contain a map of the transmission planning regions.

transmission facilities selected in the regional transmission plan for purposes of cost allocation. The method must satisfy six regional cost allocation principles.

b. Public utility transmission providers in neighboring transmission planning regions must have a common interregional cost allocation method for new interregional transmission facilities that the regions determine to be efficient or cost-effective. The method must satisfy six similar interregional cost allocation principles.
c. Participant funding of new transmission facilities is permitted but is not allowed as the regional or interregional cost allocation method.[15]

Order 1000 also contains reforms supporting the participation of non-incumbent transmission developers in regional transmission planning and the development of new transmission lines meeting certain (vague) criteria. These reforms are of most interest to me here. In particular,

a. Public utility transmission providers must remove from Commission-approved tariffs and agreements a federal right of first refusal for a transmission facility selected in a regional transmission plan for purposes of cost allocation, subject to four limitations:
    - This does not apply to a transmission facility that is not selected in a regional transmission plan for purposes of cost allocation.
    - *This allows, but does not require, public utility transmission providers in a transmission planning region to use competitive bidding to solicit transmission projects or project developers.*
    - Nothing in this requirement affects state or local laws or regulations regarding the construction of transmission facilities, including but not limited to authority over siting or permitting of transmission facilities.
b. The rule recognizes that incumbent transmission providers may rely on regional transmission facilities to satisfy their reliability needs or service obligations. The rule requires each public utility transmission provider to amend its tariff to require reevaluation of the regional transmission plan to determine if delays in the development of a transmission facility require evaluation of alternative solutions, including those proposed by the incumbent, to ensure incumbent transmission providers can meet reliability needs or service obligations.

The affected transmission utilities were given 12 months to make a compliance filing (18 months for interregional cost allocation methods). With the back and forth between the regions, FERC and the courts, the completion of the compliance process took more time than this. Most regions completed the compliance process in

[15] Participant funding refers to merchant projects that are compensated with transmission congestion revenues, the sale of transmission congestion revenue right, or contracts with specific transmission users. The developers of participant funded projects bear the full market risk of these projects. See FERC Policy Statement on participant funded transmission projects. https://www.ferc.gov/whats-new/comm-meet/2013/011713/E-2.pdf.

2014, 2015 and 2016.[16] Of most interest to me here are the reforms governing non-incumbent transmission developers and the use of the option of adopting a competitive bidding process that includes both incumbent and non-incumbent transmission developers. With or without a formal competitive solicitation process, non-incumbent utilities can participate in and, in principle, be selected to develop a project through the open regional transmission planning process. FERC seems to place a lot of faith in this open competition with ideas planning process. As discussed below, there is little evidence to support this view. The cost allocation rules are important as well, since the cost allocation rules ultimately define how transmission developers are paid and who ultimately pays for the associated transmission costs over the life of the assets. Order 1000 requires that regional cost allocation rules follow "beneficiaries pay" principles.

4 Why Competitive Procurement?

The reliance on traditional cost-of-service regulation, absent performance incentives, to determine the "revenue requirements" that a transmission owner may recover is subject to the reasonable criticism that it provides poor incentives for controlling costs. As already noted, FERC does not have a well-developed process to scrutinize the costs presented to it for inclusion in the transmission owners' revenue requirements or a history of disallowing unreasonable costs. To a first approximation FERC cost-of-service regulation is cost pass-through regulation with little scrutiny of costs. While third parties, including state public utility commissions, may challenge the costs incurred by a transmission owner or its operating performance pursuant to "prudence" or "just and reasonable" criteria, these criteria are rather vague and applied only in exceptional circumstances. In principle, a state regulator could file a complaint with FERC when the transmission owner seeks cost recovery from FERC if it believes that the costs incurred were excessive. However, I am not aware of any such cases. The ISO's regional transmission planning process may provide some natural cost containment features since cost estimates are one factor that will affect which projects are included in the Regional Transmission Expansion Plan. However, as noted by Pfeifenberger et al. (2018), roughly half of the costs of transmission investment incurred between 2013 and 2017 across ISOs are not scrutinized in any detail by the ISO or the regional stakeholder planning process. And realized costs may differ significantly from cost estimates. While FERC has accepted, in principle, the value of performance-based regulatory mechanisms and the associated rewards and penalties such as those discussed in Joskow (2014), as notes, FERC transmission incentives have focused on providing generous cost recovery rules for transmission owners to make adding transmission capacity financially attractive and for meeting certain criteria reflecting its policy goals (e.g., independent

[16] https://www.ferc.gov/industries/electric/indus-act/trans-plan/regional.asp?csrt=319600170759416567.

transmission company, merchant transmission, new technology).[17] FERC Order 679 expressed FERC's continued general interest in performance-based incentives and competitive bidding but required neither as part of its transmission incentive portfolio. Indeed, Order 679 reflects considerable skepticism about the importance of performance-based incentives and, perhaps more importantly, the ability to apply classic performance-based regulatory mechanisms to transmission in the USA given the structure of the transmission industry (e.g., balkanized transmission ownership, ISOs with no assets, continued vertical integration and state regulation of bundled transmission costs in a large fraction of the country, private, public, and municipal ownership with different regulatory regimes).

The structure of the high voltage transmission sector in the USA and the associated regulatory institutions are indeed a barrier to applying performance-based regulatory mechanisms similar to those that have been applied in other countries. Joskow (2014, pp. 326–331) provides examples of the application of incentive regulatory mechanisms in the UK. However, at that time, there was a single monopoly integrated independent transmission grid owner and operator to which incentive regulation mechanisms could be applied by a single regulator. This is not the case in the USA. In the USA, there are many transmission owners that own pieces of the regional transmission system. An ISO owns no transmission assets and does not develop or maintain transmission projects. It is not clear how one would apply classical incentive regulation mechanism to multiple owners of assets that compose a single regional grid or to the ISO which has no assets and is not really owned by anyone. The ISO could file complaints with FERC about excessive costs or poor performance by transmission owners in its region, but I am not aware that this has ever happened. FERC also has no authority to review transmission siting and environmental impacts or issue certificates of need. FERC's jurisdiction also extends only to investor-owned transmission facilities.

Laffont and Tirole (1993) have derived a variety of (second-best) efficient "incentive regulatory mechanisms" aimed at solving two fundamental problems that regulators must confront: (a) prices/revenues that exceed what is necessary to induce the developer to invest result from imperfect and asymmetric information about the firm's true costs; and (b) excessive costs result from inefficiently low firm managerial effort, or moral hazard. Very simply, this is the case because regulators have imperfect knowledge about the firm's costs while the regulated firm knows its costs and can exploit this information asymmetry to its advantage. Regulators also cannot observe the firms "effort" (managerial performance), and other things equal, managers would prefer exerting less effort to more effort. The Laffont–Tirole procurement and regulatory mechanisms can be applied both to regulation of legal monopolies, like US utilities in the USA, and to the design of procurement auctions and associated contracts

[17]FERC's policies on incentives for transmission owners were first articulated in FERC Order No. 679 (July 20, 2006). The types of incentives FERC has focused on are described in paragraphs 84–263 of Order No. 679. Paragraphs 263–279 discuss performance-based ratemaking and competitive bidding with little enthusiasm and little more than "we will continue to think about it and support its development." http://www.ferc.gov/whats-new/comm-meet/072006/E-3.pdf. Little if anything progressed on this front prior to the competitive bidding provisions of Order 1000.

for everything from firms to airplane to transmission links. More generally, "competition for the market" has long been viewed as a potential substitute for traditional commission regulation of legal monopolies (Demsetz 1968). The idea is that if there is a natural monopoly supply situation, we can satisfy the constraint of having one firm in the market by putting the monopoly franchise out for competitive bids and then rely on a long-term commercial incentive contract to manage the relationship between the firm and consumers, rather than relying on commission regulation. However, the likely practical performance of competitive bidding as a governance arrangement for long-lived sunk assets with attributes like those of electricity, gas, cable TV, railroad infrastructure, etc. has been subject to a great deal of criticism (e.g., Williamson 1976; Goldberg 1976). The problem is that the long-term contracts needed to govern the relationship with the winning bidders and consumers involve long-lived sunk investments, must deal with a large number of contingencies, are inherently incomplete and subject to contractual breakdowns of various kinds. The argument is that over time, these contractual breakdowns become so severe that an administrative agency (i.e., a regulatory agency of some kind) must be created to manage "fairly" a dynamic contractual relationship as relevant contingencies emerge over time. After all, most of the electric, gas, telephone, street railway companies, etc., started out life through a competitive franchise bidding process. Their franchise contracts broke down and they became subject to commission regulation, effectively substituting a long-term regulatory contract for a long-term commercial contract.

From the perspective of Laffont and Tirole's incentive mechanism paradigm, competitive auctions or competition for the market can be viewed as a complement or substitute for optimal regulation (Laffont and Tirole 1993; Chap. 8). The auctioning of contracts reduces the regulators' uncertainty about the firm's true costs, reducing the costs of asymmetric information (excessive rents to the firms). An incentive contract is still needed to deal with moral hazard, but the terms and conditions of the contract can be specified by the regulator or made a component of the competitive bidding program. In the context of FERC regulation of transmission, I view competitive bidding as a partial substitute for the absence of performance-based regulatory mechanisms. Because FERC is ultimately the regulator of projects and associated compensation arrangements agreed to with the winning bidder, it already is in the position to adapt the terms and conditions of the attributes of the winning bid to deal with unanticipated contingencies, holdups, bankruptcies, or other potential contractual breakdowns over time.

The analogy to competition for the market or franchise bidding should not be taken too far. As we shall see, only small segments of large transmission networks are being put out for competitive bidding. Contractual breakdown would not be too costly in this situation anyway and a backstop process for designating incumbents to invest in needed facilities and for compensating them already exists. That is, FERC provides a default regulatory contract if the commercial contract fails, the firm performs poorly or goes bankrupt. As noted earlier in the paper, the two merchant links in Australia did go bankrupt and reverted to being regulated transmission links fairly smoothly.

Accordingly, competitive procurement of the type allowed by FERC Order 1000 can be viewed constructively as a complement to FERC regulation by providing

ISOs and FERC with more information about the costs of building and operating new transmission facilities and potentially introducing performance incentives over the capital costs, operating costs, performance of new transmission facilities. This view is reflected very briefly in Order 679.

In this context, a competitive bidding program for new transmission links allows competing transmission developers effectively to propose alternative regulatory cost recovery formulas for determining annual revenue requirements. For example, bidders might agree to a cap on revenue requirements for some number of years after the transmission facility is completed, or agree to alternative annual adjustment formula like CPI-x used in the UK and Latin America, or agree to cap construction costs allowed in determining the facility's revenue requirements, or a sharing mechanism, etc. I will refer to these kinds of provisions effectively as "cost containment" provisions below. Of course, bidders could simply propose to receive traditional FERC cost-of-service recovery, expecting to win a competitive solicitation based on a project that has the lowest estimated costs, and this is how the vast majority of transmission costs are recovered. This might be the case for example where the lowest cost project requires enhancements to existing facilities (e.g., reconductoring) which can be accomplished most economically by the incumbent. Or it might be the case where there is a lot of uncertainty associated with a project due to permitting challenges.

Note, that in this discussion, I have assumed that the burden of evaluating cost estimates, trading off alternative cost commitments, risks, etc. is on the ISO when it selects projects through competitive bidding. Since the ISO is choosing the projects and Order 1000 places the burden on it to choose the most efficient or cost-effective projects identified in the regional transmission plan at first blush this makes sense. However, ISOs are not economic regulators in the traditional sense and have neither the expertise nor authority to adopt transmission ratemaking procedures. FERC is the economic regulator of transmission costs, incentives, and transmission rates and regulates the ISOs. There seems to be an institutional gap here that needs to be filled.

5 Competitive Procurement for Transmission Projects Before Order 1000

5.1 Other Countries

Before proceeding to discuss the experience with competitive procurement in the USA following Order 1000, it is useful to recognize that competitive procurement for specific transmission projects is not a new idea. As noted, competitive procurement of transmission of projects has been used in other countries for years.

Argentina initiated a competitive procurement auction for the fourth Camahue to Buenos Aires line (1700 Mw) in 1994 and a developer was selected in 1997. Much has been written about this competitive procurement process (Galetovic and Inostroza 2007; Littlechild and Skerk 2007; Littlechild and Ponzano 2008). Argentina subsequently reformed its electricity law again and continues to rely on competitive procurement for transmission (Sijm 2015). (Sijm concludes that that the Public Contest method is an interesting framework that could govern the development and operation of the EU transmission network envisioned for 2050.) Brazil (extensively), Peru, Chile, the UK, and India have also used competitive procurement (competitive tenders) for long-term contracts to support investment in new transmission facilities meeting a variety of criteria (Mountain and Carstairs 2018).

Alberta began to consider a competitive solicitation process for major regional transmission projects in 2011 and approved the details of the process in 2012 and 2013.[18] The Fort McMurray West project (500 kV) was selected through this process, though final approvals took quite a bit of time. The project was completed in March 2019. The project is supported with a 40-year contract negotiated between the developer and the Alberta ISO.[19]

5.2 *The USA*

The Long Island Power Authority (LIPA) has used a competitive procurement mechanism to select transmission interconnectors between Long Island and New England and between Long Island and PJM.[20] The New York Power Authority (NYPA) selected a transmission project (the Hudson Project completed in 2013) to connect New York City and PJM. These three projects are "participant funded" and are supported by long-term contracts with either buyers (LIPA and NYPA), or in the case of Hudson, for a large fraction of, the line's capacity. Neither the details of these contracts nor the details of the competitive procurement process and evaluation criteria are publicly available. The Hudson arrangement is interesting because a portion of the capacity was retained by the developer for "merchant" sales of point to point transmission capacity on the line.[21]

In 2009, Linden VFT, LLC, an affiliate of GE Energy Financial Services ("GE EFS"), installed a variable frequency transformer facility ("VFT") connected to the existing Linden, New Jersey, to Brooklyn, N.Y. line. This projected created an additional 315 megawatts ("MWs") of bi-directional electricity transfer capability between the control area of the PJM Interconnection, LLC ("PJM") and the

[18]https://www.aeso.ca/grid/competitive-process/.

[19]https://www.aeso.ca/grid/competitive-process/fort-mcmurray-west-500-kv-transmission-project/.

[20]The Neptune (2007) interconnector between LIPA and PJM. The Cross-Sound interconnector between New England and Long Island (2005).

[21]Unfortunately, the Hudson line incurred a series of serious system faults beginning in 2016. The developer defaulted on its loans which led to renegotiation with the New York Power Authority, lenders, and insurers. The entire cable is being replaced and new contractual arrangements negotiated.

New York Independent System Operator Inc. ("NYISO") Zone J, on an existing regulated transmission line between New Jersey and New York City. The additional capacity is participant funded and is a classic merchant project similar to the financing and contracting for new natural gas lines. Linden VFT holds open seasons where the transmission scheduling rights ("TSRs") for the project's electric transfer capacity are auctioned to market participants, as anticipated by the Merchant model.[22] While Linden VFT is an expansion of an existing regulated link and did not have to confront the licensing, cost, and construction challenges associated with an entirely new underwater link, it does clearly embody the classic merchant model.

Interconnection between PJM and New York City/Long Island and between New England and New York/Long Island are particularly attractive for supporting a merchant model as energy and capacity prices in New York City/Long Island are on average much higher than in the surrounding areas due to transmission congestion and special reliability rules applicable to New York City.[23] If the classical merchant model can work anywhere, it is here.

The Public Utility Commission of Texas approved a competitive procurement process in 2008 to select developers of about 2400 miles of new transmission lines to relieve congestion between Competitive Renewable Energy Zones (CREZ) (wind) and load in ERCOT, increasing transfer capacity by about 18,500 Mw.[24] The selection criteria and selection rationale for the chosen projects are not available to me.[25] However, these projects appear to be traditional cost-of-service regulated projects whose costs are ultimately recovered by retail consumers through their "wires" charges. Nine different transmission service providers, a mix of Texas utility incumbents and non-incumbents, were selected to build the CREZ facilities, at an estimated 2008 cost of $4.93 billion and a target completion date 2013. The projects were completed in January 2014. The final cost of the projects was about $7 billion,

[22]In April and May of 2018, Linden VFT, LLC (an affiliate of General Electric Company) will conduct an open solicitation to sell 315 MWs of transmission scheduling rights ("TSRs") pertaining to its Linden Variable Frequency Transformer Project ("Linden VFT"). The TSRs will be sold for a term beginning June 1, 2019. The term length of the TSR purchase agreements will be specified by the bidder, with a minimum term of one year. The Linden VFT TSRs allow for the withdrawal (or injection) of power at the Linden VFT switching station near Linden, NJ and the injection (or withdrawal) of power near the Goethals Substation in the Borough of Staten Island, New York City. As a result, the TSRs can be used to sell energy and capacity sourced in PJM into New York ISO ("NYISO"), as well as energy and capacity sourced in NYISO into PJM. This project is the closes to a classic merchant project that I have found in the USA.

[23]On the other hand, all of these projects had to confront technical reliability issues associated with connecting ISO-New England and PJM with New York. The two LIPA projects are DC links. The Linden VHF facility represents an investment that allowed the existing link to operate at a higher capacity by resolving reliability issues.

[24]ERCOT, the ISO that covers most of Texas, is not subject to FERC jurisdiction.

[25]In unpublished research, Stephen Littlechild and Ross Baldick (Manuscript in process, Parts I–V, 2017–2019) have studied the selection process for the CREZ projects. It was a very complex process that might best be described as competitive negotiation for the authority to build one or more regulated projects rather than the kind of competitive procurement applied by ISOs under FERC Order 1000.

though more miles of transmission were ultimately constructed than had initially been anticipated when the cost estimates were made.[26]

California adopted a competitive procurement process for certain transmission projects in 2010 before Order 1000 was issued and the process was then adjusted to conform to Order 1000. These projects will be discussed further below.

6 Early Experience with Order 1000 Competitive Procurement Programs: Overview[27]

As of 2018, five of the six RTO/ISOs subject to FERC jurisdiction have adopted and implemented competitive solicitation programs of one kind or another (New York ISO (NYISO), California ISO (CAISO), PJM, SPP, and MISO). In principle, ISO-NE has agreed to implement competitive procurement for projects that meet certain criteria and the first RFP was not issued until March 2020.[28] ISO-NE was very slow too embrace competitive procurement. Its 2017 Regional System Plan states no projects had met its criteria for competitive procurement:

> Since the effective date of the order, the ISO has completed several area needs assessments or has conducted an update to an already completed needs assessment. [footnote omitted] The results of all the needs assessments show that that time-sensitive and a few non-time-sensitive needs exist. [footnote omitted] Thus, the solutions study process has been used first to solve the time-sensitive needs, and the Competitive Solutions Process for the few non-time-sensitive needs has been placed on hold until the time-sensitive needs are addressed through the solutions study process. [footnote omitted] After the solutions have been identified for the time-sensitive needs, the ISO will begin a new needs assessment, which will include the preferred solutions for the time-sensitive needs and identify any remaining needs. [footnote omitted] The ISO will continue to review the implementation of the competitive process in New England and across the country.[29]

---

[26] Warren Lasher, "The Competitive Renewable Energy Zones Process" (presentation), ERCOT, August 11, 2014. https://www.google.com/url?sa=t&rct=j&q=&esrc=s&source=web&cd=1&cad=rja&uact=8&ved=2ahUKEwjgsr6VvJ3gAhXGnuAKHZpiAGgQFjAAegQICRAC&url=https%3A%2F%2Fwww.energy.gov%2Fsites%2Fprod%2Ffiles%2F2014%2F08%2Ff18%2Fc_lasher_qer_santafe_presentation.pdf&usg=AOvVaw1M34UYAK3OlTjTvnQH1PMO. Public Utility Commission of Texas, "Scope of Competition in Electricity Markets," January 2009. See also, Electric Transmission Texas, http://www.ettexas.com/Projects/TexasCrezand; https://www.google.com/url?sa=t&rct=j&q=&esrc=s&source=web&cd=1&ved=2ahUKEwil1aiTv53gAhWw11kKHT_5C9kQFjAAegQIChAC&url=https%3A%2F%2Fcleanenergysolutions.org%2Fsites%2Fdefault%2Ffiles%2Fdocuments%2Fjeff-billo_webinar-ercot-crez-process.pdf&usg=AOvVaw0iscL8rx-8fSIFt_OMT0T3. ERCOT, *CREZ Progress Report No. 17,* November 14 update p. 6.

[27] See the Data Appendix for a description of the process that I used to collect information for each ISO.

[28] The New England ISO's website discusses its anticipated competitive procurement process and contains information about the first RFP issued in March 2020.

[29] ISO-NE 2017 Regional System Plan, p. 69. https://www.iso-ne.com/static-assets/documents/2017/11/rsp17_final.docx.

It is also relevant that planned transmission investments mediated through ISO-NE's regional transmission planning process are expected to decline dramatically in the future. Between 2000 and 2018, over $10 billion in transmission investments were authorized, while less than $2 billion of incremental transmission investments are forecast to be made between 2018 and 2026. The expected need for reliability-related projects as load growth has stagnated, large fossil and nuclear plants are retiring, and the focus has turned to securing and integrating no-carbon generating resources. These numbers do not include "Elective Projects" which I will discuss separately below.

However, there is an alternative path to competitive procurement in New England based on state initiatives. This plays a similar role to the procurement process for public policy transmission projects as implemented by the New York ISO and the New York Public Service Commission and is expected to grow to support the expansion of no-carbon generating resources in New England. Moreover, there is growing reliance on small scale merchant investment to remove transmission constraints faced primarily by certain wind generations. Accordingly, I will discuss these developments separately below.

At this relatively early stage, it is important to recognize that the ISOs have adopted a variety of policies that significantly limit the projects that are solicited through a formal open competitive procurement. Factors that determine whether or not a project is open to competitive procurement include time until project is needed, subject to regional or local reliability criteria, type of project (reliability, public interest, market efficiency), upgrades of existing facilities, voltage, type of equipment (e.g., substations) and other considerations that are not particularly transparent. As emphasized in a recent study by Pfeifenberger et al. (2018), meaningful competitive solicitations account for a tiny fraction of transmission projects approved since Order 1000 went into effect. The ISOs have also adopted different approaches toward integrating the transmission planning process with the competitive solicitation process. CAISO, MISO, and SPP identify specific projects that they conclude are needed to meet reliability, market efficiency, and public policy needs through the regional transmission planning process. A competitive solicitation and associated RFP is then developed for a small set of these projects meeting ISO specified criteria. While specific projects are put out for competitive bidding, the details of the design of the project may vary significantly from one competitive proposal to another. FERC staff refers to this as a competitive bidding model (FERC 2017).

PJM and NYISO use the transmission planning process to identify specific reliability, market efficiency, and public policy "needs." The competitive process then allows bidders to specify proposed transmission projects that meet these needs. FERC staff refers to this as a "sponsorship" model. Arguably, this gives the competitive procurement process an even greater opportunity to attract more innovative and cost-effective solutions to a transmission need that might not have been identified through a specific project first identified by the ISO and then subject to competitive procurement. However, NYISO and PJM have applied the sponsorship model quite differently as well. In principle, the PJM Regional Transmission Expansion (RTEP) planning process and the associated competitive "windows" provide a much

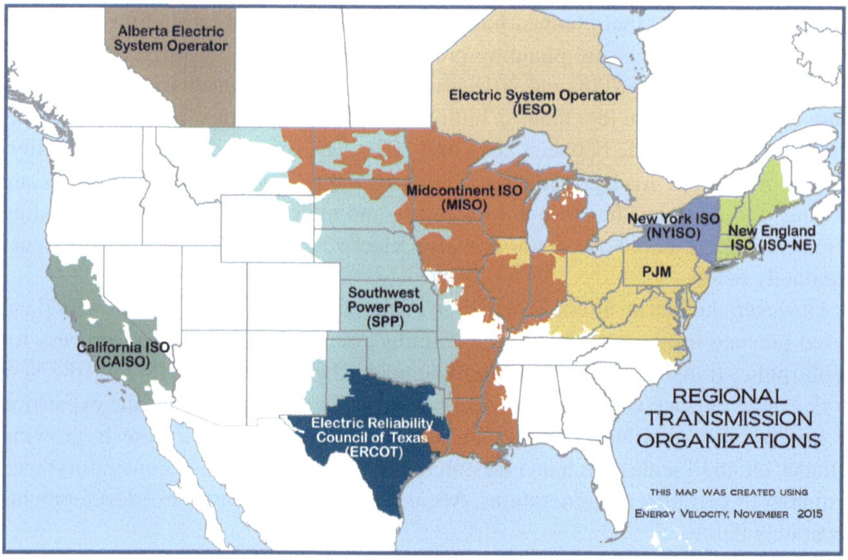

Fig. 2 Regional Transmission Organizations in the U.S. and Canada
Source FERC, https://www.ferc.gov/industries/electric/indus-act/rto.asp

larger number of opportunities for incumbents and non-incumbents to compete for reliability and market efficiency projects. As discussed further in the next section, between 2013 and 2017, about 800 proposals were received in response to 16 competitive windows that were opened.[30] About 140 project selections have been made through this process. However, only three went to non-incumbents. The NYISO has designated only two large "public interest" needs for which competitive solicitations have been initiated. PJM also has implemented a competitive process for "market efficiency" projects while NYISO has not.[31]

7 Experience in Each ISO[32]

Figure 2 is a map of the US ISOs

[30]Another seven proposals were submitted in a short-term window opened in 2018 but selections have not been announced as this is written. A 2018/19 long-term window is still open as this is written.

[31]PJM is a multi-state RTO and apparently has left it to transmission companies to work with each state to identify public policy needs.

[32]SPP issued an RFP for one project and went through a competitive solicitation and awarded the project to one of the proposed sponsors. However, the project was subsequently canceled by the regional planning organization due to declining load.

7.1 CAISO

I found ten projects selected through competitive procurement by the CAISO between 2013 and 2016.[33,34] FERC (2017, p. 22) indicates that there have been nine RFPs. The Brattle study identified ten competitive projects (Pfeifenberger et al. 2018). The ten CAISO projects that I found match those in the Brattle study.

I was able to find no additional competitive solicitations by CAISO beyond those initiated based on CAISO's 2013–2014 resource planning process. While the 2014–2015 regional planning process did authorize eight new projects, none qualified for competitive procurement. The inclusion criteria are described as follows:

> Where the ISO selects a regional transmission solution to meet an identified need in one of the three aforementioned categories that constitutes an upgrade to or addition on an existing participating transmission owner facility, the construction or ownership of facilities on a participating transmission owner's right-of-way, or the construction or ownership of facilities within an existing participating transmission owner's substation, construction and ownership responsibility for the applicable upgrade or addition lies with the applicable participating transmission owner.[35]

The same result appears to have emerged from subsequent transmission plan updates. The 2015–2016 CAISO regional plan identified 14 new regional projects. None qualified for competitive solicitation. The 2016–2017 CAISO regional transmission plan authorized two new projects neither of which met the criteria for competitive solicitation. The 2017–2018 regional transmission plan identified 17 new projects, none of which qualified for competitive procurement based on these criteria. It is unclear to me whether the CAISO continues to be interested in competitive procurement for transmission. Accordingly, we examine the ten projects authorized for competitive solicitation in the 2013–2014 transmission plan to better understand the attributes of the procurement and evaluation process.

The CAISO competitive procurement process is quite transparent and well-documented, from the identification of the project to the evaluation criteria and ultimately to the evaluation and selection of the winning proposal. A common set of evaluation criteria and a common evaluation template was applied to all ten project solicitations. This makes the solicitation and evaluation process relatively straightforward to review. The projects that are selected to be included in its competitive procurement process are developed through an annual open transmission planning process. Potential developers responding to the RFP must submit a long list of technical information, economic information, including binding cost containment commitment information, about the project, and information about the experience, financial and technical capabilities of the sponsor. The CAISO has specified about

---

[33] http://www.caiso.com/planning/Pages/TransmissionPlanning/2012-2013TransmissionPlanningProcess.aspx; http://www.caiso.com/planning/Pages/TransmissionPlanning/2013-2014TransmissionPlanningProcess.aspx.

[34] One project had only one applicant.

[35] http://www.caiso.com/Documents/Board-Approved2014-2015TransmissionPlan.pdf.

20 evaluation criteria and posts an evaluation document that discusses CAISO's evaluation of each proposal and the rationale for the project selected.

In addition to meeting the technical criteria for the project and demonstrating experience successfully developing transmission projects on budget and with good operating performance, CAISO's evaluation process has given significant weight to meaningful binding cost containment commitments. Meaningful cost containment commitments offered by some developers include construction cost caps (subject to escalation for changes in scope and other contingences), O&M cost caps for a few years, and/or future revenue requirements caps (amortization of the construction cost plus operating and maintenance expenses), subject to various contingencies. Other sponsors of proposals provide only cost estimates without firm cost containment commitments and rely on their experience and proposed construction and operating plans to convince evaluators of their merit. In this case, the default is FERC cost-of-service regulation with no performance incentives.

Table 1 provides information on all ten transmission projects that were put up for competitive solicitation by CAISO. Note the wide range of ISO cost estimates for

Table 1 California Iso Competitve Transmission Projects

Project Name	Date Approved	ISO Planning Cost Estimate	Number of Bidders	Winning Bidder	Cost Containment[1]
Imperial ValleyPolicy Element	July 11,2013	$25 million	2	Incumbent[2]	Yes
Gates-Greg	November 6, 2013	$115-$145 million	5	Incumbent	No[3]
Sycamore-Penasquitos	March 4, 2014	$111-$221 million	4	Incumbent	No
Miguel 500kV	May 1, 2014	$30-$40 million	1	Incumbent	No
Suncrest	January 6, 2015	$50-$75 million	2	Non-incumbent	Yes
Estrella	March 11, 2015	$35-$45 million	4	Non-incumbent	Yes
Wheeler Ridge Junction	March 11, 2015	$90-$140 million	4	Incumbent	No
Spring	March 11, 2015	$35-$45 million	3	Incumbent	No
Delaney-Colorado River[4]	July 10, 2015	$300 million	5	Non-incumbent	Yes
Henry-Allen To Eldorado	January 11, 2016	$144 million	3	Non-incumbent	Yes

1. Developed from project solicitation evaluation reports. http://www.caiso.com/planning/Pages/TransmissionPlanning/2013-2014TransmissionPlanningProcess.aspx
2. Project adjacent to selected proposer's territory but this territory is outside CAISO
3. Agreed only not to apply for FERC incentive rate of return on equity
4. First "economic" or "market efficiency" project

some of these projects. Incumbents were awarded six projects and non-incumbents four projects.[36] The incumbents typically do not offer binding cost commitments in their proposals, relying on their track records and natural advantages they may have as incumbents. When an incumbent proposal without cost containment commitments is selected, the costs of the project are recovered through standard FERC regulatory cost-of-service principles. The non-incumbents selected typically offer cost containment commitments—construction cost caps subject to escalation for changes in scope and other contingencies, O&M cost caps, rate of return caps, etc. Once selected and completed, these projects become regulated projects subject to cost-of-service regulation, constrained by any cost commitments that they have made.

Note that while the CAISO looks favorably on cost containment commitments, this alone will not lead to a favorable conclusion about the economics of the project. In one project evaluation (Spring in Table 1), the ISO compared a proposal with a relatively high estimated cost but with a cost cap to a project having a much lower estimated cost without a cost cap but subject to FERC cost-of-service regulation procedures. The ISO chose the second proposal. The ISO also has developed its own planning cost estimates which serve as a kind of benchmark that bidders must beat. With one exception where there was only a single bidder, all of the projects received multiple proposals, between two and five proposals with four being the median number of competing proposals in each RFP.

7.2 *MISO*

The MISO has selected only two projects eligible for competitive solicitation as of January 2018. In both cases, they are market efficiency projects ("reduce congestion costs"). In January 2016, the MISO issued its first RFP for a 345 kV transmission line between the Duff and Coleman substations, estimated to cost ~$60 million. Eleven proposals were received, of which several were from non-incumbents.[37] Bids ranged $24–$55.7 million for construction costs, a range below the ISO's pre-bid estimate. The MISO issued a selection report in December 2016.[38] The sponsors all had previous transmission construction and operating experience. The evaluation report contains a short list of technical and economic evaluation criteria. The MISO evaluation process gives weights (points) to each of the evaluation criteria: cost and design—30%; project implementation—35%; operations and maintenance -30%; transmission planning participation—5%. The proposals are evaluated against each criterion and then given an aggregate score. The MISO found all of the proposers to be highly qualified but noted significant differences in the attributes of the proposals,

[36]The Imperial Valley project winner might be reasonably classified as a non-incumbent, however.

[37]It is tricky to figure out whether a proposal is from an incumbent or a non-incumbent as the incumbents often use a subsidiary with a different name.

[38]https://cdn.misoenergy.org/Duff-Coleman%20EHV%20345kV%20Selection%20Report82339.pdf.

including wide differences in estimated costs. However, one proposal was a clear winner based on total points received. The MISO noted in particular that many of the proposals had innovative cost caps and cost containment provisions, including the sponsor awarded the project. The winner received the highest score for cost and design as well as the highest score overall. The winner was also a non-incumbent.[39]

The MISO issued an RFP for a second competitively bid proposal in July 2018 for a 500 kV line known as the Hartburg-Sabine project. The project had an estimated cost of $129 million. As noted, this too is a market efficiency project ("reduce congestion costs" by 25% more than enough to pay for the project over time in expectation). The RFP received 12 responsive proposals, including proposals from non-incumbents. The bids ranged from $95.4 to $133.9 million, with the ISO's ex ante estimate near the top of the range. An evaluation report was issued on November 27, 2018.[40] The evaluation criteria used for this second competitive MISO project are the same as for the first project. The evaluation criteria are clearly laid out; points are assigned to each project for the evaluation of its performance in each of the four evaluation "buckets." It is fairly clear from the discussion in the evaluation report that the MISO expects to see proposals that have cost caps and other cost containment commitments. The project was awarded to a non-incumbent with the highest total score (by far) as well as the highest score on cost/design and project implementation. The winning proposal capped several elements of the standard regulated annual revenue requirements as determined by FERC over the life of the project, subject to various contingencies.

It is interesting to note that market efficiency or economic projects must be justified primarily by the estimated savings in congestion costs over 15 or more years into the future. That is, the expected present discounted value of the congestion cost savings from the project must be greater (typically 25% greater) than the cost (present discounted value of revenue requirements) of the project. This is of course the situation that, in theory, would trigger a merchant investment under the classical merchant model. However, while these projects could be supported by expected congestion cost savings, the invisible hand did not lead to the development of the project. Rather they were selected through a regional planning process and the owner is compensated through FERC regulated cost recovery rule adjusted for the winning proposal's cost containment provisions. None of the proposals offered to be compensated based solely by the sale of congestion revenue rights.

[39]FERC 2017 considers the winner to be an incumbent. I would call it non-incumbent. Republic Transmission LLC, a subsidiary of LS Power Associated, a private company very active in the independent transmission space, is the primary sponsor and appears to be a private company unaffiliated with a utility. Its partner is Big Rivers Cooperative which is a G&T coop that does not have a retail service territory, though Big Rivers may be owned by retail coops. In addition, press reports indicate that Hoosier, a rural electric coop, is supporting Republic in various ways and may take an interest in Republic Transmission, but it did not have an interest when the project was awarded. https://www.elp.com/articles/2017/03/republic-transmission-wants-to-operate. It's a matter of judgment.

[40]https://cdn.misoenergy.org/Hartburg-Sabine%20Junction%20500%20kV%20Selection%20Report296754.pdf.

7.3 SPP

As of January 2018 the Southwest Power Pool (SPP) has held one competitive solicitation for the North Liberal to Walkemeyer 115 kV line with an estimated cost of $16.8 million. The substations at either end of the line also would need upgrading, but the substation components of the projects were reserved for the incumbents.[41] The project's origin was SPP's 2015 10-Year integrated transmission plan prepared in 2014. An RFP was issued on May 5, 2015.[42] The SPP board appointed an outside expert panel to review the proposals.[43] The evaluation process allocated points in five scoring categories. The scoring categories were engineering, project management, operations, rates (costs, including cost containment commitments), and financing. Incentive points could also be earned. The results of the RFP and the evaluation process were announced on April 1, 2016. Eleven (11) proposals were submitted. However, the engineering design review determined that five of the proposals did not meet minimum engineering standards. The proposals exhibited a wide range of cost estimates—from $9.5 million to $30 million—40 year NPV of revenue requirements. The process yielded a recommended proposal and an alternate proposal based on total points earned in the five evaluation categories. Both proposals have estimated construction costs of about 50% of the ISO's pre-bid estimate. The recommended proposal had the second lowest estimated cost and the alternate the third lowest. Cost containment commitments in the form of cost caps of some type were considered in the rate analysis but the estimated costs of the proposals that made such commitments were much higher than the proposals that were selected as first and alternate. The details of the cost containment commitments were unfortunately redacted from the public report and the evaluation report did not identify the names of the companies submitting the proposals, though the winner was an incumbent. The project was ultimately canceled by SPP in July 2016 when an update to the long-term transmission plan found that the project was no longer needed.[44] It does not appear that SPP has issued this type of RFP since then. Note that 86 other transmission projects were approved by the SPP board at the same time as the winner of the one competitive procurement that SPP has initiated was announced,[45]

---

[41] The CAISO competitive bid project that yielded one bidder was also a substation project owned by the sole bidder. Upgrades to substations owned and operated by an incumbent and that will continue to be owned and operated by the incumbent are probably not a good opportunity for competitive bidding.

[42] https://www.spp.org/documents/28843/spp-rfp-000001_website%20watermarked%20posting%20version_regdateupdate080315.pdf.

[43] https://www.spp.org/documents/37708/iep%20recommendation%20report%20with%20process%20and%20appendix%20public%20redacted%20041216_redacted.pdf.

[44] https://www.spp.org/documents/28843/spp-rfp-000001_website%20watermarked%20posting%20version_regdateupdate080315.pdf.

[45] https://www.spp.org/newsroom/press-releases/spp-board-votes-to-lower-planning-reserve-margins-award-first-competitively-bid-project-approve-363m-in-transmission-upgrades/.

7.4 NYISO

The NYISO manages and integrates a local transmission owner planning process, a regional reliability transmission planning process, an economic transmission planning process ("market efficiency" projects), and a public policy transmission planning process.[46] NYISO is a single state ISO and the New York Public Service Commission (NYPSC), the electric utility regulator in New York, is heavily involved in the process. The NYPSC must approve the need for transmission projects recommended by the NYISO, the RFP used by the NYISO, the evaluation criteria, and appears to be at least informally involved in the selection made by the NYISO after it runs the RFP and recommends a winner. Effectively, the NYISO and the NYPSC work cooperatively in tandem to define needs, the RFP process and selection criteria and the ultimate selection.

While CAISO, MISO, and SPP propose specific projects for competitive procurement—e.g., a line from point A to point B with a certain voltage and transfer capacity, etc., the NYISO has adopted a "sponsorship" model for projects that are put out for competitive procurement. Under this model, the ISO specifies a transmission "need" and invites proposals for projects to satisfy this need. The NYISO has only issued RFPs for two "public policy" transmission needs since Order 1000 became effective. All of the other transmission projects selected for development come out of the NYISO's regional planning process which relies heavily on local transmission plans submitted by the incumbent transmission owners. Qualified transmission developers can and do participate in this planning process and may, in theory anyway, put forward their own projects to be selected for development.

The NYISO initiated the first public policy competitive procurement in August 2014.[47] The public policy "need" is referred to as the "Western New York Public Policy Transmission Need." The NYISO identified the overloaded transmission lines in Western New York in a baseline case (later updated) as needing additional transfer capacity. On November 1, 2015 the NYISO issued an RFP soliciting proposals for the identified public policy need. The NYISO received 12 project proposals by December 31, 2015 submitted by seven unique bidders (three bidders submitted two or more proposals).[48] Several different configurations were proposed to meet the specified need to increase transfer capacity in this part of the New York transmission network. The estimated construction costs of these proposals varied from $157 million to $487 million, through the NYISO takes many other factors into account in addition

[46]https://www.nyiso.com/csppf.

[47]The public policy transmission need was identified as improved access to hydroelectric energy from the Niagara project in Western New York State and increased imports of renewable energy from Ontario, involving a 3700 Mw increase in transfer capacity. This is a big project with multiple transmission facilities. https://www.nyiso.com/public/webdocs/markets_operations/committees/bic_espwg/meeting_materials/2016-06-07/PPTPP_Update.pdf. http://www.nyiso.com/public/webdocs/markets_operations/services/planning/Planning_Studies/Public_Policy_Documents/Western_NY/NYISO_WesternNY_PPTN_VSA_2016-05-31.pdf.

[48]The sponsorship model as implemented by the NYISO and PJM allow developers to submit multiple project proposals to satisfy an identified need.

to construction costs. The winning bidder had an estimated construction cost of $181 million. In May 2016, the NYISO issued a "technical sufficiency" report designating ten of these proposals as meeting the need from a technical perspective. This report was then sent back to the NYPSC to determine whether it continued to believe that this public policy need still existed and, if so, to provide guidance to allow the NYISO to proceed with a more complete evaluation of these projects.[49] On October 13, 2016 the NYPSC confirmed the need for the Western New York Transmission expansion and sent it back to the NYISO for a more complete analysis. The NYISO then proceeded to evaluate the competing projects based on a set of quantitative and qualitative metrics. These included the technical attributes of the proposed project, whether it satisfied the identified need, total project cost, cost per MW, expandability, access to rights of way, routing, production cost savings, congestion costs savings, cost of carbon impacts, and other considerations. The review involved a lot of technical detail.

The NYISO staff recommended a winning proposal and the recommendation was approved by the NYISO Board in October 2017.[50] The proposal selected had one of the lowest cost estimates and the shortest construction schedules of those proposals submitted ($181 million). It is sponsored by a non-incumbent. While some of the proposals had cost containment commitments, the NYISO did not take them into account, noting that Order 1000 did not require that it do so. It did suggest that it would do so in future RFPs if FERC approved necessary changes to the ISO's OATT. The NYISO also indicated that it would proceed with a "lessons learned" process to improve future competitive procurements. The project is expected to go into service in June 2022.

The second competitive procurement process for public policy transmission needs grew out of the public policy needs transmission process initiated in August 2014. In an order dated December 17, 2015 the NYPSC designated a group of transmission needs in the Central East and Southeast portions of the New York State transmission network as public policy transmission needs collectively referred to as the "AC Public Policy Transmission Needs." This NYPSC order is interesting because it addresses the issue of cost containment incentives directly[51]:

"In the absence of a cost containment incentive mechanism, FERC practice is to generally allow full recovery through the NYISO Open Access Transmission Tariff of any prudently incurred costs that exceed the developer's original estimate. The Commission already ruled in these proceedings on what incentive would be appropriate to ensure accurate cost estimates.

If actual costs come in above a bid, the developer should bear 20% of the cost overruns, while ratepayers should bear 80% of those costs. If actual costs come in below a bid, then the developer should retain 20% of the savings. Furthermore, if

---

[49]FERC has now accepted a change in the ISO's tariff that eliminates this NYPSC sign-off step. *Megawatt Daily* and Platt's Market Center, February 11, 2019.

[50]New York ISO, Western New York Public Policy Planning Report, October 17, 2017.

[51]State of New York Public Service Commission, Case 12-T-0502, Case 13-E-0488, and related cases, December 17, 2015.

the developer seeks incentives from FERC above the base return on equity otherwise approved by FERC, then the developer should not receive any incentives above the base return on equity on any cost overruns over the bid price. The bid price would therefore cap the costs that may be proposed to FERC for incentives.

The Commission cannot predict at this time whether FERC will accept the Commission's preference for a cost containment incentive mechanism. The Commission also is not privy to the bidding strategies of the potential developers. Those facts raise a concern that it may be very difficult to fairly compare bids if the bids are based on different models of risk. For example, if two competing projects appear to offer equivalent value, but one offers a lower bid subject to the recovery of all actual costs, and the other offers a higher bid, but the costs are firm, it may be difficult to choose a winner.

The Commission is dedicated to a process that will ensure equity and a fair comparison. Bids should be sought from all developers in the alternative assuming both the FERC ordinary full recovery regime and the Commission's cost-overrun-sharing incentive regime. The Commission believes that this additional information as to risk assumption will be of assistance and may be crucial to discerning between close bids."[52]

In February 2016, the NYISO issued a request for proposals that included technical information and baseline analysis to meet these needs. This RFP has much more specific transmission need/project specifications, divided into two segments, than did the Western need RFP. Fifteen proposals were submitted in response to the RFP from five unique sponsors. Seven proposals were for segment A, six for segment B, and two for both segments (the segments appear to be quite independent geographically but perhaps not electrically) required to meet the specified public policy needs. An additional proposal for a distributed generation option was also submitted but failed to qualify on technical sufficiency grounds. The estimated construction costs (without contingencies) for segment A varied from $375 million to $659 million and for segment B from $275 million to $380 million. In October 2017, the NYISO issued its technical sufficiency assessment report for the proposals submitted in response to the RFP. Thirteen of the proposals met the NYISO's technical sufficiency criteria.[53] The NYISO's assessment then went to the NYPSC for confirmation that the AC Public Policy Transmission Need continued to exist, as it confirmed in an order dated January 4, 2017.[54] In March 2018, the NYISO issued a technical review report. A proposal ranking analysis was issued by the NYISO staff in June 2018. The same development consortium, a non-incumbent and the New York Power Authority, which I suppose can be considered to be an incumbent though it is not FERC or NYPSC regulated, was initially selected to build and operate both segment A and segment B specified in

[52] Ibid., pp. 48–49.

[53] NYISO, "AC Public Policy Transmission Need: Viability and Sufficiency Assessment," October 27, 2016.

[54] State of New York Public Service Commission, Case 12-T-0502, Case 13-E-0488, and related cases, January 24, 2017. I understand that the NYISO has filed tariff revisions with FERC which, among other things, eliminate this intermediate step because it takes too much time. *Megawatt Daily*, December 12, 2018.

the RFP. The Board of the NYISO requested additional analyses to address a number of issues. The NYISO issued a report "addendum" in response on December 27, 2018.[55] This led to a change in the proposal selected for segment B of the project, and this segment of the project will now be built by a consortium led by an incumbent. The Board of the NYISO subsequently approved the proposal selected initially for segment A and the revised selection for segment B.[56] The estimated construction cost for the winning bidder for segment A was $556 million and for segment B $341 million. The winners did not bid the lowest construction costs, but other economic impacts (e.g., congestion costs, capacity deferral values), technical, and social (e.g., effects on CO2 emissions) gave them the highest rankings.

The NYISO has subsequently issued two requests for suggestions for additional public policy transmission needs but these requests have not yet led to the commencement of an RFP process.[57]

7.5 PJM

PJM is by far the largest RTO/ISO in the country. Its origin can be traced back to a multi-state power pool created in 1927. PJM now covers generating and transmission facilities in 13 states plus the District of Columbia with about 180,000 Mw of generating capacity and 85,000 miles of transmission lines. About $30 billion of investment in transmission capacity has been selected in PJMs Regional Transmission Expansion Planning process (RTEP) since 2000.[58] PJM manages a set of wholesale markets for energy, ancillary services, and capacity in the PJM region, relying on for the energy and ancillary services markets security-constrained bid-based market models with nodal prices.

PJM had a comprehensive regional transmission planning process and associated procedures prior to Order 1000 the Regional Transmission Expansion Plan or RTEP. Process changes to comply with Order 1000 took effect on January 1, 2014, though these are properly viewed as enhancements to existing processes. PJM began to implement a competitive planning process consistent with Order 1000 in 2013 when

[55]https://www.nyiso.com/documents/20142/1390750/AC-Transmission-PPTN-Draft-Report-Addendum.pdf/898f1cb0-3f98-a26b-3866-0118dedafaae.

[56]https://www.nyiso.com/documents/20142/1390750/Board-Memo-on-AC-Transmission-FINAL-c.pdf/8eba9661-6ab3-0311-de12-5d54d255f11e.

[57]https://www.nyiso.com/documents/20142/1406936/Public_Policy_Needs_Solicitation_2016-08-01.pdf/c110897d-e37b-1611-d935-826124b41ab4; https://www.nyiso.com/documents/20142/1406936/2018-19-PPTPP-Needs-Solicitation-Letter.pdf/83907505-4012-3813-f3b3-f0eda17fa0dd.

[58]This excludes the costs of supplemental projects and network projects. 2017 PJM RTEP Book 1, p. 4; https://www.pjm.com/planning.aspx. Accessed January 12, 2019. "Project Statistics," January 10, 2019. https://www.pjm.com/-/media/committees-groups/committees/teac/20190110/20190110-project-statistics-2018.ashx.

it opened two competitive "windows."[59] The PJM staff works with the stakeholder Transmission Expansion Advisory Committee (TEAC) and ultimately the independent PJM Board to approve changes to the regional plan, new projects falling into certain categories, and project cancellations. PJM publishes extensive information about these processes on its website.

As noted earlier, PJM has adopted what it calls a "sponsorship" model. Rather than specifying a particular project (e.g., build a 220 kV line from A to B to increase transfer capacity by X Mw), PJM publishes a set of reliability violations (reliability projects) or a set of highly congested interfaces (market efficiency projects) and solicits proposals from incumbent and non-incumbent transmission developers to resolve the violations or to reduce forecast congestion cost sufficient to justify the investment. Market efficiency projects must have a benefit/cost ratio great than 1.25. In order to solicit projects PJM opens various competitive "windows." Projects that are five years out or more are solicited in a "long-term window." Projects that are three to five years out are solicited in a "short-term" window. Projects that are less than three years out are classified as "immediate need" projects. If there is insufficient time to open a 30-day window and evaluate proposals for immediate need projects, PJM identifies a solution and designates an incumbent to implement the solution. There are other exclusions from PJM's RTEP competitive planning projects. These include "supplemental projects," which are projects designated by transmission owners to meet local planning criteria and to replace aging infrastructure, and "network projects" which are projects associated with the interconnection of generators to meet power delivery requirements. They are reviewed by subregional committees established by PJM and the TEAC but do not go through the same level of PJM review as projects designated to meet NERC and regional reliability criteria and market efficiency projects.[60] Nor are they included in PJM's competitive procurement windows.[61] These exclusions from the competitive RTEP process are not trivial. Between 2013 and 2018, the estimated cost of new baseline RTEP projects was about $12 billion and the estimated costs of new supplemental projects

[59]2013 PJM Regional Expansion Plan, Book 1, p. 13 and p. 14, 2016. https://www.pjm.com/library/reports-notices/rtep-documents/2013-rtep.aspx.

[60]In 2016, FERC initiated an investigation of the "openness" of these regional planning processes. It found that they were not sufficiently open and violated the open planning rules established by Order 870 (enhanced by Order 1000). PJM subsequently made a compliance filing and it was accepted by FERC. 116 FERC ¶ 61,217, September 26, 2018. It is pretty clear from the record in this proceeding that FERC is much more interested in ensuring that planning processes are "open" to all stakeholders, including potential non-incumbent transmission developers, than it is in more structured competitive procurement processes.

[61]FERC issued an Order on February 15, 2018 which found that the local transmission owners and subregional planning processes associated with supplemental projects violated the transparency and openness requirements in Order 890. 162 FERC ¶ 61,129 February 15, 2018. FERC accepted PJM and transmission owner compliance filings on September 16, 2018. 164 FERC ¶ 61,217 September 26, 2018.

was about $19 billion.[62] Over a longer period of time, including the period before the first competitive window in 2013, about $29 billion has been or is estimated to be spent on RTEP Baseline projects and $26 billion on supplemental projects. Another $7 billion was spent on network projects.[63] FERC has recently approved additional exclusions from the RTEP competitive window process for projects under 200 kV and for transmission substation equipment.[64] Nevertheless, many more transmission projects (both for reliability violations and market efficiency opportunities) have been mediated through PJM's formal competitive transmission planning process than is the case for the other ISOs. PJM has also implemented an interregional planning process and identified potential interregional projects with MISO to reduce congestion between the two ISOs.[65]

Table 2 contains summary information about the proposals to develop projects submitted and selected in all of the RTEP and efficiency windows opened between 2013 and 2017. (Selections have not yet been made for the 2018 window at the top of the table.) There were 16 windows opened and completed during the 2013–2017 time period.[66] The typical window is opened with a fairly large set of reliability ("flowgate") violations on the PJM network or a fairly small set of potential market efficiency projects. Note that each reliability window "targets" related types of flowgate violations as indicated in the first column of Table 2. Depending on whether it is a short-term or long-term window, pre-qualified developers have either 60 or 120 days to submit proposals.[67] Once the window is closed the PJM staff evaluates the proposals and makes recommendations to the Transmission Expansion Advisory Committee (TEAC). The TEAC determines whether to accept these recommendations and then forwards the proposals selected to the PJM Board for final approval.

Table 2 indicates that there are 803 proposals made in response to 16 RTEP competitive windows during the 2013–17 period and 142 projects were awarded to developers based on these proposals. Opening competitive windows has certainly created a lot of interest by developers. It is evident that transmission developers ("entities") submit an average of about five proposals in a typical window, but multiple proposals may respond to different flowgate violations or market efficiency opportunities. About 45% of the proposals came from non-incumbents. However, only three of these projects were awarded to non-incumbents.[68] About 95% of the awards

[62]"Project Statistics, PJM Transmission Expansion Advisory Committee, p. 6, January 10, 2019. https://www.pjm.com/-/media/committees-groups/committees/teac/20190110/20190110-project-statistics-2018.ashx.

[63]Ibid., p. 10.

[64]2017 PJM RTEP Book 1, p. 9.

[65]2017 PJM Regional Transmission Expansion Plan Book 3, p. 240.

[66]Two additional windows were opened in 2018. As this is written one window closed in 2018 and the other closes in 2019. No information about the proposals selected had been posted as of January 12, 2019.

[67]A 30-day window may be opened for immediate need projects.

[68]The three projects are: (1) one portion of the segments awarded in the Artificial Island Solicitation awarded to a subsidiary of LS Power, (2) a market efficiency project referred to as AP-South awarded

Table 2 PJM RTEP Windows 2013-2018 (as of February 2018)

	Window Objective	Proposals	Proposal From Incumbents	Proposal From Non-Incumbent	Entities	Proposals approved by PJM Board	Approved Greenfield Projects	Approved Upgrade Projects	Approved Incumbent	Approved Non-Incumbent
2018 RTEP Proposal Window 1	2023 Summer, Winter and Light Load: N-1, Gen Deliv, Load Deliv N-1-1	7	7	0	2	N/A	N/A	N/A	N/A	N/A
2017 RTEP Proposal Window 1	2022 Summer, Winter and Light Load: N-1, Gen Deliv, Load Deliv N-1-1	51	36	15	10	10	0	10	10	0
2016/17 RTEP Long Term Proposal Window 1A	Market Efficiency	3	2	1	2	1	0	1	1	0
2016/17 RTEP Long Term Proposal Window	Market Efficiency Congestion, 15 Year Reliability Analysis	96	51	45	19	4	0	4	4	0

(continued)

Table 2 (continued)

	Window Objective	Proposals	Proposal From Incumbents	Proposal From Non-Incumbent	Entities	Proposals approved by PJM Board	Approved Greenfield Projects	Approved Upgrade Projects	Approved Incumbent	Approved Non-Incumbent
2016 RTEP Proposal Window 3 Addendum 1	Winter Reliability	6	3	3	3	3	0	3	3	0
2016 RTEP Proposal Window 3	Winter Reliability, Light Load Reliability, Short Circuit	30	15	15	7	6	0	6	6	0
2016 RTEP Proposal Window 2	N-1 Thermal and Voltage; Gen Deliv and Common Mode Outage, Load Deliv Thermal and Voltage; N-1-1 Thermal and Voltage	87	53	34	13	14	0	14	14	0

(continued)

Table 2 (continued)

	Window Objective	Proposals	Proposal From Incumbents	Proposal From Non-Incumbent	Entities	Proposals approved by PJM Board	Approved Greenfield Projects	Approved Upgrade Projects	Approved Incumbent	Approved Non-Incumbent
2016 RTEP Proposal Window 1	Generator Deliverability and Common Mode Outage Violations related to Carson-Rogers Rd 500 kV and Chesterfield-Messer Rd-Charles City Rd 230 kV and Dominion Local TO Criterion for End of Life Facilities	25	7	18	7	2	0	2	2	0
2015 RTEP Proposal Window 2	TO Thermal Criteria, TO Voltage Criteria; Light Load Thermal and Voltage	23	14	9	4	6	1	5	6	0

(continued)

Table 2 (continued)

	Window Objective	Proposals	Proposal From Incumbents	Proposal From Non-Incumbent	Entities	Proposals approved by PJM Board	Approved Greenfield Projects	Approved Upgrade Projects	Approved Incumbent	Approved Non-Incumbent
2015 RTEP Proposal Window 1	N-1 Thermal and Voltage; Gen Deliv and Common Mode Outage, Load Deliv Thermal and Voltage; N-1-1 Thermal and Voltage	91	34	57	9	20	1	19	20	0
2014/15 RTEP Long Term Proposal Window	Long Term Reliability Criteria; Long Term TO Criteria; Market Efficiency	119	67	52	22	14	1	13	13	1
2014 RTEP Proposal Window 2 Addendum 2	N-1-1 Voltage Drop	10	2	8	4	1	0	1	1	0
2014 RTEP Proposal Window 2 Addendum 1	N-1-1 Voltage Drop	4	1	3	3	0	0	0	0	0

(continued)

Table 2 (continued)

	Window Objective	Proposals	Proposal From Incumbents	Proposal From Non-Incumbent	Entities	Proposals approved by PJM Board	Approved Greenfield Projects	Approved Upgrade Projects	Approved Incumbent	Approved Non-Incumbent
2014 RTEP Proposal Window 2	Reliability Criteria Thermal and Voltage; TO Criteria	79	58	21	14	35	4	31	34	1
2014 RTEP Proposal Window 1	Reliability Criteria - Thermal	109	62	47	15	22	0	22	22	0
Market Efficiency	Market Efficiency	38	11	27	6	1	0	1	1	0
Artificial Island	Operational Performance	26	14	12	7	3	1	2	2	1
2013-2017		**804**	**437**	**367**	**147**	**142**	**8**	**134**	**139**	**3**

2018 not completed as of 2/16/19
Sources: See Data Appendix

were for upgrades to existing facilities, which under PJM rules are designated to the incumbent utility. This may help to explain why so few non-incumbents were selected. There were only seven greenfield projects awarded so non-incumbents were awarded three of the seven greenfield projects.

With a few exceptions, the details of the evaluations performed by the PJM staff and how the evaluations led to the project awards are not particularly transparent. The nature of PJM's sponsorship model makes head-to-head comparisons between proposals quite difficult since they are not "bidding" to develop a specific project but rather submit proposals for different sets of solutions to one or more flowgate violations or designated mitigation of congestion costs opportunities. While a typical PJM staff report to the TEAC for reliability projects describes in some detail the recommended solution to the flowgate violation, it does not discuss whether other equivalent proposals to solve the same violations were submitted or why they were rejected. The evaluation of market efficiency proposals is more transparent.

PJM opened the first RTEP window on August 29, 2013. In this window, PJM solicited proposals to improve operational performance of the transmission system in the Artificial Island area of Southern New Jersey. Artificial Island is the location of three nuclear generating plants owned by PSE&G. The publicly available evaluation of the proposals in response to the Artificial Island Window is more transparent than is typically the case for PJM. PJM received 26 proposals with initial cost estimates ranging from \$100 million to \$1.55 billion. The proposals put forward represent a technologically diverse set of partial and complete solutions to the reliability issues identified by PJM in the RFP.[69] The projects are not directly comparable because they include both partial and complete solutions to the Artificial Island reliability issues, though correcting for differences it appears that the incumbent made by far the most-costly proposals and did not offer to agree to cost containment commitments.[70]

This window was opened a few months before PJM's Order 1000 compliance date and PJM characterized this solicitation as a trial run.[71] Competing projects are specified, compared to one another and the rational for the proposals selected are specified fairly clearly. However, this solicitation is different from the subsequent reliability windows opened by PJM as it focused on a single set of reliability challenges in one area of the bulk transmission network around Artificial Island. In this sense, the

to Transource, a subsidiary of American Electric Power (AEP), in connection with the 2014/2015 RTEP, and (c) the Thorofare project in West Virginia also awarded to Transource in connection to the 2014 RTEP Window 2. It is not clear to me that the Thorofare project meets the PJM staff's criteria for non-incumbent as the project seems to run through the territory of an AEP subsidiary and the history of the project indicates AEP involvement with the development of the project. The Transource AP-South project was last re-evaluated by the TEAC in September 2018 and continued to show a benefit/cost ratio greater than 1.25. https://www.pjm.com/-/media/committees-groups/committees/teac/20180913/20180913-ap-south-9a-project-reevaluation-sept-2018.ashx.

[69] Artificial Island Project Recommendation White Paper, PJM, July 29, 2015. https://www.pjm.com/-/media/committees-groups/committees/teac/postings/artificial-island-project-recommendation.ashx?la=en.

[70] Ibid., pp. 12–13.

[71] Ibid., p. 1.

application of the sponsorship model in this case is more like the two solicitations conducted by the New York ISO which also claims to have a sponsorship model. The Artificial Island White Paper discusses the evaluation process in some detail.[72] The awards went to three projects that together comprise a complete solution to the Artificial Island reliability issues, one to be developed by a non-incumbent, one to be developed by the incumbent (static var compensator, substation expansion, new transformer), and one involving the installation of high-speed optical grounding wire communications on existing transmission lines owned by multiple incumbents.

The complete project configuration selected had an initial cost estimate of about $275 million.[73]The non-incumbent awardee's proposal for its segment of the project made capital cost commitments in the form of a cost cap subject to various contingencies. The contingencies included changes in scope, financing, inflation, and other factors. Capital cost commitments were a new idea for PJM and such commitments had not been, and are not now, required. Since cost commitments were a new idea to PJM (and I suspect for FERC), PJM gave the other four finalists a chance to resubmit bids with cost commitments. Three of the four submitted bids with a variety of capital cost commitment structures.[74] PJM staff adjusted the proposals to include estimates of the costs of segments not included in each proposal to make them more or less technically equivalent.

The adjusted cost estimates of the four finalists ranged from $263 million to $380 million in then current dollars.[75] The adjusted cost estimates for two of the proposals were very close, but PJM found that the proposal submitted by the non-incumbent had fewer contingencies and exclusions and the lowest expected cost (but similar to the next lowest cost estimate) It was awarded the 230 kV portion of the project from Delaware to Artificial Island (about 50% of the estimated cost of the entire project). The incumbent was awarded the portion of the project for a static var compensation, substation upgrades, and a new transformer.

The saga surrounding this project did not stop there. In 2016, PJM suspended the project for further reconsideration and then reinstated the project in 2017. The Delaware Public Service Commission granted a certificate of public need and necessity for the Delaware portion of the project in December 2018 subject to FERC approval of cost allocations to Delaware for the project meeting conditions specified by the Commission. As this is written, the ball is now in FERC's court, where cost allocation and the implementation of cost commitments are likely to be issues. Note that the solicitation, evaluation, and regulatory process started in 2013 and had not been completed by the end of 2018.

---

[72]https://www.pjm.com/-/media/committees-groups/committees/teac/postings/artificial-island-project-recommendation.ashx?la=en.

[73]Ibid. pp. 39–40.

[74]Ibid. pp. 32–35.

[75]Ibid. pp. 33.

I found no indication that the two projects awarded to non-incumbents in two other windows contained cost containment commitments.[76]

7.6 ISO-NE

As noted above, ISO-NE's FERC Order 1000 compliance filing anticipates relying on competitive procurement in certain circumstances. However, the first RFP was not issued until March 2020 beginning the process. However, at least three New England states have initiated state-sponsored renewable energy procurement programs which obligate distribution companies in these states to take competitive bids for specified supplies of renewable (no-carbon) energy pursuant to long-term contracts.[77] Two solicitations initiated by Massachusetts are perhaps the most interesting for the purposes of this paper because they effectively bundle contracts for renewable energy with contracts for dedicated transmission facilities to deliver this energy to Massachusetts customers. While the transmission projects are part of a competitive bidding process. The competitive solicitation is separate from the ISO's Order 1000 compliance process and is initiated by states rather than the ISO. The first requires Massachusetts distribution companies collectively to solicit bids to supply "clean energy" and to arrange for the transmission facilities necessary to transmit that energy without constraints to these distribution companies' interconnections with the New England transmission network. The RFP specifies general cost containment provisions such as those discussed earlier in connection with several proposals submitted in other ISOs competitive transmission solicitations. Several proposals were submitted in response to this RFP[78] offering contracts with solar, wind, and hydroelectric resources.

The winning bidder was Northern Pass transmission, a subsidiary of Eversource a distribution utility with subsidiaries in Massachusetts, Connecticut, and New Hampshire. Its winning bid proposed to build a 192 mile HVDC transmission line to connect the Hydro-Quebec network with the New England network, along with a converter station, AC transmission facilities, and substation upgrades elsewhere in New England, to support the delivery of 1090 MW of hydroelectric power to Massachusetts distribution utilities.[79] Northern Pass would be compensated for the costs of these transmission facilities through a FERC regulated tariff meeting criteria specified in the RFP and separate from either ISO-NE's regulated tariffs or regulated tariffs that apply to other transmission facilities owned by Eversource or its affiliated

[76] I would also classify one of these projects as an incumbent rather than a non-incumbent project but I have accepted PJM's categorization for counting purposes here.

[77] http://energypolicyupdate.blogspot.com/2017/06/new-england-regional-renewables.html.

[78] https://www.masslive.com/news/2017/07/transmission_hydro_and_wind_de.html.

[79] http://www.northernpass.us/project-overview.htm; http://www.northernpass.us/facilities-equipment.htm.

distribution companies. In this sense, this is no different from the revenue requirements treatments that apply to stand-alone transmission projects authorized by an ISO through a competitive procurement.

A permit for the HVDC portion of the Northern Pass project was subsequently rejected by a regulatory agency in New Hampshire. An alternative HVDC project through Maine to connect with Hydro-Quebec to access the contracted hydroelectric power—New England Clean Energy Connect[80]—that scored well in the competitive solicitation, is now going through the Maine permitting process. I anticipate that the FERC regulated tariff treatment will be similar. This project involves building 145 miles of new HVDC line, new AC lines, upgrades to existing AC lines throughout New England, a new substation, and a converter station. The developer of this project is Central Maine Power, another subsidiary of Iberdrola, but the costs of the project will be paid for by Massachusetts retail customers as the regulated transmission tariff charges are passed along to them over time. Other competing projects remain in the wings if this project does not receive the necessary permits in Maine

The second solicitation was for 400–800 MW of offshore wind generation (of an eventual 1600 MW).[81] An offshore wind developer called Vineyard Wind was selected in May 2018 as the winning bidder for 800 MW.[82] This RFP bundled the offshore wind supply with the development of the necessary transmission facilities. The original RFP provides two options for developing and paying for the associated transmission projects. The bid could include an "all in" price structure for energy and transmission or the transmission facilities could be developed separately and the costs recovered through a separate FERC regulated transmission tariff.

In some sense, these projects are conceptually similar to the public policy solicitations managed by the NYISO to bring renewable energy to load centers from Western New York and Canada. However, the NYISO is a single state ISO where the New York Public Service Commission and the NYISO can fully internalize state policies with transmission planning and development. ISO-New England covers six New England states with six public utility commissions and six sets of state electricity policies. The approach taken in New England with one or more states agreeing on renewable energy procurement policies and bundling long-term contracts for the energy with the associated transmission facilities seems to be a sensible way of resolving potential conflicts between states.

The transmission facilities associated with both of these two competitive renewable energy projects are classified by ISO-NE as "elective transmission upgrades." An elective transmission upgrade is a transmission project that has not been selected through the ISO-transmission planning process. It will not be included in the ISO's transmission tariffs or cost allocation mechanisms but will be interconnected with the

---

[80]https://www.necleanenergyconnect.org/project-overview.

[81]https://www.mass.gov/news/project-selected-to-bring-offshore-wind-energy-to-the-common wealth.

[82]A subsidiary of Iberdrola, a large international utility with a great deal of experience with both onshore and offshore wind owns 50% of Vineyard Wind.

ISO's transmission network and as a result is subject to study and approval regarding impacts on the ISO's network. According to the ISO-NE OATT:

> An entity that constructs and/or maintains an elective transmission upgrade shall be responsible for 100% of the costs and of any additions to or modifications of [the ISO-NE transmission network] that are required to accommodate the elective transmission upgrade. A request for rate treatment [regulated transmission tariff] of an elective transmission upgrade, if any, shall be determined by [FERC] in an appropriate proceeding. (ISO-NE Open Access Transmission Tariff, Section II.47.5, as of March 16, 2019)

Accordingly, while an elective transmission upgrade cannot be paid for through the ISO's standard tariff procedures, it can be paid for under a separate FERC regulated transmission tariff, as appears to be the case with these two projects, or it could be a classical merchant project that does not seek recovery pursuant a cost-of-service tariff but markets transmission rights or negotiates a contract with its customers that is not tied directly to traditional FERC cost-of-service ratemaking procedures. The latter contract would still have to be approved by FERC. The elective transmission upgrade provisions of ISO-NE's OATT can in principle used by, for example, wind generators to build or contract for their own transmission upgrades to relieve congestion on the network that is leading to curtailments of their facilities. A simpler approach would be to pay the local utility to build additional transmission upgrades beyond the standard interconnection requirements pursuant to a FERC regulated cost-of-service contract to relieve the congestion that is inhibiting the operation of their project.[83]

ISO-NE's October 2018 transmission project list contains 870 projects under construction, in development, planned, in the hopes and dreams stage, or canceled. About 75 projects are listed as "elective transmission upgrades." Most of these projects appear to be components of Northern Pass, Clean Energy Connect, and other projects of this kind that bid into the Massachusetts Clean Energy Procurement process and would ultimately become components of a FERC regulated cost-of-service tariff.

8 Discussion

It is difficult to draw strong conclusions about the performance attributes of the experience to date in the USA with competitive procurement for transmission projects. Putting PJM aside for the moment, there are only 15 projects in the USA that have been selected by the ISOs through a formal open competitive procurement process since 2014. We can add PJM's first competitive window for the Artificial Island

[83]Under the ISO's OATT, generators must pay for the costs of direct interconnections to the New England (PTF) network and incremental transmission network (PTF) upgrades deemed required to integrate them into the system. Such generators can pay for enhancements to the standard attributes of the interconnection or can build their own interconnection and pay for additional upgrades deeper in the network. To the extent that the new interconnection and any network transmission upgrades paid for by the generator create additional firm transmission rights, they are allocated to the generator. ISO-NE OATT, Schedule 11, as of March 17, 2019.

window to this total to get to 16. Many of these projects have not yet been completed, complete realized construction cost data are not readily available for those projects that have been completed, and there is no record of how the cost commitments and the associated contingency provisions contained in the winning proposals, have been applied. Turning to PJM, in principle, all of the potential projects to respond to reliability violations and potential net reductions in expected congestion costs are open for competitive bidding. Clearly, participation in the PJM competitive windows has been quite robust. There were five times more proposals submitted than projects selected through the RTEP between 2013 and 2017. Many proposals have been submitted by non-incumbents. However, 95% of the projects selected have been upgrades to existing facilities and designated to the incumbent. Supplemental projects selected through subregional planning processes are not approved by the PJM board. In February 2018, FERC also found that the subregional planning processes had not met Order 890's open planning requirements and ordered that changes be implemented.[84] Given the small number of projects that actually turn out to represent competitive opportunities for non-incumbents and for competitive offers to adopt various cost containment provisions, there may be opportunities to simplify the current process to save unnecessary time and effort. I will return to this question below.

Yet, there is quite a bit to learn from the 16 projects selected through an organized competitive procurement process by ISOs since Order 1000 went into effect. As noted earlier, one of the challenges for regulators is uncertainty about what the cost of an efficiently designed and built project should be. This is an especially important question for the USA, because FERC presently does little regulation of the reasonableness of the costs presented for inclusion in transmission operators' revenue requirement and does not apply performance-based mechanisms as have been used in other countries and other industries. It is clear from the data on ISO cost estimates and the range of cost estimates and cost commitments contained in competing proposals that there is a wide range of potential cost realizations. Indeed, perhaps the most striking thing about the proposals submitted in response to these RFPs is the wide range of estimated costs observed between the various proposals for essentially the same project or to meet the same transmission expansion need. Cost containment mechanisms aside, the wide range of cost estimates convinces me that there is substantial potential benefit in competitive procurement per se beyond non-incumbent participation in open regional planning processes unburdened by incumbent rights of first refusal. ISO evaluators and regulators can now see variations in cost estimates that they never saw when the projects were proposed and developed by a single incumbent utility. When non-incumbents have been selected their projects often have significantly lower cost estimates than the incumbent's, often combined with cost containment commitments. Competitive procurement may also induce incumbents and non-incumbents to sharpen their pencils, Artificial Island being a good example. This kind of competitive information is also necessary for an ISO to choose the most efficient or cost-effective projects as required by Order 1000.

[84] 162 FERC ¶ 61,129, February 15, 2018.

This information can also provide benchmarks for FERC if it decides to get more engaged in regulating transmission costs. While the jury is necessarily still out on whether competitive procurement leads to lower costs to meet specific transmission needs, I think that there are good reasons to believe that it likely does. The evidence from other countries, especially Argentina, is consistent with this view.

It is sometimes argued that formal competitive procurement that allows incumbents and non-incumbents to compete is not necessary because incumbent transmission owners seek competitive bids for equipment and contracts and primarily provide management oversight. This is not a compelling argument. The competitive procurements demonstrate that competing transmission developers can reduce expected costs by coming up with innovative designs to resolve transmission needs identified through the ISO regional planning process, taking on more performance risk, foregoing certain FERC revenue requirements "incentives" for which they would otherwise be eligible, etc. The cost containment commitments and related incentives (and contingencies) can be a substitute for more direct performance-based regulation of costs and could even help FERC to design and apply incentive regulation mechanisms more broadly to transmission costs. It is important to better understand how these cost containment mechanisms work in practice over time and interact with FERC's cost of service/revenue requirement recovery policies.

The costs and construction times for any developer of transmission projects, especially greenfield projects, can be very uncertain and, as a result, are not well adapted to high powered incentive schemes (e.g., a firm construction cost commitment) that do not leave a lot of expected rent on the table for the developer. Those proposals that do offer cost containment commitments also include many contingencies that would relax these cost commitments. Will the contingencies overwhelm the commitments? Time and data availability will be necessary to answer this question. On the other hand, incumbent projects regulated under traditional regulatory arrangements also can and do incur significant cost overruns. Over the period 2014–2017, the PJM RTEP reported over $1.3 billion of escalation in *estimated* construction costs.[85] The CREZ program in Texas experienced an 40% increase in realized costs from the initial estimates. Only some of the gap can be explained by input cost inflation. Pfeifenberger et al. (2018) offer additional evidence on both cost overruns and the range of costs observed in the competitive procurements that have taken place.

While the competitive procurement process may weed out projects that are not technically feasible and/or have higher projected costs than equivalent alternatives, in the end when a project is completed it becomes a FERC cost-of-service regulated project, subject to any cost containment commitments and contingencies agreed to with the ISO. These costs ultimately end up in the charges paid by transmission customers, primarily distribution utilities, and in this case passed along to retail customers in wires charges.

It would be desirable for ISOs to place more weight on cost control and performance incentives in their evaluations of proposals to lead FERC in this direction.

[85]PJM annual RTEP reports, various years; https://www.pjm.com/library/reports-notices/rtep-documents.aspx.

However, it is quite clear that the ISOs do not want to become and are not supposed to be, economic regulators in this sense and this is not where their experience lies. Order 1000 gives the ISOs the responsibility to select the least cost or most cost-efficient projects but the ISOs do not have the regulatory authority to monitor and enforce cost commitments or to evaluate whether transmission costs incurred are "just and reasonable." These regulatory responsibilities are ultimately FERC's responsibilities and, as discussed further below, FERC could take a more active role in facilitating the consideration of voluntary cost containment and performance incentives offered by developers in the project selection process. The implementation of cost containment provisions through the FERC revenue requirements process needs to be clarified as well.

It is also clear that participation in the competitive procurement process and the evaluation of competing proposals is complex, expensive, and time consuming. Transmission developers must submit a great deal of technical and financial information, past development experience, detailed development and right acquisition plans and other information to respond effectively to an RFP. The evaluation process is also quite complex, taking a wide variety of factors into consideration in evaluating competing projects. The process of developing and issuing a good RFP, proposal creation by developers, and ISO evaluation is a time-consuming process. However, some ISOs appear to be able to complete a full cycle in less than a year. Others can take five or more years. It would be helpful if the ISOs could share best practices and adopt them to streamline the process.

Participating in regional transmission planning processes and competitive transmission procurement processes is not for small inexperienced organizations. These activities require substantial financial resources, technical human resources, and technical analytical resources. In some cases, the competitive procurement processes are very burdensome and take too long (e.g., NYISO). Moreover, in evaluating proposals, ISOs place a lot of weight on engineering, operating, siting, and environmental permitting *experience*. However, we should remember that the USA has a large number of utilities with a century of transmission construction and operating experience. They can form subsidiaries and participate in planning and competitive procurement processes outside of the areas where they have retail footprints to satisfy Order 1000's criteria for being a non-incumbent. Indeed, most of the non-incumbents participating in competitive procurement processes are subsidiaries of large experience utilities, existing independent transmission companies, or independent power companies. Accordingly, there is no shortage of actual and potential non-incumbent transmission development competitors. It is clear from the competitive procurement examples that I discussed above that there are typically several competing developers that submit proposals for the same project—as many as 12.

Incumbents may have inherent advantages in some situations. They are already invested in a regional planning process, especially in the ISOs, have years of experience with it, and have no real choice but to devote resources to it. Incumbents also may have eminent domain rights, rights of way, and other soft assets that are difficult for a non-incumbent to obtain. Finally, building new transmission projects may confront community opposition of various forms. Long historical experience dealing with state and local governments and regional interest groups may convey a natural advantage. Finally, there are some types of transmission projects which may simply be easier for an incumbent to design, build, and operate. An example is the upgrade of a substation to accommodate expansion of connected transmission lines. Both New York and SPP reserved substation upgrades to the incumbent in designing a competitive procurement process for new line construction. A substation project in California that was put up for competitive procurement received one bid and it came from the incumbent.

FERC has placed great weight on open transmission planning processes, the end of any federal right of first refusal, and the participation on non-incumbents and other stakeholders in the planning process through orders 890 and 1000. But, has opening up the transmission planning process to non-incumbents and other stakeholders changed how project choices are made and expanded opportunities for non-incumbents? This is, of course, very hard to know, since there is no natural experiment or controlled trial, and associated comparative benchmark data, to rely upon. We can ask the qualitative question of whether or not these changes have opened up additional opportunities, beyond those created by formal open competitive bidding structures.

One way to get a sense for the answer to this question is to examine whether transmission companies that pursue projects through open competitive bidding are also designated as non-incumbent developers through the regional planning process when there is not a formal open competitive procurement process. NextEra Energy Transmission, a subsidiary of NextEra Energy, which is the largest electric power company in the country by market value, has been active as a competitive transmission developer. It is typically a non-incumbent. Its website notes projects won in California, New York, Texas, and Ontario where development rights were secured through competitive procurement (in the California and New York projects are in the RFPs discussed above). NextEra Energy Transmission's website lists the transmission projects it is developing in the USA. The only new development projects listed are those it secured through competitive procurement. The other projects are existing projects that it acquired from another owner. LS Power is also a major player in the competitive transmission procurement arena.[86] It is typically a non-incumbent in competitive transmission procurement processes. It was selected to develop four projects through competitive procurement processes in CAISO, PJM, MISO, and ERCOT. The fifth project is in Nevada. The Nevada project was not selected through a formal open competitive procurement process. It is being developed with a Department of Energy Loan Guarantee and the incumbent owns 25% of the project. Transource, a subsidiary of American Electric Power, also submitted bids in open competitive procurement programs. Its website lists two projects in PJM (two of the non-incumbent projects discussed above) and two additional projects in Missouri. One is pre-order 1000. The second was developed with a partner which

[86]https://www.lspower.com/project-map/.

owns an incumbent which appears to have been the initial developer of the project.[87] While this is hardly an exhaustive survey, it does not appear that merely opening up the transmission planning process to non-incumbents and removing the right of first refusal has yet led to a lot of business for non-incumbents.

9 Conclusions

What I have called the "competitive transmission procurement model" is a framework that expands the role of competition in the development and potentially operation of transmission projects. FERC Order 1000 gave new life to this model. Competitive procurement for incumbents and non-incumbents and opportunities for non-incumbents to participate in regional transmission planning and project selection have increased. The progress has been slow but promising. There is still much to be done.

It is evident from the limited evidence that we have that competitive procurement can help to resolve the adverse selection and moral hazard problems faced by regulators, in this case FERC. If we view the evidence to date as a kind of experiment, it suggests that there are potential efficiency gains from expanding open competitive solicitation opportunities meeting certain criteria. Yet, only a tiny fraction of transmission projects authorized in the USA are being selected through formal competitive procurement solicitations with transparent evaluation criteria. ISOs have adopted a wide variety of criteria to exclude and include projects from competitive procurement. PJM's more open process mostly led to projects being awarded to incumbents. It seems like a lot of time and effort for three projects out of 142 awarded to go to non-incumbents. Given the lack of experience with competitive procurement in the USA, it may have been prudent for FERC to make competitive procurement by ISOs voluntary in Order 1000 and for ISOs to limit the kinds and number of projects selected in this way. However, the experience to date is sufficiently promising to consider expanding the use of open competitive procurement solicitations for transmission projects.

How might this be accomplished? FERC can do more to encourage competitive procurement than has been the case today. It could add an incentive to the existing list of incentives to reward projects selected through an open competitive procurement process, providing both incumbents and non-incumbents with incentives to support expansions of competitive procurement by the ISOs. FERC could also take a more favorable and supporting posture toward including cost containment and other performance incentives in projects selected through competitive procurement, provide guidance to ISOs regarding evaluation of performance commitments, amend OATTs to clearly allow ISOs to take cost containment commitments into account,

[87] https://www.cfra.org/nebraska-city-maryville-sibley; https://journalstar.com/business/local/new--mile-power-line-from-nebraska-city-to-missouri/article_7413e390-d12c-5660-bf93-86d4586326c1.html.

and provide more transparent guidance for how cost containment provisions will be included in the revenue requirements calculation process.

FERC could also play a more active role in providing guidance to the ISOs for specifying criteria for transmission projects that are likely to be good candidates for competitive procurement. At this stage, excluding projects that are highly likely to be upgrades to existing facilities,[88] below a certain voltage, subject to local rather than regional planning criteria, below an estimated cost threshold, etc., would make sense if it is combined with stronger FERC support for expanding competitive procurement for the remaining types of projects. This would lead to more projects being authorized through a competitive procurement process.

Finally, as this experiment with competitive procurement evolves, it would make sense to design an evaluation program that accompanies the experiment. This would require the publication of more transparent evaluation criteria, publication of more transparent information about how these criteria were applied to support the selections, standardization of public information requirements, and information tracking of performance of projects selected through competitive procurement. With the right information in hand, both FERC staff and stakeholders would be in a position to evaluate the performance of competitive procurement and help to lead to a set of best practices.

Data Appendix

Collecting the information for this paper was not an easy task. Aside from data on the number of competitive solicitations, the numbers of bidders, and a breakdown of the number of incumbent vs. non-incumbent winners for the 2013–2016 period contained in FERC (2017) and collected from ISO websites there is no organized repository and no standard presentations of information for ISO competitive transmission solicitations. Neither FERC nor state regulators have evaluated the competitive procurement programs. Accordingly, I started with FERC (2017) and then searched the websites of all of the ISOs for relevant information on transmission planning, competitive procurement, and compliance with orders 890 and 1000. This information was supplemented by searches of contemporaneous reports in the trade press and local media. Johannes Pfeifenberger and his colleagues at the Brattle Group were kind enough to share their experience, study presentations, and information with me. This enabled me to compare the information that I found with what they found regarding competitive transmission procurement. Craig Glazer and Suzanne Glatz of PJM were kind enough to arrange for the data that I had collected for each PJM RTEP window to be checked. I went back to the source information to check

[88]This would require changes to the PJM process. The PJM staff and the TEAC would have to identify projects up-front where the most cost-effective solution is likely to be an upgrade to existing facilities. Competitive procurement would then apply when greenfield projects are identified as having a high probability of being the most cost-effective solution or where the staff is uncertain about whether a greenfield or an upgrade is likely to be the best solution.

the small number of differences that were identified and the results from this process appear in Table 2. I am confident that I have found all of the competitive procurements initiated by the ISOs and the associated available information for the period 2013–2018. Most of the primary source documents can be found in footnotes in this paper. Any remaining errors are entirely my responsibility.

References

J. Bushnell, S. Stoft, Electric grid investment under a contract networks regime. J. Regul. Econ. **10**(1), 61–79 (1996)

H.-P. Chao, S. Peck, A market mechanism for electric power transmission. J. Regul. Econ. **10**(1), 25–59 (1996)

H. Demsetz, Why regulate utilities? J. Law Econ. **11**(1), 55–65 (1968)

Federal Energy Regulatory Commission (FERC), *Transmission Metrics: Initial Results* (Staff Report, 2016). https://www.ferc.gov/legal/staff-reports/2016/03–17-16-report.pdf?csrt=5407744237430738089

Federal Energy Regulatory Commission (FERC), *2017 Transmission Metrics* (Staff Report, 2017). https://www.ferc.gov/legal/staff-reports/2017/transmission-investment-metrics.pdf?csrt=5407744237430738089

A. Galetovic, J.R. Inostroza, *A Lesson from Argentina: Setting Transmission Tolls in a Competitive Auction is Much Better than Regulating Them*.Working Paper 318 (Stanford University Center for International Development, 2007)

V. Goldberg, Regulation and administered contracts. Bell J. Econ. **7**(2), 426–448 (1976)

W. Hogan, Contract networks for electric power transmission. J. Regul. Econ. **4**, 211–242 (1992)

P. Joskow, Incentive regulation in theory and practice: electric distribution and transmission networks, in Chapter 5 in *Economic Regulation and it's Reform: What Have We Learned?* ed. by N. Rose (University of Chicago Press, 2014)

P. Joskow, J. Tirole, Merchant transmission investment. J. Industr. Econ. **53**(2), 233–264 (2005)

J.-J. Laffont, J. Tirole, *A Theory of Incentives in Procurement and Regulation* (MIT Press, Cambridge, 1993)

S. Littlechild, Merchant and regulated transmission: theory, evidence and policy. **42**(3), 308–335 (2012)

S. Littlechild, C. Skerk, Transmission expansion in Argentina 1, 2, 3, 4, 6. Energy Econ. **30**, 1367–1384, 1385–1419, 1420–1461, 1462–1419, 1527–1535 (2007), (five related articles)

S. Littlechild, E. Ponzano, Transmission expansion in Argentina 5" Energy Econ. **30**, 1491–1526 (2008)

B. Mountain, J. Carstairs, Batteries, interconnectors and institutions: the case of South Australia. Econ. Energy Environ. **7**(1), 105–126 (2018)

J. Pfeifenberger, et al., Brattle Study, in *Transmission Transitions: Potential Cost Savings Offered by Competitive Planning* (The Brattle Group, December 12, 2018)

J.P.M. Sijm, *The Governance Model of Power Transmission in Argentina* (2015). https://publicaties.ecn.nl/PdfFetch.aspx?nr=ECN-E–15–063

I. Vogelsang, Price regulation for independent transmission companies. J. Regul. Econ. **20**, 141–165 (2001)

I. Vogelsang, Can simple regulatory mechanisms realistically be used for transmission investment?: the case of H-R-G-V. Econ. Energy Environ. **7**(1), 63–88 (2018)

O. Williamson, Franchise bidding for natural monopoly—in general and with respect to CATV. Bell J. Econ. **7**(1), 73–104 (1976)

Merchant Transmission Investment Using Generalized Financial Transmission Rights

Darryl R. Biggar and Mohammad Reza Hesamzadeh

1 Introduction

The liberalization of wholesale power markets of the last few decades introduced competition into the generation sector, thereby introducing strong private, commercial incentives for efficient generation investment and operation. But the same reforms left responsibility for network operation and investment on regulated or government-owned transmission businesses. This gives rise to a somewhat awkward boundary between the private, commercial decisions of generators and the regulated, muted incentives of network operators. A key question for researchers has been whether or not it is possible to develop a mechanism which would provide efficient private, commercial decisions for network operation and investment.

It has long been observed that it is possible to allow for private, commercial investment in DC transmission links. Such links act like a combination of a generator and a load, arbitraging across differently priced locations. Merchant DC transmission investment was historically allowed in a few countries, including Australia.[1] But DC links are expensive and tend to be niche services. A more important question is

[1] Joskow et al. (2005).

The views expressed in this chapter reflect those of the authors and not the views of the ACCC or KTH.

D. R. Biggar (✉)
Australian Competition and Consumer Commission, Canberra, Australia
e-mail: darryl.biggar@accc.gov.au

M. R. Hesamzadeh
KTH Royal Institute of Technology, Stockholm, Sweden
e-mail: mrhesamzadeh@ee.kth.se

M. R. Hesamzadeh et al. (eds.), *Transmission Network Investment in Liberalized Power Markets*, Lecture Notes in Energy 79,
https://doi.org/10.1007/978-3-030-47929-9_11

whether or not it is possible to develop a mechanism which yields the correct incentives for private investment and operation in regulated AC transmission networks.

The possibility of such a mechanism has been attractive for generations of researchers and policy-makers. If such a mechanism existed, it would, in principle, allow private, commercial operators to make key transmission operation or investment decisions in a decentralized, for-profit manner, improving the efficiency of the transmission network and overcoming many of the drawbacks of regulation. But, to date, no effective mechanism has been developed.

Part of the problem is that a transmission network augmentation gives rise to both winners and losers, benefiting generators in an exporting region, and loads in an importing region, while harming generators in an importing region and loads in an exporting region. Transmission augmentations which are highly valuable for one market participant may be quite harmful to social welfare overall. Conversely, transmission augmentations may provide limited benefits to any one market participant, while producing substantial social benefits overall. Any mechanism for private transmission investment must overcome this mis-match between private and social incentives.

Fairly soon after mechanisms for efficiently pricing electricity transmission networks were proposed, it was recognised that market participants would require instruments for hedging the inter-nodal pricing risks that result. Hogan (1992) proposed the use of a now-conventional fixed-volume hedging instrument (known as a Financial Transmission Right or FTR). Almost immediately researchers explored whether FTRs could be used to signal and incentivize private or merchant transmission investment. Unfortunately, this research program achieved only limited success. In our view, the primary problem with that literature was the focus on only a limited form of inter-nodal hedging instrument—specifically, a fixed volume financial transmission right. We argue below that fixed volume financial transmission rights are inadequate as an instrument for hedging inter-nodal pricing risk, primarily because almost all market participants routinely transact electricity volumes which vary with market conditions. Instead, we have proposed a range of more general financial transmission rights which allow for hedging transactions with a variable volume of production or consumption (including a volume which may vary with the spot price). These instruments allow generators and loads at differently priced locations to achieve the same level of risk management as would arise as if they were at the same pricing node.[2]

Extending this work, in this chapter we show how the proposed generalized financial transmission rights naturally give rise to a mechanism which may allow for private incentives for operation and investment in transmission networks. Specifically, we show that a market participant (which we refer to as a 'trader') may simultaneously (a) provide hedge contracts to generators and loads, allowing them to perfectly hedge the risks they face; and (b) provide hedge contracts, in the form of generalized FTRs to the system operator, perfectly hedging the risks it faces, In doing so, the trader takes on all of the remaining risk in the market on itself. The total payoff faced

[2]Biggar et al. (2019).

by this trader is equal to the total economic welfare created in the sector. This has direct implications for the design of a merchant transmission investment mechanism.

We imagine that, prior to the augmentation, the network is in a state where the trader has provided contracts to all the generators and loads which eliminate their risk. In addition, for each of these contracts, the trader is assumed to acquire a corresponding generalized financial transmission right (defined below). This also eliminates the risk on the system operator.

We then imagine that the trader considers an upgrade to the transmission network. This may allow new players to enter the market (in which case the trader provides new hedge contracts, offset by new generalized FTRs). Overall, we show below that the change in the net payoff to the trader is equal to the change in social welfare. The trader will fund the upgrade if and only if it is socially beneficial.

In some respects, this result is striking. It comes close to the long-sought holy grail of privately funded transmission augmentation. Questions remain as to how this mechanism could be made practical. Nevertheless, we consider this an important first step, and further support for the proposed generalized financial transmission rights.

This chapter has three main sections. In the first section, we introduce the various forms of hedging instruments and the concept of the corresponding generalized financial transmission right. In the second section, we show how these generalized FTRs can place market participants (including the 'trader') in the same position as if all transactions were occurring at the same pricing node. This section also proves that the trader faces a net position equal to the total economic welfare created in the market. The third section illustrates how these principles can be used to yield efficient transmission upgrade decisions in simple networks. The final section concludes.

2 Introduction to Hedging and Generalized FTRs

Let us consider a simple wholesale electricity market comprising generators, loads, and a physical network connecting generators and loads. Without loss of generality, each generator and load can be assumed to be located at its own node in the network. To keep things simple, losses are ignored throughout this paper. In order to create a motivation for hedging, we must introduce some uncertainty into the model. Let us assume that there are different uncertain future states of the market, which we will label s.[3]

[3]Although much of the analysis that follows will depend on the state of the world s, and the point in time t, for simplicity, we will suppress the dependent on s and t in the formulae that follow. We consider that, on balance, this makes the presentation clearer but throughout the paper, this dependence on s should be kept in mind. We will have in mind a world in which all of the physical market participants (generators, loads, and the system operator) are relatively risk averse, while the financial market participants (which we refer to as traders) are close to risk-neutral. This is a special case of a more general framework in which all market participants are risk averse. However, we consider this to be a realistic starting point for electricity markets in practice.

2.1 Supply and Demand Curves

Assume we have a set of n generator pricing nodes and m load nodes. All generators and loads are assumed to be price takers. The spot price for electricity at node i is labelled p_i (these prices vary with the state s). As is conventional, we will assume that generators and loads choose a rate of production or consumption which maximizes their profit or utility.

The generator at node i is assumed to be described by a cost function $c_i(q_i)$ reflecting the rate at which costs are incurred ($/h) when producing at the rate q_i (MW). The cost function may depend on the state s, according to changes in, say, wind strength, input prices, or outages. The cost function is assumed to be upward sloping $c_i'(\cdot) > 0$ and strictly convex $c_i''(\cdot) > 0$.

When producing at rate q_i^S, the generator at node i receives profit at the rate $\pi_i(q_i^S) = p_i q_i^S - c_i(q_i^S)$ ($/h). For each value of the spot price, there is a corresponding profit-maximizing rate of production $q_i^S(p_i)$ which, by the assumptions above, is strictly increasing in the spot price. This function reflects the **supply curve** of the generator in state of the world s. We can then express the profit of the generator when facing spot price p_i (and state s) as follows:

$$\pi_i(p_i) = p_i q_i^S(p_i) - c_i(q_i^S(p_i)) \tag{1}$$

In general, the profit π_i varies with the state of the world s, so the generator is exposed to some risk.

Similarly, the load at node j is assumed to be described by a utility function $v_j(q_j)$ reflecting the rate at which utility is received ($/h) when consuming at the rate q_j (MW). This load may depend on the state s according to factors such as ambient temperature (in Australia, temperature is a major driver of air-conditioning load, a primary source of demand on hot days). The utility function is assumed to be upward sloping $v_j'(\cdot) > 0$ and concave $v_j''(\cdot) < 0$.

The rate at which the load receives utility when consuming at rate q_j^D is given by the expression $u_j(q_j^D) = v_j(q_j^D) - p_j q_j^D$ ($/h). Maximizing this expression for a given value of the spot price gives the (downward sloping) **demand curve** $q_j^D(p_j)$ in state of the world s. The utility of the load at node j facing spot price p_j can then be written:

$$u_j(p_j) = v_j(q_j^D(p_j)) - p_j q_j^D(p_j) \tag{2}$$

As is conventional, from the overall energy balance equation, the total amount of electricity produced is equal to the total amount consumed, at each point in time and in each state of the world:

$$\sum_i q_i^S = \sum_j q_j^D \tag{3}$$

2.2 The Design of Typical Hedge Contracts

As we will see, the form of effective inter-nodal hedging instruments depends in turn on the form of the instruments that generators and loads need in order to hedge their risk. So, let us first consider what forms typical hedging instruments might take.

2.2.1 Hedging for Generators

Let us start by focusing on the question of how to hedge the profit of a conventional, reliable dispatchable (e.g., thermal) generator. The profit function of a conventional generator is given in Eq. 1. Let us suppose that the generator sells a hedge contract (possibly consisting of a portfolio of hedge contracts) with the payout given by H_i^S, so that the hedged profit of the generator is:

$$\pi_i(p_i) = p_i q_i^S(p_i) - c_i(q_i^S(p_i)) - H_i^S \tag{4}$$

We will define the **implicit volume** V_i^S of the hedge contract to be the rate of change of the hedge payout with respect to the market price:

$$V_i^S(p_i) = \frac{\partial H_i^S}{\partial p_i}(p_i) \tag{5}$$

If this generator is reliable (so that its cost function is independent of the state of the world s), it only faces risk arising from variation in the spot price p_i. It can eliminate this risk by choosing a hedge contract with an implicit volume which matches its supply curve[4]:

$$\frac{\partial \pi_i}{\partial p_i} = q_i^S(p_i) - \frac{\partial H_i^S}{\partial p_i} = 0 \iff \frac{\partial H_i^S}{\partial p_i}(p_i) = q_i^S(p_i) \tag{6}$$

At this point, we can introduce one typical form of hedge contract known as a cap contract. By definition, a **cap contract** with a strike price S and a volume V pays out the difference between the spot price p and the strike price multiplied by the volume when the spot price exceeds the strike price.[5]

$$\text{Cap}(p|S, V) = (p - S)VI(p \geq S) \tag{7}$$

[4]It is important to note—here and elsewhere throughout this chapter—that this theoretical ideal hedge contract matches the forecast supply curve of the generator not the actual output. If the hedge contract paid the generator an implicit volume based on its actual output, the generator would face a moral hazard problem: it would not have an incentive to produce anything at all.

[5]Note that a swap contract is a special case of a cap contract where the strike price is below any possible realization of the spot price.

Here, $I(\cdot)$ is the indicator function[6] which takes the value one when the expression in brackets is true and zero otherwise.

Biggar et al. (2019) show that given a set of cap contracts with strike prices which are reasonably 'dense' (in the sense that, for any given price there is a cap contract in the range with a strike price close to that price), any generator with a fixed, upward-sloping marginal cost curve can construct a hedge contract with an implicit volume which approximates its supply curve. This approximation can be made arbitrarily close to the true supply curve as the strike prices of the cap contracts become arbitrarily dense (in the sense that the smallest distance between a strike price in the range and any given price tends to zero). In other words, given a dense set of cap contracts, any reliable generator with an upward-sloping marginal cost curve can come arbitrarily close eliminating all of the risk that it faces. This result is demonstrated formally in the appendix.

It is worth mentioning that, in practice, even if the portfolio of hedge contracts perfectly matches the supply curve of the generator, such a hedging strategy typically does not eliminate *all* of the risk faced by the generator. A generator may also face risks associated with changes in its cost function $c_i(\cdot)$. For example, a generator might be exposed to risk arising from plant outages (such as the loss of a generating unit) or the risk arising from variation in input-fuel cost. Hedging these risks requires additional hedging instruments, such as input-fuel price contracts.

Let us focus on a special case of a hedging contract faced by a special type of generator with a constant marginal cost, but an uncertain production capacity. We will refer to this as an **intermittent** generator. In particular, let us suppose that the cost function of the intermittent generator can be represented as a variable cost c_i (\$/MWh) up to some production capacity K_i (MW), which is uncertain (e.g., varies with the wind strength). Such a generator will produce at capacity $q_i^S = K_i$ whenever the spot price p_i exceeds the variable cost c_i. The raw or unhedged profit of the generator is therefore:

$$\pi_i(p_i) = (p_i - c_i)K_i I(p_i \geq c_i) \tag{8}$$

As before, this generator can hedge its pricing risk with a hedge contract with an implicit volume equal to its supply curve $q_i^S(p_i) = K_i I(p_i \geq c_i)$. For example, this generator could be perfectly hedged with a hedge contract which resembles a cap contract, but which has a volume whose variation matches the variation in the output of the generator:

$$H_i^S(p_i) = (p_i - c_i)K_i I(p_i \geq c_i) = \text{Cap}(p_i|c_i, K_i) \tag{9}$$

[6] Also known as the Iverson Bracket.

For example, in the case of a wind generator, the hedge contract would have a volume which is designed to reflect the forecast output of a wind farm, based on the measured wind speed. Such hedge products have recently been offered in Australia for wind and solar generators and are known as 'proxy revenue swaps.'[7] A related, more common form of hedge contract of this kind is the **Power-Purchase Agreement** or PPA.[8]

2.2.2 Hedging for Loads

Let us now consider what might be a typical hedge contract for a load. Let us suppose that the load purchases a hedge contract with a payout H_j^D so that its hedged payout is:

$$u_j(p_j) = v_j(q_j^D(p_j)) - p_j q_j^D(p_j) + H_j^D \tag{10}$$

As before, it turns out that a load can perfectly hedge the pricing risk it faces with a hedge contract with an implicit volume equal to the demand curve of the load.

$$\frac{\partial u_j}{\partial p_j} = -q_j^D(p_j) + \frac{\partial H_j^D}{\partial p_j} = 0 \iff \frac{\partial H_j^D}{\partial p_j}(p_j) = q_j^D(p_j) \tag{11}$$

Let us introduce the concept of the floor contract (which is the flip side of the cap contract). A **floor contract** with a strike price S and a volume V pays out the difference between the spot price p and the strike price multiplied by the volume when the spot price is below the strike price:

$$\text{Floor}(p|S, V) = (S - p)VI(p \leq S) \tag{12}$$

As an aside, we note that in the analogue of the well-known put-call parity result, there is a corresponding cap-floor parity, which allows a floor contract to be constructed out of a cap and a swap contract.

[7]These are described as follows: 'The project company pays the hedge provider a fixed percentage of 'proxy revenue', which is equal to the hub price multiplied by the 'proxy generation' for that settlement period. 'Proxy generation' is calculated under the hedge as the power that would have been produced by the project based on measured wind speeds and assuming pre-agreed fixed operational inefficiencies. The assumed operational inefficiencies include availability, performance and electrical losses.' https://www.projectfinance.law/publications/2017/June/hedges-for-wind-projects-evaluating-the-options.

[8]A PPA, which is based on the firm's actual output suffers from the moral hazard problem noted above.

As before, for any given supply curve, given a set of floor contracts with a sufficiently dense set of strike prices, it is possible to construct a portfolio with an implicit volume which approximates the demand curve of the load arbitrarily closely. In particular, in the case of a load which a fixed utility function $v_j(p_j)$, as the density of the set of floor contracts increases the load can create a portfolio of floor contracts which reduces the risk it faces arbitrarily close to zero.

In the literature on wholesale power markets, it is common to model loads as having a utility function which varies with factors such as the ambient temperature. One common assumption is to assume that the load has a fixed utility from consumption (which we will label M_j ($/MWh) – M_j is sometimes referred to as the 'Value of Lost Load' or VOLL) up to a varying level K_j (MW). Such a load will consume quantity K_j provided the spot price is less than M_j. The demand curve is therefore: $q_j^D(p_j) = K_j I(p_j \leq M_j)$.

Such a load can perfectly hedge the risk it faces with a variant of the floor contract which has a volume which varies with the maximum load:

$$H_j^D(p_j) = (M_j - p_j)K_j I(p_j \leq M_j) = \text{Floor}(p_j|M_j, K_j) \tag{13}$$

Such a contract is typically known as a **load-following hedge** or LFH. More generally, if the load has a downward sloping demand curve up to some maximum K_j, we need a more general form of the floor contract, which we will refer to as the FloorLFH, with a payout as follows. There is an example of the use of the FloorLFH in Sect. 4.2 below.

$$\text{FloorLFH}(p|S, V, L, K) = (S - p)VI(p \leq S, L \leq K) \tag{14}$$

2.3 *The Design of Inter-nodal Hedging Instruments*

At this point, we will introduce generalized Financial Transmission Rights. The reason for this design choice will become apparent below.

We propose that: (a) a node in the network is chosen and designated the reference node (labelled node N); and (b) for each node in the network other than the reference node, and for each hedge contract chosen by a generator or load at that node, the system operator makes available to the market a **corresponding FTR** from that node to the reference node. For each hedge contract H_i, the corresponding FTR is an FTR from node i to the reference node with the same implicit volume.

In the previous section, we introduced cap contracts, floor contracts, PPAs, and LFHs. We propose that, for each of these hedge contract types, there is made available a corresponding financial transmission right.

For example, in the previous section, we introduced the concept of the cap contract. A cap contract with a strike price S and volume V has an implicit volume equal to $VI(p \geq S)$. In exactly the same way, we propose the creation of an FTR contract which takes the form of a cap contract (which we will refer to as a CapFTR) with the same implicit volume. Specifically, a CapFTR from node i to node N, with a strike price S and a volume V pays out the following:

$$\text{CapFTR}(p_i, p_N|S, V) = (p_N - p_i)VI(p_i \geq S) \tag{15}$$

In exactly the same manner as with cap contracts, CapFTRs can be combined to form an instrument with an implicit volume which matches the supply curve of any generator with an upward-sloping supply curve. Specifically, given a set of CapFTRs with different strike prices, as the density of those strike prices increases, it is possible to form a set of CapFTRs with an implicit volume which approximates a given supply curve arbitrarily closely.

Analogously, we propose the creation of FloorFTRs, PPAFTRs, and LFHFTRs. The payout on a FloorFTR from node i to node j, with a strike price S and a volume V is as follows:

$$\text{FloorFTR}(p_i, p_N|S, V) = (p_N - p_i)VI(p_i \leq S) \tag{16}$$

The other generalized FTRs are defined in a similar way.

3 Hedging Using Generalized FTRs

To understand why these generalized FTRs might be valuable, let us first clarify the task to be solved. We will show that hedging between market participants cannot eliminate all risk. There remains a residual risk which must be borne by some party. Our objective with inter-nodal hedging, therefore, is not to enable the parties to eliminate all risk, but merely to allow them to reduce the risk down to the level that would arise if all generators and loads traded at the same pricing node.

3.1 The Theoretical Minimum Level of Risk

But what is this theoretical minimum level of risk? Let us suppose that each generator or load enters into a portfolio of financial hedge contracts which, in total, oblige the generator to pay the amount H_i^S and the load to receive the amount H_j^D in state s. In addition, we will suppose that the system operator enters into hedge contracts to hedge the risk that it faces (from variation in the merchandising surplus). In total these hedge contracts oblige the system operator to pay the amount H^{SO}. Let us define the amount H to be the total net payments made between generators, loads and the system operator. This net amount reflects the extent to which, for hedging purposes, there are payments to or from *other, outside* sources or parties (other than generators, loads, or the system operator).

$$H = \sum_i H_i^S - \sum_j H_j^D + H^{SO} \tag{17}$$

In sum, the total collective risk faced all the market participants (after hedging) is equal to the variation in the sum of the hedged profit for each generator plus the sum of the hedged utility of load plus the hedged position of the system operator. This can be written as follows:

$$\begin{aligned} W &= \sum_i \pi_i(p_i) + \sum_j u_j(p_j) + MS + H^{SO} \\ &= \sum_i (p_i q_i - c_i(q_i) - H_i) + \sum_j (v_j(q_j) - p_j q_j + H_j) + MS + H^{SO} \\ &= R + H \end{aligned} \tag{18}$$

Here $R = \sum_j v_j(p_j) - \sum_i c_i(p_i)$ is the total economic welfare of the participants in the market, H is the net payments under the hedge contracts as defined in Eq. 17, and MS is the **merchandising surplus** (also known as congestion rent) which is conventionally defined as the value of the net withdrawal of power at each node:

$$MS = -\sum_a p_a z_a \tag{19}$$

where $z_a = \sum_{i \in a} q_i^S - \sum_{j \in a} q_j^D$ is the net injection at pricing node a ($i \in a$ and $j \in a$ refers to the set of generation nodes and load nodes in pricing region a, respectively).

We will also assume that market participants do not trade hedge contracts with any other entities outside the electricity market, so that $H = 0$. Then, from Eq. 18 it follows that, no matter what hedge contracts are written, the total payoff of all the market participants is just equal to the total economic welfare $W = R$. This makes clear that although hedge contracts can shift risk around within the industry, there is a minimum economic risk which cannot be eliminated by trading in hedge contracts between generators and loads alone. That minimum risk is equal to the variation in the total economic welfare $\mathrm{Var}(R)$.[9]

[9]There is a corollary of this result which is interesting. This corollary is a parallel to the well-known Modigliani-Miller Theorem: The total value of the generators and loads in the market is independent of the trade in hedge contracts. To see this, let $V(X)$ be the present value of the uncertain future cash-flow X. This function is linear $V(X + Y) = V(X) + V(Y)$. From Eq. 18, we have that

This result is important because it establishes the theoretical minimum level of risk which must be borne in the market. The question for us now is how to package the merchandising surplus in a way which allows market participants to easily form the portfolios they need to hedge the risks they face.

3.2 *Are Fixed-Volume FTRs an Effective Inter-nodal Hedging Instrument?*

In practice, many liberalized wholesale electricity markets make available to market participants an instrument known as a fixed volume financial transmission right (FFTR). These are also known as FTR obligations to distinguish them from FTR options. A fixed-volume FTR between two locations on the network at a given point in time pays out a flow of funds equal to the difference in the nodal prices in those locations at that time multiplied by a *fixed* quantity. An FFTR of volume f_{ij} from node i to node j pays out the following amount in state s:

$$F_{i \to j} = (p_j - p_i) f_{ij} \tag{20}$$

But are fixed-volume FTRs a useful inter-nodal hedging instrument? The answer is no: A firm FTR is a hedging instrument with a fixed volume. It is therefore a useful instrument for hedging transactions which feature a *fixed volume* of production and consumption. But most conventional generators in the wholesale market have an output which varies with the state of the world, such as with changes in the wind speed, or the spot price. Such generators would prefer a hedging instrument with a hedging volume which varies with the wind speed or the spot price in a manner which mimics the production of the generator.

For example, consider the problem of hedging the output of a generic price-taking generator with a cost function $c_i(q_i)$. As noted earlier, this generator has a supply function $q_i(p_i)$ which is determined by the marginal cost function of the generator: $c_i'(q_i(p_i)) = p_i$. As the spot price p_i varies, the output of the generator $q_i(p_i)$ varies, potentially over a wide range, or the generator might shut down entirely. The risk associated with this pattern of production cannot be hedged with a *fixed volume* hedging instrument. The same, of course, applies to a wind generator. Such a generator has a pattern of production which cannot be hedged with a *fixed volume* hedging instrument.

$V(W) = \sum_i V(\pi_i) + \sum_j V(u_j) = V(R)$. The total value of the market participants is constant and independent of the trade in hedge contracts.

In our view, fixed-volume FTRs are not a satisfactory instrument for hedging inter-nodal pricing risks. The lack of effective inter-nodal hedging instruments has hindered the development of wholesale electricity markets.[10] We turn now to explain how the generalised FTRs introduced above may be used to hedge inter-nodal pricing risk.

3.3 Hedging Inter-nodal Pricing Risk Using Generalized FTRs

As we have seen, hedging between generators and loads alone cannot eliminate all risks. The remaining risks must be borne by at least one other party. For this reason, following Biggar et al. (2019), we introduce a new market participant, which we will refer to as the **trader**. The trader(s) are financial intermediaries who are assumed to behave in a manner which is close to risk neutral. The trader plays the risk-taking role, taking these residual or remaining risks on itself. Specifically, we will assume that through a process of trade and exchange in hedge contracts between generators, loads, and the trader, each generator and each load reaches a position where, due to the portfolio of hedge contracts it has acquired, it is perfectly hedged from risk. The trader takes on all of the remaining or residual risk.

Without loss of generality we can ignore trade directly between generators and loads. We therefore assume that, for each generator i, the trader purchases a hedge contract from the generator $H_i^S(p_i)$ which reduces the risk faced by the generator to zero, i.e., $\text{Var}(\pi_i + H_i^S) = 0$. This implies $H_i^S(p_i) = \pi_i(p_i) + k_i$, for some constant k_i. As we have seen, this also implies that the implicit volume in the hedge contract is equal to the supply curve of the generator.

Similarly, for each load, the trader sells a contract to the load which reduces the risk faced by the load to zero, i.e., $H_j^D(p_j) = u_j(p_j) + k_j$, for some constant k_j. Again, the implicit volume in the hedge contract is equal to the demand curve of the load.

In addition, for each hedge contract held by the trader (that is, for each hedge contract purchased from a generator or sold to a load), the trader is assumed to acquire the corresponding generalized financial transmission right from the system operator. Let us assume that node N is the reference node. For each generator at node i and load at node j, the trader acquires the corresponding generalized FTR $H_{i\to N}^S$ and $H_{j\to N}^D$.

[10] Some markets also make available FTR *options*. However this does not solve the problem identified above. FTR options payout the price difference between two nodes, but only when that price difference is positive. We have seen above that a generator or load would like an instrument which depends only on the price at one location, not on the sign of the price difference.

As we noted earlier, the corresponding generalized financial transmission right has an implicit volume which matches the implicit volume of the hedge contract.[11]

In other words, in addition to the hedge contracts H_i^S and H_j^D purchased from generators or sold to loads, the trader also acquires a portfolio of generalised FTRs with the following payoffs:

$$H_{i\to N}^S = (p_N - p_i)q_i^S(p_i) \tag{21}$$
$$H_{j\to N}^D = (p_N - p_j)q_j^D(p_j) \tag{22}$$

Immediately, we can observe that the total payoff of these generalized FTRs is equal to the merchandising surplus, eliminating the risk faced by the system operator:

$$\begin{aligned}\sum_i H_{i\to N}^S - \sum_j H_{j\to N}^D &= \sum_j p_j q_j^D - \sum_i p_i q_i^S + p_N\left(\sum_i q_i^S - \sum_j q_j^D\right)\\ &= -\sum_a p_a z_a = MS \end{aligned} \tag{23}$$

Now, let us examine the characteristics of the total hedge position of the trader. Let H^T be the total financial position of the trader. Using the results above:

$$\begin{aligned} H^T &= \sum_i H_i^S - \sum_j H_j^D + \sum_i H_{i\to N}^S - \sum_j H_{j\to N}^D \\ &= \sum_i [p_i q_i^S(p_i) - c_i(q_i^S(p_i)) + k_i] + \sum_j [v_j(q_j^D(p_j)) - p_j q_j^D(p_j) + k_j] \\ &\quad \sum_i (p_N - p_i)q_i^S(p_i) + \sum_j (p_j - p_N)q_j^D(p_j) \\ &= \sum_j v_j(q_j^D(p_j)) - \sum_i c_i(q_i^S(p_i)) + k \\ &= R + k \end{aligned} \tag{24}$$

Here, $k = \sum_i k_i + \sum_j k_j$. We conclude that $\mathrm{Var}(H^T) = \mathrm{Var}(R)$. The risk has been reduced to the minimum possible level. Market participants are placed in the same position as if there was only one pricing node.

In Sect. 4 we demonstrate how this might work in practice, but first we show how this theory has a direct application in the context of merchant transmission investment

[11] As noted earlier, this does not necessarily imply that the generalised FTR is some form of bespoke arrangement—as we noted earlier, the hedge contract required by the generator or load could consist of a portfolio of cap or floor contracts. The corresponding generalised FTRs would itself be a portfolio of the corresponding capFTR or floorFTR contracts.

3.4 Merchant Transmission Investment Using Generalised FTRs

As noted in the introduction, economists have long been interested in the possibility that locational marginal prices might, somehow, provide the correct signals for electricity network investment. If this could be achieved, private commercial incentives for network investment could improve or replace the weak, imperfect or muted incentives for investment that arise under regulatory frameworks. This is an attractive possibility.

In particular, we wish to design a mechanism under which the change in the value of the payoff to a market participant aligns with the overall economic welfare. Formally, let us suppose that the uncertain future welfare of the power system is initially R^0. A market participant (who will be the trader) is considering whether or not to make a change to power system (in this case an augmentation to the network) with an associated incremental cost C. The new uncertain future welfare of the power system after the change is assumed to be R^1. It is socially efficient for the network augmentation to be carried out if and only if

$$V(R^1) - C > V(R^0) \tag{25}$$

Here, $V(X)$ is the valuation function: For any uncertain cash-flow X, $V(X)$ reflects the present discounted value of the cash-flow.

But how can we design a mechanism in such a way that a market participant faces the change in total economic welfare following a change in the market?

The discussion above suggests that such a mechanism might be possible. The broad outline of the mechanism is as follows: The trader offers hedge contracts to generators and loads which perfectly insulate them from risk. The trader then combines those contracts with the matching generalized FTR, thereby perfectly insulating the system operator from risk. As we have seen (from Eq. 24), the trader then faces a total payoff which is equal to the total social welfare (up to a constant). Therefore, provided these hedge contracts continue to perfectly insulate the generators and loads (and any new generators and loads that enter the market), following any change in the market, the trader will face the total change in welfare following the change in the market. In particular, if the trader incurs the cost of a network augmentation, the trader will choose to augment the network if and only if the network augmentation is in the public interest.

Let us look more closely at how this might work. As noted above, we have assumed that all market participants are risk averse, except for one market participant, which we will refer to as the trader, who is risk neutral. The market participants are assumed to trade in hedge contracts. As we have seen, for each generator, the trader is assumed to purchase hedge contracts from that generator, referenced to the generator's local node, which collectively match (in volume) the supply curve of the generator (including any shifts in that supply curve resulting from factors such as changes in the wind speed). This insulates the generator from all risk. At the same time, the trader is assumed to obtain the matching or corresponding generalised FTR from that local node to the reference node. Similarly, the trader sells hedge contracts to loads, referenced to the load's local node, which match (in volume) the demand curve of the load. Again, this perfectly insulates the load from all risk. As before, the trader obtains the matching or corresponding generalised FTR from that local node to the reference node. As we have seen, (from Eq. 24) the total payoff facing the trader then matches the total welfare of the power system (up to a constant).

Now imagine that, the trader has the potential to make a change to the power system, such as a network upgrade. For the mechanism outlined above to work, it must be that the trader continues to face a payoff equal to the total welfare after the change. In the short run, the network upgrade will cause a change in prices, leading to a change in dispatch outcomes and, in the longer term, may change the entry and exit decisions of generators.

Let us start with the short-run changes. As just noted, the network upgrade will bring about a change in dispatch outcomes. This may result in generators and loads exploring new regions of their supply and demand curves, which were not previously reached. We will address this by assuming that, nevertheless the hedge contracts of the generators and loads accurately reflect these regions of the supply and demand curves (even though there were not reached before).[12] With this assumption, it follows that the generators and loads continue to be perfectly hedged after the network augmentation. This implies, in turn, that the trader continues to a cash-flow stream which matches the total welfare of the power system. The trader has an incentive to upgrade the network if and only if it is efficient to do so.

In the longer term, the network upgrade may induce generators and loads to enter or exit the market. For the mechanism to work, the change in welfare brought about by this entry and exit must be reflected back to the trader. To bring this about, we will introduce a new category of hedge contract, which we will refer to as an 'entry-contingent hedge option'.

An entry-contingent hedge call option gives the holder (in this case the trader) the right to purchase a hedge option in the future on payment of a pre-determined price f. This contract would be sold by a potential-entrant generator with the pre-determined price set equal to the fixed cost of operation of the generator f. If the call option is not exercised, the generator does not enter the market and no payments are made. If the call option is exercised the generator enters the market, incurring the fixed cost f, which is paid by the trader in exchange for exercising the call option. The generator also receives an uncertain future payment stream, which is perfectly hedged by the hedge option, passing the risk on to the trader. The call option will only be exercised if the value of the uncertain payment stream exceeds the fixed cost, which is exactly the condition for entry in an efficient, competitive market.[13]

[12]We do not consider this assumption to be unreasonable as we conjecture that it should not cost any more to provide hedge contracts to reflect parts of the supply and demand curves which are not actually reached ex ante.

[13]There is an analogous result for loads.

To be a little more precise, let us suppose that H_i is a perfect hedge for a potential-entrant generator with an uncertain future profit π_i. In a competitive market for hedge contracts, the current price for the hedge is $V(H_i)$ (where $V(\cdot)$ is the value function described above). Let us suppose the fixed cost of the generator is f_i. This generator will enter the market and sell a hedge contract if and only if $V(H_i) > f_i$. This generator may then sell an entry-contingent call option with the strike price f_i, giving the trader the right to purchase the hedge contract at the price f_i. After the trader purchases the call option, if the trader augments the network, the trader will exercise the option if and only if $V(H_i) > f_i$, which is the condition for efficient entry. In addition, the trader receives the payoff $V(H_i) - f_i$, which is the total social welfare created by the entry.

We will assume that the trader purchases such entry-contingent call options from all generators which may enter or exit the market. After the network upgrade is made, the trader invokes the call options for generators which now become profitable, bringing about new entry. At the same time, the trader does not invoke the call option for generators which now become loss-making, resulting in their exit from the market. In either case, the trader is left with a payoff which reflects the change in total economic welfare arising from the entry and/or exit decision.

In summary, the risk averse generators, loads and the system operator pass the risk they face to this central agent, referred to as the trader. The trader, making use of the generalized FTRs described above, and, if necessary, the entry-contingent hedge options described above, faces a payoff which perfectly reflects the changes in total economic welfare arising from changes in the market. The trader therefore has an incentive to upgrade the network if and only if it is efficient to do so.

4 Simple Network Examples

Let us turn now to explore how generalized FTRs might facilitate inter-locational hedging and merchant investment in practice.

4.1 Two-Node Network Example

The first network we consider has just two nodes, labelled A and B. Each node has both generators and load, as illustrated in Fig. 1.

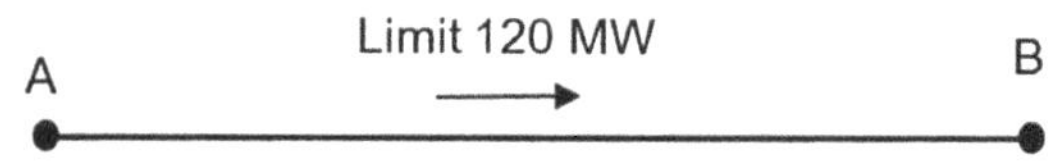

Fig. 1 Simple two-node network

Table 1 Optimal dispatch outcomes in each scenario in the network of Fig. 1. Here MS= merchandising surplus, SWF = total welfare

State	Load		Flow	Price		MS	SWF
	L_A	L_B	$A \to B$	A	B	MS	R
1	50	200	100	\$50	\$50	\$0	\$242,500
2	50	250	120	\$50	\$300	\$30,000	\$282,500
3	50	300	120	\$50	\$300	\$30,000	\$317,500
4	50	350	120	\$50	\$1000	\$114,000	\$331,500
5	85	200	100	\$50	\$50	\$0	\$275,750
6	85	250	120	\$100	\$300	\$24,000	\$315,500
7	85	300	120	\$100	\$300	\$24,000	\$350,500
8	85	350	120	\$100	\$1000	\$104,000	\$364,500

There are assumed to be three generators at node A and two at node B. Each generator is perfectly reliable and has a constant marginal cost up to a maximum capacity of 100 MW. The marginal costs (c_i) of the generators at node A are G1:\$10/MWh, G2:\$50/MWh, and G3:\$100/MWh. The marginal costs of the generators at node B are G4:\$40/MWh, G5:\$300/MWh.

There are two sources of uncertainty in the model, corresponding to uncertainty in the maximum load L_A and L_B at nodes A and B, respectively. The load at node A can take two values: $L_A = 50$ or 85. The load at node B can take four values: $L_B = 200, 250, 300$ or 350, for a total of eight different scenarios. The maximum value of consumption (sometimes referred to as the Value of Lost Load or VoLL) is assumed to be \$1000/MWh. The link $A \to B$ initially has a fixed capacity of 120 MW, but there is potential to upgrade this link to 160 MW. No generators or loads enter or exit the market following the upgrade.

The optimal dispatch outcomes (prices, flows, merchandising surplus, and overall total welfare) under each of the different load scenarios are set out in Table 1.

We will assume that each generator and load seeks to eliminate all of the risk that it faces. For the generators, this can be achieved if the trader purchases from the generator a cap contract with a strike price equal to the marginal cost of each generator and a volume of 100 MW. In the case of the loads, the elimination of risk can be achieved with a load-following floor contract, with a volume equal to the realization of the maximum load (L_A and L_B). The full list of hedge contracts is set out in Table 2. These hedge contracts completely eliminate the risk faced by the generators and loads.

Let us designate node B as the reference node. Let us suppose that, in addition, the trader acquires generalized FTRs for each generator at node A to node B. Since each generator at node A can be hedged with a cap contract, the corresponding generalized

Table 2 Hedge contracts to eliminate the risks of the market participants in the network of Fig. 1

Gen/Load	Hedge contract
G1	Cap(p_1|\$10, 100)
G2	Cap(p_1|\$50, 100)
G3	Cap(p_1|\$100, 100)
G4	Cap(p_2|\$40, 100)
G5	Cap(p_2|\$300, 100)
L1	Floor(p_1|\$1000, L_A)
L2	Floor(p_2|\$1000, L_B)

Table 3 Trader net position in the network of Fig. 1

State	Gen hedging $\sum_i H_i^S$	Load hedging $\sum_j H_j^D$	FTR Payout	Total Payoff	SWF R
1	\$5000	\$237,500	\$0	\$242,500	\$242,500
2	\$30,000	\$222,500	\$30,000	\$282,500	\$282,500
3	\$30,000	\$257,500	\$30,000	\$317,500	\$317,500
4	\$170,000	\$47,500	\$114,000	\$331,500	\$331,500
5	\$5000	\$270,750	\$0	\$275,750	\$275,750
6	\$40,000	\$251,500	\$24,000	\$315,500	\$315,500
7	\$40,000	\$286,500	\$24,000	\$350,500	\$350,500
8	\$180,000	\$76,500	\$108,000	\$364,500	\$364,500

FTR is a CapFTR contract.[14] Similarly, the trader is assumed to acquire a generalized FTR for the load at A. Since the load at A can be hedged with a load-following floor contract, we assume the trader can acquire the corresponding LfhFTR.

Now, let us consider the net position of the trader. Table 3 sets out the total hedge payout to generators, the total hedge payout to loads, and the total payout on the generalised FTRs. As table 3 shows, the total net payout on the FTRs is equal to the merchandising surplus (as shown in table 1). Importantly, the total net position of the trader matches the total welfare created in this market.

Now, let us suppose that the trader considers upgrading the link to a capacity of 160 MW. This results in a new optimal dispatch with new pricing outcomes in each scenario. It also results in a higher overall social welfare. The outcomes under optimal dispatch following the upgrade are set out in Table 9.

We will assume that the same players continue in the market, with no new entry. Moreover, the trader does not need to offer any new hedge contracts or retire any old hedge contracts. All the generators and loads can be perfectly hedged using the

[14]We will assume that the CapFTR contract pays out $H_{i,A\to B} = (P_B - P_A)V_i$ where $V_i = 100$ if $P_A > c_i$, $V_i = 0$ if $P_A < c_i$ and $V_i = Q_i^S$ if $P_A = c_i$. This last condition is required since we have violated the assumption that the supply curve is strictly upward sloping.

Table 4 Optimal dispatch outcomes in each scenario in the network of Fig. 1 with link $A \rightarrow B$ upgraded to 160 MW

State	Load		Flow	Price		MS	SWF
	L_A	L_B	$A \rightarrow B$	A	B	MS	R
1	50	200	100	$50	$50	$0	$242,500
2	50	250	150	$50	$50	$0	$290,000
3	50	300	160	$100	$300	$32,000	$327,000
4	50	350	160	$100	$300	$32,000	$362,000
5	85	200	100	$50	$50	$0	$275,750
6	85	250	150	$50	$50	$0	$312,500
7	85	300	160	$100	$300	$32,000	$358,500
8	85	350	160	$100	$300	$32,000	$393,500

Table 5 Trader net position in the network of Fig. 1, with the link $A \rightarrow B$ upgraded to 160 MW

State	Gen hedging	Load hedging	FTRs	Total	SWF
	$\sum_i H^S_{Gi}$	$\sum_j H^D_{Lj}$	Payout	Payoff	R
1	$5000	$237,500	$0	$242,500	$242,500
2	$5000	$285,000	$0	$290,000	$290,000
3	$40,000	$255,000	$32,000	$327,000	$327,000
4	$40,000	$290,000	$32,000	$362,000	$362,000
5	$5000	$270,750	$0	$275,000	$275,750
6	$20,000	$301,500	$0	$321,500	$321,500
7	$40,000	$286,500	$32,000	$358,500	$358,500
8	$40,000	$321,500	$32,000	$393,500	$393,500

contracts set out in Table 3. In addition, the set of FTRs need not change. The resulting net position of the trader after the upgrade is set out in Table 5. As before, the total net position of the trader matches the total economic welfare created in this market.

It follows immediately that the trader faces exactly the right economic incentives to upgrade this link. For example, if all of the scenarios are equally likely, the expected net position of the trader before the upgrade is $310,031 per hour and after the upgrade is $321,344 per hour—a difference of $11,313. The trader will make the upgrade if and only if the (amortized) cost per hour of the upgrade is less than $11,313, as required.

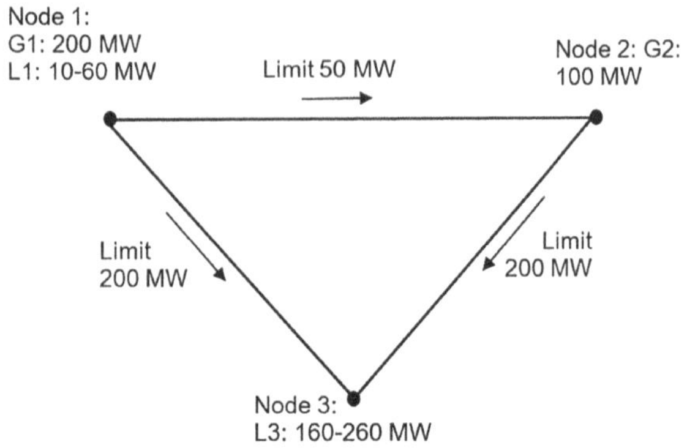

Fig. 2 Simple three-node network

4.2 Three-Node Network Example

Now, let us illustrate how the proposals might work with a simple meshed network with just three nodes.[15] This network is illustrated in Fig. 2. Each link is assumed to have identical electrical characteristics. We will also assume that the generators have quadratic cost functions up to the maximum capacity, requiring a larger portfolio of cap contracts for effective hedging.

This network features two generators and two loads. Generator G1, located at node 1, has a marginal cost function given by $c_1'(q_1^S) = 10 + 0.02 \times q_1^S$ up to a capacity of 200 MW. Generator G2, located at node 2, has a cost function given by $c\prime_2(q_2^S) = 30 + 0.04 \times q_2^S$ up to a capacity of 100 MW. Load L1 is located at node 1 and has a utility function $v\prime(q_2^D) = 1000 - 0.02 \times q_2^D$ up to a maximum load which can take values 10, 20, 30, 40, or 50 MW. Load L2 is located at node 3 and has a utility function $v\prime(q_3^D) = 5000 - 0.08 \times q_3^D$ up to a maximum load of 160, 210 or 260 MW. There are, therefore, 15 states of the world to consider. Node 3 is the reference node.

The link $1 \rightarrow 2$ initially has a fixed capacity of 30 MW, but there is potential to upgrade this link to 50 MW. Any limits on the other two links are not binding.

The optimal dispatch outcomes (prices, flows, merchandising surplus, and overall total welfare) under each of the different load scenarios are set out in Table 6.

What portfolio of hedge contracts might hedge the positions of the generators and the loads? For the generators, the risk they face can be reduced to a very low level with a portfolio of around a dozen cap cap contracts with varying strike prices contracts, as set out in Table 7. Similarly, the load utility can be effectively hedged with a set of FloorLFH contracts, as set out in Table 7.

As before, for each of the contracts set out in Table 7, the trader is assumed to acquire the corresponding generalised FTR. That is, for each cap contract purchased from generators G1 and G2 the trader acquires the corresponding CapFTR, and

[15] Biggar et al. (2019) illustrate how this might work in networks with 6 and 24 buses.

Table 6 Optimal dispatch outcomes in each scenario in the network of Fig. 2

State	Load		Flow			Price			MS	SWF
	L_1	L_2	$1 \to 2$	$1 \to 3$	$2 \to 3$	1	2	3	MS	R
1	160	10	29.9	94.1	64.3	$12.7	$31.2	$22.0	$837.9	$807329.9
2	160	20	30.0	94.2	64.2	$12.9	$31.4	$22.1	$837.4	$817212.7
3	160	30	30.0	94.2	64.2	$13.1	$31.4	$22.2	$828.6	$827077.6
4	160	40	30.0	94.2	64.2	$13.3	$31.4	$22.3	$819.8	$836938.5
5	160	50	30.0	94.2	64.2	$13.5	$31.4	$22.4	$810.5	$846794.4
6	210	10	30.0	118.9	89.0	$13.2	$32.4	$22.8	$871.0	$1055482.8
7	210	20	30.0	118.9	89.0	$13.4	$32.4	$22.9	$862.7	$1065346.5
8	210	30	30.0	118.9	89.0	$13.6	$32.4	$23.0	$853.4	$1075206.5
9	210	40	30.0	118.9	89.0	$13.8	$32.4	$23.1	$843.6	$1085061.9
10	210	50	29.6	118.7	89.2	$23.7	$32.4	$28.4	$465.2	$1094900.8
11	260	10	30.0	143.7	113.7	$13.7	$33.4	$23.6	$893.3	$1303384.2
12	260	20	30.0	143.7	113.7	$13.9	$33.4	$23.7	$884.8	$1313243.1
13	260	30	26.4	141.9	115.5	$33.6	$33.6	$33.6	$0.4	$1322991.9
14	260	40	19.8	138.6	118.8	$197.4	$197.4	$197.4	$0.0	$1332647.1
15	260	50	19.8	138.6	118.8	$998.9	$998.9	$999.0	$9.1	$1332645.2

Table 7 Hedge contract portfolios to approximately eliminate the risks of the market participants in the network of Fig. 2

Gen/Load	Hedge contract portfolio
G1	Cap(p_1|10.07, 135), Cap(p_1|12.77, 5), Cap(p_1|12.87, 5), Cap(p_1|12.97, 5), Cap(p_1|13.07, 5), Cap(p_1|13.17, 5), Cap(p_1|13.27, 5), Cap(p_1|13.37, 5), Cap(p_1|13.47, 5), Cap(p_1|13.57, 5), Cap(p_1|13.67, 5), Cap(p_1|13.77, 5), Cap(p_1|13.87, 5), Cap(p_1|13.97, 5)
G2	Cap(p_2|30.21, 10), Cap(p_2|30.61, 10), Cap(p_2|31.01, 10), Cap(p_2|31.41, 10), Cap(p_2|31.81, 10), Cap(p_2|32.21, 10), Cap(p_2|32.61, 10), Cap(p_2|33.01, 10), Cap(p_2|33.41, 10), Cap(p_2|33.81, 10)
L1	FloorLFH(p_1|$999.9, 10, 10), FloorLFH(p_1|$999.7, 10, 20), FloorLFH(p_1|$999.5, 10, 30), FloorLFH(p_1|$999.3, 10, 30), FloorLFH(p_1|$999.1, 10, 40), FloorLFH(p_1|$998.9, 10, 60)
L2	FloorLFH(p_3|$4998, 160, 160), FloorLFH(p_3|$4995.2, 50, 210), FloorLFH(p_3|$4991.2, 50, 260)

for each load-following floor contract sold to L1 and L2 the trader acquires the corresponding FloorLFHFTR.

The resulting net position of the trader is set out in Table 8. We observe as before that the total net payout on the FTRs is close to the merchandising surplus in Table 6. The remaining difference is due to the approximation of the supply and demand curves implicit in the portfolio of CapFTRs and FloorFTRs we have chosen. This difference could be made smaller by choosing a portfolio with a denser set of strike

Table 8 Trader net position in the network of Fig. 1

State	Gen hedging $\sum_i H_i^S$	Load hedging $\sum_j H_j^D$	FTR Payout	Total Payoff	SWF R
1	$376.7	$806,036.3	$925.0	$807,338	$807,330
2	$408.8	$815,876.2	$875.1	$817,160	$817,213
3	$438.9	$825,720.1	$865.9	$827,025	$827,028
4	$470.7	$835,558.1	$856.7	$836,886	$836,939
5	$505.1	$845,389.1	$758.8	$846,653	$846,794
6	$500.5	$1,054,020.3	$862.3	$1,055,383	$1,055,483
7	$532.7	$1,063,860.3	$854.3	$1,065,247	$1,065,347
8	$568.3	$1,073,694.1	$798.1	$1,075,061	$1,075,206
9	$609.7	$1,083,518.3	$696.5	$1,084,824	$1,085,062
10	$2582.2	$1,091,767.0	$372.6	$1,094,722	$1,094,901
11	$661.3	$1,301,737.6	$737.0	$1,303,136	$1,303,384
12	$697.2	$1,311,569.2	$632.7	$1,312,899	$1,313,243
13	$4655.0	$1,318,245.5	$0.3	$1,322,901	$1,322,992
14	$53787.8	$1,278,768.3	$-0.1	$1,332,556	$1,332,647
15	$294260.6	$1,038,291.8	$4.0	$1,332,556	$1,332,645

prices. In addition, Table 8 shows that the total net position of the trader is very close to the total welfare created in this market. The difference between the position of the trader and the total social welfare is less than 0.03% in each scenario.

Now, let us suppose that the trader considers upgrading the link to a capacity of 50 MW. This results in a new optimal dispatch with new pricing outcomes in each scenario. It also results in a higher overall social welfare. The outcomes under optimal dispatch following the upgrade are set out in Table 9.

As before, let us assume that the set of generators and loads remains the same. As we have seen, the generators and loads can continue to be effectively (to a close approximation) hedged using the contracts set out in table 7. As before, the set of FTRs also can remain the same. Therefore the trader can reduce its risk down to the theoretical minimum using the set of FTRs described above. The resulting net position of the trader after the upgrade is set out in Table 10. As before, the total net position of the trader closely matches the total economic welfare created in this market. The difference between the net position of the trader and the total economic welfare is less than 0.05%.

As before, we find the trader has the right incentives to fund the upgrade if and only if it is socially beneficial to do so. Specifically, if all 15 scenarios are equally likely, the trader receives a payout equal to $1,074,290 per hour, on average, when the link has a capacity of 30 MW. When the link is upgraded, the trader receives a payout equal to $1,074,590 per hour. The difference is $300 per hour, which is very close to the total economic welfare generated by the upgrade of $303 per hour. This

Table 9 Optimal dispatch outcomes in each scenario in the network of Fig. 2 with link 1 → 2 upgraded from 30 to 50 MW

State	Load		Flow			Price			MS	SWF
	L_1	L_2	1 → 2	1 → 3	2 → 3	1	2	3	MS	R
1	160	10	50.0	104.2	54.2	$13.3	$30.2	$21.7	$1276.8	$807882.1
2	160	20	50.0	104.2	54.2	$13.5	$30.2	$21.8	$1261.7	$817745.0
3	160	30	50.0	104.2	54.2	$13.7	$30.2	$21.9	$1246.5	$827603.8
4	160	40	50.0	104.2	54.2	$13.9	$30.2	$22.0	$1231.4	$837458.7
5	160	50	46.2	102.3	56.1	$30.4	$30.4	$30.4	$0.0	$847216.1
6	210	10	50.0	129.0	78.9	$13.8	$31.2	$22.5	$1314.7	$1056036.2
7	210	20	49.5	128.7	79.2	$31.2	$31.2	$31.2	$0.0	$1065881.1
8	210	30	42.9	125.4	82.5	$31.6	$31.6	$31.6	$0.0	$1075562.1
9	210	40	36.3	122.1	85.8	$32.0	$32.0	$32.0	$0.0	$1085237.1
10	210	50	29.7	118.8	89.1	$32.4	$32.4	$32.4	$0.0	$1094906.1
11	260	10	39.6	148.5	108.9	$32.8	$32.8	$32.8	$0.0	$1303664.1
12	260	20	33.0	145.2	112.2	$33.2	$33.2	$33.2	$0.0	$1313331.1
13	260	30	26.4	141.9	115.5	$33.6	$33.6	$33.6	$0.0	$1322992.1
14	260	40	19.8	138.6	118.8	$460.2	$460.2	$460.2	$0.0	$1332647.1
15	260	50	19.8	138.6	118.8	$999.2	$998.9	$999.2	$0.0	$1332647.1

Table 10 Trader net position in the network of Fig. 2, with the link 1 → 2 upgraded from 30 to 50 MW

State	Gen hedging $\sum_i H_i^S$	Load hedging $\sum_j H_j^D$	FTR Payout	Total Payoff	SWF R
1	$446.9	$806,067.5	$1221.9	$807,736	$807,882
2	$480.9	$815,911.8	$1207.4	$817,600	$817,745
3	$516.8	$825,750.0	$1110.6	$827,377	$827,604
4	$554.8	$835,582.3	$1015.9	$837,153	$837,459
5	$3853.4	$843,273.0	$0.0	$847,126	$847,216
6	$552.4	$1,054,077.9	$1171.4	$1,055,802	$1,056,036
7	$4029.2	$1,061,760.8	$0.0	$1,065,790	$1,065,881
8	$4123.1	$1,071,348.2	$0.0	$1,075,471	$1,075,562
9	$4221.0	$1,080,925.6	$0.0	$1,085,147	$1,085,237
10	$4322.9	$1,090,493.0	$0.0	$1,094,816	$1,094,906
11	$4428.8	$1,299,143.4	$0.0	$1,303,572	$1,303,664
12	$4538.7	$1,308,700.8	$0.0	$1,313,240	$1,313,331
13	$4652.6	$1,318,248.2	$0.0	$1,322,901	$1,322,992
14	$132622.7	$1,119,933.4	$0.0	$1,332,556	$1,332,647
15	$294330.5	$1,038,225.6	$0.0	$1,332,556	$1,332,647

remaining difference could be reduced through consideration of a larger set of hedge contracts and generalised FTRs in the trader's portfolio.

5 Discussion

It is worthwhile to explore the relationship between the proposal set out in this chapter, and some of the literature on merchant investment, such as the HRGV mechanism set out in Hesamzadeh et al. (2018).

Under the HRGV mechanism, an outside party (the 'regulator') allows the transmission company (referred to as a Transco) to change the fixed component of the two-part transmission tariff by an amount which does not exceed the change in total economic welfare from one period to the next. As a consequence, by construction, if the Transco upgrades the network, it receives the full change in total economic welfare in return. The Transco has the incentive to upgrade the network if and only if it is efficient to do so.

The HRGV mechanism is similar to the approach set out in this chapter. In the HRGV mechanism, the regulator ensures that the Transco receives the full gain from any change in welfare. In contrast, under the proposal in this chapter, it is the desire of generators, loads, and the system operator to be hedged which leads them to transact in hedge contracts with the trader which has the effect of leaving the trader in a position which faces the full social welfare created by the network.

But there are also key differences between the two proposals. One key difference is that, in the HRGV work, an outside party (the regulator) has to determine the change in total social welfare in order to determine how much the Transco can change the fixed fee. This requires the additional assumption that key demand and supply information is publicly available, which seems unlikely.

In contrast, in the proposal set out in this chapter, the trader is not assumed to necessarily know information about the demand and supply of generators and loads. Instead, the trader merely stands ready to transact in hedge contracts. The desire of generators and loads to hedge their risk leads them to transact in contracts in which all their risk is passed to the trader. The trader(s) then make use of generalised FTRs to trade in those contracts while taking the minimum possible risk on themselves. We have seen that the trader(s) then collectively face a payoff equal to the total social welfare. But this arises as a result of a consequence of a natural process, rather than being assumed at the outset.

There is also a deeper problem. The HRGV mechanism, by design, has the property that it expropriates the value of investments made by market participants. This undermines the incentive for market participants to make those investments in the first place.

The HRGV mechanism allow the Transco to capture the full change in economic welfare arising from any change in the market. In fact the HRGV mechanism is equivalent (in this respect) to a perfectly price discriminating monopolist who is able to extract the full surplus from all market participants. Consider the position of a generator who is considering making an investment to enhance its thermal efficiency (and thereby reduce its costs). Under the HRGV proposal, the generator immediately loses all of the benefits of that investment in the following period. The same applies to a load which is considering upgrading the electrical equipment in a factory. Again, under the HRGV proposal, all of the increase in surplus arising from that upgrade would be taken by the Transco in the following period.

The chapter by Vogelsang in this volume recognises this problem, but views it as a 'fairness' issue:

> Because the HRGV mechanism hands all social surplus increases linked to the transmission system to the Transco it provides no net benefit beyond the status quo the transmission users, which are generators and loads This can be unfair from a distributional perspective Thus the mechanism needs to be augmented by rules that lead to a fairer distribution of the social surplus increase.

We agree that this is an undesirable feature, but not just on fairness grounds. There is a strand of regulatory theory which emphasises that a fundamental objective of public utility regulation is the protection sunk investments by customers— precisely in order to ensure those customers have incentives to make socially desirable investments.[16]

One of the benefits of our proposed mechanism is that, once hedged, any change in the social surplus (e.g., a generator lowering its production cost) accrues to the individual who created that social surplus, and therefore does not undermine incentives for making such investments in the first place.[17] We consider that the approach articulated here, unlike the HRGV mechanism, is consistent with the economic foundation for public utility regulation.

In summary, the approach set out here has some similarities to the HRGV mechanism. However, we consider that the approach set out here makes an interesting and important contribution in establishing a natural link between hedging (including inter-nodal hedging using G-FTRs) and the total economic welfare in the power system. We consider that this link offers substantial promise for developing a mechanism linking private and public incentives for network augmentation in the future.

6 Conclusion

Almost since the time when locational marginal pricing of electricity networks was first proposed, researchers have explored whether or not it is possible to create a mechanism by which the incentives of private, commercial market players seeking to fund transmission augmentations would align with the overall public benefit. This chapter proposes such a mechanism, drawing on our previous work proposing the development of generalised Financial Transmission Rights.

---

[16] See Biggar (2009, 2012).

[17] The same principle could perhaps be adopted in the HRGV mechanism if the mechanism offered long-term hedge contracts in the same way as suggested here.

We consider that generalized FTRs are worthy of study, in themselves, as effective inter-nodal hedging mechanisms are becoming increasingly important. Around the world increasing penetration of Distributed Energy Resources is leading to pressure to extend current arrangements for locational marginal pricing of wholesale power markets down to lower-voltage levels. It is of critical importance that market participants have access to the tools they need to hedge those risks.

To date, liberalized wholesale power markets have only made available a strictly limited range of instruments to hedge inter-locational pricing risk. In our view this has had the effect of limiting the scope for effective inter-locational hedging. We consider this to be one of the most important weaknesses in what is otherwise one of the most important and successful sectoral liberalizations of the late twentieth century. Generalized FTRs go some way to addressing this gap.

This chapter demonstrates how, in principle, a trader using generalized FTRs faces the correct incentives to upgrade the transmission network, if and only if it is socially beneficial to do so. We envisage that the trader will trade with generators and loads, making available hedge contracts to all generators and loads which eliminate their risk. At the same time, the trader will trade with the system operator, acquiring matching generalized FTRs. We show that, if the trader is able to effectively offset the risks of generators, loads, and the system operator the trader faces a total payoff equal to the total social welfare created in the wholesale market.

Many questions remain, including whether or not the proposed mechanism can be made practical. A key question is whether or not the trader role itself can be decentralized across the market. In this case, the results set out in this chapter refer to the collective interests of the total coalition of traders in the market. The implications of this possibility are left for future research.

7 Appendix

Consider a price-taking generator with a cost function $c(g)$ facing a price p. The profit-maximising level of output of the generator is the level of output g which satisfies $c'(g) = p$ which we will write as $g(p) = (c')^{-1}(p)$. $g(p)$ represents the supply curve of the generator - for any level of the spot price it shows the corresponding profit-maximising level of output.

Up to a constant, the raw or unhedged profit of such a generator can be expressed as an integral:

$$\pi(p) = pg(p) - c(g(p)) = \int_0^{g(p)} (p - c'(g))dg \tag{26}$$

Let's suppose we have a set of cap contracts with strike prices $S_0, S_1, S_2, \ldots$. These are assumed to be ordered so that $S_0 < S_1 < S_2 \ldots$ and are assumed to span the relevant space in the sense that S_0 is below the lowest marginal cost of any generator and the largest strike price is above the largest marginal cost of any generator (or the largest marginal utility of any load). The gap between consecutive strike prices $S_{i+1} - S_i$ is assumed to be less than ΔS.

We can approximate the profit function of the generator with the following step function (here i^* is the largest value of i for which $S_i \leq p$):

$$\sum_{i=0}^{i^*} (p - S_{i+1})(g(S_{i+1}) - g(S_i)) \leq \pi(p) \leq \quad (27)$$

$$\sum_{i=0}^{i^*-1} (p - S_i)(g(S_{i+1}) - g(S_i)) + (p - S_{i^*})(g(p) - g(S_{i^*})) \quad (28)$$

We can write this as:

$$\sum_{i=0} Cap(p|S_{i+1}, g(S_{i+1}) - g(S_i)) \leq \pi(p) \leq \quad (29)$$

$$\sum_{i=0} Cap(p|S_i, g(S_{i+1}) - g(S_i)) \quad (30)$$

$$+ (p - S_{i^*})(g(p) - g(S_{i^*})) \quad (31)$$

As the spacing in the strike prices ΔS tends to zero, the last term in Eq. 31 tends to zero. The upper and lower bounds in Eq. 31 therefore approximate the profit function arbitrarily closely. Using this result we can conclude that we can approximate the optimal hedge contract arbitrarily closely with a set of cap contracts.

Theorem 1 *Given a price-taking generator with a cost function with continuous and upward sloping marginal cost $c'(\cdot)$, and a set of cap contracts with strike prices $S_0, S_1, S_2, \ldots$, as the gaps between the strike prices $S_{i+1} - S_i$ tend to zero, the generator is able to form a portfolio of cap contracts which hedges its exposure to market price risk arbitrarily closely. Specifically, suppose that, given a set of strike prices S_i and a function $g(\cdot)$ we define a hedge contract portfolio $H(g, S|P)$ as follows:*

$$H(p|g, S) = \sum_i Cap(p|S_{i+1}, g(S_{i+1}) - g(S_i)) \quad (32)$$

Then, provided we choose $g(p) = (c')^{-1}(p)$ we have that:

$$H(g, S|P) \approx \pi(p) \text{ as } \Delta S \to 0 \quad (33)$$

As before, let's suppose we have a generator with an upward sloping supply curve $g(p)$. We can write the supply curve of the generator as follows:

$$g(p) = \sum_{i=1}^{i^*} (g(S_i) - g(S_{i-1})) + g(p) - g(S_{i^*}) \quad (34)$$

Therefore, we can approximate the supply curve as follows:

$$\sum_{i=1}^{i^*}(g(S_i) - g(S_{i-1})) \leq g(p) \leq \tag{35}$$

$$\sum_{i=1}^{i^*+1}(g(S_i) - g(S_{i-1})) \tag{36}$$

This approximation becomes arbitrarily close as $\Delta S \to 0$.

From which it follows that we can approximate, arbitrarily closely, an inter-nodal hedging instrument with the required volume profile using a portfolio of CapFTRs:

$$\sum_{i=1}^{i^*} Cap(p_i, p_j|S_i, g(S_i) - g(S_{i-1})) \leq (p_i - p_j)g(p_i) \leq \tag{37}$$

$$\sum_{i=1}^{i^*+1} Cap(p_i, p_j|S_i, g(S_i) - g(S_{i-1})) \tag{38}$$

This leads to the following theorem:

Theorem 2 *Suppose a generator has an upward sloping supply curve $g(p)$. Given a set of CapFTR contracts from node i to node j with strike prices $S_0, S_1, S_2, \ldots$, it is possible to form a portfolio of CapFTR contracts with the property that, as the gaps between the strike prices $S_{i+1} - S_i$ tend to zero, the portfolio forms an internodal hedging instrument from node i to node j with a volume which matches the supply curve of the generator arbitrarily closely. Specifically, suppose that, given a set of strike prices S_i and an upward-sloping function $g(\cdot)$ we define a hedge contract portfolio $H(p_i, p_j)$ as follows:*

$$H(p_i, p_j) = \sum_{i=1} Cap(p_i, p_j|S_i, g(S_i) - g(S_{i-1})) \tag{39}$$

Then:

$$H(p_i, p_j) \approx (p_i - p_j)g(p_i) \;\; as \; \Delta S \to 0 \tag{40}$$

References

D. Biggar, Is protecting sunk investments by consumers a key rationale for natural monopoly regulation? Rev. Network Econ. **8**(2) (2009)

D. Biggar, Why regulate airports? A re-examination of the rationale for air-port regulation. J. Transp. Econ. Policy (JTEP) **46**(3), 367–380 (2012)

D.R. Biggar, M. Hesamzadeh, Generalized financial transmission rights for hedging inter-nodal pricing risks. IEEE Trans. Power Syst.

J.B. Bushnell, S.E. Stoft, Electric grid investment under a contract network regime. J. Regul. Econ. **10**(1), 61–79 (1996)

J.B. Bushnell, S.E. Stoft, Improving private incentives for electric grid investment. Resour. Energy Econ. **19**(1–2), 85–108 (1997)

M.R. Hesamzadeh et al., A simple regulatory incentive mechanism applied to electricity transmission pricing and investment. Energy Econ. **75**, 423–439 (2018)

W.W. Hogan, Contract networks for electric power transmission. J. Regul. Econ. **4**(3), 211–242 (1992)

W.W Hogan, et al., Financial transmission rights, revenue adequacy and multi-settlement electricity markets [Online]. Available Harvard University web site: http://www.hks.harvard.edu/fs/whogan/HoganFTRRevAdequacy31813

P. Joskow, J. Tirole, Merchant transmission investment. J. Ind. Econ. **53**(2), 233–264 (2005)

J. Rosellón, T. Kristiansen, *Financial Transmission Rights* (Springer, Berlin, 2013)

A Simple Merchant-Regulatory Incentive Mechanism Applied to Electricity Transmission Pricing and Investment: The Case of H-R-G-V

Ingo Vogelsang

1 Introduction

1.1 Incentive Regulation for Transcos

The traditional forms of government interference in transmission investment and operation by a private transmission company (Transco) have occurred through central planning or command and control, where the authorities tell the Transco, what it should do. This contrasts with incentive regulation, where the Transco has sometimes wide discretion to invest in and operate its network and is guided by financial rewards and penalties to act in the government's or regulator's interest. Incentive regulation takes into consideration two important features of Transco decision making. The first is that between the Transco as a private company and the government represented by the regulator there is a conflict of interest. The Transco pursues a profit goal, while the regulator pursues some different goal, which we here assume to be public welfare. The second feature is that the Transco is better informed about its technology, market environment, and demand than the regulator. If the second feature of asymmetric information between the Transco and the regulator did not hold the regulator could simply tell the Transco what to do and could achieve its own goal that way. However, because of asymmetric information the regulator actually cannot know what the best action is and the firm can deceive the regulator without the regulator knowing that he/she is being deceived.

Asymmetric information comes in two forms. The first, known as the adverse selection or hidden information problem, means that the firm has knowledge about characteristics of its environment that the regulator does not have. An example of this could be the cost function available to the Transco, where the cost would depend on

I. Vogelsang (✉)
Boston University, Boston, USA
e-mail: vogelsan@bu.edu

M. R. Hesamzadeh et al. (eds.), *Transmission Network Investment in Liberalized Power Markets*, Lecture Notes in Energy 79,
https://doi.org/10.1007/978-3-030-47929-9_12

physical inputs and the effort exerted by Transco management. The second, known as the moral hazard or hidden action problem, means that the firm can do something that the regulator cannot observe. An example could be the effort exerted by management that the regulator cannot observe.

The best-known incentive device to overcome the joint problems of conflict of interest and asymmetric information is the market. In a functioning market, sellers have to act upon their superior knowledge and exert sufficient effort in order to outcompete other firms. In the case of a monopolistic Transco, however, there is no such market competition. In this case, the regulator has to find ways to induce the Transco to use its superior information in a way that serves the benefit of the regulator. This is what incentive regulation is about.

The literature on incentive regulation has a theoretical and an applied root. The theoretical root is the principal-agent problem that characterizes many situations, in which conflicts of interest and asymmetric information exist. It is a branch of game theory with applications to many areas, such as labor markets or owner-management relations. In the regulatory area, it has among others led to regulatory proposals of menu contracts, such as the menu contracts suggested by Léautier (2000, and in Chap. 3 of this volume). The principal-agent approach is based on the assumption that the firm is of a certain type that is drawn from a subjective probability distribution of types. The firm as the agent knows its type, while the regulator as the principal only knows the type distribution but cannot observe the actual type. Usually, in the models, the type is assumed to determine a major part of the cost function, such as the main cost parameter (in case of adverse selection) or the effect of effort (in case of moral hazard). Thus, the regulator knows how types are distributed and how types affect costs but cannot observe the actual types. All other parameters, such as the demand function are assumed to be common knowledge. In the game, the regulator moves first and announces a tax/subsidy function or a menu of contracts, which are based on the firm's choice of prices. The regulator knows that the firm will choose the most profitable combination from the menu and will, therefore, determine the function or menu in such a way that the firm will reveal its type (known as the "revelation principle") and act in the regulator's interest. The main results of this approach include that the firm will not necessarily be induced to be fully efficient (as if there were no asymmetric information) and that the firm will have to receive some extra compensation for revealing its type. This so-called Bayesian approach does not easily lend itself to practical application, for example, because of the required subjective probability distribution of types.

The applied root of incentive regulation leads to a more practical approach. It is based on the notion that the firm needs to be induced to act in the regulator's interest by either providing a tax/subsidy scheme or a regulatory constraint that mimics the regulator's objective function. The main assumption here is not probabilistic as it would be under the Bayesian approach but rather based on simple asymmetry of information. Typically it is assumed that the firm knows its cost function, while the regulator does not. The regulatory constraint or tax/subsidy function is kept simple enough but can be quite drastic. This comes out clearly in the first two such mechanisms proposed by Loeb and Magat (1979) and by Vogelsang and Finsinger

(1979). Loeb and Magat simply suggested using a subsidy equal to the consumer surplus generated by the regulated firm. The firm is free to choose its prices and costs. Since in their model welfare is measured by social surplus, which is the sum of consumer surplus plus profits, this makes the firm's objective exactly equal to social welfare. Thus, the firm's and the regulator's objective are fully aligned, thereby overcoming any asymmetric information about costs. However, the regulator would have to be able to observe demand and prices in order to determine the required subsidy, and the subsidy could be substantial. Thus, the further implicit assumption is that subsidies are a pure redistribution of wealth with no welfare effects. In contrast, Vogelsang and Finsinger suggested to constrain the regulated firm's pricing in each period t such that the firm would just break even if it sold the same quantities of output in period t as it did in period t-1. This VF mechanism converges to Ramsey prices, which maximize social surplus subject to a break-even constraint. It requires the regulator to observe accounting costs, output quantities and prices, but the regulator does not have to know cost and demand functions. Also, there are no subsidies involved. The main downside of the VF mechanism is that it can be subject to strategic cost manipulations by the regulated firm and that it only converges to the optimum over time and therefore is not at the optimum at all times. Both these mechanisms, Loeb and Magat and VF have given rise to attempts for practical and realistic improvements. The H-R-G-V mechanism discussed in this chapter[1] provides such an improvement and can be traced back to both these roots.

1.2 Merchant Investment and the H-R-G-V Incentive Mechanism

The free-market alternative to central planning and government regulation of electricity transmission investment is merchant investment by competitively behaving merchant investors. Such merchant investment is characterized by private initiative, planning, ownership, and execution of the investment, which is financed by charging the transmission users, typically loads and/or generators. Merchant investment is usually associated with free-market entry, meaning that merchant investors can build new lines, as long as they view such investments to be profitable. The US merchant investors often make themselves whole through charges that are implemented via the sale of financial transmission rights (FTRs). The use of these charges mimics the merchandizing surplus. In contrast, regulated investment is usually done in a monopolistic setting, directed by the regulator and financed by a combination of regulated fixed charges and usage charges based on throughput/electricity flows. The H-R-G-V incentive mechanism discussed in this chapter provides a combination of the two approaches. Instead of featuring competitive entry, it is applied to a monopoly Transco that takes the initiative, plans, and executes the transmission investment. The Transco finances itself through a combination of a sale of FTRs based on the

[1] H-R-G-V derives from the author names Hesamzadeh et al. (2018).

merchandizing surplus and a fixed fee based on the net benefits for users created by the transmission investment. As argued below, this combination reduces some of the drawbacks of pure merchant and pure regulatory investment approaches and, at the same time, draws upon the strengths of them.

In this chapter, we first provide an overview of some known properties of the H-R-G-V mechanism. This overview leaves four important open questions about the functioning of H-R-G-V that we subsequently try to answer more specifically.

The first of these questions concerns potential shortcomings of the mechanism in the face of market power by electricity generators.

The second question deals with the application of H-R-G-V to the environmental issues associated with transmission investment in light of imperfect environmental policies.

The third question concerns the relationship of benefits conveyed to the Transco on the one hand and its users on the other. This is known as the rent extraction issue. In its original form, the H-R-G-V mechanism hands all the cumulative incremental benefits of transmission investments to the Transco, leaving users in aggregate at their original benefit level. We here look at refinements of the mechanism to let users share in the benefits provided by investment without causing too much distortion of the investment incentive.

The fourth question concerns the distribution via fixed or variable fees of the investment benefits among the different users of the transmission system. This is a fairness issue, again with incentive implications.

2 How Does H-R-G-V Work?

Before discussing other related approaches, we briefly describe H-R-G-V and its claimed properties. Under the H-R-G-V mechanism, a monopoly Transco is free to invest in transmission lines and to run and maintain them, provided it charges a two-part tariff consisting of usage fees that sum up to the merchandizing surplus and a very specific fixed fee. This fixed fee, aggregated over all transmission users, cannot exceed the sum of the previous period's fixed fee and the incremental benefit that the transmission system has generated for its users during this period compared to the previous period. Thus, the Transco's overall profit in period t is $\pi_t = MS_t + F_t - C_t$ s.t. $F_t \leq F_{t-1} + (V_t - V_{t-1})$, where MS is the merchandizing surplus, F is the revenue raised by the fixed fee, C is the annualized transmission cost, and V is the consumer surplus generated by transmission, which is the sum of surpluses generated for generators and loads as users of the transmission network.[2] Note that the regulator does not have to observe transmission cost so that the outcome is not

[2] In this chapter, we assume that the variation in transmission charges is passed on to end users in the form of price variations for the electricity they consume. In reality many end users face quite rigid electricity prices (Pollitt 2018). In that case, the surplus for the purpose of generating efficient transmission charges reflects only the cost differences for electricity generation at different nodes.

depending on, for example, the depreciation method used or on the rate-of-return on investment required by the Transco. If the regulatory constraint on the fixed fee is binding the Transco in every period t receives a total profit $\pi_t = MS_t + V_t - V_0 - C_t = W_t - V_0$, which is current social surplus minus the consumer surplus of the starting period.

Since the Transco has to pay for all costs of the transmission system, it can essentially pocket the entire social surplus generated by the transmission system minus the user surplus generated in the period before the mechanism is first implemented. In other words, the mechanism leaves the users in aggregate at their initial welfare level and gives all improvements to the Transco. As a result, the mechanism provides powerful incentives for the Transco to invest in a socially optimal way and to minimize its costs.

As described in Vogelsang (2018) the H-R-G-V mechanism is built on a two-level optimization model. At the top level of the model, the Transco maximizes profits from Transco investment (and costs of operation) based on revenues from the merchandizing surplus plus a fixed fee equal to the change in consumer surplus between the current period and an original base period. Thus, the electricity transmitted between node pairs at any measurable moment of time is taken as the output of the Transco. We will in Sects. 6 and 7 below add other outputs in order to improve the fairness of the mechanism. At the bottom level of the model, the merchandizing surplus for all node pairs is derived by an ISO based on the available transmission capacity and the generation and load bids. Thus, the function of the ISO in this approach is not to coordinate the transmission networks of various vertically integrated electric utilities, as has been common in the USA, but rather only as an organizer for the market between generators and loads and as an unbiased calculator for nodal price differences. Thus, the Transco does not negotiate variable prices with generators and loads but rather lets those be the result of the ISO's calculation based on the available capacity and the generation and load bids. The Transco takes the information as revealed by the ISO and, based on market research and its own cost information, estimates future surpluses and market profits in order to assess its investment options.

The sequencing and division of responsibilities under the proposed H-R-G-V mechanism are as follows. The regulator first sets the H-R-G-V constraint for the Transco and advises the ISO to do the short-term dispatch and to calculate the nodal price differences and merchandizing surplus. The Transco does the market research establishing projected generation and load expansions and then invests in grid capacity. Based on this grid capacity, the Transco sells point-to-point FTRs, which entitle their owners to the congestion payments calculated by the ISO based on generation and load bids. Fixed fees are then calculated based on the congestion charges and nodal price differences

The main formal results on H-R-G-V are contained in Hesamzadeh et al. (2018) and Khastieva et al. (2019), which provide theoretical and numerical derivations. These papers show that under some specified conditions a Transco investing under H-R-G-V will undertake transmission investments in a way that maximizes social surplus from such investment. In particular, generators and loads have to behave competitively and information about investment benefits to transmission users has to be common knowledge. As Hesamzadeh et al. (2018) show in numerical examples, the optimality result also holds for lumpy investments in a growth environment. Khastieva et al. (2019) and Grigoryeva et al. (2017) in addition show the optimality of investment under uncertainty if the Transco behaves in a risk-neutral way. The mechanism also induces perfectly competitive electricity generators to invest in a socially optimal way w.r.t. location and capacity.

Vogelsang (2018) shows, however, that there can be over- or underinvestment if electricity generators have market power. This holds because generators with market power distort their bidding behavior that is used for calculating the merchandizing surplus (Joskow and Tirole 2005). As we argue in Sect. 4, however, the H-R-G-V mechanism should perform well overall in the face of generators' market power.

The H-R-G-V mechanism is generally compatible with fully efficient environmental policies that affect environmental degradation associated with transmission lines or the environmental effects that new transmission lines have on the siting and running of electric generation and loads. Such efficient environmental policies would manifest themselves in the costs of building and maintaining transmission lines and in the nodal prices and user benefits used for the mechanism. However, since the mechanism does not directly deal with externalities, it can lead to imperfections in light of imperfect environmental policies. This issue is addressed in Sect. 5.

Because the H-R-G-V mechanism hands all social surplus increases linked to the transmission system to the Transco, it provides no net benefit beyond the status quo to the transmission users, which are generators and loads (and, by implication the end users of electricity). This can be unfair from a distributional perspective, because it creates an imbalance over time.[3] It can also interfere with incentives for generators and loads (a) to take actions that increase overall welfare and are at the same time associated with transmission investments and (b) to provide the necessary information (e.g., via their bidding behavior) to implement the mechanism. The main issue here is that the overall welfare benefits are the results of both, transmission and generation investment and/or load investments, and not any one of them will cause the benefits without the help of the other. Thus, the mechanism needs to be augmented by rules that lead to a fairer distribution of total surplus increases. This is the topic of Sect. 6.

Equity issues, however, also can arise between the various transmission users and user groups. The mechanism is silent with respect to the division of the fixed fee among users. This is a tricky issue, because the division of the fixed fee should not interfere with the efficiency of the mechanism. Such interference would, for example, occur if the fixed fee were distributed among users in proportion to their use of FTRs. Section 7, therefore, analyzes possibilities for the division of the fixed fee among users in a way that minimizes distortions and is fair among users. Section 8 concludes by assessing the interactions between the policy consequences discussed in Sects. 4–7 and by laying out open issues for further work.

---

[3]We assume here that the Transco is a publicly traded corporation with an ownership structure that cannot in any assignable way be linked to generators, loads, and final users. Vogelsang (1983) treats the behavior of firms that are partially owned by their users and shows that this changes their pricing behavior relative to profit maximization and also changes the distributional impact.

3 Literature Review: H-R-G-V in Contrast to Other Approaches

H-R-G-V has itself evolved from a subsidy approach originally proposed by Sappington and Sibley (1988) and traceable to Loeb and Magat (1979) and from a two-part tariff approach originally proposed by Vogelsang (1989, 2001) and traceable to Vogelsang and Finsinger (1979).

Sappington and Sibley's incremental surplus subsidy (ISS) provides the Transco with a subsidy that in each period equals the increase in consumer surplus minus the Transco's last period's market profit based on expenses (rather than costs). The Transco can keep this period's merchandizing surplus and has to bear its investment and operating costs. It, therefore, receives a total compensation of the total surplus change over last period. This approach differs from H-R-G-V in three respects. First, H-R-G-V is financed directly from consumers rather than via subsidies. Second, under H-R-G-V there is no reduction in profits by deducting the last period's profits from this period's profits. Thus, in its original form, H-R-G-V does not require the regulator to have cost data of the Transco. Third, the benefit changes conveyed to the Transco under H-R-G-V are cumulative over all periods rather than only per period. The first of these features makes H-R-G-V more compatible with the merchant approach. The second and third features provide more powerful incentives to the Transco for cost reduction and innovation, but they have the above-mentioned adverse distributional consequences to be treated in Sects. 6 and 7.

Vogelsang's (2001) two-part tariff approach was further developed by Hogan et al. (2010) and renamed HRV after the first letters of its authors. HRV uses the same institutional setup as H-R-G-V. However, the HRV fixed fee is smaller than under H-R-G-V, because HRV only uses the linear Slutsky approximation of consumer surplus change instead of the true consumer surplus change. In this case, the H-R-G-V provides immediately optimal incentives, while the HRV falls somewhat short, resulting in an imperfect investment schedule that only converges to the welfare optimum over time and only if the technical and market environments do not change too quickly.[4] On the other hand, its major advantage is that HRV requires less information about consumer benefits and can be implemented based simply on market observations of prices and quantities actually realized.

Léautier (2000) has suggested an alternative mechanism with similar incentive properties to H-R-G-V. It is described and analyzed extensively in Léautier's Chap. 3 of this volume. In some respects, it is a negative image of the H-R-G-V. Under the Léautier (2000, p. 77) approach, the Transco "is responsible for the full cost of the out-turn, plus the transmission losses, valued at the System Marginal Price."[5] Here, the (operational) out-turn is the cost of congestion created by binding transmission constraints. In other words, the total operational out-turn summed over all node

[4]If it converges HRV can only guarantee piecewise optimality, not global optimality, which H-R-G-V does (Rosellón et al. 2012).

[5]In the current paper, we neglect transmission losses. As suggested in Vogelsang (2018), they can, however, easily be incorporated in the H-R-G-V mechanism.

pairs is the potential consumer surplus that would be generated by increasing the transmission system in such a way that it is unconstrained. Thus, under this approach, the Transco is penalized for not investing in transmission capacity. This will lead to optimal investment because the Transco will invest up to the point where the cost of additional investment equals the marginal reduction in the penalty.[6]

The main issue of Léautier's approach is, therefore, the opposite of the one raised by H-R-G-V. Under H-R-G-V, the Transco is in danger of making too high profits, because it starts from a status quo, over which it will generally improve its profit position. In contrast, under Léautier's approach, the Transco starts out with a potentially huge penalty that it can reduce via investments but that will ordinarily never go away (because it is never optimal to invest so much that the system is unconstrained). Thus, under this approach, the Transco initially will have to charge a very high fixed fee with which to finance the initial penalty. From there on, the profit changes will be similar to those under H-R-G-V. In order to deal with the rent extraction issue, Léautier (2000) suggests a menu of revenue-sharing mechanisms. An alternative, closer to H-R-G-V, is to make the Transco responsible for congestion costs only relative to a preset target, as has been done in England and Wales in the 1990s. Interestingly, the process in England and Wales was stopped, when the Transcos could no longer beat the targets and incurred penalties (Léautier 2019). This may show the potential downsides of too strict rent extraction policies.

Among the suggestions from the theoretical literature, the Léautier (2000, 2019) approach is both closest to H-R-G-V and the closest one to practical application.

4 Dealing with Market Power of Electricity Generators

4.1 *The Classic Case Against Merchant Transmission by Joskow and Tirole*

According to Joskow and Tirole (2005), the tendency of merchant transmission companies to over—or underinvest in the presence of market power by generators results from the property that the nodal prices will be distorted by market power, leading either to a larger or a smaller nodal price difference under market power than under competition. Thus, if there is market power in the importing region, the larger nodal price there will make it more attractive to import electricity from low-price areas. This can represent over-investment in transmission if the generators with market power reduce their output in response to the increased imports and if the cost of producing that electricity was lower than the cost of the additional imports

[6]Léautier (2000) only shows this for totally inelastic electricity demand, while Léautier (2019) extends this to ordinary downward-sloping demand curves. The Léautier mechanism is essentially the same as the nonlinear environmental tax proposed by Kaplow and Shavell (2002), which equals the environmental damage caused by pollution. It is the negative of the subsidy proposed by Loeb and Magat (1979) for monopoly regulation.

including the transmission costs.[7] From an end-user perspective, the increased transmission capacity is good because it lowers their electricity price, but from a social welfare perspective, this comes at a too high cost.

Since this chapter only attempts to give an overview of the open issues facing an implementation of the H-R-G-V mechanism, we do not provide full-fledged modeling analyses, but rather only describe the main approaches along with conjectured results. For simplicity and to provide the main arguments, assume a transmission system with a single line connecting an area called North and an area called South with cheap, competitively supplied generation in the North and expensive generation supplied by a monopolist in the South. For simplicity, we first assume that the marginal cost of generation in the North and in the South is constant and that the marginal cost of transmission expansion is also constant.

Before analyzing the case of H-R-G-V, consider the case of pure merchant transmission. In this case, one could model the relationship between the monopolistic generation in the South and potential merchant transmission imports under a dominant firm model, where the imports are supplied from a competitive fringe of generators.[8] Now consider two cases, the first, where the combined costs of generation in the North and transmission to the South exceed the cost of generation in the South and, second, where the cost in the North plus the cost of transmission are lower or equal to the cost in the South. In the first case, the generator in the South will maximize its profit by setting an output such that the price will be slightly below the sum of marginal generation cost in the North plus marginal transmission cost.[9] As a result, the North will not export any electricity to the South and there will be no transmission and therefore no cost inefficiency.[10] Now, consider the second case. If the sum of marginal costs of generators in the North and transmission is smaller or equal to the marginal cost of generation in the South the equilibrium price will again equal these marginal costs. However, in this case, the generator in the South will produce nothing and all the market supply will come from the North. Thus, we get the welfare-optimal outcome.[11]

In contrast to the merchant transmission case, under H-R-G-V one can view the Transco as acting like a vertically integrated duopolist with the competitive supply curve of the generators in the North as part of the Transco's marginal cost curve along with the marginal cost of transmission. Conceptually, the Transco combines

[7] For a similar observation see Egerer et al. (2015).

[8] Joskow and Tirole (2005) in contrast use a generic simultaneous oligopoly model. The static dominant firm model has a dominant firm that sets its capacity or output first, while the firms in the competitive fringe set their capacity or output in the second stage along a competitive supply curve.

[9] If the marginal cost curves are upward-sloping, we can get equilibria with imports from the North. This is the case where transmission expansion exceeds the welfare-optimal amount but end-users will be better off.

[10] This is the result of a one-shot simultaneous move game. In a more dynamic setting, the generator in the South may not fully utilize its capacity and thereby may increase the price if no transmission capacity has been built.

[11] Note that this result also holds under upward-sloping marginal cost curves as long as the marginal cost condition of case 2 holds in equilibrium.

the cost of generation in the North, MC_N, and the cost of transmission to the South, MC_{TR}, resulting in marginal cost $MC_{N+T} = MC_N + MC_{TR}$, while generation in the South has marginal cost MC_S. In the following, we assume a single-period quantity-setting game that can be either Cournot or Stackelberg, in the latter case with either the Transco or the generator in the South as the Stackelberg leader.[12] Under H-R-G-V, the Transco is maximizing perceived social surplus,[13] while the (formerly) monopolistic generator maximizes its profits. Such a duopoly situation between a welfare-maximizing and a profit-maximizing firm is unusual. However, there exists some literature on the oligopolistic interaction between a public enterprise and private firms (e.g., Beato and Mas-Colell 1984; De Fraja and Delbono 1989; Cremer et al. 1989; Sappington and Sidak 2003), which has used similar constellations to the one pursued here, or between profit-maximizing private firms and private firms with a public-interest objective (Vogelsang 1983), which is similar under certain parameter values.

Independent of which of the games are played, if $MC_{N+T} < MC_S$ the Transco under H-R-G-V will invest in such a way that the monopolist in the South is fully eliminated, because the Transco will in all three games set a quantity such that $p = MC_{N+T}$, which the generator in the South cannot match. In case $MC_{N+T} = MC_S$, the generator in the South could produce a positive quantity but in equilibrium will not do so, because in all three games, it will set its quantity such that $p > MC_S$.

So, consider the case of $C_{N+T} > C_S$. The amount of electricity sold in the South is $q_S = q_C + q_M$, where q_C is the electricity imported competitively from the North and q_M is the electricity supplied by the monopolist in the South. The monopolist in the South maximizes $\pi_M = p(q_C + q_M)q_M - C_S(q_M)$, while the Transco maximizes $\pi_T = V(q_C + q_M) + p(q_C + q_M)(q_C + q_M) - C_{N+T}(q_C) - C(q_M)$.[14] This assumes that the H-R-G-V constraint is binding and it neglects the initial consumer surplus before the H-R-G-V mechanism is applied and which is deducted from the H-R-G-V fixed fee.

Now, if we have Cournot competition between the generator and the Transco we may get overinvestment in transmission, because the Transco will invest until the price in the importing region will equal its marginal cost (marginal cost of exporting generation plus marginal transmission cost) and that will be above the importing generator's marginal cost, because at the resulting price the (former) monopolist in the South will realize a Lerner index $L = s_M/\varepsilon > 0$, where s_M is the firm's market share and ε is the market demand elasticity. While such over-investment certainly is not first best, it may still be better than other alternatives achievable by the regulator.

---

[12]In a Cournot game, the players move simultaneously to set their capacities or outputs, unaware of what the other players' moves are, while in a Stackelberg game, there is one player, the leader, who moves first in setting capacity or output so that the other players, who follow simultaneously, knowing the leader's move.

[13]We neglect the deduction of the initial consumer surplus from the Transco's profit, because it is a fixed amount.

[14]Cremer et al. (1989) add a break-even constraint for the welfare-maximizing firm based on the assumption that it operates under economies of scale. This is not relevant for us, because we assume two-part tariffs that at least cover costs if the variable price equals marginal costs.

In contrast, under Stackelberg competition with the Transco as the Stackelberg leader the Transco's first-order condition is $\partial \pi_T/\partial q_C = (1 + \partial q_M/\partial q_C)p - MC_{N+T} - MC_S \partial q_M/\partial q_C$. This implies $p - MC_{N+T} = -(p - MC_S)\, \partial q_M/\partial q_C > 0$. Thus, the Transco will invest less under Stackelberg than under Cournot so that $p > MC_{N+T}$. As Beato and Mas-Colell (1984) state, this case can conventionally be viewed as a second-best outcome under the constraints of (a) a distorted merchandizing surplus and (b) the Transco investment as the only instrument.

What if instead of the Transco the generator is the Stackelberg leader, knowing the Transco's reaction function under H-R-G-V? This reaction function is given by the combined marginal cost curve of generation from the North plus the marginal cost of transmission. This corresponds to the classic static dominant firm model alluded to above for the case of pure merchant transmission. If the generation of the North and the transmission expansion are supplied at constant marginal costs, the generator in the South will fully preempt the supply from the North. However, if those supply curves are upward-sloping the result may be different. In that case, the generator will, compared to the Cournot case, lower its price and expand capacity until $p = MS_{N+T}$ in order to prevent some transmission investment. As a result, total electricity sales will be larger, the generation of South will be larger and the import from North (and therefore transmission investment) will be lower than under Cournot but, from a social surplus perspective, there can still be over-investment by the Transco. Beato and Mas-Colell (1984) have specifically looked at a similar case and have found that it may yield better or worse welfare outcomes than the case of the (welfare-maximizing) Transco as the Stackelberg leader.

We cannot simply postulate which of the three games described above will actually be played. In reality both, Transcos and generators, have to deal with major planning issues when doing their investments. Such planning takes a long time and occurs mostly in the open. Also, planned but not yet built investments can be discarded along the way. This suggests that there may be no obvious Stackelberg leader. However, the H-R-G-V approach is based on the assumption that the Transco calculates the effects of its investments and other actions on generator and load welfare. This suggests that in this game the Transco acts as a Stackelberg leader and is in line with Wolak (Chap. 4 of this volume), who states that in today's world the Transco should proactively invest.

In contrast, under pure merchant transmission, a single monopoly generator could well be a Stackelberg leader. The Cournot solution may also be quite realistic in case of many generators and merchant transmission investors. This is in line with Cremer et al. (1989), who make a strong case for the Cournot game. In addition, the game being played will likely be determined by the investment lags. As noted in Sect. 5, investment lag for conventional electricity generation, such as for nuclear, coal and hydro, is likely longer than for transmission lines, while the investment lag for renewables and gas generators is likely shorter than for transmission. This suggests that under conventional generation the generator may be the leader (meaning re-active transmission planning), while under renewable and gas generation the Transco will be the leader (meaning pro-active transmission planning).

Summing up these arguments, while the nodal prices will be distorted by generators exercising market power, under H-R-G-V the resulting tendency of overinvestment in transmission capacity will be countered by the property of the mechanism to increase transmission user surplus. There can only be too much transmission investment under H-R-G-V if the monopoly generator in the South has lower marginal cost than the marginal cost of imports including transmission costs. Even then consumers will be better off. This property may well be more important for policymakers pursuing consumer welfare than the resulting cost increase.[15]

4.2 Increasing Competition via Transmission Investment

A second important property of transmission expansion can be its effect on the level of competition between generation companies. Wolak (Chap. 4 of this volume) notes that in practice the increase in competition among generators has had particularly large welfare effects. Since the lack of competition in generation reduces user surplus and since transmission investment can increase competition among generators, the Transco has incentives under H-R-G-V to invest in such a way that it increases competition among generators. Since generator surplus change is included in the Transco's fixed fee, the Transco will have the same incentives in this respect as a social surplus maximizing social planner and will try to reduce any deadweight loss. This will, in particular, hold if the Transco acts as a Stackelberg leader in its investment game with the generating firms. For this to work, the Transco needs to have sufficient information about the generation costs of the generators with market power and must not simply be depending on observed nodal price differences or the observed bidding behavior of generators and loads.

A simple example could be the case of a generation monopoly in the North and a similar monopoly in the South, each one facing the same demand and having the same cost function and starting out with no transmission between the two nodes. Now, assume a Transco that is regulated under H-R-G-V and that is charging loads for merchandizing surplus plus fixed fees. If such a Transco builds enough transmission capacity between North and South this will lead to a duopoly for the now joint North/South market. However, the market will only be truly joint if the transmission capacity is such that there is no line congestion. Otherwise, the two markets would be differentiated. Now, if the market is truly joint the duopoly output will be associated with a lower price and larger total quantity of electricity consumed. The Transco would then be entitled to a fixed fee equal to the consumer surplus increase from the lower price minus the generators' surplus reduction from the move from two monopolies to two duopoly firms. If this net surplus increase is larger than the cost of the new transmission capacity, the Transco will make the investment. In that case,

[15] In case the monopoly generator is exporting electricity into a competitive area there can be underinvestment in transmission capacity, because the nodal price differences will be smaller than the cost differences. This could lead to higher than optimal prices in the importing area.

total surplus will increase by the reduction in deadweight loss and end-user surplus will increase by the profit reduction suffered by the generating firms.

My conjecture is that an upper bound for the necessary transmission capacity to reach this result equals the increase in total generation between the sum of the two monopoly outputs and the duopoly output plus an epsilon excess capacity in order to assure zero congestion. As Borenstein et al. (2000) have shown, however, the amount of transmission expansion necessary to sustain such increased competition can be substantially smaller in equilibrium. Also, the actual transmission capacity may not be fully used. This can lead to a violation of the conventional "used-and-useful" rule which is applied under rate-of-return regulation. Under H-R-G-V, the Transco would have to bear the cost of such "unused" capacity, but would also get the incremental benefits.

An additional important feature in this case is that generating companies can influence the market for FTRs that is relevant for the functioning of the H-R-G-V mechanism. Gilbert, Neuhoff, and Newbery (2004) have shown that the way FTRs are sold can have a strong influence on competition between generators. Thus, the sale of FTRs by the Transco has to be organized in such a way that it increases generator competition. Interestingly, an increase in competition among generators will reduce the Transco's fixed fee on account of the generators' profit reduction, but will increase it by more on account of the load surplus increase that reduces the overall deadweight loss. However, since transmission investment is an imperfect tool for increasing generator competition, the result will stop short of perfect competition.

Léautier (Léautier 2019, Chap. 8) demonstrates impressively how complex the imperfect competition issues of electricity generation are. Imperfect competition among electricity generators can take many different forms that will yield different outcomes. Government regulation of electricity transmission in view of imperfect competition among generators is therefore very hard. Similar difficulties face a Transco under H-R-G-V. Hopes that H-R-G-V will lead the Transco to implement welfare-optimal transmission investment vis-a-vis generators with market power can therefore not be too high. On the other hand, a risk-neutral Transco under H-R-G-V will have strong incentives to learn about the actual cases of imperfect generation competition it is faced with when maximizing expected surplus. However, there may be an interaction with the ISO, who organizes the auctions for the FTRs. Thus, there is a proviso that the FTR auctions are organized in such a way that they increase competition among generators.

4.3 Proactive Transmission Planning

As noted by Stoft (2007) generators will generally have incentives to sabotage transmission expansion that is not in their interest. This would hold, in particular, for lines that would increase competition between generators. The Transco under H-R-G-V could prevent such sabotage by offering generators fixed fee reductions if they would otherwise suffer profit losses from new or expanded transmission lines.

Furthermore, a Transco will be reluctant to invest in new transmission lines without being sure that new power plants will be built that these lines connect to. Stoft (2007, p. 104) in this case refers to the "embarrassment" factor that emerges if a transmission line is built and there is no power plant to connect it to. This issue will in particular come up in the permission process for new power lines. Since under H-R-G-V the Transco will receive the aggregated surplus increase of all users, i.e., of generators and loads, it should, however, be in a position to counter such tendencies. If the Transco can propose a new line that would lead to such a surplus increase provided a new power plant is built then the Transco should be able to pay the potential new generator a compensation that will leave the generator better off than without the new plant and at the same time will leave the Transco better off by building the new line. This is a tool that regulators and central planners of transmission lines usually do not have at their disposal. Typically they will face stiff opposition from generators to new transmission lines that would reduce their market power (Stoft 2007, p. 108). Including generators in the decisions on transmission investments as is done, for example, in New Zealand may lead to compromises that are not necessarily in the interest of end users.

4.4 H-R-G-V Regulation of Generators

We have so far assumed that generators with market power are not regulated and therefore simply maximize their profits. What would happen if the generators were also regulated by an H-R-G-V mechanism? This mechanism would have to be conceptually designed to be compatible with the mechanism used for the Transco. Given the interactions between all these regulated firms, a major difficulty here lies in the definition and measurement of the surplus changes that define the fixed fees for all these firms. Following Kim and Chang (1993) and Kim and Lee (1995) one could assume that the Transco and the generator(s) play a Nash game in outputs and each receives a fixed fee equal to last period's fixed fee plus the *incremental* surplus generated by each of their change in output given the output of the other firms. The incremental surplus is relevant here, because each firm maximizes its profit given the actions of the other firm(s). In the relationship between competing generators, the outputs are similar so that the competitive game between them is well defined in outputs. This may, however, not be feasible in the game between generators and the Transco. The generators produce electricity, while the Transco transports it. Both may be measured in the same units, such as kwh, but they are very different services. Thus, the Nash condition may have to refer to capacities/investments rather than to outputs.

Within competitive constraints under H-R-G-V, the variable price charged by each generator could be freely selected by the generator, because it is directly linked to the fixed fee, if the generator charges a high variable fee that would (a) reduce its sales and (b) reduce its fixed fee. Thus, the generator would have every incentive to choose a variable fee equal to its marginal cost without the necessity of the regulator to measure the generator's costs. For this purpose, the generator's contribution to the relevant consumer surplus is the incremental change in the end users'/loads' gross consumer surplus minus the amount paid by end users/loads to the generator (excluding the fixed fee) and minus the cost of transmission (neglecting other generators). The Transco surplus (again excluding the Transco's fixed fee) is already captured here, because the revenues of the Transco are paid by the end users and therefore cancel out when measuring gross rather than net consumer surplus. Thus, in the application of H-R-G-V to generators, the regulator has to know the change in the cost of transmission. Now, the H-R-G-V mechanism for the Transco provides the Transco with a fixed fee that is equal to last period's fixed fee plus the incremental change in the consumer/load surplus and the generator surplus (excluding the generators' fixed fees) resulting from the Transco's performance in the current period.

While an H-R-G-V mechanism that is only applied to the Transco leaves the users (generators and loads/end users) in aggregate at their original surplus level, the simultaneous use of the mechanism for the Transco and the generators could actually burden the loads/end users. This could happen in spite of the fact that the fixed fees of the oligopolistic generators under H-R-G-V would add up being smaller than the aggregate incremental surplus provided by them. The reason is that generation and transmission are complementary services, while the services of oligopolistic generators are substitutes for each other. The incremental surplus from generation resulting from an additional generator is the surplus resulting from all generators minus the surplus without the additional generator. With downward-sloping demand curves, the incremental surplus is always lower than the average surplus. Thus, the sum of all incremental surpluses is less than the total surplus. In contrast, in the relationship between all generators and the Transco the surplus without the generators is zero and (assuming the generators require transmission to deliver their services) the surplus without the Transco is zero. Thus, the incremental surplus in this case equals the total surplus. Thus, if there were only one generator and one Transco, the sum of the increase in fixed fees from one period to the next would exceeds the surplus increase generated by the two firms. If there were several generators, this effect could vanish, but this remains an empirical issue.

A further issue arising when applying the H-R-G-V mechanism to generators is that under H-R-G-V generators compete with each other using two prices, the usage fee and the fixed fee. In such competition, rational generation customers will choose the generator who provides them with their highest consumer surplus. While ordinarily under H-R-G-V a more efficient firm would be rewarded by a higher fixed fee, the consumers would not benefit and would, therefore, have no reason to choose the more efficient supplier. There appear to be two possibilities for the H-R-G-V mechanism to yield efficient outcomes nevertheless. The first would be that the fixed fees of the different generators are aggregated by the regulator and paid by all users on an average basis. This approach has the advantage that one can stay in a quantity-setting framework, such as Cournot competition, rather than moving the game to pure price competition. Pure price competition, even without any two-part tariffs, would already yield marginal cost prices, assuming that electricity generation is a homogeneous service. This result would also extend to two-part tariffs (Griva and Vettas 2015). The second possibility would be to allow two-part tariff competition in both the fixed fee and the usage charge, where the firms can charge fixed fees at levels below the binding regulatory constraint. This would again lead to the Bertrand result if electricity is viewed as a homogeneous service. Thus, the question is if one could use a sequential approach, where generators set their capacities/quantities first and then compete in two-part tariffs.

What makes the combined use of H-R-G-V for the Transco and the generators so appealing is that the generators will have incentives to bid at their marginal costs. The congestion prices will therefore not be distorted and the Transco will invest in a welfare-optimal fashion. The difficulty is that the transmission costs have to be known by the regulator in order to apply the H-R-G-V constraint on generators. However, if that is the case H-R-G-V regulation loses its major purpose, which is to overcome information asymmetries.

5 Environmental Problems and Opportunities for Transmission Investment

5.1 Environmental Problems from Siting Transmission Lines

The siting and building of electricity transmission lines are associated with direct environmental degradation from cutting down trees, changing the landscape, potentially affecting the physical health of abutters, etc. These environmental problems are usually addressed in the transmission line approval process through conditions imposed on the transmission investor during the permit phase. This resembles the negligence approach to civil liability. As long as the Transco fulfills the siting conditions imposed, it is not negligent. As far as I know, it is unusual for the transmission investor to directly compensate abutters and others for this kind of environmental impact, which would resemble a strict liability approach. Provided the negligence standard is set correctly, both a negligence standard and strict liability can lead to optimal environmental precaution. They have, however, very different distributional consequences if and when environmental damages occur. In that case, the negligence standard leaves the victims worse off, while under strict liability, they would be fully compensated. Since the H-R-G-V mechanism gives all the benefits of transmission investments to the Transco, it appears to be only fair that the Transco also covers all the social costs of transmission. Under the current approach, it is hard to include this issue automatically in the H-R-G-V mechanism, because the information for both the variable fee and for the benefits that determine the fixed fee is derived from the bidding behavior of the transmission users. However, in order to get building permits, the Transco under H-R-G-V should during the permit process be willing to offer compensation to the stakeholders affected by the environmental problems caused by the transmission lines. Since the Transco can pocket all the incremental benefits resulting from a new transmission line, it should also have a greater willingness and ability to pay compensation for any environmental degradation associated with it. A downside of this greater willingness and ability to pay could be that the Transco could also engage in rent-seeking activities by bribing officials or by paying for the election of politicians that would favor the building of transmission lines.[16]

A different regulatory approach would be to directly include in the H-R-G-V mechanism the environmental degradation caused by transmission lines. Based on the Loeb-Magat mechanism, Kim and Chang (1993) have suggested a combined tax/subsidy mechanism to solve the environmental and market power issues simultaneously. Under their mechanism, the Transco would receive a subsidy equal to the consumer surplus generated by the Transco and would have to pay a tax equal to the environmental damage caused by transmission lines. Applying the same idea to H-R-G-V would mean that the Transco would as before receive a two-part tariff consisting of the merchandizing surplus and the consumer surplus increase caused by the Transco but now the Transco would also have to pay for the additional environmental degradation it has caused. In this case, the Transco's overall profit in period t is $\Pi_t = MS_t + F_t - C_t$ s.t. $F_t \leq F_{t-1} + (V_t - V_{t-1}) - (D_t - D_{t-1})$, where D is the environmental damage caused by transmission. Assuming the constraint is binding and solving backwards for F_{t-1} we get $\Pi_t = MS_t + V_t - D_t - C_t - V_0 + D_0$. The first four right-hand-side terms here represent current welfare, while the last two terms are given at the outset at $t = 0$. A welfare-maximizing policy instrument against environmental degradation would maximize the difference between the environmental benefits and the costs of abatement. Since in this case of expanded H-R-G-V the Transco would receive all the benefits and carry all the costs of abatement, it would fully internalize the welfare problem and strive for the welfare-maximizing outcome.

There are two problems with this approach. First, the regulator has to be able to calculate the environmental damage in such a way that all stakeholders can agree on it. Second, if the amount is paid out to the state it is really a tax and the victims of the environmental degradation are worse off. They suffer damages and get nothing in return. Thus, ideally, those victims should be identified and compensated. In this case, the approach would resemble strict liability with some extra enforcement by the regulator. The advantage of this approach over a Pigou tax is that the regulator does not have to know the Transco's cost of abatement.

[16]An interesting issue not treated here is transmission expansion guided by "market integration" (which does not differentiate between connecting fossil-fuel generation or "green" generation) as opposed to transmission expansion guided by "green renewable integration." This has been a hot topic in Germany, where the planning of transmission expansion (Netzentwicklungsplan) has not been able to build the links to bring wind generation from the north to consumption areas in the south of Germany. See Kemfert et al. (2016).

5.2 *Environmental Effects of Electricity Generation Affected by the Transmission System*[17]

5.2.1 Coordinated Investment of Transco and Generators

Transmission investment today is often the result of changes in the set of generation technologies used. In particular, large wind farms and utility-sized solar installations are typically not close to large electricity consumption centers and so require new transmission lines. At the same time, old transmission lines may become superfluous because of the shutdown of old generation plants and because of new distributed generation facilities based on wind and solar power. Thus, the build-out of the transmission system is strongly affected by the development of the electricity generation facilities. At the same time, a well-adapted transmission system will affect the rate and direction of generation expansion and retirement. Most of the new low environmental impact generation will be based on solar and wind energy and will, therefore, be intermittent and to some extent decentralized. Madrigal and Stoft (2011) notice that building such generation facilities usually takes much less time than conventional generating plants. As a result, very often new transmission lines will take longer times to plan and build than those generation facilities. Therefore, Madrigal and Stoft (pp. 7, 50 and 82ff) suggest that transmission expansion should occur proactively rather than reactively in response to new renewable electricity generation.[18] Note that by the same argument the case can be made for reactive transmission planning for conventional generation, because it takes longer to build than transmission lines.[19] Will under H-R-G-V the Transco have incentives to undertake the right choice between proactive and reactive transmission investment?

Under a reactive approach, the Transco only builds new connections if it is sure that the new generation facilities are being or have been built. Since building transmission lines take time, this may lead to delays in the start-up of the new generation projects. Such delays can be shortened if the transmission project starts, once the generation project is sufficiently committed. However, the generation project can only be sure about transmission and its cost if the Transco is committed. What is the risk of the Transco under proactive transmission planning? It could be stranded without any new generation. Groppi and Fumagalli (2014) argue that there is some cost in proactive planning, which they put at 10% of the investment. However, in their case study in Italy, this still makes proactive transmission planning preferable with a high probability. If the Transco under H-R-G-V is risk neutral he/she will take the socially optimal decision by basing it on expected values. If he/she is not risk neutral, he/she can pay generators to commit investing in new generation if that is

[17]For the issues raised in this section, see also Chap. 14 by Rudnick and Velásquez in this volume.

[18]See, however, Sect. 4.3.

[19]See also Groppi and Fumagalli (2014).

welfare enhancing. If one includes the consumers' costs from outages, some excess transmission capacity may be optimal and that may be achievable under H-R-G-V.[20]

5.2.2 Interaction of Transco with Imperfect Environmental Regulation

When environmental regulation is perfect, implemented by Pigou taxes or by cap-and-trade, it directly or indirectly affects the costs of transmission and generation in such a way that the Transco under H-R-G-V will automatically make the right environmental decisions, when investing in transmission projects. In case of cap-and-trade, this presupposes that the surplus which generators/loads gain from emissions trading can be clearly separated from the surplus caused by the transmission system.

In reality, environmental regulation tends to be imperfect. In the case of renewables, regulators and governments tend to either set quantitative targets in terms of the renewables percentage in electricity generation or set subsidies for renewables on a per kwh basis. Since such policies are decidedly second best, the question arises whether H-R-G-V does well relative to these second-best policies. It is important to find out under what circumstances second-best policies lead to distorted information that either leads to too little renewable generation or to too much renewable generation.

If environmental policy prescribes a certain percentage of electricity generation to come from renewables, the Transco cannot influence this percentage (although it can, in principle via the price paid by loads for renewables and for transmission, influence the total amount generated by renewables). Thus, it has to take the environmental effect of generation as given. However, it can influence the combined cost of generation and transmission arising from renewable generation. Assuming that the percentage renewables mandate exactly refers to a Transco's territory and is tradable among loads and assuming the Transco is fully informed about generation costs in its territory the Transco under H-R-G-V could be incentivized to build transmission lines in such a way that renewables are installed in a cost-minimizing fashion. This could hold, because a cost-minimizing trade of renewables obligations will minimize generation costs, which will lead to higher generator surplus and/or load/end-user surplus.

However, since the Transco under H-R-G-V leaves transmission users at their status quo surplus, there is a potential conflict between the trading system for renewables obligations and H-R-G-V. The loads will only trade renewables obligations if that increases their surplus, which means if the cost differential between renewables and non-renewables including the transmission cost differences is such that trading is worthwhile. Now, if the Transco is rewarded for transmission investment that it undertakes for renewables with the resulting increase in load surplus the load's gains from trade may be taken away. Thus, in principle, there needs to be a differentiation between gains due to the market trades for renewables obligations and gains due to transmission investment. This difficulty and the resulting tensions are quite similar

[20]See Brito and Rosellón (2011) for a related argument about excess pipeline capacities.

to the difficulty of assigning surplus increases for renewable generation investment between the generators and the Transco that invests in transmission in order to enable the renewable generation to be sold to loads. Thus, while the H-R-G-V mechanism can help implement standards-based policies in a cost-minimizing manner if they are already quite efficiently designed there is a conflict that needs to be resolved.

What happens if the percentage obligations cannot be traded among loads? In that case, each load will have to fulfill its percentage obligation and purchase that percentage of renewable energy. The Transco under H-R-G-V will still have every incentive to allow the loads to receive this electricity at minimum costs but the total cost will in general be higher than in the case of tradable obligations.

6 The Rent Extraction Issue

The H-R-G-V mechanism is a two-part tariff. The fixed fee of this tariff conveys rents to the Transco, who becomes the main beneficiary of transmission investment. Such a rent transfer could make it politically difficult to implement the mechanism, which may be opposed by generators and loads. Laboratory experiments have shown that people usually do not accept offers that provide them with low benefits if the offering side receives large benefits that they do not share. This is best known from studies about the "ultimatum game," where players routinely reject offers that provide them with much less benefits than the other side even if the rejection means that they will receive no benefit at all (Güth et al. 1982). It is therefore important to refine the H-R-G-V mechanism by allowing a sharing of the Transco's rents with the transmission users without jeopardizing the investment incentives. This is difficult, because the strength of the H-R-G-V mechanism rests on the property that the Transco can capture all the welfare improvements from its investments (or from other of its acts, such as maintenance or innovation). Thus, giving the Transco less profit opportunity might jeopardize these incentives. We will consider three approaches for dealing with this issue.

The first approach for reducing any excessive rents going to the Transco is to reposition the starting point of the mechanism, but leaving the mechanisms otherwise as before. As shown in Hesamzadeh et al. (2018), the Transco itself has incentives to change the starting point in its favor by charging excessive prices (and investing too little) before the mechanism starts. Thus, in order to change the starting point in favor of generators and loads without affecting the behavior of the Transco, the regulator would have to surprise the Transco or must have independent information about the Transco's ability to improve welfare by investment. If the Transco starts de novo the regulator could organize an auction for the Transco license as, for example, suggested by Sharkey (1979) for the Loeb and Magat (1979) scheme. In that case, there have to exist enough potential bidders, who can fill the position of the Transco. The problem with this approach is that any change in the starting position gets outdated over time and therefore has to be renewed from time to time.

A refinement of this approach is the RPI-X formula first proposed by Baumol (1982) and made popular by Littlechild (1983). Under this approach, a price index of the Transco's services (including fixed fees) would be adjusted each period by an inflation factor (RPI) and a rent extraction factor (X). This formula has been especially designed to deal with the dynamic adjustment of regulation to changing circumstances. It is usually applied multiplicatively to the prices charged by a regulated entity. Unfortunately, if applied to H-R-G-V a multiplicative RPI-X-factor leads to investment distortions (Vogelsang 2018). Foreseeing the (usually downward) adjustment of its profits by the X-factor, the Transco compensates by withholding investment and thereby increasing the price. Thus, instead of leading to usage prices equal to marginal costs, the resulting Lerner index will in the absence of inflation be X times the monopoly Lerner Index or $L = (p - \mathrm{MC})/p = X/\varepsilon$, where ε is the demand elasticity. This can be a small distortion if X is only a few percent but it can be large if X is large. A less distortive approach is to apply the RPI-X formula additively. Provided the formula remains constant this will eliminate investment distortions but may result in excessive rents or losses over time if the RPI-X formula does not well trace the Transco's input prices, productivity changes, and market environment. Overall both the multiplicative and the additive RPI-X formulas nevertheless are good approaches, as long as the X-factor is small.

A second method is to deduct a fixed percentage share of the Transco's total H-R-G-V profit and hand it over to transmission users in the form of a fixed fee reduction. This would lead to no change in investment incentives compared to the original H-R-G-V scheme, provided the fixed fee reduction can be done without affecting user incentives and provided the Transco's total costs are captured in its measured profits. User incentives are unlikely to be affected if users are small and many. The effect of profit sharing on cost-reducing incentives, however, can be substantial if the sharing percentage is high and if costs depend on unobservable effort. In the case of sharing of the Transco's total H-R-G-V profit, the regulator has to be able to do some cost measurement, thereby reducing one of the motivating advantages of the method.

For the case that some cost measurement can be done, a third approach would be to use a Bayesian regulatory mechanism such as proposed by Léautier (2000). For this purpose, he assumes that the Laffont–Tirole cost pricing dichotomy holds (Laffont and Tirole 1993, Chap. 3). This means that cost minimization decisions are independent of pricing decisions. Thus, Léautier's (2000) transmission pricing mechanism can hold independently from cost-reducing incentives provided. In this case, the Transco would be presented with a menu of revenue-sharing combinations, from which it could choose. In principle, this could also apply to the H-R-G-V mechanism. However, it is not clear that the cost pricing dichotomy holds for transmission systems.[21]

Summing up, the rent extraction problem can be solved with fairly little distortions of the H-R-G-V mechanism. However, some cost measurement is necessary,

---

[21] While there has been little practical application of menu regulation, the UK has been a frontrunner of water, gas, and electricity distribution systems. See https://www.oxera.com/agenda/menu-regulation-is-it-here-to-stay-revisited/.

because otherwise the Transco's rents cannot be determined and overshooting of rent extraction leading to losses or undershooting leading to excessive profits can result.

7 The Fairness Issue Among Transmission Users[22]

In the previous section, we addressed the rent extraction issue at an aggregate level, combining all transmission users in a single entity that receives its share in the transmission benefits. In actuality, a sharing of the benefits between the different users has to be arranged. This sharing and the overall allocation of the fixed fee under H-R-G-V, however, has to be accomplished in such a way that it is viewed as fair by the users and that it does not distort the investment decisions. In particular, the users should be induced by their purchase or bidding decisions to provide truthful information about their specific preferences.

We see two distinct fairness issues, in particular, that need to be addressed.

7.1 Fairness Between Electricity Generators and Loads

The first such fairness issue relates to the distribution of fixed fees between electricity generators and loads. If only loads pay for the fixed fees they may actually be worse off under H-R-G-V than under the status quo. The reason is that H-R-G-V leaves all users in total at their status quo benefit level. Thus, in cases where generators benefit, loads in total will be worse off. This issue will be less pronounced if, as suggested in Sect. 6, H-R-G-V is adjusted in such a way that transmission users in aggregate share in the net benefits created by the mechanism.

However, another important issue is that generators actually cause major, partly individually assignable, transmission costs and convey major welfare benefits by, for example, providing new facilities that are less costly and more environmentally friendly. The cost causation justifies a distinction between *connection assets* assignable to generators or new loads and *network assets* common to all users (see, e.g., Madrigal and Stoft 2011, p. 14, and, in particular, Rivier's and Olmos' Chap. 5 in this volume). My interpretation of the logic behind the Rivier and Olmos approach to transmission network costing is that besides producing congestion relief as multiple outputs (since all the node pairs generate different outputs) a Transco is providing many more outputs. In particular, all the connections to generators and loads can be seen as many different outputs. Generators and loads would then have to pay directly assignable connection costs, which they cause or have caused. That would justify assigning costs to these outputs and would go a long way toward filling out the complementary charge, leaving quite little for a genuine assignment of common

[22]For a more extensive treatment of the fairness issues, see Chap. 5 by Rivier and Olmos in this volume.

fixed costs. One can interpret the connection charges as access fees for the option to send or receive electricity over the grid. In contrast, the not directly assignable network costs are common and can, in my view, be interpreted as usage costs that should be assigned to the usage-related services. Nevertheless, under H-R-G-V they would for the most part be paid as fixed fees. Although both generators and loads would be causing the associated services, it probably is best to have loads pay for them, because in terms of the value chain they are closer to end users, who ultimately have to bear them.

In practice, connection costs are interpreted quite differently by different jurisdictions, ranging from super-shallow connections, which only include enabler facilities, to deep connections, which include network upgrades necessitated by accommodating the new generation facilities (Madrigal and Stoft 2011, pp. 16ff). Having new generators pay only super-shallow connection costs will favor such new generation facilities (Madrigal and Stoft 2011, p. 15). In case of H-R-G-V, the regulator could influence the build-out of renewables by prescribing such shallow cost allocation for new renewable generation and a deeper cost allocation for other new generation. It is questionable, however, if such policy systematically provides good environmental policy, because the resulting implicit subsidies for renewables appear to be arbitrary and will vary by distance of renewable generation from load centers. Such policy would also affect competition between generators in a nonsystematic and arbitrary fashion.

In cases of new renewables, net benefits improve from new transmission lines connecting such facilities with loads, but the responsibility for those benefits is shared between the generators and the Transco. This joint benefit creation could therefore also justify a special sharing mechanism between generators and the Transco. A major question here is if the Transco would offer such sharing on a voluntary basis in order to attract new generation that would provide such benefits. This would depend on the distribution of information between the Transco and potential new generators. Furthermore, if the Transco and generators are not fully informed about the benefits it is even less likely that the regulator will be well-informed.

In contrast, when transmission investment increases competition between imperfectly competitive generators, the generators in aggregate will see reduced surplus. This will in general be justified to the extent that their original surplus included excess profits.

Although transmission charges are ultimately passed on to end users, the distribution of those charges between generators and loads can have major influences, both in equity and efficiency terms. This will be the case because cost pass-on will not be 100% and it will vary by type and location of economic agents. For example, postage pricing, where all kwh sent from all locations are priced the same will favor remote locations that cause higher costs for connection/transmission assets.

7.2 *Fairness Among Loads*

The second fairness issue among transmission users concerns differently sized loads that from a fairness perspective should not all pay the same fixed fees. The large industrial entity should not pay the same fixed fee as the small home. The fixed fees for loads should be allocated in such a way that they are perceived as fair but should have no effect on the loads' incentives to buy electricity. Such undesirable effects have been shown by Vogelsang (2001) for the case of fixed fees that were allocated according to the individual user's peak electricity consumption. Furthermore, high fixed fees should not induce end users to disconnect from the electricity network. Such disconnection is rare under the systems currently used in practice (Pollitt 2018) but could happen under the H-R-G-V mechanism, where variable fees would tend to be low and fixed fees high. The resulting fixed fees should also provide no incentives for inefficient behavior, such as inefficient mergers. Thus, equal fixed fees for all loads are out of question. This also holds, because different loads can cause very different connection costs. Furthermore, they can be small or large and therefore small loads would have incentives to merge in order to reduce the impact of fixed fees.

An allocation of fixed fees by population size could work as a simple proxy for size, but that would treat loads serving a large share of commercial and/or industrial activities better than purely residential loads. It would also treat suburban areas with low density better than urban areas with high density. A potentially better approach would, therefore, allocate fixed fees by total net income of residential users plus the value added created by business. While this seems to be quite fair and does not seem to affect electricity purchases by the loads, it does create measurement problems for regulators. Thus, it may be best to observe how currently loads actually charge their customers, which are residential end users, commercial entities, and industrial users. Typically, end users would be charged a uniform fixed fee with some reductions for specific households, such as the poor. This is important from a distributional and political economy perspective, because two-part tariffs have regressive properties (Florio 2013) and affect competition between non-residential users (Hoernig and Vogelsang 2013). Commercial and industrial entities would be charged different fixed fees that may be differentiated by voltage and size of the entity. This charging could then be retranslated into the method for the aggregate fixed fee charged by the Transco. There could also be connection fees for various types of users.

8 Conclusions

In this chapter, we have tried to bring the H-R-G-V mechanism closer to an actual application in practice. H-R-G-V is a regulatory incentive mechanism aimed at overcoming the dual problem of asymmetric information and conflict of interest. It uses a non-Bayesian approach, combining properties of subsidy-based and constraint-based mechanisms by using two-part tariffs. The welfare optimality properties of this mechanism have only been shown for restrictive environments with competitive generators and loads, an absence of environmental problems, and a lack of distributional concerns. The current chapter addresses these issues without formal modeling and provides a number of conjectures that suggest that the mechanism will work well if these assumptions are relaxed.

While being related to merchant investment, the H-R-G-V mechanism differs by two important features. The first is that it applies to a monopoly situation of the transmission network and therefore includes all the complementarities and externalities of the network build-out. As shown in Chap. 13 of this volume by Papadaskalopoulos et al., merchant investment is generally not feasible for meshed networks. The second is that its pricing is not simply built on the merchandizing surplus but includes a fixed fee that adjusts in opposite direction to the congestion prices. This allows one to deal with economies of scale in the transmission build-out, again something that according to the Papadaskalopoulos et al., modeling the merchant approach cannot deal with. Both of these properties eliminate many of the concerns raised against merchant investment. Nevertheless, the H-R-G-V mechanism shares some of the merchant transmission issues.

In particular, H-R-G-V can lead to socially too much investment in transmission connecting an exporting region with competitive generation and an importing region with monopolistic generation. However, these issues are reduced by (a) the Transco acting as a Stackelberg leader vis-à-vis the monopolistic generator and (b) by helping create more competition in generation. Simultaneously using H-R-G-V regulation for the Transco and monopolistic generators can lead to consistency problems and requires the regulator to have more information than would be compatible with the incentive regulation approach.

While H-R-G-V is fully compatible with perfect environmental regulation, it could interact less well with imperfect regulation. This is part of the second-best problem, according to which perfect and imperfect regulations do not fully mix. However, our analysis suggests that H-R-G-V could in some cases include environmental variables and can be converted into an instrument for both market power regulation and environmental regulation.

Since H-R-G-V gives all the incremental surplus to the Transco, its distributional impact is less benign than its efficiency impact. Correcting this leads to tradeoffs the solution of which will inevitably reduce some of the efficiency properties. It is well known from Laffont and Tirole (1993) that this is a property of all mechanisms under distributional concerns. Thus, the task is to find the fairest solution with the least negative efficiency impact. The same holds for the distribution of the fixed fees among users.

In Sects. 4–7, we have discussed several practical problems with the mechanism and offered solutions. Some of these solutions interact with each other. This holds, in particular, for the interaction between generator market power, environmental effects of generation and rents accruing to generators. Clearly, reducing generator market power via increased transmission capacity will lower rents accruing to those generators but will also expand outputs. If such generation causes pollution, adverse environmental effects will result. Such effects may be prevented if the increase in competition will reduce the generation by polluting generators and benefit competitive renewables.

We also found that the H-R-G-V mechanism potentially interacts negatively with other incentive schemes for generators and/or loads, because the H-R-G-V scheme gives all surplus increase to the Transco. In order for other incentive schemes to work side-by-side with the H-R-G-V for the Transco either an appropriate sharing mechanism for total benefits between the Transco and generators and loads has to be devised or the improvements caused by each incentivized party have to be measured and assigned separately.

As Léautier in Chap. 3 remarks, the proposed H-R-G-V scheme requires an adaptation of the institutional setup for transmission regulation. This holds for all incentive mechanisms that try to address Transco performance in total rather than only certain aspects of it. In the USA, for example, our proposal would require a change in responsibilities between the Transco and the ISO. Under H-R-G-V, the ISO would only be responsible for the dispatch and for the calculation of nodal prices and the merchandizing surplus, while maintenance of the grid would move to the Transco. Under this new division of labor, the ISO would only be responsible for tasks that can be verified by others.

The current paper has only developed conjectures. Future work should deepen the theoretical analyses of these issues and do numerical simulations to the extent that analytical solutions are not achievable.

Acknowledgement The author would like to thank Lew Evans and his co-editors Mohammad Hesamzadeh and Juan Rosellón for many valuable suggestions.

Bibliography

W. Baumol, Productivity-incentive clauses and rate adjustment for inflation. Public Utilities Fortnightly **110**(2), (1982)

P. Beato, A. Mas-Colell, The marginal cost pricing rule as a regulation mechanism in mixed markets, in *The Performance of Public Enterprises*, ed. by M. Marchand, P. Pestieau, H. Tulkens (North-Holland Elsevier, Amsterdam, 1984), pp. 81–100

S. Borenstein, J. Bushnell, S. Stoft, The competitive effects of transmission capacity in a deregulated electricity industry. RAND J. Econ. **31**, 294–325 (2000)

D. Brito, J. Rosellón, Lumpy investment in regulated natural gas pipelines: an application of the theory of the second best. Netw. Spat. Econ. **11**, 533–553 (2011)

H. Cremer, M. Marchand, J.-F. Thisse, The public firm as an instrument for regulating an oligopolistic market. Oxford Economic Papers **41**, 283–301 (1989)

G. De Fraja, F. Delbono, Alternative strategies for a public enterprise in oligopoly. Oxford Economic Papers **41**, 302–311 (1989)

J. Egerer, J. Rosellón, W.-P. Schill, Power system transformation toward renewables: an evaluation of regulatory approaches for network expansion. Energy J. **36**, 105–128 (2015)

M. Florio, *Network Industries and Social Welfare* (Oxford University Press, Oxford, 2013)
R. Gilbert, K. Neuhoff, D. Newbery, Allocating transmission to mitigate market power in electricity networks. RAND J. Econ. **35**, 691–709 (2004)
A. Grigoryeva, M.R. Hesamzadeh, J. Rosellón, I. Vogelsang, *Analysis of the impact of Transco's risk aversion and market expectations on HRGV regulatory mechanism.* Working paper (2017)
K. Griva, N. Vettas, On two-part tariff competition in a homogeneous product duopoly. Int. J. Ind. Organ. **41**, 30–41 (2015)
A. Groppi, E. Fumagalli, Network expansion by a proactive transmission system operator: a case study. Energy Policy **69**, 610–623 (2014)
W. Güth, R. Schmittberger, B. Schwarze, An experimental analysis of ultimatum bargaining. J. Econ. Behav. Organ. **3**, 367–388 (1982)
M.R. Hesamzadeh, J. Rosellón, S.A. Gabriel, I. Vogelsang, A simple regulatory incentive mechanism applied to electricity transmission pricing and investment. Energy Econ. **75**, 423–439 (2018)
S. Hoernig, I. Vogelsang, The ambivalence of two-part tariffs for bottleneck access, in *Wettbewerb und Regulierung in Medien, Politik und Märkten*, ed. by R. Dewenter, J. Haucap, C. Kehder (Nomos, Baden-Baden, 2013), pp. 63–80. Available at http://papers.ssrn.com/sol3/papers.cfm?abstract_id=2176095
W. Hogan, J. Rosellón, I. Vogelsang, Toward a combined merchant-regulatory mechanism for electricity transmission expansion. J. Regul. Econ. **38**, 113–143 (2010)
P.L. Joskow, & J. Tirole, Merchant transmission investment. J. Ind. Econ. **LIII**, 233–264 (2005)
L. Kaplow, S. Shavell, On the superiority of corrective taxes to quantity regulation. Am. Law Econ. Rev. **4**, 1–17 (2002)
C. Kemfert, F. Kunz, J. Rosellón, A welfare analysis of electricity transmission planning in Germany. Energy Policy **94**(C), 446–452 (2016)
D. Khastieva, M. Hesamzadeh, I. Vogelsang, J. Rosellón, M. Amelin, Value of energy storage for transmission investments. Energy Strat. Rev. **24**, 94–110 (2019)
J.-C. Kim, K.-B. Chang, An optimal tax/subsidy for output and pollution control under asymmetric information in oligopoly markets. J. Regul. Econ. **5**, 183–197 (1993)
J.-C. Kim, S.-H. Lee, An optimal regulation in an intertemporal oligopoly market: the generalized incremental surplus subsidy scheme. Inf. Econ. Policy **7**, 225–249 (1995)
J.-J. Laffont, J. Tirole, *A theory of incentives in procurement and regulation* (The MIT Press, Cambridge, MA, 1993)
T. Léautier, Regulation of an electric power transmission company. Energy J. **24**, 61–92 (2000)
T. Léautier, Transmission constraints and imperfect markets for power. J. Regul. Econ. **19**, 27–54 (2001)
T. Léautier, *Imperfect markets and imperfect regulation: An introduction to the microeconomics and political economy of power markets* (MIT Press, Cambridge, MA, 2019)
S.C. Littlechild, *Regulation of British Telecommunication's Profitability*. Report (The Secretary of State, Department of Industry, London, 1983)
M. Loeb, W. Magat, A decentralized method for utility regulation. J. Law Econ. **22**, 399–404 (1979)
M. Madrigal, S. Stoft, *Transmission Expansion for Renewable Energy Scale-Up*. Energy and Mining Sector Board Discussion Paper No. 26 (The World Bank Washington, 2011)
M.G. Pollitt, Electricity network charging in the presence of distributed energy resources: principles, problems and solutions. Econ. Energy Environ. Policy **7**(1), 89–103 (2018)
J. Rosellón, I. Vogelsang, H. Weigt, Long-run cost functions for electricity transmission. Energy J. **33**, 131–160 (2012)
D.E.M. Sappington, D. Sibley, Regulating without cost information: the incremental surplus subsidy scheme. Int. Econ. Rev. **29**, 297–306 (1988)
D.E.M. Sappington, J.G. Sidak, Incentives for anticompetitive behavior by public enterprises. Rev. Ind. Organ. **22**, 183–206 (2003)
W.W. Sharkey, A decentralized method for utility regulation: a comment. J. Law Econ. **22**, 405–407 (1979)

S. Stoft, Problems of transmission investment in a deregulated power market, in *Competitive Electricity Markets and Sustainability*, ed. by F. Lévêque (Edward Elgar, Cheltenham, UK, 2007), pp. 87–130

I. Vogelsang, Public interest firms as leaders in oligopolistic industries, in *Public Sector Economics*, ed. by J. Finsinger (Macmillan, London, 1983), pp. 86–99

I. Vogelsang, Two-part tariffs as regulatory constraints. J. Publ. Econ. **39**, 45–66 (1989)

I. Vogelsang, Price regulation for independent transmission companies. J. Regul. Econ. **20**, 141–165 (2001)

I. Vogelsang, Can simple regulatory mechanisms realistically be used for electricity transmission investment? The case of H-R-G-V. Econ. Energy Environ. Policy **7**(1), 63–87 (2018)

I. Vogelsang, J. Finsinger, A regulatory adjustment process for multiproduct monopoly firms. Bell J. Econ. Manag. Sci. **10**, 157–171 (1979)

Game-Theoretic Modeling of Merchant Transmission Investments

Dimitrios Papadaskalopoulos, Ying Fan, Antonio De Paola, Rodrigo Moreno, Goran Strbac, and David Angeli

Nomenclature

(A) Indices and sets

$i \in I$	Index and set of merchant investors
$l \in L$	Index and set of network loops
$m \in M$	Index and set of network branches
$n \in N$	Index and set of network nodes
$t \in T$	Index and set of time periods in the operational timescale

(B) Parameters

n_m^s	Reference sending node of branch m
n_m^r	Reference receiving node of branch m
$\mathbf{\Phi}$	Matrix of sensitivities $\varphi_{n,m}$ for power outflow from node n with respect to power flow on branch m

D. Papadaskalopoulos (✉) · R. Moreno · G. Strbac · D. Angeli
Imperial College London, London, UK
e-mail: d.papadaskalopoulos08@imperial.ac.uk

Y. Fan
State Grid, Beijing, China

A. De Paola
University of Bath, Bath, UK

R. Moreno
University of Chile, Santiago, Chile

M. R. Hesamzadeh et al. (eds.), *Transmission Network Investment in Liberalized Power Markets*, Lecture Notes in Energy 79,
https://doi.org/10.1007/978-3-030-47929-9_13

$\boldsymbol{\Psi}$	Matrix of sensitivities $\psi_{l,m}$ for voltage drop of loop l with respect to power flow on branch m
F_m^0	Existing capacity of branch m (MW)
T_m^F	Fixed investment cost of branch m (£/h)
T_m^V	Variable investment cost of branch m (£/MWh)
a_n^G	Quadratic cost coefficient of generation company of node n (£/MW^{2}h)
b_n^G	Linear cost coefficient of generation company of node n (£/MWh)
$G_n^{\max}$	Maximum generation limit of generation company of node n (MW)
a_n^D	Quadratic benefit coefficient of demand company of node n (£/MW^{2}h)
b_n^D	Linear benefit coefficient of demand company of node n (£/MWh)
$D_n^{\max}$	Maximum demand limit of demand company of node n (MW)
w_t	Weighting factor of period t

(C) Variables

$\boldsymbol{u}$	Vector of binary variables u_m expressing whether new capacity is added on branch m ($u_m = 1$ if it is $u_m = 0$ if it is not)
$\boldsymbol{F}$	Vector of continuous variables F_m expressing the total capacity addition on branch m (MW)
$\boldsymbol{F}(\boldsymbol{i})$	Vector of continuous variables $F_m(i)$ expressing the capacity addition by merchant investor i on branch m (MW)
$f_{m,t}$	Power flow on branch m and period t (MW)
$G_{n,t}$	Power generated at node n and period t (MW)
$D_{n,t}$	Power consumed at node n and period t (MW)
$p_{n,t}$	Net power outflow from node n at period t (MW)
$\lambda_{n,t}$	Locational marginal price at node n and period t (£/MWh)

(D) Functions

$T_m(\cdot)$	Investment cost of branch m (£/h)
$C_{n,t}(\cdot)$	Operating cost of generation company of node n at period t (£/h)
$B_{n,t}(\cdot)$	Benefit of demand company of node n at period t (£/h)
$J_i(i,\cdot)$	Surplus of merchant investor i (£/h)

1 Introduction

1.1 Motivation

During the last decades, deregulation of the electricity industry has been observed worldwide, involving the unbundling of vertically integrated monopoly utilities, the introduction of competition in the generation and supply sectors, and the open access to the electricity networks. In this deregulated environment, two general approaches are adopted for transmission network planning (Kirschen and Strbac 2004; Joskow and Tirole 2005; Shrestha and Fonseka 2007; Strbac et al. 2014).

Under the first approach, known as regulated transmission investment planning and investigated in Part 2 of this book, planning is centrally carried out by a regulatory authority, the system/network operator or the regulated transmission company, which realizes under regulatory supervision the optimal transmission expansion plan that maximizes the social welfare while satisfying security of supply requirements. The required capital cost plus a suitable rate of return for the transmission company is recovered from the network users. In this context, quantitative research efforts have focused on the solution of the centralized optimal transmission planning problem (Latorre et al. 2003), as well as the allocation of transmission costs among the network users (Lima et al. 2009).

Under the second approach, known as merchant transmission investment planning (Joskow and Tirole 2005) and investigated in Part 3 of this book, transmission planning relies on competitive market forces and decentralized, profit-driven decisions of self-interested players. These players may generally include merchant transmission companies (companies aiming at making profits through investing in transmission) as well as generation and demand users of the network, who are rewarded on the basis of the collected congestion revenues created by their network investments. This paradigm is gaining continuously ground as it accounts for the interests of the different market agents and the resulting competition in transmission planning is advocated as a further step toward the deregulation and liberalization of the electricity industry (Joskow and Tirole 2005; Shrestha and Fonseka 2007; Strbac et al. 2014; Gil et al. 2002). The first instances of merchant transmission planning can be found in the USA, Australia, Argentina, Brazil, and Chile, although the adopted frameworks constitute a mix of centralized and merchant planning where the regulator still determines the final expansion plan, reconciling the conflicting interests of the different entities (Joskow and Tirole 2005; Shrestha and Fonseka 2007; Littlechild 2003; Federal Energy Regulatory Commission 2002; Electric Light & Power 2016; DUKE American Transmission Co 2017; StarWood Energy Group et al. 2007).

However, critical open questions need to be answered before the widespread application of this merchant transmission planning paradigm:

(i) Which entities are likely to undertake network investments under this planning paradigm?

(ii) Is this planning paradigm able to achieve the same (maximum) social welfare as the traditional centralized planning approach?

1.2 Relevant Work

A few recent papers have developed quantitative models of this new transmission planning paradigm in order to answer these questions. Employing *Lagrangian relaxation* (LR) principles, authors in Gil et al. (2002) demonstrate that this decentralized paradigm leads to the same planning solution as the one obtained by the centralized paradigm, concluding that introduction of competition in network planning is plausible. However, this outcome is subject to two simplifying assumptions. First of all, the fixed costs of network assets are neglected because LR is generally unable to produce the centralized solution in the presence of non-convexities (Bazaraa et al. 2006); as a result, the undeniable economies of scale associated with transmission investment are not properly considered (Kirschen and Strbac 2004; Joskow and Tirole 2005). More importantly, players participating in merchant transmission planning are assumed to be competitive, price-taking entities, considering the *locational marginal prices* (LMP) as exogenous signals that cannot be influenced by their individual actions. In reality, however, in a similar fashion as strategic behavior observed in energy markets, participants will attempt to exercise market power and manipulate the LMP to increase their profits beyond the competitive equilibrium levels, through strategic network investments (Joskow and Tirole 2005; Shrestha and Fonseka 2007; Sauma and Oren 2007; Bushnell and Stoft 1996; Bushnell and Stoft 1997; Molina and Rudnick 2010).

In Bushnell and Stoft (1996), Bushnell and Stoft (1997) and Joskow and Tirole (2005), this competitive behavior assumption is removed. Authors in Bushnell and Stoft (1996), Bushnell and Stoft (1997) show that under certain conditions (neglect of fixed costs of network assets, congestion rights satisfying certain feasibility constraints, no imperfections in the energy market), merchant investments are socially efficient. In the seminal work (Joskow and Tirole 2005), the authors demonstrate through theoretical discussion and illustrative examples that this conclusion does not hold when the above simplifying conditions are relaxed. However, these papers investigate the social efficiency of investments by a single merchant company, neglecting that the very essence of the merchant planning paradigm lies in the introduction of competition in transmission planning, through the participation of multiple players. In fact, authors in Joskow and Tirole (2005) recognize through a simple 3-node example that gaming interactions between multiple merchant investors are likely in reality (pages 54–55), but they do not provide a comprehensive modeling framework capturing these interactions.

Authors in Shrestha and Fonseka (2007) make the first attempt to consider a setting with multiple strategic merchant companies and analytically derive the relation between the procured transmission capacities under the centralized and the merchant planning paradigm. The results indicate that decentralized planning by merchant companies leads to under-investment with respect to the centralized approach but the extent of this under-investment is reduced as the number of merchant companies is increased. Extending the analysis to the theoretical case where the number of merchant companies approaches infinity, the authors in Shrestha and Fonseka (2007) demonstrate that the differences between the planning solutions of the two paradigms in this case tend to zero. Although this theoretical scenario with an infinite number of merchant companies does not correspond to a realistic setting, this result is of great significance as it implies that under a "sufficiently large" number of competing merchant companies, the socially optimal transmission planning solution can be approached. However, this paper carries out simplifying assumptions that sacrifice the generality of the obtained results; transmission branches are presumed congested at the optimal solution and fixed costs of network assets are neglected. More importantly, the multiple merchant companies are assumed to make investment decisions sequentially, without accounting for the reactions of the competing players. In other words, the adopted approach does not comprehensively model the decision-making interactions between multiple investors.

1.3 Chapter Contributions

As mentioned in Molina and Rudnick (2010), a *non-cooperative game-theoretic* modeling framework is required to accurately capture the strategic behavior and interactions of multiple merchant investors, discussed in Sect. 1.2. This Chapter aims at developing such a novel framework and exploiting it to answer the research questions outlined in Sect. 1.1. More specifically, two different models, both based on non-cooperative game theory, are developed.

The first model adopts an *equilibrium programming* approach. The decision-making problem of each player is formulated as a *bi-level optimization* problem, accounting for the impacts of its own actions on LMP as well as the actions of all competing players. The upper-level problem represents the surplus maximization of the player, and the lower-level problem represents the energy market clearing process subject to the network constraints. This bi-level problem is formulated for different types of players (merchant transmission companies, generation companies, and demand companies) and solved after converting it to a *mathematical program with equilibrium constraints* (MPEC). An iterative *diagonalization* method is employed to search for the likely outcome of the strategic interactions between multiple players, i.e., *Nash equilibria* (NE) of the game.

Case studies on a simple 2-node system provide the following answers to the identified research questions:

(i) Which entities are likely to undertake network investments under the merchant planning paradigm?

Networks investments will be mostly undertaken by generation companies in areas with low LMP and demand companies in areas with high LMP *(higher-motivated players)*, as apart from collecting congestion revenue they also increase their energy surpluses. Merchant transmission companies, generation companies in areas with high LMP, and demand companies in areas with low LMP *(lower-motivated players)* could also be motivated to invest by the collection of congestion revenue under certain circumstances. Case studies illustrate the interdependencies between the different players' decisions; in certain cases, the large network capacity desired by higher-motivated players reduces the obtainable congestion revenue by lower-motivated players and thus prevents the latter from investing in capacity.

(ii) Is the merchant planning paradigm able to achieve the same (maximum) social welfare as the traditional centralized planning approach?

The merchant planning solution approaches the centralized one as the number of competing players increases. The largest deviations from the centralized solution are observed in the case where the set of participating players includes only merchant transmission companies, as they procure significantly lower capacity in order to increase their profits through higher LMP differentials.

However, because of its iterative nature, this first model cannot guarantee convergence to existing NE, especially as the number of players and the size of the network increase; as a result, the examined case studies are limited to a 2-node system with up to 10 players. In other words, although this model captures the strategic decision-making interactions between competing merchant investors and accounts for fixed costs of transmission assets [aspects not captured by the modeling framework of Shrestha and Fonseka (2007)], it cannot establish whether the important finding of Shrestha and Fonseka (2007) (i.e., that merchant planning yields the same solution as centralized planning under the participation of a "sufficiently large" number of competing investors) is valid or not, as it cannot deal with a large number of players, especially in large networks.

In order to address this challenge and validate this important finding of Shrestha and Fonseka (2007), a second model is developed, where the set of merchant investors is approximated as a continuum. The proposed approximation makes the impact of each infinitesimal player's decisions on system quantities negligible, allowing us to derive mathematical conditions for the existence of a NE solution in an analytical fashion.

Based on this model, this Chapter investigates the validity of the finding of Shrestha and Fonseka (2007), through analytical and numerical comparison of the merchant planning solution against the one obtained through the traditional centralized paradigm. This comparison demonstrates that merchant planning can achieve the same (maximum) social welfare as the centralized planning approach only when the following conditions are satisfied:

(a) fixed investment costs are neglected, and
(b) the network is radial and does not include any loops.

As these conditions do not generally hold in reality, our findings suggest that even a fully competitive merchant transmission planning framework, involving the participation of a very large number of competing merchant investors, is not generally capable of maximizing social welfare, as implied by the previous work (Shrestha and Fonseka 2007). Numerical simulations supporting these findings are carried out on a 2-node, a 3-node, and a 24-node system, while the largest case study examined in the previous relevant works discussed in Sect. 1.2 corresponds to a 6-node system.

1.4 Chapter Outline

The rest of this Chapter is organized as follows. Section 2 outlines a basic model of traditional centralized transmission planning, against which the merchant planning approach will be later compared. Sections 3 and 4 detail the two developed game-theoretic models and present results of relevant case studies. Finally, Sect. 5 discusses conclusions and future extensions of this work.

2 Centralized Transmission Planning Model

Under the centralized transmission planning paradigm, a regulatory authority, the system/network operator or the regulated transmission company, determines the capacity to be added in the existing network, so as to maximize the long-term social welfare or, equivalently, minimize the long-term system cost (Kirschen and Strbac 2004). The latter is given by the sum of two terms: the difference between generation operating cost and demand benefit, plus the investment cost required for delivering the new capacity. Employing a DC load flow model, the optimization problem determining the centralized planning solution is formulated as follows:

$$\min_{\substack{u_m,\, F_m,\, f_{m,t} \\ G_{n,t},\, D_{n,t},\, p_{n,t}}} {}_{,\forall m, \forall n, \forall t} \quad S = \sum_m T_m(u_m, F_m) + \sum_t \sum_n w_t\big[C_{n,t}(G_{n,t}) - B_{n,t}(D_{n,t})\big] \tag{1}$$

where:

$$T_m(u_m, F_m) = u_m\big(T_m^F + T_m^V F_m\big), \forall m \tag{2}$$

Subject to:

$$0 \le F_m, \forall m \tag{3}$$

$$D_{n,t} + p_{n,t} - G_{n,t} = 0 : \lambda_{n,t}, \forall n, \forall t \tag{4}$$

$$-\left(F_m^0 + u_m F_m\right) \leq f_{m,t} \leq F_m^0 + u_m F_m, \forall m, \forall t \tag{5}$$

$$p_{n,t} = \sum_m \varphi_{n,m} f_{m,t}, \forall n, \forall t \tag{6}$$

$$\sum_m \psi_{l,m} f_{m,t} = 0, \forall l, \forall t \tag{7}$$

$$0 \leq G_{n,t} \leq G_n^{\max}, \forall n, \forall t \tag{8}$$

$$0 \leq D_{n,t} \leq D_n^{\max}, \forall n, \forall t \tag{9}$$

Accounting for the realistic economic properties of network investments, the network investment cost T_m for branch m includes (i) a fixed component, which does not depend on the procured capacity but only on the binary decision u_m of whether new capacity will be added on branch m or not and (ii) a variable component, which is incurred when this binary decision is positive ($u_m = 1$), and is proportional to the procured capacity F_m, as expressed by (2). System operation constraints are expressed by (4)–(7); the Lagrangian multipliers $\lambda_{n,t}$ associated with the nodal demand–supply balance constraints (4) express the LMP at the respective node n and period t. Generation and demand limits are enforced by (8) and (9).

For presentation clarity and without loss of generality, the above model (as well as the rest of the models in this Chapter) involves the following assumptions:

(i) the addition of transmission capacity on branch m does not affect its reactance; in other words, the matrices $\boldsymbol{\Phi}$ and $\boldsymbol{\Psi}$ are constant and do not depend on capacity additions, and
(ii) a one-to-one mapping between nodes and generation/demand participants; in other words, each generation/demand participant corresponds to the whole generation/demand at a particular node n.

3 Modeling Merchant Transmission Planning: Equilibrium Programming

3.1 Setting and Assumptions

Under the merchant paradigm investigated in this Chapter, transmission planning relies on competitive market forces and decentralized, profit-driven decisions of self-interested players. Specifically, different market entities participate in network planning by making network expansion proposals, including merchant transmission companies as well as generation and demand companies. These entities are assumed to be rational players and determine their network expansion proposals so as to maximize their own economic surpluses. In the case of merchant companies, the surplus is given by the difference between the congestion revenue and the investment cost associated with the network capacity they procure. In the case of generation and demand companies, apart from the congestion revenue and investment cost of the procured capacity, their proposals are also driven by the impact of network capacity additions on their profits from selling energy and on their utilities from buying energy, respectively.

The expansion proposals made by the different players are interdependent, since the power flows, the LMP and the generation/demand dispatch affecting their surpluses will be driven by the aggregation of the individual network expansion decisions. Therefore, each player needs to account for the decisions of the rest of the players. Furthermore, in a similar fashion as behavior observed in energy markets, each of these players will not act as a price-taker but will rather attempt to manipulate the LMP through its expansion decisions in order to increase its surplus beyond the competitive equilibrium levels. These interactions can be described through a non-cooperative game among the players involved in the planning process.

3.2 Bi-Level Optimization Model of Merchant Investor

In the non-cooperative game-theoretic setting outlined in Sect. 3.1, the decision-making of a single player i can be formulated as a bi-level optimization problem, a modeling approach widely adopted in literature investigating the strategic behavior of generation companies in electricity markets (Hobbs et al. 2000; Weber and Overbye 2002; Ruiz and Conejo 2009). The upper-level (UL) problem determines the optimal individual transmission expansion decisions maximizing the surplus of player i and is subject to the lower level (LL) problem representing the energy market clearing process. These two problems are coupled, since the expansion decisions made by the UL problem affect the power flow constraints of the LL problem, while the power flows, the LMP and the generation/demand dispatch determined by the LL problem affect the objective function of the UL problem.

According to Sect. 3.1, the formulation of this bi-level problem depends on whether the considered player i is a merchant transmission company, a generation company, or a demand company. In case of a merchant transmission company, this problem is formulated as follows:

(Upper level)

$$\max_{F_m(i),\forall m} J(i) = \sum_m u_m \left[\sum_t w_t \left(\lambda_{n_m^r,t} - \lambda_{n_m^s,t} \right) f_{m,t} \frac{F_m(i)}{F_m + F_m^0} \right] - \sum_m u_m \left[T_m^F \frac{F_m(i)}{F_m} + T_m^V F_m(i) \right] \tag{10}$$

where:

$$F_m = \sum_i F_m(i), \forall m \tag{11}$$

$$u_m = \begin{cases} 0 \text{ if } F_m = 0 \\ 1 \text{ if } F_m = 1 \end{cases}, \forall m \tag{12}$$

Subject to:

$$0 \leq F_m(i), \forall m \tag{13}$$

(Lower level)

$$\min_{G_{n,t}, D_{n,t}, f_{m,t}, p_{n,t} \forall m, \forall n, \forall t} \sum_t \sum_n w_t \left[C_{n,t}(G_{n,t}) - B_{n,t}(D_{n,t}) \right] \tag{14}$$

Subject to:

$$D_{n,t} + p_{n,t} - G_{n,t} = 0 : \lambda_{n,t}, \forall n, \forall t \tag{15}$$

$$-\left(F_m^0 + u_m F_m\right) \leq f_{m,t} \leq F_m^0 + u_m F_m, \forall m, \forall t \tag{16}$$

$$p_{n,t} = \sum_m \varphi_{n,m} f_{m,t}, \forall n, \forall t \tag{17}$$

$$\sum_m \psi_{l,m} f_{m,t} = 0, \forall l, \forall t \tag{18}$$

$$0 \leq G_{n,t} \leq G_n^{max}, \forall n, \forall t \tag{19}$$

$$0 \leq D_{n,t} \leq D_n^{max}, \forall n, \forall t \tag{20}$$

The objective function (10) of the UL problem constitutes the surplus of the merchant transmission company *i* and is given by the difference between the congestion revenue (first term) and the investment cost (second term) associated with the network capacity it procures on each branch. The share of the total congestion revenue belonging to player *i* is equal to the share of the total capacity it owns, as expressed by the first ratio in (10). Likewise, the share of the total fixed investment cost paid by player *i* is equal to the share of the total capacity addition it procures, as expressed by the second ratio in (10). This total capacity addition is given by the sum of the individual players' capacity additions (11). The UL problem is subject to procured capacity limits (13) and the LL problem. The latter represents the market clearing process, maximizing the short-term social welfare (14), subject to system operation constraints (15)–(18) and generation/demand limits (19)–(20).

In the case that the considered player i is the generation company of node n, the objective function of the UL problem is given by (21). Apart from the congestion revenue (first term) and the investment cost (second term) associated with the network capacity player i procures, this objective function also includes its revenue from selling energy at the LMP of node n (third term) and its operating cost (fourth term).

$$\begin{aligned}\max_{F_m(i),\forall m} J(i) = &\sum_m u_m\left[\sum_t w_t\left(\lambda_{n_m^r,t} - \lambda_{n_m^s,t}\right) f_{m,t}\frac{F_m(i)}{F_m + F_m^0}\right] \\ &-\sum_m u_m\left[T_m^F\frac{F_m(i)}{F_m} + T_m^V F_m(i)\right] \\ &+\sum_t w_t\lambda_{n,t}G_{n,t} - \sum_t w_t C_{n,t}\left(G_{n,t}\right)\end{aligned} \tag{21}$$

In the case that the considered player i is the demand company of node n, the objective function of the UL problem is given by (22). Apart from the congestion revenue (first term) and the investment cost (second term) associated with the network capacity player i procures, this objective function also includes its payment for buying energy at the LMP of node n (third term) and its perceived benefit (fourth term).

$$\begin{aligned}\max_{F_m(i),\forall m} J(i) = &\sum_m u_m\left[\sum_t w_t\left(\lambda_{n_m^r,t} - \lambda_{n_m^s,t}\right) f_{m,t}\frac{F_m(i)}{F_m + F_m^0}\right] \\ &-\sum_m u_m\left[T_m^F\frac{F_m(i)}{F_m} + T_m^V F_m(i)\right] \\ &-\sum_t w_t\lambda_{n,t}D_{n,t} + \sum_t w_t B_{n,t}\left(D_{n,t}\right)\end{aligned} \tag{22}$$

In order to solve this bi-level optimization problem in a mathematically rigorous fashion, following the approach adopted in literature investigating the strategic behavior of generation companies in electricity markets (Hobbs et al. 2000; Weber and Overbye 2002; Ruiz and Conejo 2009), we convert it to a mathematical program with equilibrium constraints (MPEC). This is achieved through the replacement of the LL problem by its *Karush–Kuhn–Tucker* (KKT) optimality conditions, which is enabled by the continuity and convexity of the LL problem. The MPEC formulation is omitted but follows the logic detailed in Hobbs et al. (2000), Weber and Overbye (2002), Ruiz and Conejo (2009).

3.3 Determining Nash Equilibrium

The above bi-level optimization/MPEC model expresses the decision-making problem of a single player *i*. Our interest, however, lies in determining the likely outcome of the strategic interactions between multiple merchant investors, given that the very essence of the merchant planning paradigm is the introduction of competition in transmission planning (Sect. 1.1). According to game theory (Fudenberg and Tirole 1991), this likely outcome corresponds to a Nash equilibrium (NE) of the non-cooperative game, which expresses a condition where none of the players can increase its surplus by unilaterally modifying its decisions.

In order to determine a NE solution of the merchant planning game, the iterative diagonalization method, which was in introduced the mathematical paper (Pang and Chan 1982) and was employed in Hobbs et al. (2000), Weber and Overbye (2002), is adopted. This iterative procedure involves three steps:

(1) The players' expansion decisions are initialized, the iteration counter is set to 1 and the convergence tolerance ε is determined.
(2) At every iteration r, each player determines its expansion decisions by solving its respective MPEC, accounting for the decisions of the rest of the players as fixed parameters, equal to their values at iteration $r - 1$.
(3) The vector of all players' decisions at iteration r is compared to the one at iteration $r-1$. If their distance is lower than ε, the iterative procedure terminates. As discussed in Hobbs et al. (2000), Weber and Overbye (2002), the resulting outcome after convergence corresponds by definition to a *pure strategy NE* of the game, since none of the players can increase its surplus by unilaterally modifying its decisions.

It should be noted at this point that existence and uniqueness of NE are not generally guaranteed and that the iterative diagonalization approach is not generally guaranteed to converge, even if NE exist (Hobbs et al. 2000; Weber and Overbye 2002).

3.4 Case Studies on 2-Node System: Analyzing Which Entities Undertake Network Investments

The relevant generation and demand data (Kirschen and Strbac 2004) of the 2-node system considered in the case studies of this Section is illustrated in Fig. 1. It is assumed that the existing capacity of the single branch is zero and that the operational timescale of the planning problem includes a single time period. Generation costs are assumed to be quadratic functions of the respective power productions. The demands in the two nodes are assumed inelastic and equal to constant values, i.e., their benefit functions are constant and can thus be omitted from the two optimization problems.

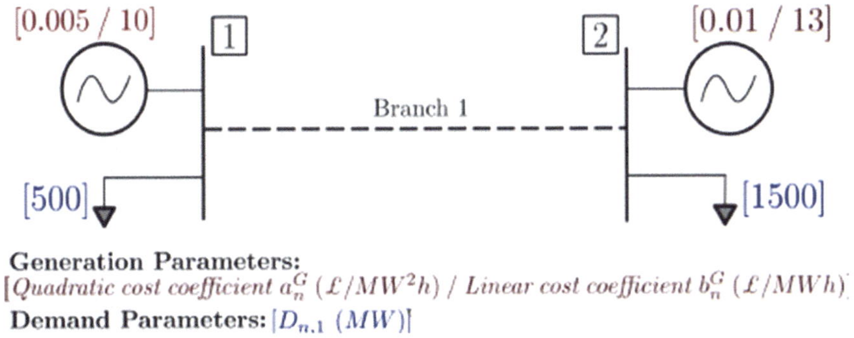

Fig. 1 Topology and parameters of 2-node system

The investment cost includes only a variable component $T_1^V = 4$£/MWh while fixed costs are neglected ($T_1^F = 0$).

In order to comprehensively analyze the outputs of the merchant planning model, the following five cases with different sets of participating players have been examined. The NE determined by the diagonalization approach of Sect. 3.3 in each of these cases are presented in Table 1. Due to the existence of a single network branch in the considered system, the subscript $m = 1$ is omitted from the representation $F_m(i)$ of the expansion decisions in Table 1.

Case 1.1 The set of players includes two merchant transmission companies M_1 and M_2. The two companies procure equal capacities in the NE solution. Table 2 justifies the expansion decision of each of the two players given the decision of its competitor as determined by the NE solution of Table 1. By procuring a capacity of 267 MW, each company collects a congestion revenue which is higher than the incurred investment cost and therefore makes a positive profit of £2131. It is noted that energy surplus is not defined in the case of merchant companies, as they are not involved in the energy market.

Case 1.2 The set of players includes the two generation companies G_1 and G_2 of nodes 1 and 2, respectively. Only G_1 procures capacity in the NE solution. The addition of network capacity increases the power exported from the lower-priced (due to the combination of cheaper generation and lower demand) node 1 to the

Table 1 Merchant planning solutions in 2-node system

Case	Players' expansion decisions in NE (MW)
1.1	$F(M_1) = 267$, $F(M_2) = 267$
1.2	$F(G_1) = 580$, $F(G_2) = 0$
1.3	$F(D_1) = 0$, $F(D_2) = 900$
1.4	$F(G_1) = 345$, $F(G_2) = 0$, $F(D_1) = 0$, $F(D_2) = 588$
1.5	$F(G_1) = 345$, $F(G_2) = 0$, $F(D_1) = 0$, $F(D_2) = 588$, $F(M_1) = 0$, $F(M_2) = 0$

Table 2 Justification of expansion decision of each player given its competitor's decision in Case 1.1

	Energy surplus (£)	Congestion revenue (£)	Investment cost (£)	Total surplus (£)
M_1: surplus change by procuring 267 MW instead of 0 MW	N/A	3199	−1068	2131
M_2: surplus change by procuring 267 MW instead of 0 MW	N/A	3199	−1068	2131

higher-priced (due to the combination of more expensive generation and higher demand) node 2. Therefore, it also reduces the LMP differential between the two nodes, by increasing the LMP in node 1 and reducing the LMP in node 2. As a result, G_1 invests in network capacity not only to collect congestion revenue (like the merchant companies of Case 1.1) but also to increase its energy surplus [given by the difference between its revenue from selling energy and its operating cost in (21)], as demonstrated in Table 3. On the other hand, G_2 does not invest despite the potential congestion revenue, due to the adverse effect of the interconnection on its energy surplus; this is justified by Table 3, which presents the impact of a small capacity procurement by G_2 (10 MW) on its surplus.

Case 1.3 The set of players includes the two demand companies D_1 and D_2 of nodes 1 and 2, respectively. Only D_2 procures capacity in the NE solution. Since the addition of network capacity reduces the LMP in node 2, D_2 invests in high network capacity in order to increase its energy surplus (i.e., reduce its energy payment, since demand is assumed inelastic), despite the fact that the congestion revenue it collects does not cover the incurred investment cost (Table 4). On the other hand, D_1 does not invest due to both the adverse effect of the interconnection on its energy surplus and

Table 3 Justification of expansion decision of each player given its competitor's decision in Case 1.2

	Energy surplus (£)	Congestion revenue (£)	Investment cost (£)	Total surplus (£)
G_1: surplus change by procuring 580 MW instead of 0 MW	4582	6148	−2320	8410
G_2: surplus change by procuring 10 MW instead of 0 MW	−183	103	−40	−120

Table 4 Justification of expansion decision of each player given its competitor's decision in Case 1.3

	Energy surplus (£)	Congestion revenue (£)	Investment cost (£)	Total surplus (£)
D_1: surplus change by procuring 10 MW instead of 0 MW	−50	7	−40	−83
D_2: surplus change by procuring 900 MW instead of 0 MW	27,000	900	−3600	24,300

the fact that the potential congestion revenue does not cover the required investment cost, given the high capacity procured by D_2 (Table 4).

Case 1.4 The set of players includes the two generation companies G_1 and G_2 and the two demand companies D_1 and D_2. Following the analysis of Cases 1.2 and 1.3, G_1 and D_2 procure capacity in the NE solution in order to increase their energy surpluses (increase their energy profit and reduce their energy payment, respectively), despite the fact that they do not collect any congestion revenue in this case (Table 5). The reason behind this zero congestion revenue is that the total capacity procured in the NE solution of this case is so high that it eliminates congestion and therefore the price differential between the two nodes. It is also worth noting that this value of the total capacity ($F(G_1) + F(D_2) = 933$ MW) constitutes the minimum value for

Table 5 Justification of expansion decision of each player given its competitors' decisions in Case 1.4

	Energy surplus (£)	Congestion revenue (£)	Investment cost (£)	Total surplus (£)
G_1: surplus change by procuring 345 MW instead of 0 MW	4350	0	−1380	2970
G_2: surplus change by procuring 10 MW instead of 0 MW	0	0	−40	−40
D_1: surplus change by procuring 10 MW instead of 0 MW	0	0	−40	−40
D_2: surplus change by procuring 588 MW instead of 0 MW	17,655	0	−2352	15,303

which congestion is eliminated (Kirschen and Strbac 2004). In other words, no player has motivation to invest in further capacity, as this action will not affect the dispatch and the LMP, and consequently they will have to incur the additional investment cost without improving their energy surplus and congestion revenue. This also explains why G_2 and D_1 experience no change in their energy surplus and congestion revenue from potential investments (Table 5) and thus do not invest in network capacity.

Case 1.5 The set of players includes the two generation companies G_1 and G_2, the two demand companies D_1 and D_2, and two merchant transmission companies M_1 and M_2. As in Case 1.4, G_1 and D_2 procure capacity in the NE solution. Given that the total capacity they procure is so high that it eliminates congestion and the price differential between the two nodes, the two merchant companies do not have motivation to invest in further capacity (Table 6), in contrast with Case 1.1 where they constitute the only participating players. This result demonstrates the interdependencies between the different players' decisions in the merchant planning framework.

Table 6 Justification of expansion decision of each player given its competitors' decisions in Case 1.5

	Energy surplus (£)	Congestion revenue (£)	Investment cost (£)	Total surplus (£)
G_1: surplus change by procuring 345 MW instead of 0 MW	4350	0	−1380	2970
G_2: surplus change by procuring 10 MW instead of 0 MW	0	0	−40	−40
D_1: surplus change by procuring 10 MW instead of 0 MW	0	0	−40	−40
D_2: surplus change by procuring 588 MW instead of 0 MW	17,655	0	−2352	15,303
M_1: surplus change by procuring 10 MW instead of 0 MW	N/A	0	−40	−40
M_2: surplus change by procuring 10 MW instead of 0 MW	N/A	0	−40	−40

Table 7 Merchant planning solutions in 2-node system with higher generation cost differential between the two nodes

Case	Players' expansion decisions in NE (MW)
1.1	$F(M_1) = 456$, $F(M_2) = 456$
1.2	$F(G_1) = 900$, $F(G_2) = 50$
1.3	$F(D_1) = 11$, $F(D_2) = 1177$
1.4	$F(G_1) = 558$, $F(G_2) = 0$, $F(D_1) = 0$, $F(D_2) = 904$
1.5	$F(G_1) = 558$, $F(G_2) = 0$, $F(D_1) = 0$, $F(D_2) = 904$, $F(M_1) = 0$, $F(M_2) = 0$

Next, we analyze the outputs of the merchant planning model in each of the above five cases when the linear cost coefficient of generation company G_2 is increased to 30£/MWh (Table 7).

In every case, the total and individual capacity additions are increased with respect to Table 1, due to the higher generation cost differential (and therefore higher LMP differential) between the two nodes, which motivates further capacity investments.

Furthermore, the particularly interesting difference in the results is that G_2 and D_1 invest in network capacity in Cases 1.2 and 1.3, respectively. This is because the higher LMP differential makes the collected congestion revenues more significant than the adverse effect of these investments on their energy surpluses and the required investment costs (Tables 8 and 9). However, the capacity procured by G_2 and D_1 is still lower than the one procured by G_1 and D_2, respectively, who are motivated to invest in capacity not only by the congestion revenues but also by the improvement of their energy surpluses.

Furthermore, in Cases 1.4 and 1.5, the total capacity procured by the higher-motivated players G_1 and D_2 reduces the obtainable congestion revenue by G_2 and D_1, preventing the latter from investing in capacity (Table 10), in contrast with Cases 1.2 and 1.3 where fewer players participate. This result again demonstrates the interdependencies between the different players' decisions in the merchant planning framework.

Table 8 Justification of expansion decision of each player given its competitor's decision in Case 1.2 with higher generation cost differential between the two nodes

	Energy surplus (£)	Congestion revenue (£)	Investment cost (£)	Total surplus (£)
G_1: surplus change by procuring 900 MW instead of 0 MW	9000	14,850	−3600	20,250
G_2: surplus change by procuring 50 MW instead of 0 MW	−575	825	−200	50

Table 9 Justification of expansion decision of each player given its competitor's decision in Case 1.3 with higher generation cost differential between the two nodes

	Energy surplus (£)	Congestion revenue (£)	Investment cost (£)	Total surplus (£)
D_1: surplus change by procuring 11 MW instead of 0 MW	−55	103	−44	4
D_2: surplus change by procuring 1177 MW instead of 0 MW	35,310	11,017	−4708	41,619

Table 10 Justification of expansion decision of each player given its competitors' decisions in Case 1.4 with higher generation cost differential between the two nodes

	Energy surplus (£)	Congestion revenue (£)	Investment cost (£)	Total surplus (£)
G_1: surplus change by procuring 558 MW instead of 0 MW	9391	636	−2232	7795
G_2: surplus change by procuring 10 MW instead of 0 MW	−7	8	−40	−39
D_1: surplus change by procuring 10 MW instead of 0 MW	−50	8	−40	−82
D_2: surplus change by procuring 904 MW instead of 0 MW	27,120	1031	−3616	24,535

3.5 Case Studies on 2-Node System: Comparing Centralized and Merchant Planning Solutions

In this Section, we make an attempt to validate the important findings of the previous work (Shrestha and Fonseka 2007), i.e., that the merchant planning solution approaches the centralized one as the number of participating players increases, and the two solutions become identical under the participation of a "sufficiently large" number of players. To this purpose, we apply the developed equilibrium programming model to the same 2-node system presented in Sect. 3.4.

In this context, we have executed the developed model for different scenarios regarding the number of participating players as well as the centralized planning model of Sect. 2, and compared their solutions. Figures 2, 3, and 4 present the total network capacity in the merchant planning solution and the percentage deviation of the long-term system cost of this solution from the respective cost of the centralized solution, when the set of players participating in merchant planning includes:

- From 1 to 10 merchant transmission companies (Fig. 2).
- From 1 to 10 generation companies per node (Fig. 3). Each identical company at node 1 and 2 owns an equal share of the total generation capacity at the respective node.
- From 1 to 10 demand companies per node (Fig. 4). Each identical company at node 1 and 2 supplies an equal share of the total demand at the respective node.

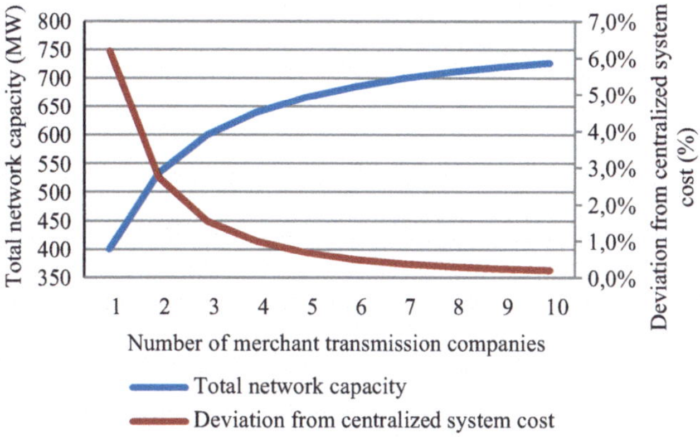

Fig. 2 Merchant planning solution in 2-node system for different numbers of participating merchant transmission companies

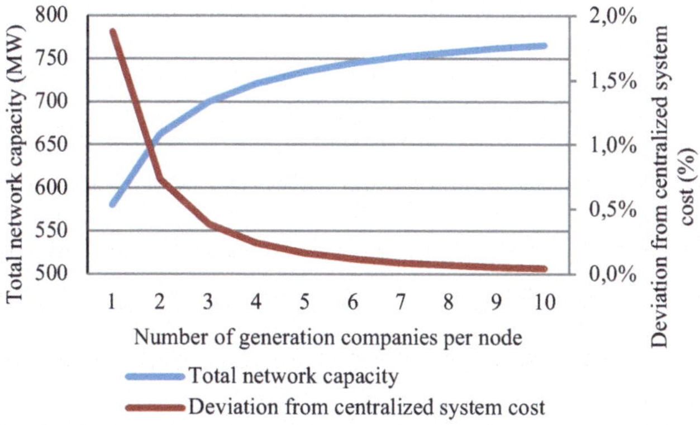

Fig. 3 Merchant planning solution in 2-node system for different numbers of participating generation companies

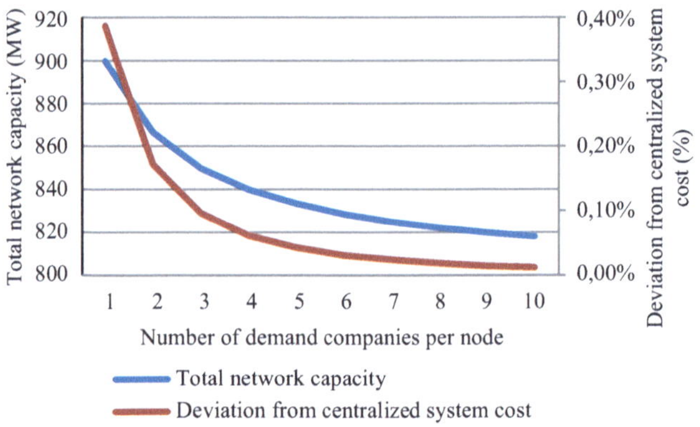

Fig. 4 Merchant planning solution in 2-node system for different numbers of participating demand companies

The above results seem to suggest that the findings of Shrestha and Fonseka (2007) are valid, since the total network capacity and system cost of the merchant planning solution approach the respective capacity (800 MW) and system cost of the centralized planning solution, as the number of participating players increases. The largest deviations from the centralized solution are observed in the case where the participating players are merchant transmission companies (Fig. 2), as they procure significantly lower capacity in order to increase their surpluses through higher LMP differentials.

However, these results are not sufficient to comprehensively validate the findings of Shrestha and Fonseka (2007). First of all, the examined studies include up to 10 "active" players (i.e., players procuring positive capacity, unlike the generation companies at node 2 and the demand companies at node 1 in the cases of Figs. 3 and 4, respectively). Therefore, although the merchant planning solution approaches the centralized one, we cannot guarantee that this trend will be still valid for a larger number of players and that the two solutions will eventually become identical under the participation of a "sufficiently large" number of players. Furthermore and more importantly, the examined studies are carried out on a very simple 2-node system; we cannot guarantee that the observed trend will be still valid in larger, more realistic systems.

Unfortunately, the developed equilibrium programming model cannot effectively deal with such larger numbers of players and larger networks. This is because the employed diagonalization approach (Sect. 3.3) cannot guarantee convergence to an existing NE, since it is very sensitive to the initialization of the players' decisions and its iterative nature often results in an oscillatory behavior, with these problems being aggravated when the number of players and the size of the network increase.

4 Modeling Merchant Transmission Planning: Continuum Approximation

4.1 Setting and Assumptions

In order to overcome the limitations of the equilibrium programming model presented in the previous Section and comprehensively validate the findings of Shrestha and Fonseka (2007), a second model is developed in this Section. This model also adopts a non-cooperative game-theoretic framework. However, in order to deal effectively with a large number of players, the set of players is approximated as a continuum (Khan and Sun 2002). Similar approaches have been previously considered in other economic (Aumann 1964, 1966) and smart grid (Couillet et al. 2012; De Paola et al. 2016) applications. The proposed approximation makes the impact of each infinitesimal player's decisions on system quantities negligible, allowing us to derive mathematical conditions for the existence of a NE solution in an analytical fashion, and therefore avoid the limitations of the iterative diagonalization approach, discussed in Sect. 3.5.

Before proceeding to a detailed description of this second model, it should be noted that it has its own limitations. First of all, the proposed continuum approximation implies that the number of players approaches infinity, which does not correspond to realistic settings where the number of players is always finite. Nevertheless, this theoretical scenario constitutes a good approximation of a setting with a "sufficiently large" number of players, and, more importantly, it is also examined in Shrestha and Fonseka (2007), the very findings of which we aim at validating. Secondly, this continuum approximation implies that the considered players are identical. Therefore, this model cannot provide an answer to the first research question of Sect. 1.1 (which entities are likely to undertake network investments under merchant planning), for which the equilibrium programming model of Sect. 3 is more suitable. For this reason, the players we consider in this second model are merchant transmission companies which are similar in practice, rather than generation or demand companies which have distinct characteristics (such as generation operating costs, generation capacities, and demand sizes.).

Under the proposed continuum approximation, the set of merchant transmission companies is not described as a finite collection $I = \{1, 2, .., |I|\}$ (as in the equilibrium programming model of Sect. 3), but rather as a closed interval $I \subset \mathbb{R}$. With this approximation, system quantities such as investment decisions u_m and F_m are not impacted by each infinitesimal player's decisions, but only depend on the aggregation of all players' decisions. In this context, the total capacity addition on branch m is not expressed by the sum (11) but rather as the integration:

$$F_m = \int_I F_m(i)\mathrm{d}i \tag{23}$$

The surplus function (10) of player i can be expressed as $J(i, \boldsymbol{F}(\boldsymbol{i}), \boldsymbol{u}, \boldsymbol{F})$ where $\boldsymbol{F}(\boldsymbol{i}) = \{F_m(i), \forall m\}$ denotes the vector of the investment decisions of player i, and $\boldsymbol{u} = \{u_m, \forall m\}$ and $\boldsymbol{F} = \{F_m, \forall m\}$ denote the vectors of the aggregate binary investment decisions and total capacity investment decisions of all players, respectively. Given the above approximation, player i can only modify $\boldsymbol{F}(\boldsymbol{i})$ but cannot impact $\boldsymbol{u}$ and $\boldsymbol{F}$.

4.2 Determining Nash Equilibrium

Consider feasible vectors $\boldsymbol{F}^*(\boldsymbol{i}), \boldsymbol{u}^*$, and $\boldsymbol{F}^*$. Based on the definition of NE (Sect. 3.3), and given that player i can only modify $\boldsymbol{F}(\boldsymbol{i})$, these quantities constitute a NE of the merchant planning game if, for any feasible vector $\boldsymbol{F}(\boldsymbol{i})$, the following holds:

$$J\left(i, \boldsymbol{F}^*(\boldsymbol{i}), \boldsymbol{u}^*, \boldsymbol{F}^*\right) \geq J\left(i, \boldsymbol{F}(\boldsymbol{i}), \boldsymbol{u}^*, \boldsymbol{F}^*\right) \tag{24}$$

It is thus critical to analyze which values of $\boldsymbol{F}(\boldsymbol{i})$ maximize the surplus function (10) of player i for fixed values of $\boldsymbol{u}$ and $\boldsymbol{F}$. Note that this surplus function is linear with respect to each individual capacity addition $F_m(i)$ and can alternatively be written as:

$$J(i, \boldsymbol{F}(\boldsymbol{i}), \boldsymbol{u}, \boldsymbol{F}) = \sum_m \Lambda(u_m, F_m) \cdot F_m(i) \tag{25}$$

where the term $\Lambda(u_m, F_m)$ is expressed as:

$$\Lambda(u_m, F_m) = u_m \frac{\sum_t w_t \left(\lambda_{n_m^r,t} - \lambda_{n_m^s,t}\right) f_{m,t}}{F_m + F_m^0} - u_m \left(\frac{T_m^F}{F_m} + T_m^V\right) \tag{26}$$

Three different conditions need to be examined for each term of the sum in (25):

- $\Lambda(u_m, F_m) > 0$: the function J is monotonically increasing with respect to $F_m(i)$. It follows that the surplus of player i can always be increased by selecting a higher value of $F_m(i)$ and therefore a NE can never be reached.
- $\Lambda(u_m, F_m) < 0$: the function J is monotonically decreasing with respect to $F_m(i)$. Therefore, the surplus of player i can always be increased by choosing a lower value of $F_m(i)$. As a result, a NE could potentially be reached if and only if $F_m(i) = 0, \forall i$. This is never the case, as the mentioned conditions would lead to $u_m = 0$ and $\Lambda(u_m, F_m) = 0$, contradicting the initial hypothesis.
- $\Lambda(u_m, F_m) = 0$: the function J does not depend on $F_m(i)$. If this is true for all $m \in M$, (24) holds as equality and a NE is reached. It should be noted that, in this case, the marginal value [first term of (26)] and the marginal cost [second term of (26)] of an additional unit of network capacity investment by player i are equal.

Based on the three conditions examined above, the following result can be deduced:

Theorem 1 *The vectors* $\boldsymbol{F}^*(\boldsymbol{i}), \boldsymbol{u}^*$, *and* $\boldsymbol{F}^*$ *constitute a NE of the merchant planning game if and only if*:

$$u_m^* F_m^* \sum_t w_t \left(\lambda_{n_m^r,t} - \lambda_{n_m^s,t}\right) f_{m,t} = u_m^* \left(F_m^* + F_m^0\right)\left(T_m^F + T_m^V F_m^*\right), \forall m \tag{27}$$

Proof The three above conditions for $\Lambda(u_m, F_m)$ are considered. When $\Lambda(u_m, F_m) > 0$, we have established that a NE does not exist. This is consistent with the theorem statement, as (27) does not hold in this case. In fact, since $\Lambda(u_m, F_m) > 0$, the term in the left-hand side of (27) is strictly larger than the term in the right-hand side of (27). A similar procedure can be followed for the case $\Lambda(u_m, F_m) < 0$: having established that a NE is never reached, it is sufficient to note that the left-hand side of (27) is strictly smaller than its right-hand side. When $\Lambda(u_m, F_m) = 0$, it has been shown that a NE is reached and (27) always holds, thus concluding the proof.

Theorem 1 provides the necessary and sufficient conditions (27) for existence of a NE of the merchant planning game in an analytical fashion. However, as mentioned in Sect. 3.3, according to game-theory literature (Fudenberg and Tirole 1991), uniqueness of NE is generally not guaranteed. Therefore, it is possible that multiple different investment solutions fulfill (27). Since the focus of this work is not on identifying all possible NE of the merchant planning game but rather on investigating whether merchant planning can yield the same social welfare maximizing solution as centralized planning, we will seek for the NE solution yielding the largest social welfare.

Therefore, the optimization model we will employ for determining the merchant planning solution is formulated as follows:

$$\min_{\substack{u_m, F_m, f_{m,t} \\ G_{n,t}, D_{n,t}, p_{n,t}}, \forall m, \forall n, \forall t} \quad S = \sum_m T_m(u_m, F_m) + \sum_t \sum_n w_t \left[C_{n,t}(G_{n,t}) - B_{n,t}(D_{n,t})\right] \tag{28}$$

Subject to (3)–(9),

$$u_m F_m \sum_t w_t \left(\lambda_{n_m^r,t} - \lambda_{n_m^s,t}\right) f_{m,t} = u_m \left(F_m + F_m^0\right)\left(T_m^F + T_m^V F_m\right), \forall m \tag{29}$$

This problem is similar to the one determining the centralized planning solution (Sect. 2), but it also considers the NE condition (29) of Theorem 1, to be verified on each network branch.

4.3 Theoretical Comparison of Centralized and Merchant Planning Solutions

Given the analytical formulations of the optimization problems determining the centralized solution (CS) (Sect. 2) and the merchant solution (MS) under the participation of a "sufficiently large" number of players (Sect. 4.2), this Section aims at theoretically analyzing under which conditions the two solutions are identical. Based on our analysis, we claim that this equivalence holds if the following conditions are satisfied:

(A1) Fixed investment costs are neglected, i.e., $T_m^F = 0, \forall m$.
(A2) The network is radial and does not include any loops, i.e., $L = \emptyset$.

The sufficiency of the aforementioned conditions A1 and A2 is theoretically proved through Theorem 2 below. This theorem claims that if A1 and A2 hold, then the CS and MS coincide. In order to simplify the theoretical analysis, two auxiliary conditions are introduced:

(B1) The operational timescale of the planning problem includes a single period, i.e., $|T| = 1$.
(B2) The existing capacity of every branch is zero, i.e., $F_m^0 = 0, \forall m$.

Theorem 2 *The CS determined by problem* (1), (3)–(9) *and the MS determined by problem* (28), (3)–(9), (29) *coincide if conditions A1, A2, B1, and B2 hold.*

Proof Without loss of generality, it is assumed that the capacity addition of the CS is positive for all branches, i.e., $F_m > 0, \forall m$. If this is not the case for some branches, the following analysis can be performed on the subset of branches $\tilde{M} \subset M$ for which this assumption holds, i.e., $\tilde{M} = \{m \in M : F_m > 0, u_m = 1\}$. If this is not the case for any branch, i.e., $F_m = 0, u_m = 0, \forall m$, it can be shown that the CS and MS coincide as both sides of the NE conditions (29) are zero.

Given condition B1, the subscript t is omitted in the remainder of this proof. Under the current assumptions, a simplified expression can be derived for the problem (1), (3)–(9) determining the CS. Given condition A1, the investment cost term (2) in the objective function (1) can be rewritten as:

$$T_m = T_m^V F_m, \forall m \tag{30}$$

Regarding the constraints, (7) is omitted as a result of condition A2. Assuming without loss of generality a "positive" power flow on each branch (i.e., power flows from the reference sending node to the reference receiving node), we implicitly account for constraints (5) by imposing:

$$f_m = F_m, \forall m \tag{31}$$

These equations hold since (i) $f_m > F_m$ violates (5) given that $F_m^0 = 0$ from condition B2 and ii) $f_m < F_m$ is suboptimal as the unused capacity $F_m - f_m$ increases the objective function (1). As a result of the above, by combining (4) and (6) and by rewriting each of the (8) and (9) as two separate constraints, the problem determining the CS can be reformulated as:

$$\min_{\substack{F_m, \forall m \\ G_n, D_n, \forall n}} \quad S = \sum_m T_m^V F_m + \sum_n [C_n(G_n) - B_n(D_n)] \tag{32}$$

Subject to:

$$D_n + \sum_m \varphi_{n,m} F_m - G_n = 0 : \lambda_n, \forall n \tag{33}$$

$$-G_n \leq 0 : \mu_n^-, \forall n \tag{34}$$

$$G_n - G_n^{max} \leq 0 : \mu_n^+, \forall n \tag{35}$$

$$-D_n \leq 0 : \nu_n^-, \forall n \tag{36}$$

$$D_n - D_n^{max} \leq 0 : \nu_n^+, \forall n \tag{37}$$

The Lagrangian function associated with this optimization problem is expressed as:

$$\begin{aligned} L = & \sum_m T_m^V F_m + \sum_n [C_n(G_n) - B_n(D_n)] \\ & + \sum_n \lambda_n \left(D_n + \sum_m \varphi_{n,m} F_m - G_n \right) \\ & - \sum_n \mu_n^- G_n + \sum_n \mu_n^+ (G_n - G_n^{max}) \\ & - \sum_n \nu_n^- D_n + \sum_n \nu_n^+ (D_n - D_n^{max}) \end{aligned} \tag{38}$$

Derivation of the Lagrangian with respect to F_m yields the following set of necessary conditions for optimality:

$$\frac{\partial L}{\partial F_m} = T_m^V + \sum_n \varphi_{n,m} \lambda_n = 0, \forall m \tag{39}$$

The term $\varphi_{n,m}$ in (39) denotes the element in the nth row and mth column of the sensitivity matrix $\mathbf{\Phi}$, describing the network topology. For each column m of $\mathbf{\Phi}$, we have $\varphi_{n_m^s,m} = 1$ and $\varphi_{n_m^r,m} = -1$, while $\varphi_{n,m} = 0$ for all nodes n not connected to branch m. Therefore, (39) can be rewritten as:

$$T_m^V + \lambda_{n_m^s} - \lambda_{n_m^r} = 0, \forall m \tag{40}$$

Regarding the MS, as a result of conditions A1, B1, B2, and (31), the necessary and sufficient conditions (29) for achieving NE can be rewritten as:

$$\lambda_{n_m^r} - \lambda_{n_m^s} = T_m^V, \forall m \tag{41}$$

The optimality conditions (40) of the CS are equivalent to the NE conditions (41) of the MS. This implies that the CS and the MS coincide, concluding the proof.

It should be noted that although the above theoretical analysis considers the auxiliary simplifying hypotheses B1 and B2, the case studies presented in the following Sections will numerically demonstrate that B1 and B2 are not necessary. In other words, it will be shown that the CS and the MS coincide even when B1 and B2 do not hold. On the other hand, these case studies indicate that A1 and A2 are not only sufficient but also necessary: the CS and MS are in principle different when one of the conditions A1 or A2 does not hold.

The physical significance behind the sufficiency and necessity of conditions A1 and A2 is particularly interesting and is discussed below:

Condition A1: As demonstrated in Kirschen and Strbac (2004), under the CS, the total congestion revenue in the whole network covers exactly the variable component of the total investment cost but does not cover fixed costs. On the other hand, the NE condition (29) of the MS requires that the total congestion revenue covers exactly the total investment cost (i.e., both variable and fixed components), as the rational merchant transmission companies do not accept economic losses. Therefore, as demonstrated by the case studies in the following Sections, when fixed costs are accounted for, the total network capacity procured under the MS is lower than the respective capacity procured under the CS, in order to increase the collected congestion revenue and thus cover the fixed costs.

Condition A2: Under the CS, although the total congestion revenue in the whole network is equal to the variable component of the total investment cost, this equality does not necessarily hold for each individual network branch when the network is meshed; in such cases, some branches may generate higher congestion revenue than their variable investment cost, while other branches may generate lower congestion revenue, as demonstrated in Kirschen and Strbac (2004). On the other hand, the NE condition (29) of the MS requires that this equality holds on an individual branch basis, as demonstrated by the case studies in the following Sections. This requirement makes sense since the impact of each infinitesimal player's decisions on system conditions is negligible. As a result, each of these players assesses its decision for each branch individually, ignoring the impact of this decision on the congestion revenue associated with other branches; it will strive to increase its procured capacity on a branch m if the obtainable congestion revenue from m is higher than the required investment cost and decrease it if the obtainable congestion revenue from m is lower than the required investment cost. Therefore, as demonstrated by the following case studies, the CS and MS do not coincide when the network is meshed.

Table 11 Centralized and merchant planning solutions in 2-node system

	Case 1.1		Case 1.2	
	CS	MS	CS	MS
F_1 (MW)	800	800	800	690
Congestion revenue (£/h)	3200	3200	3200	5042
Investment cost (£/h)	3200	3200	5483	5042

4.4 Case Studies on 2-Node System

The 2-node system (Fig. 1) and the relevant assumptions considered in the case studies of this Section are the same with the ones considered in Sects. 3.4 and 3.5. The only difference is that two different cases regarding the transmission investment cost have been examined in this Section, with the respective CS and MS presented in Table 11.

Case 1.1: The investment cost includes only a variable component $T_1^V = 4£/\text{MWh}$ while fixed costs are neglected ($T_1^F = 0$). In this case, the CS and MS are identical and involve investment on a line of 800 MW (Table 11). This result follows from Theorem 2, as conditions A1, A2, B1, and B2 hold and therefore the CS and MS must coincide.

Case 1.2: The investment cost includes both a variable component $T_1^V = 4$ £/MWh and a fixed component $T_1^F = 2283$ £/h. The capacity procured under the CS does not change with respect to Case 1.1 and, as discussed in Sect. 4.3, the congestion revenue does not cover the full investment cost, due to the existence of fixed costs. On the other hand, the capacity procured under the MS is now reduced to 690 MW, to ensure that the congestion revenue covers the full investment cost (Sect. 4.3). This result suggests that condition A1 (zero fixed investment costs) is a necessary condition for the CS and MS to coincide.

4.5 Case Studies on 3-Node System

The considered 3-node system along with its relevant generation and demand data (Kirschen and Strbac 2004) is illustrated in Fig. 5. It is assumed that the existing capacity of the three branches is zero, their investment costs are equal and their reactances after any capacity addition are equal. The operational timescale of the planning problem includes two time periods with weighting factors $w_1 = 0.25$ and $w_2 = 0.75$. Generation costs are assumed to be quadratic functions of the respective power productions and demands are assumed inelastic. Four different cases have been examined, with the respective CS and MS presented in Table 12.

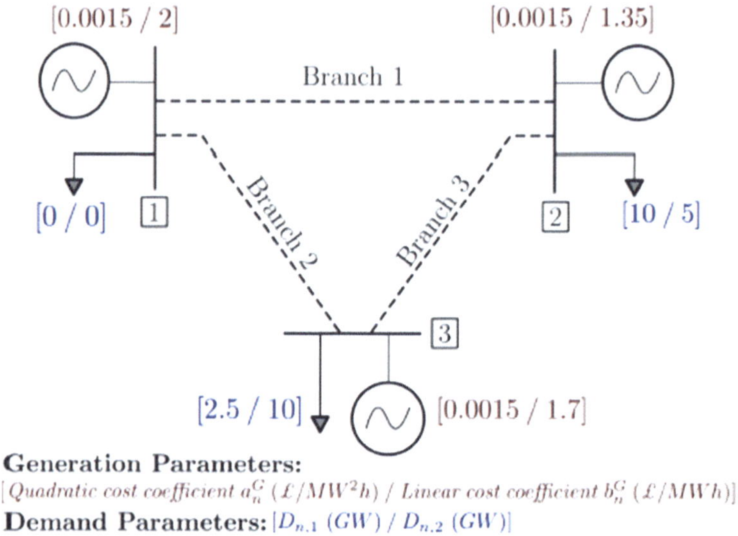

Fig. 5 Topology and parameters of 3-node system

Case 2.1: The investment cost includes only a variable component $T_m^V = 3.42$ £/MWh, $\forall m$, while fixed costs are neglected. In contrast to Case 1.1, the CS and MS are different (Table 12). As discussed in Sect. 4.3, while the equality between congestion revenue and investment cost holds for each individual network branch under the MS, the same does not hold under the CS. This result suggests that condition A2 (no network loops) and/or condition B1 (single period in the operational timescale) is/are necessary condition(s) for the CS and MS to coincide.

Case 2.2: In order to investigate which of the conditions A2 and B1 is critical for the equivalence between the CS and MS, we consider a case where capacity can be added only on branches 1 and 2, imposing $F_3 = 0$ in the two optimization problems. All the other parameters remain the same as in Case 2.1. In this scenario, the CS and MS are identical (Table 12). This suggests that A2 is a necessary condition for the CS and the MS to coincide, since in this scenario the network is radial and does not include loops. On the other hand, it also demonstrates that condition B1 is not necessary for the CS and MS to coincide.

Case 2.3: In order to further explore this interesting result, we consider a theoretical scenario where capacity can be added on all three branches but *Kirchhoff's voltage law* (KVL), expressed through (7), is neglected in both optimization problems. All the other parameters remain the same as in Case 2.1. In this theoretical scenario, the CS and MS are again identical (Table 12). This result suggests that the physical reason behind the necessity of condition A2 lies in the unavoidable consideration of the KVL in meshed networks. As already noted for Case 2.2, it seems that condition B1 is not necessary for the equivalence between the CS and MS.

Table 12 Centralized and merchant planning solutions in 3-node system

	Case 2.1		Case 2.2		Case 2.3		Case 2.4	
	CS	MS	CS	MS	CS	MS	CS	MS
F_1 (MW)	1963	2193	2678	2678	2044	2044	2044	1686
F_2 (MW)	2887	2808	3991	3991	2089	2089	2089	2211
F_3 (MW)	1387	1609	0	0	2156	2156	2156	1811
Congestion revenue—Branch 1 (£/h)	6690	7511	9172	9172	7002	7002	7002	8057
Investment cost—Branch 1 (£/h)	6723	7511	9172	9172	7002	7002	9285	8057
Congestion revenue—Branch 2 (£/h)	8337	9615	13,667	13667	7154	7154	7154	9855
Investment cost—Branch 2 (£/h)	9887	9615	13,667	13,667	7154	7154	9437	9855
Congestion revenue—Branch 3 (£/h)	6333	5510	–	–	7382	7382	7382	8485
Investment cost—Branch 3 (£/h)	4750	5510	–	–	7382	7382	9665	8485
Congestion revenue—Total (£/h)	21,360	22,636	22,839	22,839	21,538	21,538	21,538	26,397
Investment cost—Total (£/h)	21,360	22,636	22,839	22,839	21,538	21,538	28,387	26,397

Case 2.4: The KVL is neglected as in Case 2.3 but the investment cost also includes a fixed component $T_m^F = 2283$ £/h, $\forall m$. In contrast to Case 2.3, the CS and the MS do not coincide. As discussed in Sect. 4.3, while the total congestion revenue covers the total investment cost under the MS, the same does not hold under the CS due to the existence of fixed costs. Like in Case 1.2, this result suggests that A1 is necessary for the CS and MS to coincide.

4.6 *Case Studies on IEEE 24-Node System*

Although the case studies of Sects. 4.4 and 4.5 validate the theoretical analysis of Sect. 4.3, demonstrating the criticality of conditions A1 and A2 for the equivalence between the CS and MS, they are carried out on very simple 2-node and 3-node systems, respectively. In order to establish that these insights are still valid in larger, more realistic systems, the IEEE 24-node system is examined in this Section. This system along with its relevant network, generation and demand data (Conejo et al. 2010), is illustrated in Fig. 6. All lines represent existing branches that can be expanded. The operational timescale of the planning problem includes a single time period. Generation costs and demand benefits are assumed to be quadratic functions of the power productions and consumptions, respectively. Three different cases have been examined, with the respective CS and MS presented in Table 13. For compactness reasons, branches with zero capacity additions in all three cases have been omitted from Table 13.

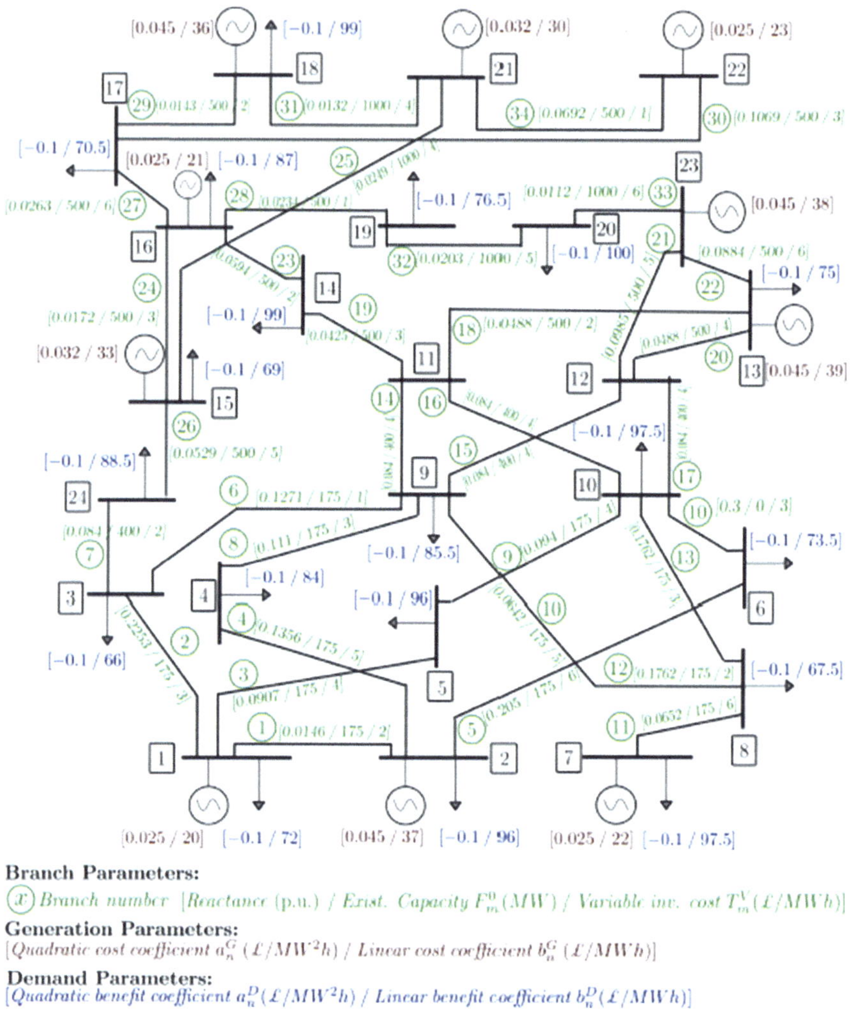

Fig. 6 Topology and parameters of IEEE 24-node system

Table 13 Centralized and merchant planning solutions in IEEE 24-node system

	Case 3.1		Case 3.2		Case 3.3	
	CS	MS	CS	MS	CS	MS
F_1 (MW)	322.02	323.03	241.53	241.53	241.53	246.39
F_3 (MW)	185.06	179.81	93.91	93.91	93.91	88.39
F_{11} (MW)	140.21	144.91	146.45	146.45	146.45	142.36
F_{23} (MW)	311.42	310.99	117.54	117.54	117.54	82.12
F_{26} (MW)	209.62	202.36	0	0	0	0
F_{28} (MW)	234.70	240.27	548.91	548.91	548.91	567.43
F_{34} (MW)	4.33	0	0	0	0	0
Congestion revenue (£/h)	3928	4108	2521	2521	2521	2682
Investment cost (£/h)	4135	4108	2521	2521	2771	2682

Case 3.1: The investment cost includes only a variable component (presented in Fig. 6 for each branch), while fixed costs are neglected. As in Case 2.1, the CS and MS are different (Table 13), suggesting that condition A2 (no network loops) and/or condition B2 (zero existing capacity on every branch) is/are necessary condition(s) for the CS and MS to coincide.

Case 3.2: In order to investigate which of the conditions A2 and B2 is critical for the equivalence between the CS and MS, following the rationale of Case 2.3, we consider a theoretical scenario where the KVL is neglected. As in Case 2.3, the CS and MS are identical (Table 13). This result again supports the idea that condition A2 is necessary for the CS and MS to coincide. On the other hand, it also demonstrates that condition B2 is not necessary for the CS and MS to coincide.

Case 3.3: The KVL is neglected as in Case 3.2 but the investment cost also includes a fixed component $T_m^F = 50\,£/\text{h}, \forall m$. In contrast to Case 3.2, the CS and the MS are different. As in Cases 1.2 and 2.4, this suggests that condition A1 is necessary for the CS and MS to coincide.

5 Conclusions and Future Work

Merchant transmission investment planning has recently emerged as a promising alternative or complement to the traditional centralized planning paradigm and it is considered as a further step toward the deregulation and liberalization of the electricity industry. However, its widespread application requires addressing two fundamental research questions: which entities are likely to undertake merchant transmission investments and whether this planning paradigm can maximize social welfare as the traditional centralized paradigm. Unfortunately, previously proposed approaches to quantitatively model this new planning paradigm do not comprehensively capture the strategic behavior and decision-making interactions between multiple merchant investors.

This Chapter has proposed a novel non-cooperative game-theoretic modeling framework to capture these realistic aspects of merchant transmission investments and provide insightful answers to the above research questions. More specifically, two different models, both based on non-cooperative game theory, have been developed.

The first model adopts an equilibrium programming approach. The decision-making problem of each merchant investing player is formulated as a bi-level optimization problem, accounting for the impacts of its own actions on locational marginal prices (LMP) as well as the actions of all competing players. This bi-level problem has been formulated for different types of players that can potentially participate in merchant investments (merchant transmission companies, generation companies, and demand companies) and solved after converting it to a mathematical program with equilibrium constraints (MPEC). An iterative diagonalization method is employed to search for the likely outcome of the strategic interactions between multiple players, i.e., Nash equilibria (NE) of the game.

Case studies on a simple 2-node system have provided the following answers to the above research questions:

(i) Which entities are likely to undertake network investments under the merchant planning paradigm?

Networks investments will be mostly undertaken by generation companies in areas with low LMP and demand companies in areas with high LMP *(higher-motivated players)*, as apart from collecting congestion revenue they also increase their energy surpluses. Merchant transmission companies, generation companies in areas with high LMP and demand companies in areas with low LMP *(lower-motivated players)* could also be motivated to invest by the collection of congestion revenue under certain circumstances. Case studies have illustrated the interdependencies between the different players' decisions; in certain cases, the large network capacity desired by higher-motivated players reduces the obtainable congestion revenue by lower-motivated players and thus prevents the latter from investing in capacity.

(ii) Is the merchant planning paradigm able to achieve the same (maximum) social welfare as the traditional centralized planning approach?

The merchant planning solution approaches the centralized one as the number of competing players increases. The largest deviations from the centralized solution are observed in the case where the set of participating players includes only merchant transmission companies, as they procure significantly lower capacity in order to increase their profits through higher LMP differentials.

However, because of its iterative nature, this first model cannot guarantee convergence to existing NE, especially as the number of players and the size of the network increase; as a result, the examined case studies are limited to a 2-node system with up to 10 players. In other words, although this model provides insightful answers to the first question, it cannot establish whether the merchant planning solution yields the same solution as centralized planning under the participation of a "sufficiently large" number of competing investors, as it cannot deal with a large number of players, especially in large networks.

In order to address this challenge and provide insightful answers to this second research question, a second model has been developed, where the set of merchant investors is approximated as a continuum. The proposed approximation makes the impact of each infinitesimal player's decisions on system quantities negligible, allowing us to derive mathematical conditions for the existence of a NE solution in an analytical fashion.

Based on this model, we have performed an analytical comparison of the merchant planning solution under the participation of a "sufficiently large" number of competing investors against the one obtained through the traditional centralized paradigm, as well as a numerical comparison through case studies on a 2-node, a 3-node, and a 24-node system. These comparisons have demonstrated that merchant planning can achieve the same (maximum) social welfare as the centralized planning approach only when the following conditions are satisfied:

(a) fixed investment costs are neglected, and
(b) the network is radial and does not include any loops.

As these conditions do not generally hold in reality, our findings suggest that even a fully competitive merchant transmission planning framework, involving the participation of a very large number of competing merchant investors, is not generally capable of maximizing social welfare, as implied by previous work.

This conclusion implies that some sort of regulatory interventions will be required to align the outcome of merchant transmission investment planning with social optimality. However, these interventions need to remain at a minimum level, in line with the vision of deregulation. The analytical design of such regulatory measures constitutes a significant challenge for future research.

Furthermore, the two models of merchant planning developed in this Chapter—as well as the rest of the models in the relevant literature—assume a fixed generation mix and do not consider generation expansion decisions. In reality, however, transmission and generation expansion decisions are interdependent. In this context, future work aims at developing an integrated transmission and generation planning framework and comparing the impacts of centralized and merchant transmission planning on generation expansion decisions.

References

R.J. Aumann, Markets with a continuum of traders. Econometrica **32**(1/2), 39–50 (1964)

R.J. Aumann, Existence of competitive equilibria in markets with a continuum of traders. Econometrica **34**(1), 1–17 (1966)

M.S. Bazaraa, H.D. Sherali, C.M. Chetty, *Nonlinear Programming: Theory and Algorithms* (Wiley, Hoboken, 2006)

J.B. Bushnell, S.E. Stoft, Electric grid investment under a contract network regime. J. Regul. Econ. **10**(1), 61–79 (1996)
J.B. Bushnell, S.E. Stoft, Improving private incentives for electric grid investment. Resource Energy Econ. **19**(1–2), 85–108 (1997)
A.J. Conejo, M. Carrion, J.M. Morales, *Decision Making Under Uncertainty in Electricity Markets* (Springer, Berlin, 2010)
R. Couillet, S.M. Perlaza, H. Tembine, M. Debbah, Electrical vehicles in the smart grid: a mean field game analysis. IEEE J. Sel. Areas Comm. **30**(6), 1086–1096 (2012)
A. De Paola, D. Angeli, G. Strbac, Distributed control of microstorage devices with mean field games. IEEE Trans. Smart Grid **7**(2), 1119–1127 (2016)
DUKE American Transmission Co, *Zephyr Power Transmission Project a Transmission Project to Connect Wyoming Wind to the Southwestern U.S.* (2017). Available: http://www.datcllc.com/projects/zephyr/
Electric Light & Power, *Arizona Regulators Approve $2 billion SunZia Project in Close Vote* (2016). Available: http://www.elp.com/articles/2016/02/arizona-regulators-approve-2-billion-sunzia-project-in-close-vote.html
Federal Energy Regulatory Commission, *Remedying Undue Discrimination Through Open Access Transmission Service and Standard Electricity Market Design* (2002). Available: https://www.gpo.gov/fdsys/pkg/FR-2002–08-29/pdf/02-21479.pdf
D. Fudenberg, J. Tirole, *Game Theory* (MIT Press, 1991)
H.A. Gil, E.L. da Silva, F.D. Galiana, Modeling competition in transmission expansion. IEEE Trans. Power Syst. **17**(4), 1043–1049 (2002)
B.F. Hobbs, C. Metzler, J.S. Pang, Strategic gaming analysis for electric power systems: an MPEC approach. IEEE Trans. Power Syst. **15**(2), 638–645 (2000)
P. Joskow, J. Tirole, Merchant transmission investment. J. Industr. Econ. **53**(2), 233–264 (2005)
M.A. Khan, Y. Sun, Non-cooperative games with many players, in *Handbook of Game Theory*, ed. by R. Aumann, S. Hart (Elsevier Science, 2002), vol. 3, pp. 1761–1808
D. Kirschen, G. Strbac, *Fundamentals of Power System Economics* (Wiley, Hoboken, 2004)
G. Latorre, R.D. Cruz, J.M. Areiza, A. Villegas, Classification of publications and models on transmission expansion planning. IEEE Trans. Power Syst. **18**(2), 938–946 (2003)
D.A. Lima, A. Padilha-Feltrin, J. Contreras, An overview on network cost allocation methods. Electr. Power Syst. Res. **79**(5), 750–758 (2009)
S. Littlechild, *Transmission Regulation, Merchant Investment, and the Experience of SNI and Murraylink in the Australian National Electricity Market* (Electricity Policy Research Group, University of Cambridge, 2003). Available: https://sites.hks.harvard.edu/hepg/Papers/Littlechild.Transmission.Regulation.Australia.pdf
J.D. Molina, H. Rudnick, Transmission expansion investment: cooperative or non-cooperative game?, in *IEEE Power Energy Soc. Gen. Meet.*, 2010, pp. 1–7
J.-S. Pang, D. Chan, Iterative methods for variational and complementarity problems. Math. Program. **24**(1), 284–313 (1982)
C. Ruiz, A.J. Conejo, Pool strategy of a producer with endogenous formation of locational marginal prices. IEEE Trans. Power Syst. **24**(4), 1855–1866 (2009)
E.E. Sauma, S.S. Oren, Economic criteria for planning transmission investment in restructured electricity markets. IEEE Trans. Power Syst. **22**(4), 1394–1405 (2007)
G.B. Shrestha, P.A.J. Fonseka, Optimal transmission expansion under different market structures. IET Gen. Transm. Distr. **1**(5), 697–706 (2007)
StarWood Energy Group, *Neptune Transmission System, Financed by Energy Investors Fund and Star-Wood Energy Group, Begins Delivering Power to Long Island* (2007). Available: http://starwoodenergygroup.com/wpcontent/uploads/2014/06/6 NeptuneAnnouncement.pdf
G. Strbac, M. Pollitt, C. Vasilakos Konstantinidis, I. Konstantelos, R. Moreno, D. Newbery, R. Green, Electricity transmission arrangements in Great Britain: time for change? Energy Pol. **73**, 298–311 (2014)
J.D. Weber, T.J. Overbye, An individual welfare maximization algorithm for electricity markets. IEEE Trans. Power Syst. **17**(3), 590–596 (2002)

Transmission Investment Coordination and Smart Grid

Transmission Investment and Renewable Integration

Hugh Rudnick and Constantin Velásquez

1 Introduction: Conditions of Integration of Renewables in Modern Power Markets

The world needs renewables, and renewables need transmission. Renewable electricity generation is a key element for the much-needed transition to low-carbon economies of the future. According to the IPCC, renewables are projected to supply 70–85% (interquartile range) of electricity in 2050 across the future pathways for limiting the hazards of climate change (IPCC 2018). Cleaner power generation enables effective climate change abatement through electrification of other fossil fuel reliant sectors, such as heating and transportation. Electrification would lead to more carbon emissions if the power sector remains heavily reliant on fossil fuel generation.

The world has witnessed a tremendous growth of variable renewable energy (VRE) generation over the past decade. Indeed, renewable energy generation accounted for 9.3% of global power generation in 2018, up from only 3% a decade ago (BP 2019). Initially driven by support mechanisms and subsidies, pure economics and evolving market-based regulations are now driving the growth of VREs around the world, partly due to the plunge in investment costs over recent years (IEA 2018; Kavlak et al. 2018). For the near future, the IEA expects a 46% growth of renewable power capacity between 2018 and 2023 in its main case forecast. This expansion would be mostly in VRE resources, with more than half coming from solar PV generation, and wind remaining the second-largest contributor (IEA 2018). Partly due to the recent growth of renewables, the power sector is often depicted as a success for

H. Rudnick (✉) · C. Velásquez
Pontificia Universidad Católica de Chile, Santiago, Chile
e-mail: hrudnick@ing.puc.cl

C. Velásquez
e-mail: csvelasq@puc.cl

M. R. Hesamzadeh et al. (eds.), *Transmission Network Investment in Liberalized Power Markets*, Lecture Notes in Energy 79,
https://doi.org/10.1007/978-3-030-47929-9_14

decarbonization of energy and sometimes as the only energy sector for which the future path seems clear (compared to heat and transportation).

Despite the incredible growth of VREs over the past decade, a challenging path lies ahead for the power sector. The world is still highly reliant on fossil fuels for power generation. As of 2018, coal remains the dominant worldwide fuel for power generation with 38%, the same share as 20 years ago (BP 2019). Increased and sustained efforts are thus clearly needed to accelerate and deepen the integration of VREs, to quickly lower carbon emissions of the electricity sector.

The disruption of renewable generation poses new challenges and opportunities for the transmission system, both from the system and the investor perspective.

- Variable and uncertain generation profiles of VRE require flexibility in operations, planning and regulation of the power sector. New transmission capacity is a primary source of flexibility for the power system. Adequate transmission capacity and flexible operation procedures (such as transmission line switching) allow for sharing of the most economic and flexible resources across the power system, key for a secure operation under varying flow patterns.
- Unlike coal and gas, wind and sun cannot be transported to more convenient locations. Renewable power plants need to be located in resource-rich regions. Regions with high-quality renewable resources are often far away from load centers. Therefore, transmission infrastructure is needed to transport the electricity from renewable power plants to the main grid and to final customers.
- Smaller power plants require flexible expansion and smooth coordination of investments across the generation, transmission and distribution segments. Renewable power plants are often much smaller than fossil-fueled and hydro-dams power plants. Moreover, large VRE projects (more than 300 MW) can be developed quickly and flexibly in small incremental stages, given the modular nature of wind turbines and solar PV panels. Smaller power plants require lower direct investments, shorter construction times and are often widely dispersed geographically. Therefore, common transmission infrastructure might be beneficial to economically harness the potentials of renewable generation hubs. Moreover, coordination is needed between transmission and distribution investments and operations, given the increasing role of distributed generation and other flexible resources such as storage.

Therefore, rapidly achieving high shares of renewable energy in the electricity generation mix requires timely and efficient development of the transmission system. Such development includes transmission investment in both the main grid and in locations with high renewable generation potential. New challenges for planning, pricing and regulation of the transmission system are arising due to the disruption of VRE generation, which is rapidly re-shaping power systems.

This chapter highlights some of these challenges associated to transmission investments needed for integration of renewable generation, as well as the approaches to deal with these issues. Some of the issues and lessons analyzed in the chapter draw on the Chilean experience (Velásquez 2017; Watts and Rudnick 2014). Chile has seen record growth of VREs in recent years, increasing the share of electricity generation from wind and solar renewables from 0% in 2006 to 5% in 2015 and 12% in 2018 (8.7 TWh). Renewables were initially driven by quota mechanisms which aimed at 10% of renewable generation participation by 2024. Although a later law increased the target renewable share, it is now clear that these targets will be vastly surpassed and far in advance, due to the explosive economic-driven growth of renewables.

The explosive growth of renewables required significant transmission investments, and massive new projects will also be needed in the future. Chile is a long country with widely dispersed renewable resources. Indeed, the renewable potential considered for the Ministry's strategic long-term planning highlights the availability of high-quality solar generation potential concentrated in the north, in contrast with hydro- and wind generation potential in the south (see Fig. 1a). Given high concentration of demand in the center zone where the capital is located, transmission investments to harness the renewable potential widespread across Chile will be significant in the medium- and long-term (see Fig. 1b, c). This trend will intensify given the recently announced plan to decommission all coal-fired power plants by 2040. Coal generation accounted for 38% of total power generation in Chile during 2018, and 872 MW of coal-fired capacity are located in the center zone. The first decommissioning stage comprises 1,047 MW of generation capacity (322 MW of which are located in the center) by 2025, which will be replaced by renewable projects located in the far-north (mostly solar PV) and the south (wind farms).

Recent experiences in Chile highlight the complexities of the planning, permitting and siting process for new transmission systems. Commissioning of a major 500 kV line between the center and the north suffered a 17-month delay (further discussed below). This experience raises concerns regarding the timely development of future expansion, such as the planned HVDC line between the north and the center, much needed for harnessing the solar generation potential. Complexities may result in delays and a development time between 10 and 15 years for such a large HVDC transmission project. Moreover, uncertainty in the expected location of future renewable generation may result in excess transmission toward the north and insufficient transmission toward the south.

Renewable investors in Chile have already suffered the impacts of inadequate transmission capacity and delays of important transmission projects. Figure 2 depicts the evolution since 2017 of hourly locational spot prices in selected nodes of the northern and center zones. Before November 2017, the Chilean power market was composed of two independent power grids: the SING in the north, mainly composed by large mining companies and coal-fired generation, and the SIC in the center and south zones, with a mix of industrial, commercial and household customers as well as thermal and hydro-generation. Over the past decade, renewable generation projects in the northern SIC zone were built far more quickly than the required transmission infrastructure. Price decoupling reveals that transmission congestions between the northern and center SIC provoked curtailment of wind and solar PV generation during daylight hours. The lack of timely transmission capacity meant significant foregone revenues for renewable generators due to both curtailed electricity production and lower spot prices.[1]

[1]The lack of operational flexibility is also a significant contributor to renewable generation curtailment, given the high shares of inflexible coal-fired capacity in the north, as well as gas-fired plants

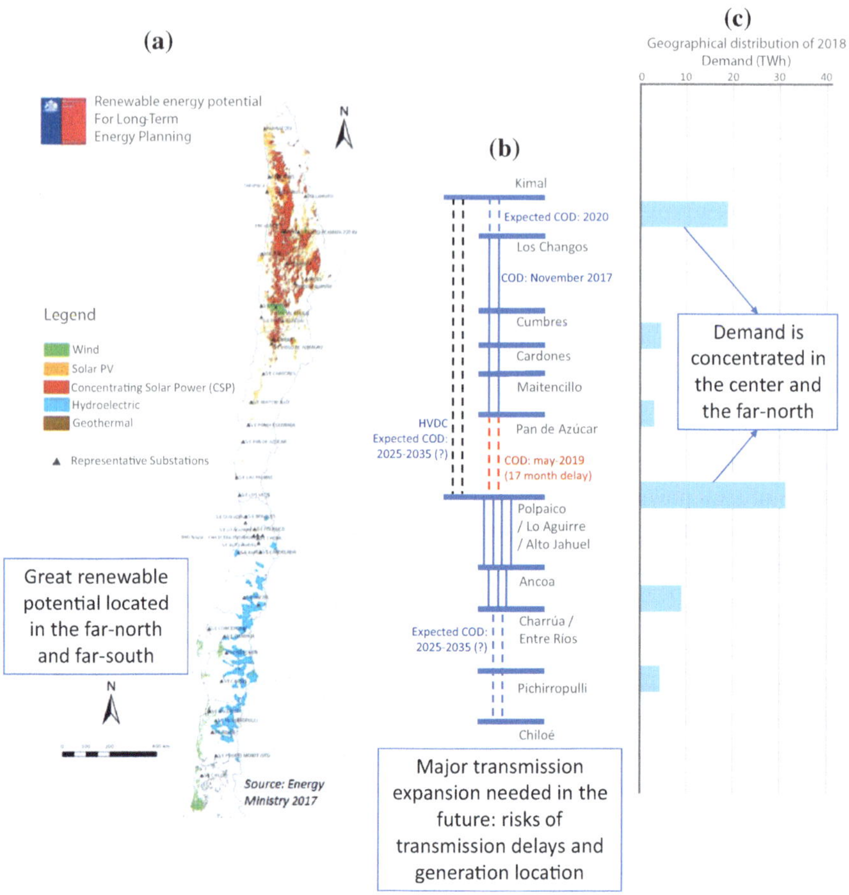

Fig. 1 Location of renewable energy potential across Chile (**a**) main transmission system (**b**) and geographical distribution of demand (**c**). The chart presents a simplified depiction of the main transmission system (500 kV and HVDC) and the geographical distribution of yearly energy demand for 2018, for illustration purposes only. Several substations were grouped or omitted for presentation purposes. HVAC transformers and HVAC/HVDC transformation omitted for illustration. Different zones are grouped for the geographical demand distribution. Commercial Operation Date (COD) provided for new transmission lines is referential, based on the author's experience, to illustrate the potential challenges of new transmission projects. *Source* Own, based on Energy Ministry (2017) and data published by National Electric Coordinator (CEN)

Curtailment fell drastically following the commissioning of the transmission interconnection project between the SIC and the SING in late November 2017, leading to

---

with inflexible LNG supply agreements which are given priority dispatch. However, the single most relevant contributing factor to renewable curtailment in Chile is the lack of adequate transmission capacity.

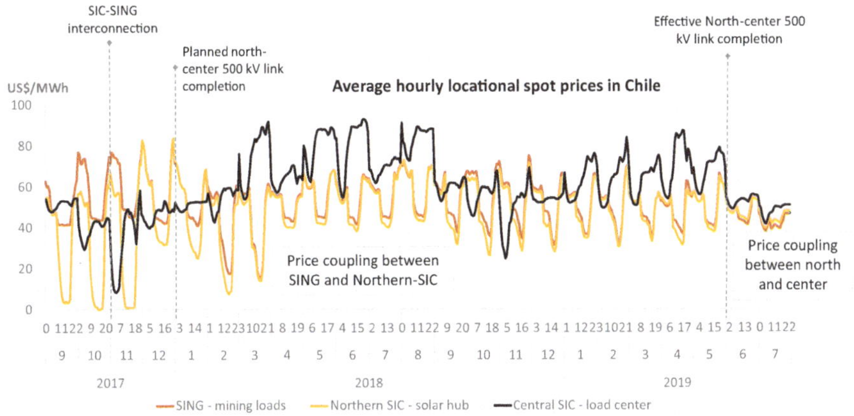

Fig. 2 Average hourly locational spot prices in Chile. The marginal cost of three representative busbars is presented to illustrate price difference across geographic regions of center–northern Chile. SING is represented by Crucero 220 kV, northern SIC by Cardones 220 kV and center SIC by Alto Jahuel 220 kV. For each month, the arithmetic average of hourly marginal costs was calculated across all days of the month. Planned and effective completion dates for transmission expansions are depicted at hour 0 of each month, given that no daily resolution is presented. *Source* Own, based on ISO data (the National Electrical Coordinator, or CEN for its Spanish acronym)

increased price coupling between the SING and the northern SIC. However, decoupling persisted between these northern zones and the center SIC. Although such decoupling would have been solved from February 2018 onwards due to the planned commissioning of a 500 kV link between the north and the center (connected only through 220 kV lines), the last tranche of the 500 kV link was delayed for 17 months, partly due to the complicated right-of-way negotiation process and intense public opposition. Commissioning of the full 500 kV link occurred in May 2019, leading to price coupling across the north and center zones of the interconnected grid in recent months. The prolonged delay of this project meant sustained foregone revenues for renewable generators, mostly due to lower-than-expected prices, rather than curtailment.

To tackle the challenges brought about by the disruption of renewables, as well as wider weaknesses in transmission regulation, the Chilean regulator conducted over the course of two years (2014–2016) a widely participatory process to develop a new legal framework for transmission expansion and operation. The new transmission law was enacted in June 2016, introducing deep reforms to expansion and operation of transmission systems. The key elements of the Chilean transmission law related to the accommodation of renewables can be summarized as follows (Ferreira et al. 2016):

- Governance of power system operations and interconnection procedures was strengthened through increased independence of the ISO. Moreover, the two previously independent ISOs were merged in a single ISO for the interconnected national grid.

- Beyond reliability and least-cost production, a wider set of benefits must be explored to justify transmission expansion plans, including competitive and resiliency benefits.
- Transmission planning must address long-term uncertainty through scenario analysis, incorporating a variety of long-term visions of the evolution of the energy sector to guide transmission planning.
- The planner is explicitly granted the ability to consider spare transmission capacity for possible future developments of supply and demand fundamentals.
- Transmission for renewable hubs can now be proactively developed through centrally planned investment in the required transmission infrastructure. Renewable hubs are conceived as zones with high potential of renewable generation and relatively far from the existing transmission networks, for which proactive transmission expansion may be required to harness the full renewable potential.
- The cost allocation method, based on locational signals and cost sharing between generators and demand, was simplified by transferring costs of the main grid to demand through a simple postage stamp method (with no locational signal), through a 15-year transition period for transferring these costs.
- Some responsibilities of the siting process were transferred from transmission concessionaire companies to the state. While routing of new transmission projects was previously the responsibility of the concessionaire, for complex transmission projects, the authority must now conduct a strip study to determine the spatial route in which the project must be developed, considering a variety of environmental and societal criteria. The resulting strip will be subject to a strategic assessment process and the approval of the Council of Ministries.

The foundation of these reforms ranges from theoretically sound arguments to primarily practical considerations. These theoretical and practical foundations, as well implementation issues that have already emerged, will be further discussed in this chapter.

The rest of the chapter is structured as follows. Section 2 overviews planning and expansion of the main transmission grid to accommodate high levels of renewable generation, from scenario generation to planning studies and the difficulties with approval and siting of new projects. Section 3 analyzes the alternatives for efficiently harnessing renewable generation hubs, discussing the economics of transmission and generation coordination. Section 4 concludes this chapter.

2 The Backbone for Low-Carbon Power Systems: Developing Transmission Grids for High Levels of Renewable Generation

Developing the transmission grid that renewables require is no easy task. At the system level, planning the optimal transmission network is a highly complex engineering and regulatory challenge, which requires dealing with uncertainty and

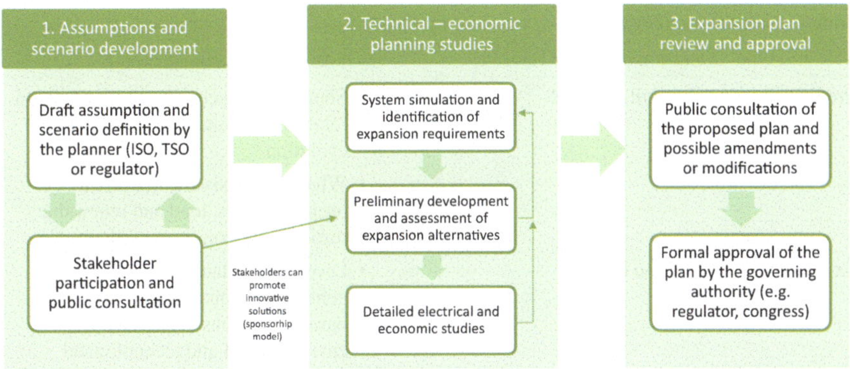

Fig. 3 Generic transmission expansion planning process in practice. *Source* Watts and Rudnick (2014)

multiple objectives (as previously discussed). Moreover, once transmission expansions are defined, their cost must be allocated among market participants, and the works must be financed and successfully completed within reasonable times. However, the development of new transmission lines has become more challenging due to growing environmental and social concerns, yielding longer and more uncertain lead times for new transmission projects due to siting and permitting difficulties. This section will discuss some of the issues related to planning, allocating costs and executing much-needed transmission expansions for the renewable scale-up.

Transmission planning in practice is developed in three stages: assumptions, technical–economic studies and approval process, as depicted in Fig. 3 (Watts and Rudnick 2014). Each of these stages has its own set of challenges, for which a variety of possible solutions have emerged worldwide, as further discussed below.

2.1 Scenario Generation for Transmission Planning

In the first stage of the transmission planning process, key assumptions and multiples scenarios are developed and agreed upon by the planner and the stakeholders. The precise definition of assumptions and scenarios shapes the results of the planning process. Therefore, this is a key early participation tool for stakeholders, which they can use to express their interests and expectations of the process, whether it is profit maximization (for generation companies) or sustainability (for communities). Therefore, this process often entails extensive stakeholder participation and public consultation.

Scenarios can be broadly classified under three categories: predictive, explorative and normative (Börjeson et al. 2006). Each scenario type attempts to answer a different kind of question about the future, and thus, different examples exist for transmission planning (see Table 1). The motivation and generation techniques for

Table 1 Scenario typology

Scenario type	Question about the future	Transmission planning example
Predictive	What will happen?	• Short-term baseline scenario based on relatively certain supply and demand evolution
		• What-if analysis and sensitivities of projects delays, load and renewable resource pattern, among others
Explorative	What can happen?	• Diverse and plausible scenarios, generated by quantitative models based on various economic, environmental and technological assumptions
		• Higher and lower estimates of renewable generation integration, demand growth and other key uncertainties
Normative	How can a specific target be reached?	• Goals for renewable generation and fuel diversification
		• Envisioned energy mix in the long-term

Source Own, based on Börjeson et al. (2006)

each of these three scenario types are outlined below, given the diversity of transmission planning approaches and scenario techniques employed across different countries.

Predictive scenarios which attempt to forecast future conditions have historically been the basis for transmission planning, specially under vertically integrated utilities. These scenarios can simplify the representation of uncertainties of relatively low complexity, such as **k**nown and **u**nknown uncertainties (Diebold et al. 2010; Gomory 1995; Velasquez et al. 2016). Predictive or case-driven scenarios are often sensitivities or limited deviations from base case assumptions. Case-driven scenarios describe many possible combinations of outcomes of some set of uncertainties such as winter/ summer peak, generation expansion or load growth rates (Bustamante-Cedeño and Arora 2008; Buygi et al. 2006; Gorenstin et al. 1993; Mejia-Giraldo and McCalley 2014; PJM 2015a).

As previously mentioned, power system operations and planning in the vertically integrated regime have been historically driven by least-cost engineering analyses and computer simulation models (Stoll 1989). Industry restructuring and the introduction of competition make economics and value-based transmission become more important (Buygi et al. 2004; Kirschen and Strbac 2005; Oliveira et al. 2007). This trend has led many countries to devise transmission planning processes primarily around theoretically sound quantitative models for generating and analyzing predictive scenarios. For example, this has been historically the case of PJM's reliability and market efficiency studies for transmission planning.

As the uncertainty and complexity of the power market increase with more competitors and new renewable technologies, **explorative and normative scenarios** become more important for the medium- and long-term. As PJM puts it, for the first ten years of its transmission planning process, uncertainties were limited, and a single set of assumptions was enough for reliability and market efficiency planning. However, market and policy developments in more recent years required PJM to undertake scenario planning (PJM 2017).

Explorative scenarios can be used to represent a broad spectrum of plausible future evolutions of the power system, such as different levels of policy-driven renewable generation. These scenarios ensure internal consistency by analyzing interactions among several uncertainties, and selecting between three and six scenarios to broadly represent plausible uncertainty realizations (Gu and McCalley 2010; Linares 2002; Munoz et al. 2014; National Grid 2015a, b; PJM 2015b; Sanchis et al. 2015; van der Weijde and Hobbs 2012).

In turn, normative scenarios can be used to guide the planning process toward a strategic long-term vision (often the government's vision), such as resource adequacy levels or fuel diversification goals (e.g., for heavily hydro-reliant countries such as Colombia and Peru). These normative scenarios often portray the authority's vision for the future of the power sector. While normative scenarios can be part of the planning process, the set of considered scenarios should also be diverse to represent a wide range of possible futures (Schoemaker 1993).

Contrasting with the quantitative approach for generating predictive scenarios, explorative and normative scenarios are often built by the intuitive or qualitative approach (van Notten et al. 2003). The intuitive approach conceives scenarios as a mean to bound, understand and communicate uncertainty, rather than accurately predicting or forecasting future outcome (Bradfield et al. 2005; Myers and Kitsuse 2000; van der Weijde and Hobbs 2012; Watts and Rudnick 2014). Each scenario should present a trajectory to some future state in a narrative and compelling fashion, outlining the interaction between the most important uncertainties in an internally consistent manner.

The process for strategic scenario definition should be designed for building stakeholder consensus while promoting variety of outcomes and incorporating policy guidance. Various qualitative techniques for intuitive scenario generation have been developed, including surveys, workshops, the think-tank model—back-office scenario development by team of experts—and the Delphi method—based on multiple rounds of expert panel questionnaires (Börjeson et al. 2006). These and other techniques, integrated in a strategic scenario generation process with participation from different individuals (whether experts or not), can achieve a richer variety of future possibilities and help overcome psychological biases (Schoemaker 1993; Tversky and Kahneman 1974).

Transmission planning should draw techniques from both the intuitive and formal approaches to generate scenarios. Quantitative models for developing scenarios are a must for detailed modeling of the power market and transmission expansion plans. However, purely quantitative scenarios can result in future possibilities that are too narrow or lack internal consistency. Qualitative processes allow for more diverse scenarios which are also easier to communicate and discuss among both expert and non-expert stakeholders. A mixed approach is probably best for long-term transmission planning under complex uncertainties. Such an approach was developed by the IPCC to generate its 2000 emissions scenarios, based on a mix of expert consultation, results across different models and elaboration of storylines (IPCC 2000). A similar approach is employed by the European Network of TSOs, whose scenario generation process combines storylines, rounds of stakeholders' participation and quantitative modeling (ENTSO 2018).

It is worth noting that these mixed approaches for scenario generating process go beyond the combination of multiple isolated uncertainties (or qualitative case-driven scenarios). Such a process would qualitatively assess the range of plausible values for each of the individual uncertainties. Then, extreme scenarios are generated based on all the possible combinations of these individual uncertainties. The resulting scenarios should be assessed for internal consistency and plausibility, to eliminate impossible scenarios. Nonetheless, these scenarios would lack the compelling narrative and storyline of the intuitive approach described above. Therefore, scenario generation by combination of uncertainties could be insufficient for long-term strategic transmission planning.

This combinatorial scenario generation process was employed in Chile for the first Long-Term Energy Planning Process (Energy Ministry 2017). The scenario planning process was introduced in 2016 by the new transmission law, to generate long-term energy and electricity scenarios that are later used for transmission planning. The new process was a major success in introducing strategic visions, qualitative techniques and expert panels to scenario generation. Resulting scenarios span a rich set of future possibilities which should yield more robust transmission plans, compared to the relatively simpler scenarios historically used for transmission planning (which only considered supply uncertainties). However, the process fell short of producing compelling and credible storylines for a reduced set of long-term scenarios. These storylines are a core component of participative scenario generation processes developed in USA and Europe.

2.2 *Assessing Expansion Projects and Elaborating the Transmission Plan*

Once scenarios and assumptions are completed, the second stage of transmission planning proceeds with technical and economic planning studies. These studies are often conducted in an iterative fashion between technical–economic optimization and detailed electrical simulations, given the computational complexity of the transmission expansion problem. This stage involves both the identification of expansion needs (e.g., identifying reliability violations) and the assessment of alternative solutions to these needs, primarily through quantitative modeling. The outcome of this stage is a set of recommended transmission system expansion works and their development schedule, with a general outline of the technical, economic and environmental specifications of the project.

Modeling and optimization techniques for transmission planning are increasingly complex in modern power markets. Rapidly growing renewable generation and other technological disruptions (such as distributed resources) impose the need for increased flexibility in power system operations and planning (Ela et al. 2014; Holttinen et al. 2013; Milligan et al. 2016). Therefore, the evolution of power systems requires enhanced modeling through higher spatial and temporal resolution (Munoz et al. 2015). Moreover, more precise representation of the underlying dynamics of the power market is needed, including scheduling, ancillary services, corrective operational measures, spatial and temporal correlation of renewable resources, weather phenomena and market-based forces (Dillon et al. 2014; Jin et al. 2014; Moreira et al. 2018; Moreno et al. 2013; Munoz et al. 2012, 2015; Neuhoff et al. 2013; Orfanos et al. 2013; Pérez Odeh and Watts 2019; Sauma and Oren 2006; Watts et al. 2016).

Based on these evolving modeling techniques, facilitating the integration of renewables requires the full range of benefits of transmission to be considered in planning studies. Conventional planning methodologies in the integrated utility regime aimed at reliability as high as necessary and design as economical as possible (Schlabbach and Rofalski 2008; Stoll 1989). Economic efficiency benefits beyond production cost savings are also commonly analyzed, including the effect of transmission expansions on market prices, increased competition and market power mitigation (Awad et al. 2006; Sauma and Oren 2006). However, transmission expansion projects simultaneously offer a number of benefits (Joskow 2005), ranging from operational, to environmental and investment benefits. These benefits include enhanced reserve scheduling, alleviation of reliability-must-run dispatch, economic valuation of increased reliability, emissions benefits and fuel diversification (Inzunza 2014). While assessing these benefits is challenging, relying solely on easily quantifiable production cost savings would often lead to the rejection of otherwise beneficial investments (Hou and Pfeifenberger 2012).

Capital-intensive transmission projects of strategic value may seem sub-optimal if these additional benefits are dismissed. A practical example is the transmission interconnection project between the two Chilean power systems (Bustos-Salvagno and Fuentes 2017). Simple production cost analyses estimated net benefits between US$ 0.5 and US$ 1.5 bn (Synex–Mercados 2012), in net present value. Another wider economic assessment of the interconnection project assessed, among others, benefits of increased competition in the contracts market and resiliency against shocks (e.g., fuel disruption and project delays). Such wider economic assessment found benefits from the interconnection project between US$ 3.2 and US$ 9.1 bn, in addition to direct production cost benefits (Bustos-Salvagno and Fuentes 2017; CNE 2013). That is, the economic benefit of these additional assessments is a staggering 2–18 times higher than the benefits suggested by conventional planning studies. As mentioned in the introduction, the interconnection between both power systems was commissioned in November 2017 and ever since it has fostered competition (e.g., allowing generators in the south to compete for the supply of large mining companies) and reduced curtailment of renewable generation due to transmission congestion, among many other benefits.

A key trade-off in transmission planning is the desired level of transmission congestions versus acceptable levels of spare transmission capacity. Under expected future demand growth, spare transmission capacity results due to the fundamental properties of transmission infrastructure. Indeed, lumpy investments preclude small incremental investments for the required capacity in each moment, while economies of scale determine that it is better to build a little bigger to begin with, to accommodate future demand growth (Hirst and Kirby 2001). Therefore, there is a trade-off between congestion risks due to lack of adequate and timely expansions and the risks of capacity underutilization for far too long or even over-investment. Fast-growing economies such as Alberta and Chile tend to emphasize the need of robust planning transmission through spare capacities, to avoid the country-wide economic impacts of transmission congestion due to under-investment and delays. Moreover, the complete elimination of transmission congestions is pursued in some countries, despite it being a sub-optimal planning strategy (Stoft 2006). Alberta's transmission plan focused on achieving an unconstrained system until recent years (AESO 2014; Watts and Rudnick 2014). Transmission planning in Germany has also been historically guided by a copper plate standard aiming at unconstrained power markets, although the possibility of 3% renewable curtailment was introduced to the planning process in 2015 (Von Hirschhausen et al. 2018).

Spare capacity for robustness of transmission expansion plans was one of the key components of the Chilean transmission law. Transmission planning at the time was perceived to inadequately address uncertainty through scenarios of low diversity, short planning horizons and lack of strategic long-term vision. The authority argued that the regulation of the transmission planning process precluded enough spare capacity to be considered, thus resulting in prolonged transmission congestions, price decoupling within the power system and curtailment of renewable generation. Such spare capacity would be a key planning tool under uncertainty, given expectations for high demand growth and long lead times for new transmission projects. Much of the legislative discussion focused on the risks of over-investment due to speculative planning (Baldick and Kahn 1993). Although congress granted the regulator the ability to consider spare transmission capacities for future expected uses, implementation problems emerged relating primarily to the cost allocation reforms that were also introduced, as further discussed below.

When it comes to solutions, however, new wires are not everything. While spare capacities are needed for long-term planning, flexibility is paramount for short- and medium-term horizons. Flexibility can be defined as "*the ability to adapt the planned development of the transmission system, quickly and at a reasonable cost, to any change, foreseen or not, in the conditions that were considered at the time it was planned*" (Latorre et al. 2003). Flexibility encompasses many components of transmission planning, including the following:

- Optimization techniques for balancing robustness and flexibility of expansion plans under uncertainty (Mejia-Giraldo and McCalley 2014).
- Flexible solutions for addressing transmission needs, ranging from operational measures to non-wire investments. Flexible solutions require lower capital investment and lead times than new transmission lines, thus enabling deferral of transmission investments until uncertainty diminishes, as well as management of incremental variations in flow patterns in the meantime (e.g., due to quicker-than-expected completion of expected renewable projects). Solutions include optimization of the existing infrastructure through improved system operation (MIT 2011); repowering (Tejada et al. 2015); transmission switching (Fisher et al. 2008; Khodaei et al. 2010); dynamic line ratings (Douglass and Edris 1996; Fernandez et al. 2016); asset management (Brown and Humphrey 2005; Shahidehpour and Ferrero 2005); flexible equipment such as FACTS, phase shifters and storage (Blanco et al. 2011; Konstantelos and Strbac 2015) and other non-wire solutions such as demand response, energy efficiency and distributed generation solutions analyzed in California (CAISO 2013), Denmark (Weber et al. 2013) and UK (National Grid Plc 2014). Moreover, the proposition of additional solutions by independent project sponsors should be encouraged to foster innovation (Herling et al. 2016).
- Project management of planned capacity additions. More precisely, this includes timing of investments (Garcia et al. 2010), real options approach (Chamorro et al. 2012), decision trees (Buygi et al. 2003; RTE 2014) and staged project development to allow adaptation. Staged development should establish adaptability-enabling milestones for the complete expansion process, from conceptual design to spatial layout and permitting. For example, the UK TSO can recommend pre-construction studies to start outlining projects that could be necessary in the future. Moreover, projects under development can be postponed or even canceled in case of major changes in the market (National Grid Plc 2014).
- The flexibility of the process itself can be improved through higher frequency of the scenario and planning process (at least on a yearly basis for planning). Moreover, projects can be grouped in clusters of similar properties or complexities, to allow for expedited approval processes for the less controversial projects.

One illustrative example of plan flexibility pertains to the interaction of spare capacities and repowering. In its 2013–2014 transmission plan, the Chilean energy regulator (CNE) proposed a 500 kV line to supply the southernmost zone of the power system. A generation company presented a discrepancy against this project to the conflict resolution body of the Chilean power market, the Panel of Experts. The generator argued that the project should be developed in stages by deferring some branches and initially powering the line in 220 kV. In turn, the regulator's arguments included frustration with insufficient expansions from previous transmission planning processes and the need of a long-term vision for harnessing the wind generation potential in the southern zone. After careful analysis, in its Resolution N°3 of 2014, the Panel of Experts accepted the generators' proposal to develop the project in stages, notwithstanding the relative agreement regarding the long-term need of the project. This resolution made it clear that a wider variety of benefits and scenarios should be considered for these kinds of expansion projects to be feasible.

However, the lack of flexibility persisted as one of the weaknesses of the transmission process in Chile. Although the previously mentioned experience motivated profound and positive changes in the transmission law aiming at the long-term development of the market, medium-term transmission expansion was largely left unchanged. Crucially, the regulation did not establish the ability to postpone, modify or cancel complex expansion projects. In its 2017 expansion plan, the first under the recently enacted transmission law, the regulator proposed the biggest transmission project ever in Chile, a massive 1500 km/3000 MW HVDC transmission line with US$ 1.8 bn of referential investment. Such project would connect the north and center zones of the system, enabling the long-term development of solar generation in the north to supply growing demand in the center. Although the need of the project conveyed widespread agreement, the accelerated planning process with incomplete information and insufficient time for comments prompted a discrepancy to the Panel of Experts, this time by a mining company, arguably because the transmission law allocated expansion costs entirely to final customers (with no costs borne by generators). Panel's Resolution N° 7 of 2018 delayed the HVDC line to the 2018 plan for further analysis and specification. Unfortunately, given the inflexibilities of the process, this lengthy conflict resolution process also delayed by several months all the other expansion projects (many of which raised no opposition in the first place). In the 2018 planning process, the regulator finally achieved approval for a smaller 2000 MW HVDC line with US$ 1.3 bn of referential investment.

Given the importance, size and complexity of this HVDC project, staged development might be useful, particularly given the extreme difficulty that is expected from the overall siting process of this project (further discussed below). Moreover, the basic properties of the project are already defined, and the respective right-of-way shall be planned for the smaller approved project. This inflexibility precludes a later decision to build a higher capacity line which requires a wider strip of land, in case the renewable potential turns out to develop faster than expected. The need to incorporate more flexibility in the planning process has already been acknowledged by the authority and is a key part of a transmission planning improvement law currently being prepared by the Ministry.

2.3 Cost Allocation, Plan Approval and Project Development

Governance of the transmission approval and development process is pivotal to the success of the transmission expansion framework for renewable integration. After the optimal transmission network has been planned, such plan is subject to several stages of regulatory approval, administrative permitting and siting processes. Notwithstanding the need for careful planning, difficulties in the approval and development stages can result in severe delays, rerouting and redefinition of new transmission projects. Severe difficulties in this process can result in large opportunity losses for renewable generation and final customers due to curtailment and price decoupling. Moreover, inadequate processes undermine the confidence of new investors on the transmission expansion framework, slowing the pace of renewable generation investment.

A core issue in the transmission approval process is the trade-off between efficiency and simplicity of cost allocation methods. A beneficiaries-pay cost allocation methodology promotes market efficiency. However, application of a beneficiaries-pay rule is difficult for large-scale transmission projects with various benefits spread throughout wide geographic areas and different market participants (Hogan 2011). Socialization of transmission costs is relatively simple in practice, but would reduce the incentives for efficient expansion of the combined generation and transmission infrastructure.

Cost allocation in Chile for the main transmission grid (trunk grid) historically relied on a complex usage-based methodology. Congestion rents are assigned to Transcos, and transmission costs not covered by these congestion rents were shared among generators and loads. For the "common influence area" (defined by engineering criteria as the grid used by both generators and loads across the entire grid), 80% of the costs were allocated to generators and 20% to loads. Allocation between generators and between loads was based on approximate usage factors derived from simulations of the system's operation. Results were highly dependent on hydrological conditions, and the overall cost allocation framework was deemed too complex for new investors to understand and manage.

To facilitate and accelerate renewable investment, the Chilean transmission reform drastically simplified the transmission cost allocation method. After the law, costs of the trunk transmission grid (now called "national" grid) are allocated entirely to final customers through a simple postage stamp methodology. A 15-year transition period was established to gradually transfer transmission costs from generators to final customers for supply contracts signed before enactment of the law. Nonetheless, new generation projects would be automatically exempt from bearing transmission costs of the trunk grid (although generation interconnection costs are still borne by generators). These regulatory changes were expected to facilitate renewable investment by new investors, smaller than incumbent Gencos in Chile. Moreover, allocating costs of the trunk grid to final customers was consistent with the most common international practices (PJM 2010).

The simplification of cost allocation method means that locational and efficiency signals for investment were reduced (Matamala et al. 2019). Locational signals remain at the core of the Chilean market since the power pool still operates on short-run locational marginal prices. The lack of a locational signal in transmission cost allocation may in the long-term reduce the efficiency of the combined generation—transmission investment, incentivizing too much generation away from load centers. However, high-quality renewable resources cannot be transported to more convenient locations near demand centers. Thus, consensus emerged among participants of the transmission law discussion regarding the idea that locational signals in transmission cost allocation are meaningless for the transition to a highly renewable energy mix.

Allocating transmission costs directly to final customers was expected to increase transparency in final customer's bills. This would also reduce the risk premium that generators may be charging their customers due to the uncertainty embedded in transmission tolls, given their strong dependence on hydrological conditions. Indeed, supply contracts in Chile often pass through transmission costs to final customers. Thus, the argument goes, tolls should be directly allocated to final customers, since they end up paying for it anyway (Baldick et al. 2007).

This also increases the risk of overbuilding transmission if demand is too passive in the planning process, compared to generators. Historically, generators have been actively involved in the transmission planning process, arguably because the costs of new transmission projects would be mostly borne and managed by generators. Generators participation thus provided a market-based loop for transmission expansion efficiency. In turn, customers are often deemed to be less interested in the details of their electricity bill and the transmission planning process. However, Chilean experience suggests otherwise, since the new HVDC line planned by the authority was delayed and its capacity reduced due to opposition of large mining customers. Although such level of participation cannot be expected from residential customers, the conflict around the HVDC line highlights the need for final customer participation in early planning stages, with adequate time allowed to review and comment the expansion plan. Moreover, a benefit-based cost allocation procedure for new transmission projects could make issues and opposition to transmission expansion more transparent (Baldick et al. 2007).

After conflicts due to cost allocation are resolved and the expansion plan is approved, the complex permitting and siting process begins. Delays of major transmission expansion projects due to public opposition, permitting and siting processes can have significant impacts on the power market. These delays have proven very difficult to manage since they are largely locational specific, depending on the communities and administrative divisions involved. One related Chilean experience is the Cardones–Polpaico 2 × 500 kV line, which suffered a delay of 17 months with sizable economic impacts for renewable generators located in the northern zone (solar PV and wind). The delay was partly related to intense public opposition in the center zone, where electricity demand is concentrated. Rights-of-way negotiation was slow in the center zone given the large number of land owners. After negotiations finished, public opposition intensified with a few extreme acts that delayed completion of the last line segments. Public opposition to this transmission project emerged despite its need for renewable generation integration, which has more social support than the thermal generation that renewables replace.

Siting difficulties highlight the need for early stakeholder participation in transmission planning. Policy-makers and TSOs should acknowledge the large body of the literature addressing the underlying factors of public acceptance of new transmission projects, primarily the consensus that concerns of inhabitants and organized stakeholders go well beyond the Not-In-My-Backyard (NIMBY) phenomenon. Easily accessible information and a better representation of a project's impacts—beyond pure economics and cost-based analysis—are thus required (Devine-Wright 2012; Komendantova and Battaglini 2016a; Schmidt and Lilliestam 2015).

Further research is required to propose and compare participative solutions to the transmission siting conundrum (Cohen et al. 2014). Although participation and stakeholder empowerment from the need definition and spatial planning stages are ways of avoiding project delays, participation is project-tailored and does not automatically eliminate conflicts (Späth and Scolobig 2017). Participation is a dynamic process which requires optimal engagement time (not too late and not too early) and addressing new concerns that appear due to increased stakeholder awareness (Komendantova and Battaglini 2016b). While participation is a key means for social acceptance of new lines, some authors argue for participation to become a goal in itself. A siting approach based on an open dialog, with the possibility to co-decide and shape the project's definition, can foster societal acceptance of large interconnection projects (Ciupuliga and Cuppen 2013). Despite growing experience with transmission siting processes and practices around the world, much of the available literature is focused on European countries where one or a few TSOs plan and develop new transmission projects. In the Chilean framework, new transmission projects are planned by the authority and built by a transmission concessionaire. Other country-specific aspects make it difficult to successfully transfer lessons and best practices (Consorcio Centro Cambio Global UC—Centro de Energía U. de Chile y Teco Group 2018).

The Chilean transmission law introduced new instruments for spatial transmission planning and siting with a stronger role of the state as a "*guarantee of social welfare*". Before the law, new transmission projects were auctioned with little information on its route and awarded to the least-cost proposal (Ferreira et al. 2016). A broad consensus emerged in the public discussion of the transmission law, regarding the need of an increased role of the state in route definition of new transmission projects. In the new regime, the state formulates alternative strips with early public participation and a wider set of criteria beyond economic efficiency, including social and environmental sustainability. A strategic environmental assessment is developed in parallel to inform this strip study. The outcome of this process is a strip of land subject to approval by the Sustainability Council of Ministries. After the strip is approved, a public auction is conducted, and the awarded transmission concessionaire will be responsible for detailed route definition and project construction, as well as obtaining environmental and administrative permits (which also require public participation).

Despite broad consensus on the direction of these reforms, effective implementation will be crucial to success of the new transmission siting framework. Recent experiences suggest that new transmission lines will face intense public opposition if the siting process is not implemented effectively. The first strip study should start in 2020 for two tranches of a new 500 kV transmission system in the far-south zone of the Chilean grid, with an estimated length of 421 km. However, potentially conflictive projects will not be subject to this state-led process, given the authorities' criteria for determining project complexity. In fact, a 25 km line in the center-south zone will not be subject to a strip study, despite being more complex from a social–environmental perspective than one of the tranches of the 500-kV southern project. Given the Ministry's methodology, a strip study is not justified for the center-south project given its low technical complexity (in turn due to its low extension), despite its high social–environmental complexity. Hence, in practice, the transmission reform tackled the siting process for large transmission projects without addressing the siting process for smaller projects, which are often located near cities and communities (whereas large projects are often located farther from big cities). Siting of these small transmission projects will continue being managed by a variety of transmission concessionaires according to their own procedures for obtaining environmental and administrative permits.

Another potential weakness of the regulatory framework is the inability to plan a common strip for multiple transmission lines, nor for a possibly larger transmission line in the future. This flexibility is important, for example, for the planned HVDC between the center and northern zones required for scaling-up solar generation in the north. Seeking to ensure its approval, the authority adjusted the HVDC project by lowering its capacity and investment requirements below what was initially considered necessary. A strip study adjusted to the approved capacity will be conducted, probably precluding a larger project to be built in the future in case the approved capacity is later found to be too small to accommodate solar generation. Also, the strip cannot be wider than necessary to accommodate other transmission projects in the future. If poorly managed, this situation could result in two parallel HVDC projects which occupy a land strip much larger than necessary and which are developed in times much longer than necessary.

3 Reaching Out: Transmission for Harnessing Renewable Generation Hubs

Coordination of transmission and renewable generation investment is the key for market efficiency. Such coordination is especially challenging for renewable energy "hubs", or zones with a concentrated high-quality potential for renewable generation, for three reasons.

- First, renewable energy hubs can be located far from main demand centers or the transmission grid, thus requiring new dedicated transmission systems to cover widespread areas in order to harness the renewable potential.
- Second, renewable resources can be dispersed in large geographic areas, where independent connections designed for each possible generation project might be inefficiently costly, leading to less-than-optimal investment in renewable generation. Moreover, such independent connections can have significant socio-environmental impact, due to the multiplicity of transmission lines.
- Third, many smaller generation projects (i.e., less than 20 MW) are unable to finance investment in new dedicated transmission lines required to transport generation for large distances to the main transmission grid. Therefore, less projects than the efficient level would be developed due to the failure to coordinate and share the transmission costs among multiple developers.

Given these difficulties, anticipative and proactive transmission expansion approaches have been proposed. The anticipative approach would anticipate to the development of new generation projects to plan an optimized transmission network. A proactive planner would expand transmission considering the effect of these expansions on competing generators, seeking maximum efficiency by "guiding" generation investment through transmission investments.

Mixed outcomes have resulted from the implementation of these approaches to coordinate renewable generation investment in Texas, Brazil, Australia and recently Chile (Chattopadhyay 2011; Hasan et al. 2013; Madrigal and Stoft 2011; Porrua et al. 2009; Rudnick et al. 2012). The Texas experience with proactive transmission planning and staged coordination of new generation projects has been very successful, achieving integration of over 19 GW of wind generation and reducing curtailment to low, economic levels around 0.5% by 2014, down from 17% in 2009 (Billo 2017). The Brazilian experience with coordination of transmission for renewable projects that participate in supply auctions was also initially successful (Porrua et al. 2009). However, coordinated transmission expansion was later abandoned, partly due to transmission project delays and the difficulties of risk allocation between customers, planners and generators whose construction was uncertain (Bayer et al. 2018). Finally, to this date there have not been major successful experiences in Australia and Chile, despite their efforts to coordinate transmission for clusters or hubs of renewable generation.

However, many issues curb the development of coordinated transmission systems for renewable hubs. First, anticipative and proactive planning is by itself a challenging task, requiring new optimization methodologies and institutional arrangements compared to traditional planning. Second, planning transmission to incentivize optimal generation investment inevitably risks transmission asset stranding, in case some of the new generation projects get canceled. Third, the timing and development times of multiple generation firms can vary widely. Fourth, competing generation firms might be unwilling to trust their direct competitors with commercially sensitive information regarding their generation project portfolios. Moreover, generation firms might be wary of depending on the decisions of their direct competitors, let alone helping them achieve lower transmission costs and shorter time-to-market, since the potential competitive loss may outweigh transmission cost reductions due to coordination.

The coordination between transmission and generation investment for renewable energy hubs will be discussed in further detail below, by using simple examples to illustrate the underlying competitive and market forces. First, coordination between two similar generators by private initiative will be analyzed, highlighting the reasons that curb such private coordination (Sect. 3.1). Second, centralized coordination of transmission interconnection will be analyzed, highlighting the potential risks of such a solution (Sect. 3.2). Third, Sect. 3.3 overviews open access and its practical implications for renewable integration.

3.1 Voluntary Coordination of Two Similar Generators

Consider two similar new generation projects located nearby each other, but far from the main transmission grid (about 50 km). Suppose two alternative connection solutions are available, as depicted in Fig. 4. The first solution ("uncoordinated" solution) would leave the full development of transmission interconnection to generation developers, thus probably resulting in one dedicated interconnection transmission system for each generation project. Instead, the second solution ("coordinated" solution) would optimize the transmission interconnection system considering both generation projects, thus building a bigger common line and two smaller and independent interconnection lines.

The coordinated solution reaps the benefits of scale economies in transmission systems, thus achieving lower overall costs when compared to the uncoordinated interconnection solution. Using standard investment costs for the Chilean transmission systems, the example above would result in a total interconnection cost of MUS$ 24.5 in the uncoordinated solution (two independent transmission lines of MUS$ 12.3 each) and MUS$ 23.1 under a coordinated solution (composed of a common 15.4 MUS$ line and two independent lines that cost MUS$ 3.8 each). Thus, coordination would yield transmission interconnection savings for MUS$ 1.5, or about 6% of transmission interconnection costs, given the lower cost per MW of capacity for bigger transmission lines.

However, strategic considerations may deter a coordinated interconnection solution to be agreed upon and executed by the two generators. The first strategic concern is related to the increased risk that the potential cost saving implies. Consider the "payoff matrix" for both generators under the uncoordinated and coordinated solutions presented in Table 2, depicting the transmission interconnection costs borne by each generator (in each of the four columns), depending on the investment decision taken by the first generator (rows) and by the second generator (columns). In the uncoordinated solution, each generator would pay MUS$ 12.3 for transmission interconnection if (and when) it finally decides to invest, and would pay nothing otherwise.

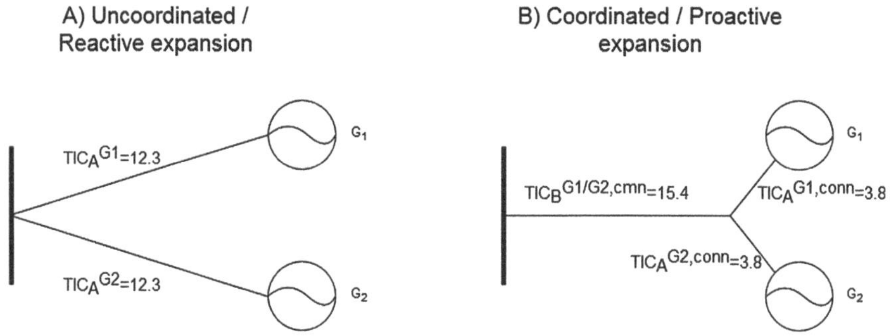

Fig. 4 Uncoordinated and coordinated transmission connection solutions

Table 2 Generators-game payoff matrix under both expansion alternatives

G1\G2	Invest		Not invest	
(A) Uncoordinated expansion				
Invest	−12.3	−12.3	−12.3	0
Not invest	0	−12.3	0	0
(B) Coordinated expansion				
Invest	−11.5	−11.5	−19.2	0
Not invest	0	−19.2	0	0

The outcome for each generator in the uncoordinated solution is independent of the other generator's investment decision and its timing.

In turn, the coordinated solution results in lower interconnection cost of MUS$ 11.5 for each generator, but only if both generators follow through with their pre-commitments to invest in a coordinated transmission interconnection project. However, consider that only G1 decides to invest in the coordinated system, but G2 decides not to participate of the coordinated system (whether because its project is canceled, or because it decides to develop an independent connection). Then G1 would bear the full cost of the common interconnection line (MUS$ 15.4), as well as its independent line (MUS$ 3.8), thus resulting in an interconnection cost of MUS$ 19.2, which is a MUS$ 7 loss with respect to the uncoordinated solution.

Considering that both generators invest in their respective power plants, but can either coordinate transmission or pursue an independent connection, the resulting game is an instance of the well-known stag hunt game (see Table 3). The stag hunt is a common example of coordination failures among individuals. Theoretically, this simple game has two pure Nash equilibrium strategies: both players coordinate or both fail to do so (Fudenberg and Tirole 1991). If both players coordinate, none of them has the incentive to unilaterally change their strategy, since payoffs would be lower. However, this is also true if both players do not coordinate.

Therefore, game theory does not predict a single pure strategy for rational players. In turn, the mixed strategy equilibrium (i.e., where each player has a probability distribution over his set of actions) depends on the probability of coordinating. For this example, coordinating would be better only if the probability that the other generator coordinates is over 90% (due to the large downside risk of building a larger than needed line). The probability of coordination required for it to be the optimal strategy needs to be higher with more players. In theoretical terms, the "both

Table 3 Generators-game payoff matrix if both invest in their power plant and choose whether or not to coordinate transmission

G1\G2	Coordinate	Not coordinate
Coordinate	−11.5, −11.5	−19.2, −12.3
Not coordinate	−12.3, −19.2	−12.3, −12.3

fail to coordinate" equilibrium is said to *risk-dominate* "both coordinate", despite coordination being *payoff dominant* over non-coordination (Harsanyi and Selten 1988).

Of course, if G2 withdraws from the coordinated system, G1 could drop the coordinated solution altogether and develop an independent interconnection instead. Then, the transmission cost of the coordinated solution is always lower than the coordinated expansion (the worst-case cost would be MUS$ 12.3 instead of MUS$ 19.2). Rational generators would thus pursue the coordinated solution first, and fall back to the uncoordinated option should coordination fail. Moreover, the credibility and feasibility of a coordinated solution may increase with enforceable contracts and the establishment of a special independent firm for developing the coordinated transmission project.

However, the risks of project delay can easily outweigh the benefits due to interconnection cost savings. Indeed, this optionality would require the transmission project to be postponed until enough certainty regarding the development of both generation projects is achieved.[2] Waiting for such certainty may significantly delay revenues from the generation project, since the other generator could fail in an advanced stage of project development. While the direct costs associated to the delay may be very small (e.g., restarting interconnection studies and permits), the opportunity costs could be tremendous due to foregone revenue from energy sales (especially if the generation project is committed for supplying a contract with final customers). Another option would be the parallel development of a coordinated and an uncoordinated solution. However, parallelism does not completely eliminate delay risks and would probably undermine confidence on the viability of the coordinated solution.

The importance of time-to-market over minor cost savings, and the properties of the strategic game between both generators described above, could well be enough to impede a coordinated solution to be pursued by competing generators in many real cases. However, other wider commercial considerations come at play against a coordinated solution. Attempting coordination would inevitably mean disclosing confidential information regarding the generator's project portfolio and commercial strategy. Moreover, the coordinated solution could ultimately improve the commercial position of a direct competitor and could thus be negative for a profit-maximizing generator. In the long-term, the portfolio-wide competitive loss due to a couple of additional competing renewable projects can be very small for large companies with several GWs of installed generation capacity. However, a few hundred MWs of additional renewable capacity can result in enormous short-term losses due to curtailment and price decoupling in constrained transmission networks such as the Chilean grid, further reducing incentives for a coordinated connection.

---

[2]A joint venture between the two generation firms could allow for full certainty by bundling the two generation projects as well as the common transmission system. However, that would not be a case of coordination among different firms. Moreover, joint ventures may be far less likely in zones with many generation projects (instead of only two).

For the reasons outlined above and given the relatively low level of potential cost savings due to coordination (when compared to total project costs and incomes), it seems rather unlikely that generators would voluntarily pursue a coordinated interconnection solution. It is worth noting that the issues with the coordinated interconnection (namely risks of projects' delay and slower time-to-market, risks of higher transmission costs, disclosing commercially sensitive information and helping a direct competitor) are also present in more general cases with multiple generation projects of different technologies and installed capacity. Indeed, while the potential cost savings of a coordinated solution could be higher for many small projects, the potential risks could also increase (with worse possible outcomes), as well as the difficulty of coordinating and establishing trust among many competing firms.

3.2 Centralized Coordination of Two Similar Generators

Transmission cost savings and cost sharing among multiple power plants can be key for developing new generation projects. Unlike thermal power plants, neither the wind nor the sun can be transported to a more convenient location. Given that high-quality renewable potential can be located far from the existing transmission networks, long transmission lines may be required to harness the full potential of the hub. In that case, the higher transmission cost associated to an uncoordinated solution can deter or even preclude altogether the development of some generation projects. The appeal of the coordinated interconnection solution is thus twofold: it lowers overall transmission costs, thus enabling more renewable potential to be harnessed.

Centralized coordination of transmission investment for facilitating renewable generation development is therefore an appealing policy. A coordinated transmission project would be designed, developed and executed with the state's direction. Renewable projects could then use the transmission capacity provided they pay their "fair share" for the coordinated system, so that an appropriate efficiency signal is preserved for the location and volume of renewable projects.

To illustrate the potential benefits of transmission coordination, assume, for the sake of simplicity, that two 200 MW solar PV power plants could be developed in a nearby location far from the main grid. Each power plant requires an investment of MUS$ 200 (consistent with public information of PV projects under development in Chile). Considering a capacity factor of 31% and an average spot price of 50 US$/MWh, each power plant would roughly save MUS$ 27.2 in total system operation costs on a yearly basis, or a present value of MUS$ 231.2 over a 20-year lifespan (with a 10% discount rate).

If coordination between both projects is implemented perfectly, coordination would yield savings of MUS$ 1.5 compared to the uncoordinated solution. However, should one of the projects fail to reach completion and pay for the common transmission infrastructure, the TSP would have to bear the stranded cost of the common line, equal to MUS$ 7.7 (see Table 4).

Table 4 Transmission interconnection costs with Transco-driven coordination

G1	G2	Transco (Tx)					
		Not invest			Invest		
		G1	G2	Tx	G1	G2	Tx
Invest	Invest	−11.5	−11.5	0	−11.5	−11.5	0
	Not invest	−19.2	0	0	−11.5	0	−7.7
Not invest	Invest	0	−19.2	0	0	−11.5	−7.7
	Not invest	0	0	0	0	0	−15.4

If excess costs due to stranded assets are not transferred away from the TSP and generators, and through to final customers or the state, it is unlikely that the coordinated solution is financed and executed. Indeed, if the TSP is allowed to increase charges to the completed generator, then prospective generators would refrain from participating in the coordinated solution, preferring an independent solution instead. If the TSP bears the risks of stranded assets, the project is unlikely to achieve financing due to the significant risks involved, which are not rewarded through higher expected returns.

However, transferring risks away from project developers could ultimately harm system efficiency. While facilitating the development of renewable projects through coordinated or proactive transmission can allow for increased investment in renewable generation, inefficient transmission or generation projects could end up developing at the cost of final customers.

As previously mentioned, centralized coordination could enable increased investment in generation projects which would otherwise be inviable. The benefit of higher and more efficient overall investment is far greater than the benefits due to transmission interconnection cost savings for large, high-quality generation projects which would be developed anyway (i.e., with or without coordination). The potential benefit related to guiding more efficient generation investments has been studied in the literature through proactive transmission planning models. Such models are formulated as sequential strategic games where the transmission planner has perfect information regarding the cost structure of individual generators, thus allowing the planner to determine optimal investment decisions for each generation firm, given the transmission expansion plan under assessment (Sauma and Oren 2006). While these assumptions could be adequate for the long-run planning of the overall transmission system, practitioners may find it difficult to determine optimal investment decisions for private generation firms in particular zones of the system.

Given the uncertainty involved in estimating generator's investment decisions, a probabilistic approach could be more adequate for analyzing the development of renewable hubs. The simplest approach is assuming that building a transmission system for a particular hub increases the probability of generation projects' completion. For example, based on the simple example presented above, consider that one generation project's completion (G1) is absolutely certain (i.e., with or without a coordinated transmission system), while the second-generation project's completion

(G2) is deemed uncertain. Further, assume that the planner estimates a 30% chance that G2 reaches completion if no coordinated system is developed. If a coordinated system is developed, lower interconnection costs would increase the probability that G2 is completed.

In this setting, building the coordinated system would only be beneficial (in expected present value) if the probability of G2's completion increases from 30% without coordination, to over 53%. Instead, if the probability is lower than 53% then, in expected present value, the planner would be better-off leaving transmission interconnection solely to generator developers. This sensitivity to perceived project completion likelihood highlights the risks involved in proactive transmission development for renewable generation hubs.

The general formulation of the related optimization problems would be as follows:

$$\min \sum_{l} \mathrm{CTIC}_l \cdot y_l + \sum_{g} E\left[\mathrm{GIC}_g + \mathrm{GTIC}_g + \mathrm{VC}_g | \{y_l\}_{\forall l}\right]$$

Subject to production, demand, transmission flows and variable nature constraints. Where

y_l Binary decision variable for investment in coordinated transmission systems.
CTIC_l Coordinated transmission investment costs, shared among coordinated generators.
GIC_g Generation investment costs.
GTIC_g Transmission interconnection costs borne by each generation firm.
VC_g Total variable costs for each generator of the system.

The model presented above minimizes the total expected investment and operation costs, considering the impact that a centralized coordinated solution has on the likelihood of project completion. Unfortunately, the formulation above is nonlinear, since the probabilities required for calculating expected generation costs depend on the decision variables for coordinated transmission expansion. Nonetheless, for assessing rather small generation hubs, the problem can be solved by comparing the solutions with and without coordination, as depicted above.

The model captures the uncertainty associated with generation project development but fails to appropriately represent the underlying market forces in liberalized power markets, primarily, profit-maximizing generation firms. A more theoretically sound (albeit complicated) model could be posed as a Bayesian game where players (i.e., generators and the transmission planner) have uncertain information regarding the game itself (e.g., regarding the cost structure of each firm and generation project).

3.3 Open Access and Governance of the Transmission System

Open access to the transmission system is a core component of competitive power markets (Hogan 1998, 2002; Hunt 2002; Rudnick et al. 1997). Open access to the transmission grid means open access to dispatch, thus enabling short-run competition among generators and efficient transmission capacity allocation. Moreover, open access allows investment and the entry of new competitors through interconnection of new power plants.

Achieving effective open access in practice can be tricky. Investment in new renewable generation projects is particularly sensitive to the many components of the interconnection process and its regulation. Such components include cost allocation policies for new connections (deep vs. shallow), times and costs of the interconnection process, process standardization, governance and independence of the TSO/ISO, among others (Ellery et al. 2013; Madrigal and Stoft 2011).

For renewable generation in Chile, the scope of open access to dedicated transmission systems is particularly important. Dedicated systems are those that connect a single generation project or large customer to the main transmission grid. As explained above, the usage of the existing transmission lines located near high-quality renewable zones often poses significant benefits for small renewable power plants. These benefits are twofold. First, the power plant would bear lower interconnection costs, since it is usually cheaper to reinforce or expand existing transmission systems (although not always possible depending on the sizes of the new projects), rather than building a new transmission line. Second, the power plant would also face lower construction times and lower delay risks, since right-of-way and many other administrative permits are needed for shorter strips. Furthermore, there are many societal and environmental benefits since less transmission lines are built.

However, lacking appropriate institutions, new renewable project developers can find it hard to connect to these dedicated transmission systems. This was perceived to be the case in Chile by 2015, given the ownership structure and regulation of the transmission system. Transmission ownership in Chile is not solely allocated to a single or few transmission companies. Instead, transmission system ownership is dispersed among transmission, generation and distribution companies and even large customers (mostly mining companies). Furthermore, regulations for open access to dedicated transmission systems were not clear in many respects. Indeed, there was no bylaw regulation for dedicated systems, the regulation allowed some dedicated systems to be exempt from open access and there were no established procedures and referential costs for connection of new projects to different transmission systems.

If not standardized, connection procedures and costs can be discretionally set by the transmission system owner. This gives the owner the power to deter new connections by imposing high connection costs (whether through complicated and lengthy procedures and studies or through excessive tolls). The owner could be compelled to do so, for example, if he plans to develop new projects in the future making use of the existing system, if he wishes to deter entrance of new competitors (for generators) or if he simply wishes to avoid the trouble of multiple connections and toll agreements (for large customers such as mining or manufacturing companies, whose primary business is not electricity). Even if the transmission owner does not wish to deter new connections, different procedures and criteria for connection to each transmission system make it harder for new generators to enter the market.

The Chilean experience suggests that institutions for open access to dedicated transmission systems may need revisiting to allow for the efficient integration of renewable resources. Four lessons from the Chilean experience are outlined next. First of all, regulations must clarify the scope of open access. The best scenario for new renewable generators would be that every transmission system is subject to open access administered by the ISO. This was not the case in Chile, where the lack of open access to some dedicated systems did not seem to be a problem in the past, given the existence of only few market participants, who developed mostly large generation projects with dedicated transmission lines.

Second, the roles and functions regarding open access must be clarified and ideally allocated to a single entity (such as the ISO), provided other opportunities for dispute resolution (with the regulator or a dedicated expert entity, before leaving it to the courts). Connection procedures should be standardized across the power market administered by the ISO, independently of the owner of each individual transmission system. Such procedures should clearly specify the timing and general process for connection, including deadlines for yearly connection windows, in case connection applications are not received and processed all year long. Moreover, the process for determining and allocating Available Transfer Capacity (ATC) should be standardized and publicly available, and the process should also be administered by an independent ISO instead of individual market players, to avoid doubts regarding its fairness.

Third, the criteria applied by the ISO for approving new connections should be as transparent and standardized as possible, seeking to preserve reliability and system security above all, as well as fostering competition. For example, in Chile, the connection of new transmission systems to the main grid was not fully standardized. Procedures allowed for up to one tap-off connection to only one circuit of a transmission line in the main grid. However, when there were two or more connections to a single point of the main grid, new sectioning substations for all the circuits were required, with additional costs borne by the owner of the second or third connection. This standard made it far cheaper for the first-generation project to connect to the main grid. It also placed most of the financial burden of a secure connection on the second- or third-generation projects to connect to the main grid in a given location. With the advent of renewable generation, tap-offs quickly spread throughout the Chilean transmission grid and as of 2019 are still being replaced by fully functioning sectioning substations.

Fourth, property rights should be clearly allocated, and coordination thereof should be the responsibility of the ISO. In Chile, the transmission law and bylaws successfully clarified the scope of open access to electricity transmission infrastructure but failed to clarify property rights and open access to the communication lines bundled within power transmission lines. This led to some delays in the interconnection of new transmission projects and is therefore a key part of the refinement law currently under development by the Energy Ministry.

4 Conclusions

Timely and adequate transmission capacity is key for renewable energy integration. Short-run impacts for investors due to curtailment and market decoupling can be significant. However, the risks of overbuilding transmission should also be managed. A carefully designed and implemented framework for transmission expansion can support long-run efficiency and sustainability in the evolution of power systems. This chapter discussed some of the elements that could improve transmission planning, thus facilitating renewable integration.

A coordination problem between transmission and generation investments arises for harnessing hubs of high renewable potential. Small and geographically dispersed renewable projects could benefit from economies of scale stemming from coordinated transmission expansions. However, economic risks and strategic considerations curb the development of such coordinated transmission solutions. The Texas experience with proactive coordination of wind farms is a major success of transmission–generation coordination for efficient wind farm accommodation. However, mixed experiences have emerged in Brazil, Chile and Australia, given the difficulties associated to planning, coordination and allocation of the costs and risks of proactive transmission investments. Further research and analyses are needed to shed lights on possible solutions to the coordination conundrum. These solutions should aim both at economic efficiency and practical feasibility.

Regarding planning studies, practitioners should resort to the wide variety of optimization models developed to support transmission planning. Quantitative modeling should be a core component of the transmission planning process. Although many optimization models are computationally challenging to solve, incremental improvements of transmission planning modeling should be continually pursued by practitioners. For example, planning could be improved by modeling the temporal and spatial correlation of renewable resources, as well as flexible expansion alternatives such as FACTS and storage. Practitioners should also consider general policy recommendations stemming from sophisticate models which may be difficult to solve directly for real case studies. For example, competitive benefits of transmission should be considered at least approximately, given that competitive equilibrium models might be difficult to calibrate and solve in practice.

The whole process of transmission expansion deals with various uncertainties and complexities due to multiple conflicting criteria. Primarily, quantitative techniques guide the planning process in the short- to medium-term. Mostly, qualitative participatory processes for generating long-term scenarios have emerged worldwide as a primary tool to model and communicate more complex uncertainties. In any case, optimization approaches should be used more often by practitioners to determine the optimal expansion plan. Although academic research has made tremendous advances, many planners rely on simple heuristics for selecting "optimal" projects, not relying on optimization techniques such as stochastic programming, robust optimization and multi-objective optimization. In particular, tremendous uncertainty on the future location and technological mix of renewable generation calls for increased reliance on optimization techniques to guide the transmission planning decision-making process. A simple example is provided by Chile, where both the solar potential of the north and the wind and hydro-potential of the south could develop in the future. The decision of building lines from the center (where demand is concentrated) to the north or to the south ultimately depends on both the qualitative visions regarding plausible futures and the optimality of the expansion plan under various kinds of uncertainties and multiple criteria.

Long-run reliability, efficiency and sustainability of the power sector require a holistic approach to transmission expansion. In this sense, the most complex process within transmission planning is the approval and siting of new overhead transmission lines. Lack of early and effective participation of communities and stakeholders in the decision-making process increases the risks of later delays due to opposition, judicialization and even redefinition of the projects. Transmission siting and therefore the whole expansion process are shaped by a confluence of regulation, technical–economic theory, underlying market fundamentals, social and environmental dynamics and ultimately good governance and institutional capacity to ensure the practical effectiveness of the transmission expansion framework. Convergence of approaches and analyses from all these disciplines is required to facilitate a smooth transition towards the much-needed low-carbon economies of the future.

References

AESO, *AESO 2013 Long-term Transmission Plan* (2014)

M. Awad, S. Broad, K.E. Casey, A.S. Geevarghese, J.C. Miller, A.J. Perez, et al., The California ISO transmission economic assessment methodology (TEAM): principles and application to Path 26. *2006 IEEE Power Engineering Society General Meeting*, 8 p (2006). https://doi.org/10.1109/PES.2006.1709178

R. Baldick, A. Brown, J. Bushnell, S. Tierney, T. Winter, *A national Perspective: On Allocating the Costs of New Transmission Investment: Practice and Principles* (2007). Retrieved from https://hepg.hks.harvard.edu/files/hepg/files/rapp_5-07_v4.pdf

R. Baldick, E. Kahn, Network costs and the regulation of wholesale competition in electric power. J. Regul. Econ. **5**(4), 367–384 (1993). https://doi.org/10.1007/BF01065403

B. Bayer, L. Berthold, B.M.R. de Freitas, The Brazilian experience with auctions for wind power: an assessment of project delays and potential mitigation measures. Energy Policy **122**, 97–117 (2018). https://doi.org/10.1016/j.enpol.2018.07.004

J. Billo, *The Texas Competitive Renewable Energy Zone Process* (2017). Retrieved from https://cleanenergysolutions.org/sites/default/files/documents/jeff-billo_webinar-ercot-crez-process.pdf

G. Blanco, F. Olsina, F. Garces, C. Rehtanz, Real option valuation of FACTS investments based on the least square Monte Carlo method. IEEE Trans. Power Syst. **26**(3), 1389–1398 (2011). https://doi.org/10.1109/TPWRS.2010.2094211

L. Börjeson, M. Höjer, K.-H. Dreborg, T. Ekvall, G. Finnveden, Scenario types and techniques: towards a user's guide. Futures **38**(7), 723–739 (2006). https://doi.org/10.1016/j.futures.2005.12.002

BP, *BP Statistical Review of World Energy 2019* (2019). https://doi.org/10.2307/3324639

R. Bradfield, G. Wright, G. Burt, G. Cairns, K. Van Der Heijden, The origins and evolution of scenario techniques in long range business planning. Futures **37**(8), 795–812 (2005). https://doi.org/10.1016/j.futures.2005.01.003

R.E. Brown, B.G. Humphrey, Asset management for transmission and distribution. IEEE Power Energ. Mag. **3**(3), 39–45 (2005). https://doi.org/10.1109/MPAE.2005.1436499

E. Bustamante-Cedeño, S. Arora, Stochastic and minimum regret formulations for transmission network expansion planning under uncertainties. J. Oper. Res. Soc. **59**(11), 1547–1556 (2008). Retrieved from http://www.jstor.org/stable/20202237

J. Bustos-Salvagno, F. Fuentes, Electricity interconnection in Chile: prices versus costs. Energies **10**(9), 1438 (2017). https://doi.org/10.3390/en10091438

M.O. Buygi, G. Balzer, H.M. Shanechi, M. Shahidehpour, Market-based transmission expansion planning. IEEE Trans. Power Syst. **19**(4), 2060–2067 (2004). https://doi.org/10.1109/TPWRS.2004.836252

M.O. Buygi, H.M. Shanechi, G. Balzer, M. Shahidehpour, Transmission planning approaches in restructured power systems, in *Power Tech Conference Proceedings, 2003 IEEE Bologna*, p. 7, vol. 2 (2003). https://doi.org/10.1109/PTC.2003.1304666

M.O. Buygi, H.M. Shanechi, G. Balzer, M. Shahidehpour, N. Pariz, Network planning in unbundled power systems. IEEE Trans. Power Syst. **21**(3), 1379–1387 (2006). https://doi.org/10.1109/TPWRS.2006.873016

CAISO, *Consideration of Alternatives to Transmission or Conventional Generation to Address Local Needs in the Transmission Planning Process* (2013). Retrieved from https://www.caiso.com/Documents/Paper-Non-ConventionalAlternatives-2013-2014TransmissionPlanningProcess.pdf

J.M. Chamorro, L.M. Abadie, R. de Neufville, M. Ilic, Market-based valuation of transmission network expansion. A heuristic application in GB. Energy **44**(1), 302–320 (2012). https://doi.org/10.1016/j.energy.2012.06.028

D. Chattopadhyay, Scale efficient network development to support renewable generation development. IEEE Trans. Sustainab. Energy **2**(3), 329–339 (2011). https://doi.org/10.1109/TSTE.2011.2138173

A.R. Ciupuliga, E. Cuppen, The role of dialogue in fostering acceptance of transmission lines: the case of a France-Spain interconnection project. Energy Policy **60**, 224–233 (2013). https://doi.org/10.1016/J.ENPOL.2013.05.028

CNE, *Impacto Económico y Social de Interconexión SIC-SING* (2013). Retrieved from https://www.cne.cl/wp-content/uploads/2015/07/Impacto-Económico-y-Social-de-Interconexión-SIC-SING.pdf

J.J. Cohen, J. Reichl, M. Schmidthaler, Re-focussing research efforts on the public acceptance of energy infrastructure: a critical review. Energy **76**, 4–9 (2014). https://doi.org/10.1016/J.ENERGY.2013.12.056

Consorcio Centro Cambio Global UC—Centro de Energía U. de Chile y Teco Group, *Estudio para la Implementación del Proceso de Determinación de Franjas Preliminares* (2018). Retrieved from http://franjas.minenergia.cl/sites/default/files/estudio_para_la_implementacion_del_proceso_de_determinacion_de_franjas_preliminares.pdf

P. Devine-Wright, Explaining "NIMBY" objections to a power line: the role of personal, place attachment and project-related factors. Environ. Behav. **45**(6), 761–781 (2012). https://doi.org/10.1177/0013916512440435

F.X. Diebold, N.A. Doherty, R.J. Herring, Introduction, in *The Known, the Unknown, and the Unknowable in Financial Risk Management: Measurement and Theory Advancing Practice*, ed. by F.X. Diebold, N.A. Doherty, R.J. Herring (2010), pp. 1–30. Retrieved from http://www.jstor.org/stable/j.cttq9s3f.4

J. Dillon, D.J. Burke, M.J. O'Malley, A. Tuohy, Should unit commitment be endogenously included in wind power transmission planning optimisation models? IET Renew. Power Gener. **8**(2), 132–140 (2014). https://doi.org/10.1049/iet-rpg.2011.0272

D.A. Douglass, A.-A. Edris, Real-time monitoring and dynamic thermal rating of power transmission circuits. IEEE Trans. Power Delivery **11**(3), 1407–1418 (1996). https://doi.org/10.1109/61.517499

E. Ela, M. Milligan, A. Bloom, A. Botterud, A. Townsend, T. Levin, *Evolution of Wholesale Electricity Market Design with Increasing Levels of Renewable Generation* (2014)

E. Ellery, A. Brown, E. Crousillat, AF-Mercados EMI, V. Loksha, R. Tan, B. Dilli, *International Experience with Open Access to Power Grids: Synthesis Report* (2013), pp. 1–56. Retrieved from http://documents.worldbank.org/curated/en/2013/11/18699545/international-experience-open-access-power-grids-synthesis-report

Energy Ministry, *Proceso de planificación energética de largo plazo* (2017). Retrieved from http://pelp.minenergia.cl/files/100

ENTSO, *TYNDP 2018: Scenario Report* (2018). Retrieved from https://docstore.entsoe.eu/Documents/TYNDP documents/TYNDP2018/Scenario_Report_2018_Final.pdf

E. Fernandez, I. Albizu, M.T. Bedialauneta, A.J. Mazon, P.T. Leite, Review of dynamic line rating systems for wind power integration. Renew. Sustain. Energy Rev. **53**, 80–92 (2016). https://doi.org/10.1016/J.RSER.2015.07.149

R. Ferreira, H. Rudnick, L. Barroso, The expansion of transmission—the challenges faced in South America. IEEE Power Energy Mag. **14**(4) (2016). Retrieved from http://ieeexplore.ieee.org/document/7491387/

E.B. Fisher, R.P. O'Neill, M.C. Ferris, Optimal transmission switching. IEEE Trans. Power Syst. **23**(3), 1346–1355 (2008). https://doi.org/10.1109/TPWRS.2008.922256

D. Fudenberg, J. Tirole, *Game theory* (The MIT Press, Cambridge, 1991)

R.C. Garcia, J. Contreras, P.F. Correia, J.I. Munoz, Transmission assets investment timing using net present value curves. Energy Policy **38**(1), 598–605 (2010). https://doi.org/10.1016/j.enpol.2009.10.012

R.E. Gomory, The known, the unknown and the unknowable, in *Scientific American* (1995). Retrieved from http://www.scientificamerican.com/article/the-known-the-unknown-and-the-unkno/

B.G. Gorenstin, N.M. Campodonico, J.P. Costa, M.V.F. Pereira, Power system expansion planning under uncertainty. IEEE Trans. Power Syst. **8**(1), 129–136 (1993). https://doi.org/10.1109/59.221258

Y. Gu, J. McCalley, Market-based transmission expansion planning under uncertainty. North Am. Power Sym. (NAPS) **2010**, 1–6 (2010). https://doi.org/10.1109/NAPS.2010.5619975

J.C. Harsanyi, R. Selten, *A General Theory of Equilibrium Selection in Games* (MIT Press, Cambridge, 1988)

K.N. Hasan, T.K. Saha, M. Eghbal, D. Chattopadhyay, Review of transmission schemes and case studies for renewable power integration into the remote grid. Renew. Sustain. Energy Rev. **18**, 568–582 (2013). https://doi.org/10.1016/j.rser.2012.10.045

S. Herling, F. Koza, P. McGlynn, The sponsorship model: competitive construction of transmission facilities in PJM interconnection. IEEE Power Energy Mag. **14**(4) (2016)

E. Hirst, B. Kirby, Key transmission planning issues. Electr. J. **14**(8), 59–70 (2001). https://doi.org/10.1016/S1040-6190(01)00239-1

W.W. Hogan, *Competitive Electricity Markets: A Wholesale Primer* (1998). Retrieved from http://www.hks.harvard.edu/fs/whogan/empr1298.pdf

W.W. Hogan, Electricity Market Restructuring: Reforms of Reforms. J. Regul. Econ. **21**(1), 103–132 (2002). https://doi.org/10.1023/A:1013682825693

W.W. Hogan, *Transmission Benefits and Cost Allocation* (2011). Retrieved from http://www.hks.harvard.edu/fs/whogan/Hogan_Trans_Cost_053111.pdf

H. Holttinen, A. Tuohy, M. Milligan, E. Lannoye, V. Silva, S. Muller, L. Soder, The flexibility workout: managing variable resources and assessing the need for power system modification. IEEE Power Energ. Mag. **11**(6), 53–62 (2013). https://doi.org/10.1109/MPE.2013.2278000

D. Hou, J.P. Pfeifenberger, Transmission's true value: adding up the benefits of infrastructure investments. Public Utilities Fortn. **7** (2012). Retrieved from http://westpointproject.com/wp-content/uploads/2010/10/Transmissions-True-Value2.pdf

S. Hunt, *Making Competition Work in Electricity* (Wiley, New York, 2002)

IEA, *Renewables 2018: Executive Summary* (2018). Retrieved from https://webstore.iea.org/download/summary/2312?fileName=English-Renewables-2018-ES.pdf

Inzunza, A., *CVAR Constrained Planning of Renewable Generation with Consideration of System Inertial Response, Reserve Services and Demand Participation* (Pontificia Universidad Católica de Chile, 2014). Retrieved from http://web.ing.puc.cl/~power/paperspdf/AndresInzunza.pdf

IPCC, *Emissions Scenarios* (2000). Retrieved from https://www.ipcc.ch/site/assets/uploads/2018/03/emissions_scenarios-1.pdf

IPCC, Summary for policymakers, in *Global Warming of 1.5°C. An IPCC Special Report on the Impacts of Global Warming of 1.5°C Above Pre-industrial Levels and Related Global Greenhouse Gas Emission Pathways, in the Context of Strengthening the Global Response to the Threat of Climate Change* ed. by V. Masson-Delmotte, P. Zhai, H.-O. Pörtner, D. Roberts, J. Skea, P.R. Shukla, et al. (2018), p. 32. Retrieved from https://www.ipcc.ch/site/assets/uploads/sites/2/2019/05/SR15_SPM_version_report_LR.pdf

S. Jin, A. Botterud, S.M. Ryan, Temporal versus Stochastic Granularity in thermal generation capacity planning with wind power. IEEE Trans. Power Syst. **29**(5), 2033–2041 (2014). https://doi.org/10.1109/TPWRS.2014.2299760

P.L. Joskow, *Patterns of Transmission Investment* (2005). Retrieved from http://economics.mit.edu/files/1174

G. Kavlak, J. McNerney, J.E. Trancik, Evaluating the causes of cost reduction in photovoltaic modules. Energy Policy **123**, 700–710 (2018). https://doi.org/10.1016/J.ENPOL.2018.08.015

A. Khodaei, M. Shahidehpour, S. Kamalinia, Transmission switching in expansion planning. IEEE Trans. Power Syst. **25**(3), 1722–1733 (2010). https://doi.org/10.1109/TPWRS.2009.2039946

D. Kirschen, G. Strbac, *Fundamentals of Power System Economics* (2005). Wiley, New York

N. Komendantova, A. Battaglini, Beyond Decide-Announce-Defend (DAD) and Not-in-My-Backyard (NIMBY) models? Addressing the social and public acceptance of electric transmission lines in Germany. Energy Res. Soc. Sci. **22**, 224–231 (2016a). https://doi.org/10.1016/J.ERSS.2016.10.001

N. Komendantova, A. Battaglini, Social challenges of electricity transmission: grid deployment in Germany, the United Kingdom, and Belgium. IEEE Power Energ. Mag. **14**(4), 79–87 (2016b). https://doi.org/10.1109/MPE.2016.2550399

I. Konstantelos, G. Strbac, Valuation of flexible transmission investment options under uncertainty. IEEE Trans. Power Syst. **30**(2), 1047–1055 (2015). https://doi.org/10.1109/TPWRS.2014.2363364

G. Latorre, R.D. Cruz, J.M. Areiza, A. Villegas, Classification of publications and models on transmission expansion planning. IEEE Trans. Power Syst. **18**(2), 938–946 (2003). https://doi.org/10.1109/TPWRS.2003.811168

P. Linares, Multiple criteria decision making and risk analysis as risk management tools for power systems planning. IEEE Trans. Power Syst. **17**(3), 895–900 (2002). https://doi.org/10.1109/TPWRS.2002.800991

M. Madrigal, S. Stoft, *Transmission Expansion for Renewable Energy Scale-Up Emerging Lessons and Recommendations* (2011). Retrieved from http://siteresources.worldbank.org/EXTENERGY2/Resources/Transmission-Expansion-and-RE.pdf

C. Matamala, R. Moreno, E. Sauma, The value of network investment coordination to reduce environmental externalities when integrating renewables: case on the Chilean transmission network. *Energy Policy* **126**(October 2018), 251–263 (2019). https://doi.org/10.1016/j.enpol.2018.10.065

D. Mejia-Giraldo, J.D. McCalley, Maximizing future flexibility in electric generation portfolios. IEEE Trans. Power Syst. **29**(1), 279–288 (2014). https://doi.org/10.1109/TPWRS.2013.2280840

M. Milligan, B.A. Frew, A. Bloom, E. Ela, A. Botterud, A. Townsend, T. Levin, Wholesale electricity market design with increasing levels of renewable generation: revenue sufficiency and long-term reliability. Electr. J. **29**(2), 26–38 (2016). https://doi.org/10.1016/j.tej.2016.02.005

MIT. *The Future of the Electric Grid* (2011). Retrieved from http://mitei.mit.edu/system/files/Electric_Grid_Full_Report.pdf

A. Moreira, G. Strbac, R. Moreno, A. Street, I. Konstantelos, A five-level MILP model for flexible transmission network planning under uncertainty: a min–max regret approach. IEEE Trans. Power Syst. **33**(1), 486–501 (2018). https://doi.org/10.1109/TPWRS.2017.2710637

R. Moreno, D. Pudjianto, G. Strbac, Transmission network investment with probabilistic security and corrective control. IEEE Trans. Power Syst. **28**(4), 3935–3944 (2013). https://doi.org/10.1109/TPWRS.2013.2257885

C. Munoz, E. Sauma, J. Contreras, J. Aguado, D. La Torre, Impact of high wind power penetration on transmission network expansion planning. IET Gener. Transm. Distrib. **6**(12), 1281–1291 (2012). https://doi.org/10.1049/iet-gtd.2011.0552

F.D. Munoz, B.F. Hobbs, J.L. Ho, S. Kasina, An engineering-economic approach to transmission planning under market and regulatory uncertainties: WECC case study. IEEE Trans. Power Syst. **29**(1), 307–317 (2014). https://doi.org/10.1109/TPWRS.2013.2279654

F.D. Munoz, J.-P. Watson, B.F. Hobbs, Optimizing your options: extracting the full economic value of transmission when planning under uncertainty. Electr. J. **28**(5), 26–38 (2015). https://doi.org/10.1016/j.tej.2015.05.002

D. Myers, A. Kitsuse, Constructing the future in planning: a survey of theories and tools. J. Plan. Educ. Res. **19**(3), 221–231 (2000). https://doi.org/10.1177/0739456X0001900301

National Grid, *Electricity Ten Year Statement 2015* (2015a). Retrieved from http://www2.nationalgrid.com/UK/Industry-information/Future-of-Energy/Electricity-Ten-Year-Statement/

National Grid, *Future Energy Scenarios 2015* (2015b). Retrieved from http://www2.nationalgrid.com/WorkArea/DownloadAsset.aspx?id=41952

National Grid Plc, *Electricity Ten Year Statement 2014* (2014). Retrieved from https://www.nationalgrideso.com/document/46941/download

K. Neuhoff, J. Barquin, J.W. Bialek, R. Boyd, C.J. Dent, F. Echavarren et al., Renewable electric energy integration: quantifying the value of design of markets for international transmission capacity. Energy Econ. **40**, 760–772 (2013). https://doi.org/10.1016/j.eneco.2013.09.004

G.C. Oliveira, S. Binato, M.V.F. Pereira, Value-based transmission expansion planning of hydrothermal systems under uncertainty. IEEE Trans. Power Syst. **22**(4), 1429–1435 (2007). https://doi.org/10.1109/TPWRS.2007.907161

G.A. Orfanos, P.S. Georgilakis, N.D. Hatziargyriou, Transmission expansion planning of systems with increasing wind power integration. IEEE Trans. Power Syst. **28**(2), 1355–1362 (2013). https://doi.org/10.1109/TPWRS.2012.2214242

R. Pérez Odeh, D. Watts, Impacts of wind and solar spatial diversification on its market value: a case study of the Chilean electricity market. Renew. Sustain. Energy Rev. **111**, 442–461 (2019). https://doi.org/10.1016/J.RSER.2019.01.015

PJM, *A Survey of Transmission Cost Allocation Issues, Methods and Practices* (2010). Retrieved from https://core.ac.uk/download/pdf/21721132.pdf

PJM, *2014 Regional Transmission Expansion Plan Report—Book 1 RTEP in Review* (2015a). Retrieved from http://www.pjm.com/~/media/documents/reports/2014-rtep/2014-rtep-book-1.ashx

PJM, *2014 Regional Transmission Expansion Plan Report—Scenario and Interregional Study Results* (2015b). Retrieved from http://pjm.com/~/media/documents/reports/2014-rtep/2014-rtep-book-4.ashx

PJM, *2017 RTEP Process Scope and Input Assumptions* (2017). Retrieved from https://www.pjm.com/-/media/library/reports-notices/2017-rtep/20170731-rtep-input-assumptions-and-scope-whitepaper.ashx?la=en

F. Porrua, R. Chabar, L.M. Thome, L.A. Barroso, M. Pereira, Incorporating large-scale renewable to the transmission grid: technical and regulatory issues. IEEE Power Energy Soc. Gen. Meet. **2009**, 1–7 (2009). https://doi.org/10.1109/PES.2009.5275657

RTE, *Schéma décennal de développement du réseau Édition 2014* (2014). Retrieved from https://www.rte-france.com/sites/default/files/sddr_2014.pdf

H. Rudnick, L.A. Barroso, D. Llarens, D. Watts, R. Ferreira, Flexible connections: solutions and challenges for the integration of renewables in South America. IEEE Power Energ. Mag. **10**(2), 24–36 (2012). https://doi.org/10.1109/MPE.2011.2178300

H. Rudnick, R. Varela, W. Hogan, Evaluation of alternatives for power system coordination and pooling in a competitive environment. IEEE Trans. Power Syst. **12**(2), 605–613 (1997). https://doi.org/10.1109/59.589615

G. Sanchis, B. Betraoui, T. Anderski, E. Peirano, R. Pestana, B. De Clercq et al., The corridors of power: a Pan-European, "Electricity Highway" system for 2050. IEEE Power Energ. Mag. **13**(1), 38–51 (2015). https://doi.org/10.1109/MPE.2014.2363528

E. Sauma, S. Oren, Proactive planning and valuation of transmission investments in restructured electricity markets. J. Regul. Econ. **30**(3), 358–387 (2006). Retrieved from http://econpapers.repec.org/RePEc:kap:regeco:v:30:y:2006:i:3:p:358–387

J. Schlabbach (Jürgen), K.-H. Rofalski, *Power System Engineering: Planning, Design, and Operation of Power Systems and Equipment* (Wiley-VCH, Weinheim, 2008)

P. Schmidt, J. Lilliestam, Reducing or fostering public opposition? A critical reflection on the neutrality of pan-European cost–benefit analysis in electricity transmission planning. Energy Res. Soc. Sci. **10**, 114–122 (2015). https://doi.org/10.1016/j.erss.2015.07.003

P.J.H. Schoemaker, Multiple scenario development: its conceptual and behavioral foundation. Strateg. Manag. J. **14**(3), 193–213 (1993). https://doi.org/10.1002/smj.4250140304

M. Shahidehpour, R. Ferrero, Time management for assets: chronological strategies for power system asset management. IEEE Power Energ. Mag. **3**(3), 32–38 (2005). https://doi.org/10.1109/MPAE.2005.1436498

L. Späth, A. Scolobig, Stakeholder empowerment through participatory planning practices: the case of electricity transmission lines in France and Norway. Energy Res. Soc. Sci. **23**, 189–198 (2017). https://doi.org/10.1016/J.ERSS.2016.10.002

S. Stoft, Transmission investment in a deregulated power market, in *Competitive Electricity Markets and Sustainability*, ed. by F. Lévêque (2006), p. 34. Retrieved from http://stoft.com/metaPage/lib/Stoft-2005-07-Tx-Invest-EU-final.pdf

H.G. Stoll, *Least-Cost Electric Utility Planning* (Wiley-Interscience, New York, 1989)

Synex–Mercados. *Análisis Técnico y Económico de una Interconexión SING–SIC* (2012). Retrieved from https://www.cne.cl/wp-content/uploads/2015/07/Informe-Final-Interconexión-Sing-Sic.pdf

D. Tejada, J. Lopez-Lezama, M.J. Rider, G. Vinasco, Transmission network expansion planning considering repowering and reconfiguration. Int. J. Electr. Power Energy Syst. **69**, 213–221 (2015). https://doi.org/10.1016/j.ijepes.2015.01.008

A. Tversky, D. Kahneman, Judgment under uncertainty: heuristics and biases. Science **185**(4157), 1124–1131 (1974). Retrieved from http://www.jstor.org/stable/1738360

A.H. van der Weijde, B.F. Hobbs, The economics of planning electricity transmission to accommodate renewables: using two-stage optimisation to evaluate flexibility and the cost of disregarding uncertainty. Energy Econ. **34**(6), 2089–2101 (2012). https://doi.org/10.1016/j.eneco.2012.02.015

P.W. van Notten, J. Rotmans, M.B. van Asselt, D.S. Rothman, An updated scenario typology. Futures **35**(5), 423–443 (2003). https://doi.org/10.1016/S0016-3287(02)00090-3

C. Velásquez, *Robust and Efficient Decision Making for Transmission Expansion Planning Under Uncertainty* (2017). Retrieved from http://hrudnick.sitios.ing.uc.cl/paperspdf/ConstantinVelasquez.pdf

C. Velasquez, D. Watts, H. Rudnick, C. Bustos, A framework for transmission expansion planning: a complex problem clouded by uncertainty. IEEE Power Energy Mag. **14**(4) (2016). https://doi.org/10.1109/MPE.2016.2547278

C. Von Hirschhausen, C. Gerbaulet, C. Kemfert, C. Lorenz, P.-Y. Oei, *Energiewende "Made in Germany": Low Carbon Electricity Sector Reform in the European Context* (Springer, Cham, 2018)

D. Watts, N. Oses, R. Pérez, Assessment of wind energy potential in Chile: a project-based regional wind supply function approach. Renew. Energy **96**, 738–755 (2016). https://doi.org/10.1016/j.renene.2016.05.038

D. Watts, H. Rudnick, *Expansión de la Transmisión Informe Final* (2014)

A. Weber, T. Beckers, P. Behr, N. Bieschke, S. Fehner, C. von Hirschhausen, *Long-term Power System Planning in the Context of Changing Policy Objectives—Conceptual Issues and Selected Evidence from Europe* (2013). Retrieved from http://www.sefep.eu/activities/publications-1/SEFEP_gridplanning_June2013.pdf

The Impact of Transmission Development on a 100% Renewable Electricity Supply—A Spatial Case Study on the German Power System

Jens Weibezahn, Mario Kendziorski, Hendrik Kramer, and Christian von Hirschhausen

Abbreviations

AC	Alternating current
DC	Direct current
FLH	Full load hour
HVDC	High-voltage direct-current
lib	Lithium-ion battery
lib	Lithium-ion batteries
NDP	Network development plan
NGO	Non-governmental organization
NUTS2	Nomenclature of Territorial Units for Statistics, Level 2
p2g	Power-to-gas
phes	Pumped hydroelectric energy storage
pv-open	Open-space photovoltaics
pv-roof	Rooftop photovoltaics
PV	Photovoltaic
RES	Renewable energy sources
rfb	Redox flow battery

J. Weibezahn (✉) · M. Kendziorski
Technische Universität Berlin, Berlin, Germany
e-mail: jew@wip.tu-berlin.de

M. Kendziorski
e-mail: mak@wip.tu-berlin.de

H. Kramer
Technische Universität Berlin and Universität Duisburg-Essen, Berlin, Germany and Essen, Germany
e-mail: hendrik.kramer@uni-due.de

C. von Hirschhausen
Technische Universität Berlin and German Institute for Economic Research (DIW Berlin), Berlin, Germany
e-mail: cvh@wip.tu-berlin.de

M. R. Hesamzadeh et al. (eds.), *Transmission Network Investment in Liberalized Power Markets*, Lecture Notes in Energy 79,
https://doi.org/10.1007/978-3-030-47929-9_15

rfb	Redox flow batteries
ror	Run-of-river
TSO	Transmission system operator
TYNDP	Ten-Year Network Development Plan
wind-off	Offshore wind
wind-on	Onshore wind

1 Introduction

The role of electricity transmission infrastructure for an energy system has, is, and continues to be an issue of greatest importance, and of controversy, too. Traditionally, the structure of transmission networks has followed the pattern of generation and load, serving mainly as a backup service in cases of larger discrepancy, or connecting power plants to the grid that had been planned without consideration to "optimal" location, such as nuclear power plants (usually far away from load centers). Thus, in the old days, transmission planning was considered to be fuel-neutral, and the role of the transmission network was limited to fuel-neutral backup service. Foci of research were incentives in transmission planning, for example, high- or low-powered incentives (Hogan et al. 2010; Olmos and Pérez-Arriaga 2009; Rosellón and Kristiansen 2013), and financing, for example, merchant vs. regulated financing (Gerbaulet and Weber 2018; Joskow and Tirole 2005, 2006).

However, with the need to phase out fossil generation, and the arrival of massive amounts of distributed renewable energy, the assessment of the link between the electricity mix and the transmission requirements has fundamentally changed. The technology shift from carbon-intense conventional power plants toward largely renewable technologies has strong implications for the power system's operation and investment decisions. A crucial characteristic of renewable energy sources (RES) is their increased distributed structure, among others due to the smaller power ratings of such technologies. In particular, it has become evident that transmission planning is *not* neutral vis-à-vis the electricity mix, but that there is a direct link between the design of the transmission system and the resulting electricity flows, be they driven by carbon-intense, nuclear, or renewable generation. Thus, it is now commonly agreed that in a carbon-intense electricity system, transmission expansion leads to *more* carbon emissions: Transmission expansion does affect the electricity mix and leads to rising CO_2 emissions in the European context, for example, in the thesis by Carlo (2013) and the theory-based numerical assessment of Abrell and Rausch (2016).

Transmission expansion cannot be analyzed independently from the specific institutional context in which it takes place, that is, the form of regulation, financing, etc. In this chapter, we focus on the German case, a typical case in which transmission planning was considered to be a "standard" activity given to the transmission companies until recently, but where the interdependence with the electricity fuel mix has come out clearly once the *energiewende*, the no-carbon, no-nuclear transformation of the energy system, has taken off. In the old days of vertical integration and regional monopolies, the transmission companies assured a congestion-free network with ample overcapacity, which is, by the way, the basis why transmission constraints are not really binding until today (Gerbaulet 2018; Kunz et al. 2013). After vertical unbundling (in 1998) and first targets for renewables in the network development plan (2005), transmission system operators (TSOs) adapted a rhetoric of linking the share of renewables to additional transmission expansion, even though the real level of congestion was, and still is, negligible. As generously rate-of-return regulated companies with a significant information advantage vis-à-vis the regulator, TSOs tried to maximize expansion plans at the cost of economic efficiency (Kemfert et al. 2016).

On top of this restructuring, the objective of the energy reform (*energiewende*) in Germany puts additional stress onto the system. As part of the larger package, the *energiewende* includes strict greenhouse gas reduction targets (−80 to −95% by 2050, basis 1990), over 80% of electricity from renewables (by 2050), the disconnection of nuclear power plants by 2022, and strict targets for economy-wide and sectoral energy efficiency (Hirschhausen 2018). The impact of the *energiewende* on transmission planning was discussed controversially, with two major issues. TSOs continue their quest for large-scale transmission expansion, whereas from a climate perspective, it had become clear that, given the carbon-intense fuel mix, this would still mainly benefit the coal plants in East Germany (Mieth et al. 2015). The few hours of network congestion were indeed correlated to high use of coal plants, mainly lignite in East Germany. In 2015, the network regulator integrated carbon constraints in network planning for the first time and has tightened these since (Mieth et al. 2015). Today, there is a general consensus on phasing out coal in the 2030s (Göke et al. 2018).

Thus, TSOs are in a key position not only to determine the amount of network expansion, but also to affect the electricity mix indirectly. Based on the assumption of massive network congestion, renewables expansion was linked to an "appropriate" level of transmission expansion, particularly large-scale transmission lines between the North/East of Germany and the South/West, in the national Ten-Year Network Development Plans (TYNDP) (50Hertz Transmission 2018); the plan also included several high-voltage direct-current (HVDC) lines. With the rate-of-return on capital of over 6% (and interest rates of about 0%), no wonder TSOs tried to maximize expansion plans. On the other hand, our own work has confirmed the detrimental effect of transmission expansion in a fossil-fuel basin, that is, the lignite basin in East Germany (Lusatia), on the dispatch of lignite plants in the region: when expanding the East–South high-voltage corridor, about 30 GWh of additional lignite would be produced, corresponding to almost the entire electricity deficit of Bavaria.

While this conversation is still ongoing, it seems necessary to look ahead and consider the longer term, that is, when the objective of the low-carbon, no-nuclear *energiewende* will be attained, including a largely renewables-based electricity generation. That is why we adopt a different perspective in this chapter, analyzing the link between the electricity mix, with a focus on centralized and distributed renewables, and the nature of the transmission system. Our hypothesis is that the geographical distribution of the renewable electricity mix interacts with different transmission architectures. To test this hypothesis, we develop a stylized model of transmission and generation investment, and operation based on the traditions of electricity network modeling (Leuthold et al. 2012; Weibezahn and Kendziorski 2019), but add a high degree of technical and spatial detail in the spirit of the DIETER model (Zerrahn and Schill 2015).

The remainder of this chapter is structured as follows: Sect. 2 analyzes the current literature concerning distributed resources and their effects on transmission requirement. Section 3 provides a description of the investment and dispatch model that is developed to compare the different locations of renewables, the data, and the scenarios. Section 4 provides the results of the scenario runs and discusses them, and Sect. 5 concludes.

2 Literature Review

Many studies analyze optimal renewable power plant placement in electricity distribution grids. 50Hertz Transmission (2019) assess the literature published on the optimal placement of renewable energy sources, discuss the drivers of increased interest, and compare different optimization approaches. They state that most studies focus on distribution grids for a given network setup without considering network extension. Besides renewable power plant siting and sizing, optimization studies also focus on system flexibility to integrate RES, such as storage systems (Lund et al. 2015). Sophisticated models such as written by Kayal and Chanda (2015) consider secure grid operation and different weather conditions, but treat the transmission network as static.

Fewer studies are performed to assess cost optimal capacity extension of wind and photovoltaic (PV) power plants in combination with transmission network extension options. Schlachtberger et al. (2017) propose a cost-minimizing optimization problem. A case study for Europe is elaborated. In one scenario, no energy exchange between countries is allowed, whereas in another scenario the effect of international electricity trade is highlighted. They conclude that it is important to consider spatial and temporal scales when performing research on the integration of high shares of renewable power in the given grid infrastructure.

Likewise, Grams et al. (2017) show that spatial deployment of wind power over a large region allows minimizing renewable energy output variability. For Europe, they conclude that large-scale spacial deployment could be a strategic response to the multi-day volatility challenge of the common weather regimes on the European continent. Based on a nodal approach, Abrell and Rausch (2016) point out that increased inter-European cross-border transmission capacities allow for more renewable power usage. Furthermore, the European climate targets could be reached at cheaper cost, if national climate mitigation plans and thus their view on transmission adequacy would be matched more in a cooperative fashion. On a national scale, Drechsler et al. (2017) conclude that a spatially even mix of wind and solar power is preferable for the German national electricity system. They highlight that the current tender mechanism for wind power plant subsidies incorporate a regional correction factor to support regional distribution to some extent, whereas such a factor is missing for

PV tender auctions. This so-called reference yield model (*Referenzertragsmodell*) balances wind power over the whole territory of Germany, based on geographical characteristics defined for each postal code. It neither considers present network infrastructure information nor distance to regions of high electricity demand. Back in 2010 in Germany, strategic planning and support of erecting wind turbines was absent, such that investors faced obstacles to install wind power plants at the most beneficial locations (Ohl and Eichhorn 2010). Likewise, Ohlhorst (2015) found that federal state government targets for wind and PV power plants are not in line with national top-down climate mitigation ambitions. In particular, the current energy policy is characterized by a separated planning approach of grid infrastructure extension and power generation dispatch planning, resulting in higher total cost. An integrated approach combining both aspects would result in welfare gains (Kemfert et al. 2016). Clearly in the case of Germany, there are incentives for overinvesting into transmission infrastructure.

There are no simple answers to resolve the issue of "optimal" transmission planning for a largely renewables-based electricity system. Clearly, the higher the oversupply of transmission, the easier it is to feed in surplus renewables, but this holds for surplus fossil-fuel electricity, too—as currently practiced in Germany, which has a 50 TWh export surplus, mainly based on coal and lignite sources. Thus, while Fürsch et al. (2013) favor grid extension to integrate RES, there needs to be a compromise between different flexibility options, in particular in a dynamic perspective where the spatial distribution of renewable electricity is endogenous. Not only environmental non-governmental organizations (NGOs) argue against overinvestment into the transmission network, and that issues of sustainable generation should be prioritized vis-à-vis transmission issues[1]; this has also been shown, again, recently in techno-economic research on the German electricity grid (in the European context), such as Grimm et al. (2016a, b).

In this chapter, we add a spatial component of distributed resources and also integrate (spatially differentiable) storage capacities, to assess the relation between different transmission designs and the optimal allocation of generation and storage.

3 Model, Data, and Scenarios

3.1 Dispatch and Investment Model with Linearized Power Flow

The analysis is methodically based on an investment model minimizing the sum of the costs of installed infrastructure investments and operational power generation cost. The model is inspired by ELMOD (Egerer, 2016), Joulia.jl (Weibezahn and Kendziorski, 2019), dynELMOD (Gerbaulet and Lorenz, 2017) and DIETER (Zerrahn and Schill, 2015). Combining elements of the models mentioned before, the

[1] Naturschutzbund Deutschland e. V. (2019): "Stellungnahme zum NEP Strom 2030"

following equations account for investment and dispatch activities, while also considering the network topology. The model does not account for the current power plant fleet (greenfield model approach), as infrastructure has to be renewed until 2050 anyway.

The model sets are technologies $\mathcal{T}$ (with subsets for dispatchable units $\mathcal{T_D}$, non-dispatchable units $\mathcal{T_N}$, and storage technologies $\mathcal{T_S}$), regional zones $\mathcal{Z}$ (as subsets of countries $\mathcal{C}$, alternating current (AC) transmission lines $\mathcal{L}$, hours $\mathcal{H}$ and seasons $\mathcal{W}$. Decision variables for the dispatch are power output by generation units G^{gen} (including storage discharge), storage charge G^{ch}, storage state of charge E^{soc}, transmitted power through power injection in one region F^{ni}, high-voltage direct-current (HVDC) line usage F^{dc}, and lost load LL. Investment relevant variables are installed power output P^{inst}, installed charging power P^{ch}, and installed storage energy capacity E^{inst}. Model parameters are power demand p^{load}, the availability factor ζ additionally restricting power availability for non-dispatchable technologies, the autarky factor ϕ reducing international electricity exchange, and investment and generation cost factors c^p, c^{ch}, c^e and c^{mc}, as well as c^{ll} reducing unserved electricity demand. Further network parameters are power and energy restrictions on network elements, that is, generation (p^{max}), storage (e^{max}, η, ρ) and transmission ($ptdf$, f^{max}). The time scaling factor γ allows for cost comparisons on a yearly scale.

minimize

$$\gamma^{\text{year}}\left[\sum_{\substack{t\in\mathcal{T}\\z\in\mathcal{Z}\\h\in\mathcal{H}}} c_t^{mc} G_{t,z,h}^{gen} + c^{ll}\sum_{\substack{z\in\mathcal{Z}\\h\in\mathcal{H}}} LL_{z,h}\right] + \sum_{\substack{t\in\mathcal{T}\\z\in\mathcal{Z}}} c_t^{p} P_{t,z}^{inst} + \sum_{\substack{t\in\mathcal{T_S}\\z\in\mathcal{Z}}} c_t^{ch} P_{t,z}^{ch} + \sum_{\substack{t\in\mathcal{T_S}\\z\in\mathcal{Z}}} c_t^{e} E_{\text{t,z}}^{inst}$$

subject to

(1)

$$\sum_{t\in\mathcal{T}} G_{t,z,h}^{gen} - \sum_{t\in\mathcal{T_S}} G_{t,z,h}^{ch} - load_{z,h} + \sum_{zz\in\mathcal{Z}} F_{zz,z,h}^{dc} - \sum_{zz\in\mathcal{Z}} F_{z,zz,h}^{dc} + LL_{z,h} = F_{z,h}^{ni} \quad (\forall z\in\mathcal{Z}, h\in\mathcal{H}) \tag{2}$$

$$G_{t,z,h}^{gen} \leq \zeta_{t,z,h} P_{t,z}^{inst} \quad (\forall t\in\mathcal{T_N}, z\in\mathcal{Z}, h\in\mathcal{H}) \tag{3}$$

$$G_{t,z,h}^{gen} \leq P_{t,z}^{inst} \quad (\forall t\in\mathcal{T_D}, z\in\mathcal{Z}, h\in\mathcal{H}) \tag{4}$$

$$G_{t,z,h}^{gen} \leq P_{t,z}^{inst} \quad (\forall t\in\mathcal{T_S}, z\in\mathcal{Z}, h\in\mathcal{H}) \tag{5}$$

$$G_{t,z,h}^{ch} \leq P_{t,z}^{ch} \quad (\forall t = p2g, z\in\mathcal{Z}, h\in\mathcal{H}) \tag{6}$$

$$E_{t,z,h}^{soc,h} + \gamma^{\text{season}} E_{t,z,w}^{soc,w} \leq E_{t,z}^{inst} \quad (\forall t\in\mathcal{T_S}, z\in\mathcal{Z}, h\in\mathcal{H}) \tag{7}$$

$$\gamma^{\text{year}} \sum_{\substack{z \in \mathcal{Z} \subseteq \mathcal{C} \\ h \in \mathcal{H}}} G^{gen}_{t,z,h} \leq p^{max,gen}_{t,c} \qquad (\forall t \in \mathcal{T}, c \in \mathcal{C}) \tag{8}$$

$$P^{inst}_{t,z} \leq p^{max,inst}_{t,z} \qquad (\forall t \in \mathcal{T}, z \in \mathcal{Z}) \tag{9}$$

$$E^{inst}_{t,z} \leq e^{max,inst}_{t,z} \qquad (\forall t \in \mathcal{T_S}, z \in \mathcal{Z}) \tag{10}$$

$$\begin{aligned} E^{soc,h}_{t,z,h} &= \rho_t E^{soc,h}_{t,z,h-1} + \sqrt{\eta^{ch}_t} G^{ch,h}_{t,z,h} - \frac{1}{\sqrt{\eta^{dch}_t}} G^{gen}_{t,z,h} \\ &+ G^{ch,w}_{t,z,h \text{ if first h of } \mathcal{W}} - G^{dch,w}_{t,z,h \text{ if last h of } \mathcal{W}} \\ &\qquad (\forall t \in \mathcal{T_S}, z \in \mathcal{Z}, h \in \mathcal{H}) \end{aligned} \tag{11}$$

$$E^{soc,w}_{t,z,w} = \rho_t^{\gamma^{\text{season}}} E^{soc,w}_{t,z,w-1} + G^{ch,w}_{t,z,w} - G^{dch,w}_{t,z,w} \quad (\forall t \in \mathcal{T_S}, z \in \mathcal{Z}, w \in \mathcal{W}) \tag{12}$$

$$-f^{max}_l \leq \sum_{z \in \mathcal{Z}} ptdf_{l,z} F^{ni}_{z,h} \leq f^{max}_l \qquad (\forall l \in \mathcal{L}, h \in \mathcal{H}) \tag{13}$$

$$F^{dc}_{z,zz,h} \leq f^{max}_{z,zz} \qquad (\forall z, zz \in \mathcal{Z}, h \in \mathcal{H}) \tag{14}$$

$$\sum_{z \in \mathcal{Z}} F^{ni}_{z,h} = 0 \qquad (\forall h \in \mathcal{H}) \tag{15}$$

$$\sum_{\substack{z \in \mathcal{Z} \subseteq \mathcal{C} \\ h \in \mathcal{H}}} \left(LL_{z,h} + \sum_{t \in \mathcal{T}} G^{gen}_{t,z,h} - \sum_{t \in \mathcal{T_S}} G^{ch}_{t,z,h} \right) \geq \phi \sum_{\substack{z \in \mathcal{Z} \subseteq \mathcal{C} \\ h \in \mathcal{H}}} load_{z,h} \qquad (\forall c \in \mathcal{C}) \tag{16}$$

The objective function minimizes the overall system costs that are represented by the sum of the power plant fleet investment cost, the storage investment cost, the power generation cost, and penalty costs for lost load (Equation 1). Market clearing in each region implies that power generation, demand, storage power interaction, power exchange through HVDC lines, and lost load equal the net exchange over conventional transmission lines between regions (Equation 2). Power output from dispatchable and non-dispatchable power plants has to be within the installed capacity, power output from fluctuating technologies might be lower due to lacking availability (Equations 3 and 4). The power output from storage entities is limited by its installed power rating and – particularly for the power-to-gas technology – the power infeed rate can be set differently from its outflow rate (Equations 5 and 6). The state of charge of the storage units including the seasonal storage energy flows need to be within the installed storage capacity (Equation 7). If technologies have a restriction on the energy that can be provided per year, it may not exceed that limit (Equation 8). The amount of installed storage power and storage capacity cannot be increased above its exogenously given potentials (Equation 9 and 10). Storage interaction exhibits losses when being charged, discharged, or when energy is kept within the storage device (self-discharge). If applicable, at the time slices defining a season start or ending, the energy injection or withdrawal, respectively, is possible (Equation 11). For each season, seasonal storage balance is defined by last seasons storage state of charge, inter-seasonal losses, and the interaction with storage at the season's start and end (Equation 12). The transmission line capacities are limited by their thermal limits (Equations 13 and 14). The net power balance in all modeled regions has to be zero (Equation 15). Each country's cross-border energy exchange is restricted to an autarky factor, such that a country's energy balance to the neighboring countries does not exceed an exogenously set percentage (Equation 16).

The model is written in the programming language Julia v1.1 in combination with the modeling tool JuMP v0.19 and uses the solver Gurobi v8.1.

3.2 Data

A regional split-up is implemented based on the Nomenclature of Territorial Units for Statistics, Level 2 (NUTS2) information of the European Union. This framework divides Germany into 38 regions. Each neighboring country is mapped as one model region, thus obtaining nine further network nodes (Netherlands, Belgium, France, Switzerland, Austria, Czech Republic, Poland, and Denmark). Figure 1 illustrates the geographical zone setup for this analysis. The names of the regions are attached in Table 4 in the appendix. The model includes the transmission lines of 380 kV and 220 kV, whereas lower voltage levels are neglected. The set of transmission lines is reduced to "system-relevant" lines following the methodology of Weinhold and Mieth (2020) so that only those lines are considered, whose bounds directly constrain the DC power flow solution, as they reach their thermal limits first. For the scenarios with transmission network expansion, data from the network development plan (NDP) (50Hertz Transmission 2018) is taken as reference. Based on this source, HVDC transmission lines are also modeled in the future scenarios as proposed by the NDP (see Fig. 2).

The potentials for renewable energies are calculated based on an approach put forward by Nahmmacher et al. (2014). Firstly, data on land area categorized into agriculture, forest, continuous urban fabric, and discontinuous urban fabric are taken from the European Environment Agency (2019). As the dataset is reported on a NUTS3 level, the values are then dis-aggregated to meet the defined NUTS2 zones. The land area that is available for wind turbines and PV panels is calculated according to factors in Table 1. This amount of land area is then multiplied with the energy density in order to obtain the potential generation capacity in megawatt (MW). Wind offshore is assumed to have an installation limit of 75 GW in Germany. Also no distinction is made in terms of investment costs or full load hours. However, in reality, costs will increase the further away from the coast the offshore wind park is being built.

Electricity demand data for each NUTS2 region is taken from Kunz et al. (2017). Availability time series of renewable energy sources—wind and PV—is provided by

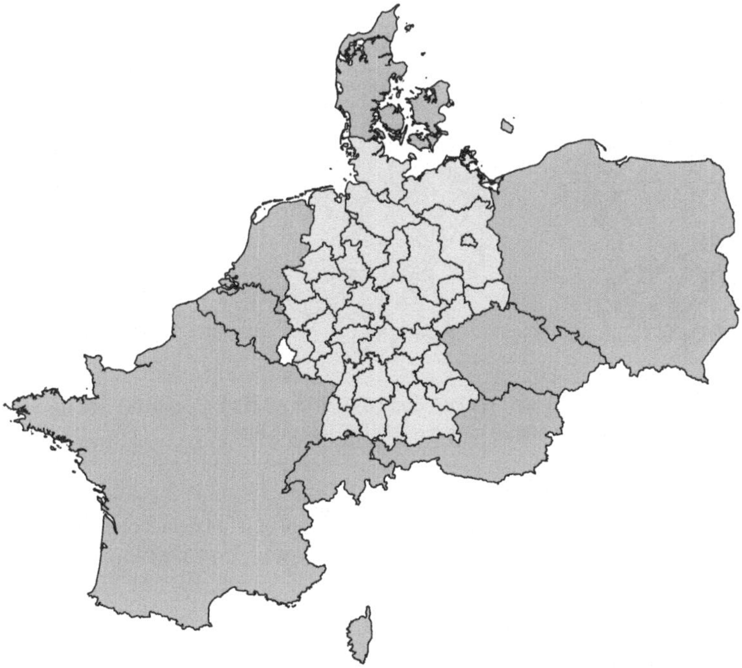

Fig. 1 NUTS2 zones in Germany (light gray) and neighboring countries (dark gray)

Table 1 Parameters used in the calculation of the renewable potentials

Source	Wind onshore	Open-space PV	Rooftop PV commercial	Rooftop PV residential
	Nahmmacher et al. (2014)	Nahmmacher et al. (2014)	own assumptions	own assumptions
Agriculture	30%	2%	–	–
Forest	5%	–	–	–
Continuous urban fabric	–	–	8%	25%
Discontinuous urban fabric	–	–	2%	1%
Energy density (MW/km^2)	4	30	0.16	0.16

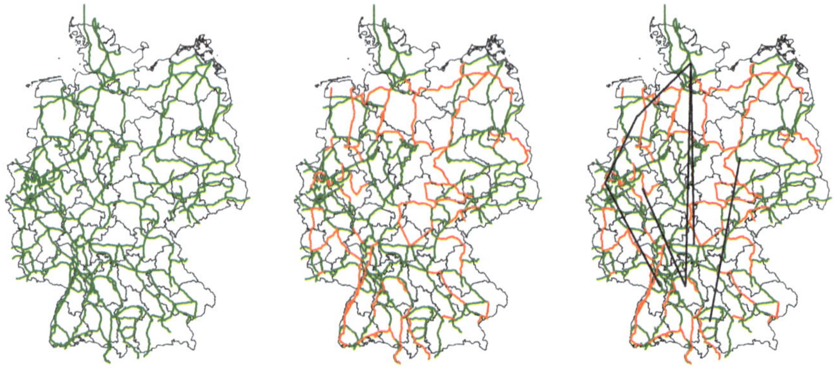

Fig. 2 Grid 2022, Grid 2035 w/o HVDC, and Grid 2035 w/ HVDC: existing AC lines in green, AC lines built or reinforced until 2035 in red, HVDC lines in black

Table 2 Technology data

Technology	Fixed cost (EUR/MW)	Overnight cost (EUR/MW)	Overnight cost (EUR/MWh)	Marginal cost (EUR/MWh)	Lifetime (years)
pv-open	6.375	246.000	0	0	20
pv-roof	6.375	467.000	0	0	20
wind-on	21.500	900.000	0	0	20
wind-off	80.000	2280.000	0	0	20
lib	1.960	75.000	164.000	1.3	12
rfb	2.000	550.000	122.000	1.3	20
phes	10.000	3000.000	10.000	0.5	60
p2g	20.000	2287.000	300	8	10
ror	30.000	3000.000	0	0	50

Pfenninger and Staffell (2016). Offshore wind data is assigned to the three coastal regions Weser-Ems, Schleswig-Holstein, and Mecklenburg-Vorpommern in Northern Germany, depending on where the submarine cables are linked to the onshore transmission grid.

The model contains the following generation and storage technologies: open-space photovoltaics (pv-open), rooftop photovoltaics (pv-roof), onshore wind (wind-on), offshore wind (wind-off), lithium-ion batteries (lib), redox flow batteries (rfb), pumped hydroelectric energy storages (phess), power-to-gas (p2g), and run-of-river (ror). Cost assumptions of each generation and storage technology are listed in Table 2.

Due to the computational complexity of the investment model, the time series method from Poncelet et al. (2016) is applied to obtain a time period of four weeks suitable for analysis.

3.3 Scenarios

The objective of this paper is to relate distributed renewable generation portfolios to different scenarios of transmission topologies and congestion patterns. To that end, we define different representative scenarios, in order to assess different investment patterns resulting thereof. These (exogenously defined) transmission scenarios are the following:

(i) Copper Plate A corner solution is to allow an unrestricted flow of electricity within the transmission grid. Thus, the geographical location of generation would become irrelevant;
(ii) Grid 2022 is a scenario in which the existing transmission network in the year 2022 is taken as the basis for the analysis. In this setting, a little network congestion does occur since the optimal locations for renewable sources are not identical to load centers;
(iii) Grid 2035 w/o HVDC describes a transmission expansion scenario that corresponds to the official network development plan by the TSOs, but without engaging into high-voltage direct-current (HVDC) lines;
(iv) Grid 2035 w/ HVDC corresponds to the full-fledged version of the official network development plan until 2035, including the HVDC lines.

Assuming a copper plate system setup, geographical distances on a national basis are neglected. Power flow between different regional zones is unlimited. Another analysis is performed with the existing/planned transmission grid of 2022. In the real world, network congestion occurs as renewable energy sources were installed far from the demand centers. As a consequence, the NDP quests for additional transmission capacity until 2035.[2] This enforced grid then also is going to contain DC transmission links that allow for a more flexible energy dispatch, as transmitted power can be directly controlled. The analysis of these different scenarios allows to draw conclusions about what would be the cost minimal solution for designing an efficient energy system starting from scratch or starting from today's setup. Differences in the obtained results thus indicate long-term lock-in costs from today's network topology.

4 Results and Discussion

Since this paper is primarily concerned with the link between transmission and distributed generation, we focus on the changes that different transmission designs, that is, the scenarios, have on generation, both in a static and a dynamic perspective. In particular, a 100% renewable electricity mix requires a fine balance between

[2]In this analysis, we refer to scenario B 2035 of the NDP.

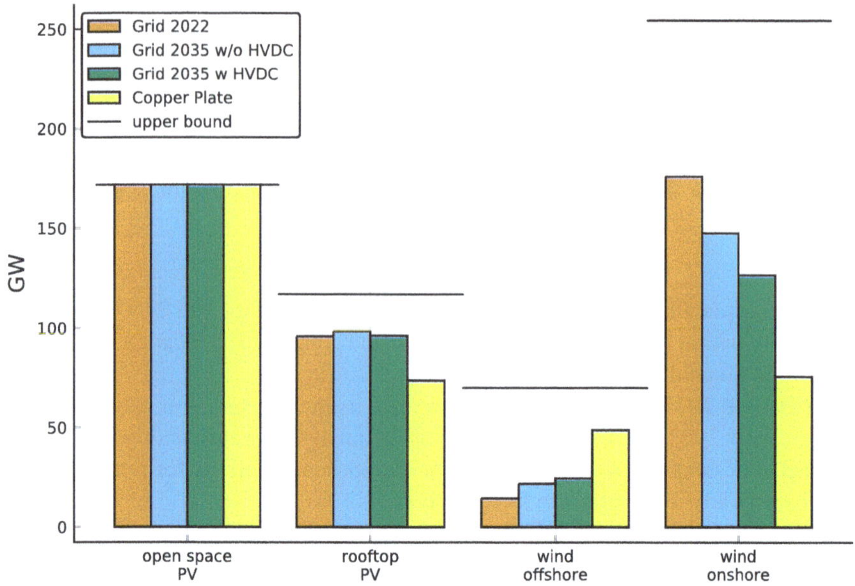

Fig. 3 Installed power in GW in Germany

flexibility options, among them transmission grids and storage.[3] The results and the discussion therefore focus on generation, storage, and transmission congestion, respectively.

4.1 Distributed Electricity Mix

Results show that the existing transmission grid has a huge impact on the optimal investment decisions. Figures 3 and 6 show the optimal investments into generation and storage power for the different scenarios. The solid black line in Fig. 3 indicates the upper bound for the respective technology given its installation potentials.

While utilizing the maximum installable potential of open-space PV is worthwhile in every scenario, other decisions vary depending on the level of grid expansion. The largest trade-off exists between onshore wind and offshore wind. In the copper plate scenario, only 76 GW of onshore wind is being built. In contrast, in the Grid 2022 scenario—having the lowest level of grid expansion—the investment into onshore wind increases to 176 GW. On the other side, the installed power of offshore wind decreases from 49 to 15 GW. Due to Germany's geographical location, the offshore wind sites are positioned in the north only while a significant share of the electricity demand lies in the southern regions. As a result, high investments into offshore

[3]Demand-side management, another important flexibility option, is not covered in this chapter.

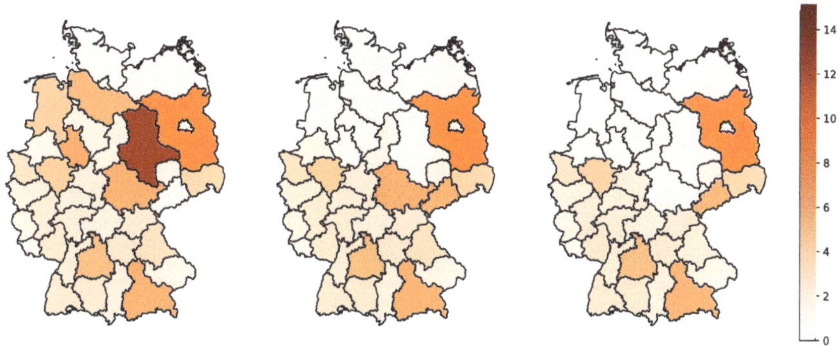

Fig. 4 Installed generation power: rooftop PV in Grid 2022, Grid 2035 w/o HVDC, Grid 2035 w/ HVDC (GW)

power are only feasible if enough transmission capacity is available. However, the lack of transmission capacity is offset by higher and more diversified investments into onshore wind and rooftop PV capacities. These two technologies are not bound to the coast and hence can be placed in a system-friendlier fashion.

Albeit the amount of installed power of rooftop PV does not differ between the scenarios that consider grid constraints, the spatial pattern changes (depicted in Fig. 4). In the scenario Grid 2022, rooftop PV panels are relatively evenly distributed among the regions, while in the other scenarios the investments are more concentrated in the southern regions where the full load hours are higher.

A similar pictures can be seen in Fig. 5. The onshore wind turbines are also present in the south even though yields are lower. With an an increasing grid expansion, the installed wind power diminishes almost completely in the very south.

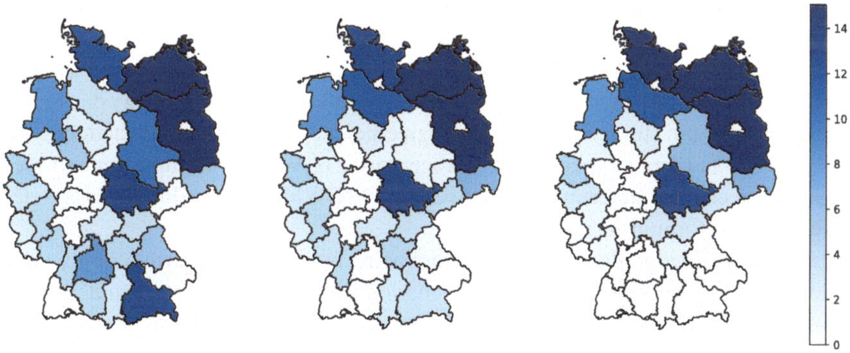

Fig. 5 Installed generation power: onshore wind in Grid 2022, Grid 2035 w/o HVDC, Grid 2035 w/ HVDC (GW)

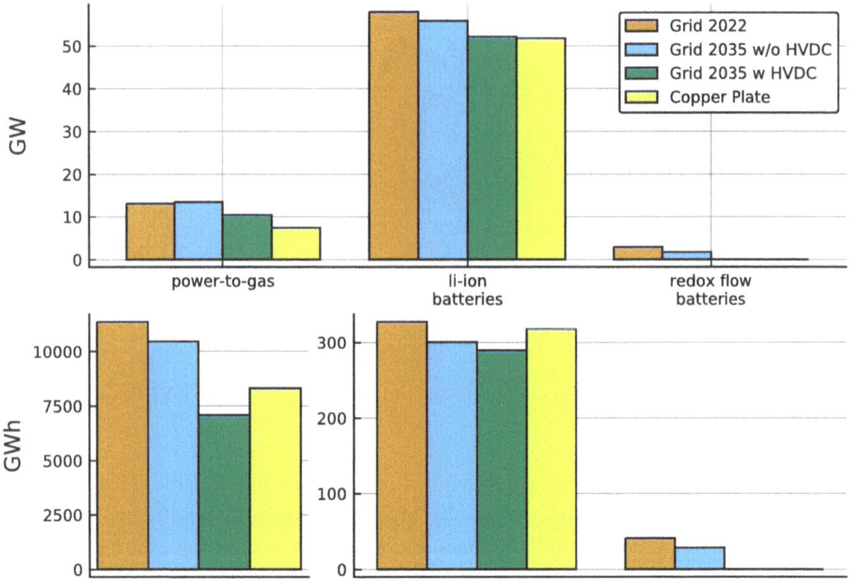

Fig. 6 Installed storage power and capacity in GWh in Germany

4.2 *Storage Capacities*

The flexibility option "storage" is particularly important in a system dominated by distributed generation. The storage capacity in GWh is depicted in Fig. 6. The left part of the plot shows the investment into power-to-gas storage. This large-scale storage is used for seasonal storing due to its low investment costs in capacity (GWh), while the conversion from electricity to gas and the re-electrification is inefficient and associated with high investment costs into power (GW). On the right hand side, the installed storage capacity of the lithium-ion and the redox flow batteries are plotted (note that the scale of the y-axis is different). Lithium-ion batteries serve as a short-term storage that is frequently in use for shifting smaller amounts of energy due to low investments cost in storage power and a high efficiency. The redox flow batteries are less efficient and more expensive in storage power (GW) than lithium-ion batteries but in return cheaper in terms of storage capacity (GWh). Thus, redox flow batteries are best suited for mid-term flexibility.

Interestingly, redox flow batteries are not being built in the scenarios copper plate and Grid 2035 w/ HVDC while in the scenarios with higher grid constraints investments in redox flow storage capacity is advantageous. Contrarily, the seasonal storage capacity declines with a higher level of grid expansion which is not true for the copper plate scenario. The redox flow batteries in combination with the lithium-ion batteries are used to resolve grid congestion. Since grid bottlenecks can occur for more than a several hours (e.g., the wind is blowing strongly in the north for a couple of days, and a longer period of cloudy days occurs in the south), a mid-term storage is sufficient.

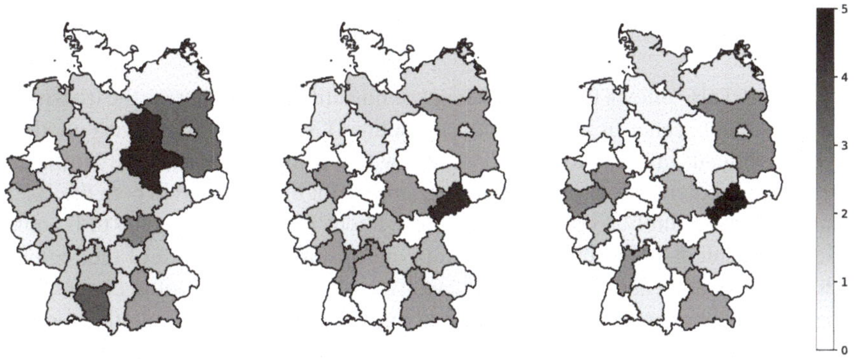

Fig. 7 Installed storage power: lithium-ion batteries in Grid 2022, Grid 2035 w/o HVDC, Grid 2035 w/ HVDC (GW)

These bottlenecks do not exist in the copper plate scenario where the excessive energy is stored in a seasonal storage immediately. While the scenarios including the grid have an investment pattern that diversifies stronger into different technologies and locations, the copper plate scenario can harvest the best spots without considering any grid limitations. As discussed in the previous section, investments into onshore wind and rooftop PV are significantly higher in the scenario Grid 2035 w/ HVDC than in copper plate, resulting in a lower need for seasonal storage.

In Fig. 7, the locations of the lithium-ion batteries also shift from a more distributed pattern to a slightly stronger concentrated pattern in the case of Grid 2035. The effect is more distinct for the locations of the power-to-gas storage capacity, which are mainly focused in the north-west (Fig. 8).

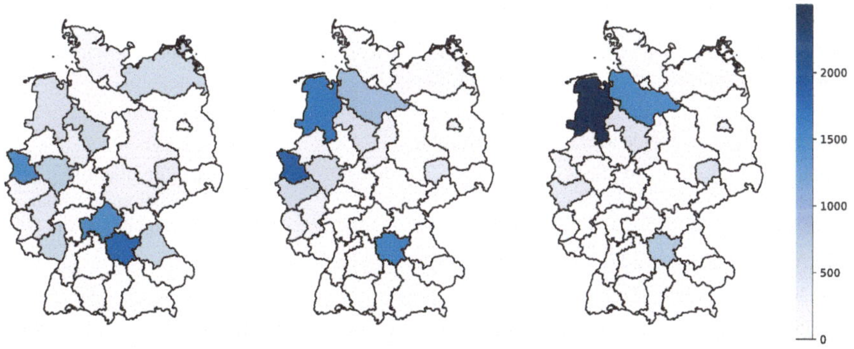

Fig. 8 Installed storage capacity: power-to-gas in Grid 2022, Grid 2035 w/o HVDC, Grid 2035 w/ HVDC (GWh)

4.3 Transmission Congestion

The model also allows for an assessment of line utilization (whereas the transmission investment scenarios are exogenously defined). Figure 9 shows a very modest level of overload: In scenario Grid 2022, a few line connections are highly utilized. However, this potentially congested lines disappear for the most part in both the 2035 scenarios. Interestingly, even the 2035 grid without HVDC lines seems to be able to accommodate the distributed energy mix quite comfortably. Most of the regions with high generation and/or high load are well equipped and do not suffer congestion. When HVDC lines are added, they are highly utilized, though. However, the number of average line congestions per hour does not decrease in the Grid 2035 w/ HVDC scenario. This explains the higher investments in storage capacity of power-to-gas and redox flow batteries. The HVDC lines can transport the electricity right away, while in the other scenarios, the energy has to be stored until enough transmission capacity is available (Table 3).

Fig. 9 Average utilization of AC lines [% of thermal limit] in Grid 2022, Grid 2035 w/o HVDC, and Grid 2035 w/ HVDC: categorized into high (>70%, red), medium (>30%, yellow), and low (<30%, green)

Table 3 Total system wide average line utilization

Scenario	AC lines (%)	DC lines (%)	Avg. number of congestions per hour (binding constraints)
Grid 2022	22.0	–	19.8
Grid 2035 w/o HVDC	21.2	–	16.2
Grid 2035 w/ HVDC	20.9	71.4	16.3

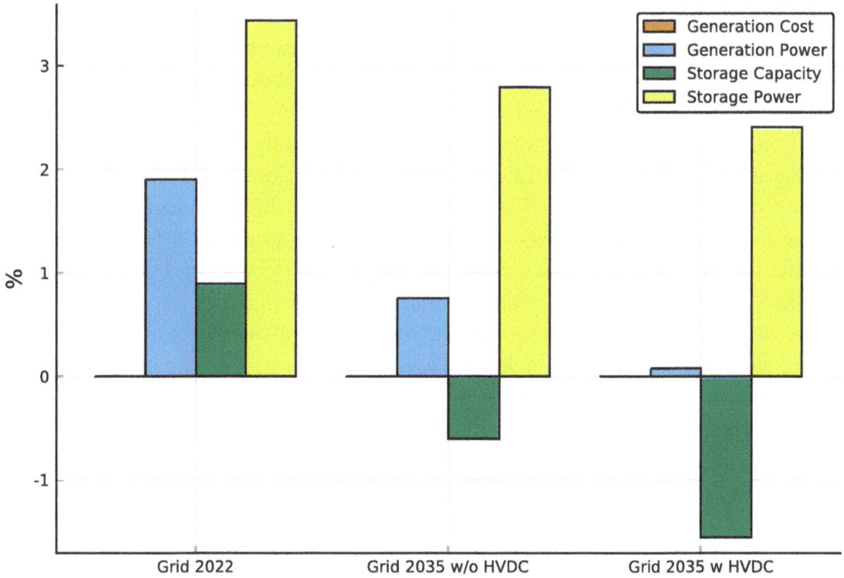

Fig. 10 Relative costs to the copper plate scenario

4.4 Cost Considerations

Last but not least, in Fig. 10, we compare the costs of electricity generation and of storage in the different transmission scenarios relative to the costs in the copper plate scenario: Clearly, more transmission leads to less generation investment, resulting from less capacity installed, as shown above. The same results hold for storage power and storage capacity, which is inversely related to the level of grid development. The generation costs do not play a role since renewable energies are assumed to not have any variable costs.

Investment costs in generation power increase less than 2% between the full grid of 2035 and the existing grid of 2022. The increase for investment costs in storage power amounts to about 1%. A peculiar result emerges for storage capacity: costs compared to the copper plate scenario even decrease in the Grid 2035 w/ HVDC scenario. These changes in costs have to be counted against the changes in transmission investments.

5 Conclusions

Transmission investment is an important element of the low-carbon energy policy agenda that many industrialized countries have embraced over the last years, and that will shake the sector upside down with respect to the (almost entirely de-carbonized) generation mix and the interaction with storage and other flexibility options. In this chapter, we have chosen an extreme scenario, 100% distributed renewable generation resources, to analyze the interdependence between transmission development and other elements of the electricity system. To this end, we have developed a stylized model in the spirit of previous electricity sector models in our research group, adding technical detail on generation and storage. The application to a 38 region representation of the German electricity system allows for a spatially dis-aggregated analysis, yielding new insights into the interaction between the design of the transmission system, the generation mix, and storage infrastructure. In the model, transmission expansion is exogenous, whereas we differentiate into four possible designs, ranging from the status quo in 2022, an extended network in 2035 with and without HVDC lines, and a full-fledged copperplate without any congestion.

The model runs yield some expected results, but also offer some challenges. Among the expected results, a higher level of transmission expansion leads to lower requirements of distributed renewable capacities, mainly for rooftop solar and onshore wind. On the other hand, offshore wind benefits from more transmission capacities.

A similar, though less strong effect can be observed for long-term power-to-gas and short-term lithium-ion storage, which play an important role in the new, renewables-based energy system: storage capacity requirements tend to decrease with more transmission, though this relation is not quite robust with respect to the scenarios (particularly, the copper plate scenario). Due to the assumption of high national autarky, that is the requirement that countries have to fulfill their own energy demand netted over the course of the year, long-term storage capacity (in GWh) turns out to be quite important, whereas short-term capacities are very small. The model allows a spatial representation of the installation, indicating a trend to less storage requirements in the north as transmission is expanded.

Simplified network analysis suggests that Germany would be well on its way moving to a 100% distributed renewable generation portfolio: though some network congestion is observed in the 2022 and the 2035 w/o HVDC scenarios, those seem to be of minor importance, and represent only a marginal share of the electricity transported. Even in the 2035 w/o HVDC scenario, line overruns can hardly be identified, and if so these occur mainly at the margins of the network. We conclude that a 100% distributed renewables world leads indeed to a major overhaul of the system, but—given the simplified, aggregated level of modeling deployed here—it seems that transmission grids are unlikely to be a critical factor of that pathway. Future research should extend the analysis to a fully European level and consider stochasticity (mainly of distributed generation) and other flexibility options, mainly demand-side management.

Acknowledgements This work was carried out as part of the project "Long-term planning and short-term optimization of the German electricity system within the European framework: further development of methods and models to analyze the electricity system including the heat and gas sector," funded through grant "LKD-EU," FKZ 03ET4028A, German Federal Ministry for Economic Affairs and Energy. The authors would like to thank Alexander Roth for comments. The usual disclaimer applies.

Model Nomenclature

Model Nodes

Table 4 gives an overview of all NUTS2 area codes in Germany used as nodes in the model.

Table 4 List of NUTS2 area codes for Germany

Code	Region	Code	Region
DE11	Stuttgart	DE91	Braunschweig
DE12	Karlsruhe	DE92	Hannovr
DE13	Freiburg	DE93	Lüneburg
DE14	Tübingen	DE94	Weser-Ems
DE21	Oberbayern	DEA1	Düsseldorf
DE22	Niederbayern	DEA2	Köln
DE23	Oberpfalz	DEA3	Münster
DE24	Oberfranken	DEA4	Detmold
DE25	Mittelfranken	DEA5	Arnsberg
DE25	Unterfranken	DEB1	Koblenz
DE27	Schwaben	DEB2	Trier
DE30	Berlin	DEB3	Rheinhessen-Pfalz
DE40	Brandenburg	DEC0	Saarland
DE50	Bremen	DED2	Dresden
DE60	Hamburg	DED4	Chemnitz
DE71	Darmstadt	DED5	Leipzig
DE72	Gießen	DEE0	Sachsen-Anhalt
DE73	Kassel	DEF0	Schleswig-Holstein
DE80	Mecklenburg-Vorpommern	DEG0	Thüringen

Table 5 Model sets

Sets	
$c \in \mathcal{C}$	Set of countries
$z \in \mathcal{Z}$	Set of zones/regions
$h \in \mathcal{H}$	Set of hours
$w \in \mathcal{W}$	Set of seasons
$t \in \mathcal{T}$	Set of technologies
$t \in \mathcal{T_D} \subseteq \mathcal{T}$	Subset of dispatchable technologies
$t \in \mathcal{T_N} \subseteq \mathcal{T}$	Subset of non-dispatchable technologies
$t \in \mathcal{T_S} \subseteq \mathcal{T}$	Subset of storage technologies
$l \in \mathcal{L}$	Set of transmission lines

Table 6 Model variables

Variables	
$P^{inst}_{t,z}$	Installed generation power in MW
$P^{ch}_{t,z,h}$	Storage charge power in MW
$E^{inst}_{t,z}$	Installed storage capacity in MWh
$G^{gen}_{t,z,h}$	Generation power in MW
$LL_{z,h}$	Lost load in MW
$G^{ch,w}_{t,z}$	Storage charge for first hour of a season in MW
$G^{dch,w}_{t,z}$	Storage discharge for first hour of a season in MW
$E^{soc,h}_{t,z,h}$	Storage hourly state of charge in MWh
$E^{soc,w}_{t,z,w}$	Storage seasonal state of charge in MWh
$F^{dc}_{z,zz}$	Flow on DC lines from zone z in MW
$F^{dc}_{zz,z}$	Flow on DC lines to zone z in MW
$F^{ni}_{z,h}$	Net input in MW

Table 7 Model parameters

Parameters	
c^{p}_{t}	Investment cost for generation technologies in EUR/MW
c^{ch}_{t}	Investment cost for charging technologies in EUR/MW
c^{e}_{t}	Investment cost for storage technologies in EUR/MWh
c^{mc}_{t}	Marginal generation cost for generation technologies in EUR/MWh
c^{ll}	Penalty cost for lost load in EUR/MWh
$load_{z,h}$	Load in MW
$p^{max,inst}_{t,z}$	Maximum installed generation power in MW
$p^{max,gen}_{t,c}$	Maximum provided energy per year in MWh
$e^{max,inst}_{t,z}$	Maximum installed storage capacity in MWh
$\zeta_{t,z,h}$	Availability factor for non-dispatchable technologies
ρ_{t}	Self-discharge rate of storage
η^{ch}_{t}	Storage charge efficiency
η^{dch}_{t}	Storage discharge efficiency
f^{max}_{l}	Thermal limit of AC line in MW
$f^{max}_{z,zz}$	Thermal limit of DC line in MW
$ptdf_{l,z}$	Power transfer distribution factor matrix
γ	Scaling factors for reduced time horizon in one year
ϕ	Autarky factor

References

50Hertz Transmission, Amprion, TenneT, and TransnetBW, Szenariorahmen für den Netzentwicklungsplan Strom 2030. VERSION 2019, Entwurf der Über- tragungsnetzbetreiber (2018)

J. Abrell, S. Rausch, Cross-country electricity trade, renewable energy and European transmission infrastructure policy. J. Environ. Econ. Manag. **79**, 87–113 (2016)

B.M.-A. Carlo, Electricity without borders - the need for cross-border transmission investment in Europe. Proefschrift / Dissertation, Technische Universiteit Delft, (2013)

M. Drechsler, J. Egerer, M. Lange, F. Masurowski, J. Meyerhoff, M. Oehlmann, Efficient and equitable spatial allocation of renewable power plants at the country scale. Nature Energy **2**(9), 17124 (2017)

J. Egerer, Open Source Electricity Model for Germany (ELMOD-DE). DIW Data Documentation 83. Berlin: DIW Berlin - Deutsches Institut für Wirtschaftsforschung e. V. (2016)

European Environment Agency, CORINE Land Cover - Copernicus Land Monitoring Service (2019)

M. Fürsch, S. Hagspiel, C. Jägemann, S. Nagl, D. Lindenberger, E. Tröster, The role of grid extensions in a cost-efficient transformation of the European electricity system until 2050. Appl. Energy **104**, 642–652 (2013)

C. Gerbaulet, C. Lorenz, dynELMOD: A Dynamic Investment and Dispatch Model for the Future European Electricity Market. DIW Data Documentation 88. Berlin: DIW Berlin - Deutsches Institut für Wirtschafts- forschung e. V. The Impact of Transmission Development on a 100 (2017)

C. Gerbaulet, The role of electricity transmission infrastructure, in *Energiewende "Made in Germany": Low Carbon Electricity Sector Reform in the European Context*, ed. by C. von Hirschhausen, C. Gerbaulet, C. Kemfert, C. Lorenz, P.-Y. Oei (Springer International Publishing, Cham, 2018), pp. 193–216

C. Gerbaulet, A. Weber, When regulators do not agree: are merchant interconnectors an option? Insights from an analysis of options for network expansion in the Baltic Sea region. Energy Policy **117**, 228–246 (2018)

L. Göke, M. Kittel, C. Kemfert, P.-Y. Oei, C. von Hirschhausen, Scenarios for the Coal Phase-out in Germany - A Model- Based Analysis and Implications for Supply Security. DIW Wochenbericht 28. Berlin: DIW Berlin - Deutsches Institut für Wirtschaftsforschung e. V (2018)

C.M. Grams, R. Beerli, S. Pfenninger, I. Staffell, H. Wernli, Balancing Europe's wind-power output through spatial deployment in- formed by weather regimes. Nat. Clim. Change **7**(8), 557 (2017)

V. Grimm, A. Martin, M. Schmidt, M. Weibelzahl, G. Zöttl, Transmission and generation investment in electricity markets: the effects of market splitting and network fee regimes. Euro. J. Oper. Res. **254**(2), 493–509 (2016)

V. Grimm, B. Rückel, C. Sölch, G. Zöttl, Reduction of network expansion through redispatch and efficient feed-in management: A model-based assessment. List Forum für Wirtschafts- und Finanzpolitik **1–34** (2016)

Hertz Transmission, TenneT Amprion, TransnetBW,. Netzentwicklungsplan Strom 2030 (2019). VERSION 2019, 2. E. Abdmouleha, A.G. Zeineb, L. Ben-Brahima, M. Haouarib, N.A. Al-Emadia, Review of optimization techniques applied for the integration of distributed generation from renewable energy sources. Renew. Energy **113**, 266–280 (2017)

C. Hirschhausen, German energy and climate policies: a historical overview, in *Energiewende "Made in Germany": Low Carbon Electricity Sector Reform in the European Context*, ed. by C. von Hirschhausen, C. Gerbaulet, C. Kemfert, C. Lorenz, P.Y. Oei (Springer International Publishing, Cham, 2018), pp. 17–44

W. Hogan, J. Rosellón, I. Vogelsang, Toward a combined merchant-regulatory mechanism for electricity transmission expansion. J. Regul. Econ. **38**(2), 113–143 (2010)

P. Joskow, J. Tirole, Merchant transmission investment. J. Ind. Econ. **2**, 233–264 (2005)

P. Joskow, J. Tirole, Retail electricity competition. RAND J. Econ. **37**(4), 799–815 (2006)

P. Kayal, C.K. Chanda, Optimal mix of solar and wind distributed generations considering performance improvement of electrical distribution network. Renew. Energy **75**, 173–186 (2015)

C. Kemfert, F. Kunz, J. Rosellón, A welfare analysis of electricity transmission planning in Germany. Energy Policy **94**, 446–452 (2016)

F. Kunz, C. Gerbaulet, C. von Hirschhausen, Mittel- fristige Strombedarfsdeckung durch Kraftwerke und Netze nicht gefährdet. DIW Wochenbericht 48. Berlin: DIW Berlin - Deutsches Institut für Wirtschafts- forschung e. V (2013)

F. Kunz, M. Kendziorski, W.-P. Schill, J. Weibezahn, J. Zepter, C. von Hirschhausen, P. Hauser, M. Zech, D. Möst, S. Heidari, J. Felten, C. Weber, Electricity, Heat and Gas Sector Data for Modelling the German System. DIW Data Documentation 92. Berlin: DIW Berlin - Deutsches Institut für Wirtschaftsforschung e. V (2017)

F. Leuthold, H. Weigt, C. von Hirschhausen, A large-scale spatial optimization model of the European electricity market. Networks Spat. Econ. **12**(1), 75–107 (2012)

P.D. Lund, J. Lindgren, J. Mikkola, J. Salpakari, Review of energy system flexibility measures to enable high levels of variable renewable electricity. Renew. Sustain. Energy Rev. **45**, 785–807 (2015)

R. Mieth, C. Gerbaulet, C. von Hirschhausen, C. Kemfert, F. Kunz, R. Weinhold, Perspektiven für eine sichere, preiswerte und umweltverträgliche Energieversorgung in Bayern. Politikber- atung kompakt 97. Berlin: DIW Berlin - Deutsches Institut für Wirtschafts- forschung e. V (2015)

R. Mieth, R. Weinhold, C. Gerbaulet, C. von Hirschhausen, C. Kemfert, Electricity grids and climate targets: new approaches to grid planning. DIW Econ. Bull. (Berlin) **5**(6), 75–80 (2015)

P. Nahmmacher, E. Schmidt, B. Knopf , Documentation of LIMES-EU - A long-term electricity system model for Europe (2014)

C. Ohl, M. Eichhorn, The mismatch between regional spatial planning for wind power development in Germany and national eligibility criteria for feed-in tariffs-A case study in West Saxony. Land Use Policy **27**(2), 243–254 (2010)

D. Ohlhorst, Germany's energy transition policy between national targets and decentralized responsibilities. J. Integ. Environ. Sci. **12**(4), 303–322 (2015)

L. Olmos, I.J. Pérez-Arriaga, A comprehensive approach for computation and implementation of efficient electricity transmission network charges. Energy Policy **37**(12), 5285–5295 (2009)

S. Pfenninger, I. Staffell, Long-term patterns of European PV output using 30 years of validated hourly reanalysis and satellite data. Energy **114**, 1251–1265 (2016)

K. Poncelet, H. Höschle, E. Delarue, A. Virag, W. D'haeseleer, Selecting representative days for capturing the implications of integrating intermittent renewables in generation expansion planning problems. IEEE Trans. Power Syst. **32**(3), 1936–1948 (2016)

J. Rosellón, T. Kristiansen (eds.), *Financial transmission rights: analysis, experiences and prospects*, vol. 7 (Lecture notes in energy (Springer, London, New York, 2013)

D.P. Schlachtberger, T. Brown, S. Schramm, M. Greiner, The benefits of cooperation in a highly renewable European electricity network. Energy **134**, 469–481 (2017)

J. Weibezahn, M. Kendziorski, Illustrating the benefits of openness: a large-scale spatial economic dispatch model using the Julia language. Energies **12**(6), 1153 (2019)

R. Weinhold, R. Mieth, Fast Security-Constrained Optimal Power Flow through Low-Impact and Redundancy Screening. IEEE Transactions on Power Systems (2020)

A. Zerrahn, W.-P. Schill, A greenfield model to evaluate long-run power storage requirements for high shares of renewables. DIW Discussion Paper 1457. Berlin: DIW Berlin - Deutsches Institut für Wirtschaftsforschung e. V (2015)

Coordination of Gas and Electricity Transmission Investment Decisions

Seabron Adamson, Drake Hernandez, and Herb Rakebrand

1 Introduction

The increasing penetration of gas-fired generation in some liberalized power markets has created new challenges in coordinating planning and investment in gas and electric transmission infrastructure. When gas-fired generation is critical to meeting electric demand, the regional availability of natural gas supplies can be a key driver of reliability and may have a large impact on consumer power prices. This is especially true in regions (such as New England) where there are coincident high winter gas and electric demand. While there is a substantial literature on investment in the electric grid, there is a more limited literature on policy issues associated with gas–electric interactions. In this chapter, we explore some of the economic and policy issues arising from differing investment models for electric and natural gas transmission, with a focus on New England, a region of the United States that has been substantially impacted by lack of new investment in regional pipeline infrastructure.

The opinions expressed in this chapter are solely those of the authors and do not reflect the opinion of Charles River Associates or its clients.

S. Adamson (✉) · D. Hernandez · H. Rakebrand
Charles River Associates, Boston, MA, USA
e-mail: sadamson@crai.com; seabron.adamson@bc.edu

S. Adamson · H. Rakebrand
Carroll School of Management, Boston College, Chestnut Hill, Newton, MA, USA

M. R. Hesamzadeh et al. (eds.), *Transmission Network Investment in Liberalized Power Markets*, Lecture Notes in Energy 79,
https://doi.org/10.1007/978-3-030-47929-9_16

2 Investment Models in U.S. Electric and Gas Transmission

Electric and natural gas transmission are largely federally regulated in the United States, but the mechanisms for investment in new transmission assets are substantially different.[1] Planning in the electric sector is generally centralized, with most regional and inter-regional expansion planning conducted by regional transmission organizations (RTOs) and independent system operators (ISOs), which operate the transmission grid in large regions. While not all transmission-operating utilities are in an ISO, FERC's Order 1000 (issued in 2011) required other federally regulated transmission utilities to join some form of regional transmission planning group.

The interstate natural gas industry is quite different, as transmission expansion is not planned by any central authority. Natural gas pipelines do not demonstrate the network external costs issues affecting integrated electric grids, making decentralized operations economically possible (Makholm 2012). FERC approval is needed to build an interstate gas pipeline, under the terms of the *Natural Gas Act of 1938*. A pipeline company is required to obtain a certificate of public convenience and necessity to build a new pipe, but no regional or federal agency oversees pipeline system planning or coordinates investment decisions between pipeline companies. Instead, pipeline companies are often in vigorous competition to build new pipelines to serve customers willing to sign the requisite long-term gas transportation contracts.

2.1 *Regional Electric Transmission Planning*

An ISO typically performs three primary roles:

1. Short-term operation of the regional power grid,
2. Managing wholesale electric markets and,
3. Future transmission system planning.

Power system planning requires the ISO to make certain that the region has adequate resources and transmission capacity to meet the demand for electricity over a planning horizon—this includes the interchange of power between adjacent ISOs. To initiate this process, the ISO performs detailed network studies to determine where system improvements and/or upgrades are required. The results of this analysis are made public to provide market signals as to where investments are required.

An ISO will typically develop and maintain a regional transmission system plan, which looks at system adequacy and reliability over a longer horizon. Under FERC's Order 1000 requirements, ISOs and other regional transmission planning groups

[1]Electric transmission in Alaska, Hawaii and much of Texas are state regulated, as these systems are loosely (or not at all) connected to other grids and hence do not fall under federal interstate regulation. Most long-distance pipelines cross state borders and hence are regulated by the Federal Energy Regulatory Commission (FERC), but intrastate pipelines may be subject to state-level regulation.

must consult with stakeholders, consider state and federal public policy requirements (such, for example, state renewal portfolio standards), and coordinate with other regions, if necessary. Order 1000 also changed long-standing U.S. policy and removed any "right of first refusal" for existing transmission owners to build new lines or facilities in their service territories. Historically, a transmission-owning utility could build any new required transmission projects within its service territory. The elimination of the right of first refusal created the scope in some cases for competition to build new projects within a region needed under the ISO's regional plan. While transmission developers may compete to build these projects, the costs are recovered in transmission rates. ISOs (or other FERC-regulated regional transmission planning groups) are responsible for evaluating proposals for the required system improvements. Allocating the costs of these projects among different users of the regional transmission system has proved controversial in some cases, despite broad FERC policy guidance on the topic (Adamson 2018).

2.2 *The Decentralized Pipeline Expansion Model*

In stark contrast to the electric planning process, the process for new gas infrastructure investment begins as a discussion between an interstate natural gas pipeline—which seeks to sell transportation capacity on the pipeline under contract—and potential shippers of gas on the pipeline. A shipper may be a natural gas local distribution company (LDC), a gas marketer, large industrial customer, a gas producer or a power generator. These discussions are generally non-public and do not address regional issues, but focus on the needs of the potential shipper customers. In most cases, especially for larger projects, the discussions may include several shippers. Once the pipeline and shippers conclude their discussions and arrive at a project structure that meets their needs, the pipeline will hold an "open season." US pipelines are subject to open access regulations, and an open season is required by FERC to assure all parties have an equal opportunity to acquire capacity on the proposed project. Often the open season announcement is the first time there is any public notice of such discussions or that a project is being considered. In some cases the open season is largely a formality. In other cases, the open season process is used to identify additional shippers for the pipeline. Upon the completion of the open season process, the pipeline will typically begin the public process of permitting the project. FERC administers the approval and permitting process. In most cases, the pipeline will present contracts, known as precedent agreements, it has executed with the shippers as proof of need for the project.

Historically, FERC policy has deemed the existence of commercial counterparties willing to sign long-term contracts for natural gas transportation as a strong market signal that such capacity is needed, and FERC will often not undertake a market needs analysis.

Typically there has been limited or no discussion or consideration of coordination or broader regional needs for a natural gas pipeline project. As part of the FERC approval process, there is a requirement to investigate alternatives to the proposed new infrastructure. The primary purpose of this exercise is to minimize potential environmental impacts of the new project. At this point, FERC may consider coordinated grid improvements between existing pipelines, but this path is rarely taken given the established contractual agreements that exist between the pipelines and their shippers. Upon receipt of a FERC certificate and associated permits, the pipeline must secure appropriate financing to construct the facilities. Historically, pipeline companies have been the sole equity holders for new pipeline projects, leaving little room for outside investment. This has changed recently as new shippers have often been able to secure equity positions in the pipeline assets—this is particularly true for larger greenfield pipelines.

3 National and Regional Transmission Investment Experience

The differences between regulatory models for investment in natural gas pipelines and electric transmission have helped produce differing levels of investment over the last few decades.

3.1 National Experience in Transmission Investment

U.S. interstate natural gas pipeline companies had have substantial success building new pipeline infrastructure over the past fifteen years. Demand for new natural gas pipelines grew sharply around 2007 as gas needed to be moved from new shale producing regions. As shown in Fig. 1, pipeline developers were able to respond quickly to shippers' need for new transportation capacity. Investment more than tripled in a few years, much of it into new "producer-push" pipelines—that is, a pipeline largely contracted by gas producers that takes gas from producing fields to market. The large amount of investment and new capacity in 2017 was largely associated with new pipelines bringing gas from the Marcellus and Utica shale regions to market.

Investment on the electric side has grown much more modestly, even with the need to integrate substantial quantities of new renewable generation in many regional markets. Figure 2 shows the trend in electric transmission investment over the decade to 2015.

Many of the investment dollars represented in Fig. 2 do not reflect new transmission capacity. Rather, the investments have been primarily used for the rehabilitation of existing assets (Edison Electric Institute 2016). Since transmission-owning utilities may have incentives to capitalize large rehabilitation and rebuilding of existing

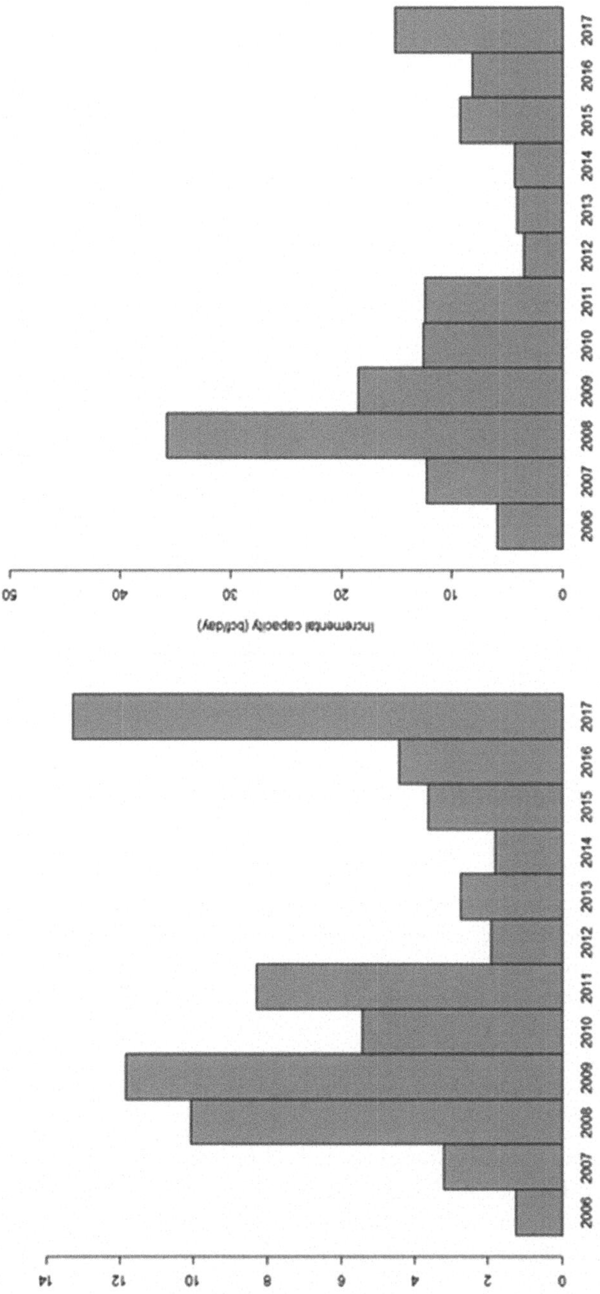

Fig. 1 New interstate pipeline investment and capacity (2006–2017). *Source* Authors' analysis of EIA database

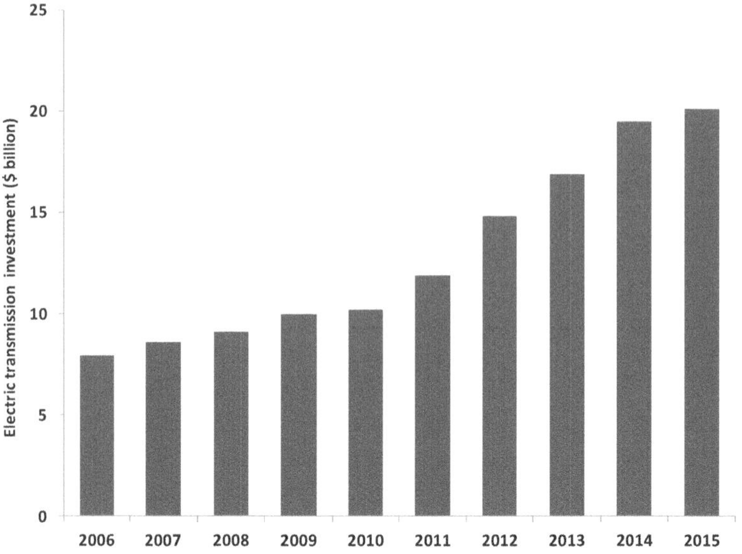

Fig. 2 U.S. electric transmission investment by year ($ billions nominal). *Source* Data from Edison Electric Institute

transmission asset, these often are reflected in new electric transmission capital expenditure data.

3.2 *The New England Regional Experience*

The New England experience over the last 10–15 years differs substantially from the national experience. ISO New England (ISO-NE), the regional transmission and market operator, has seen relatively significant levels of investment, while new gas pipeline additions have been minimal despite a significant need for new gas transportation capacity into the region.

3.2.1 ISO New England and Regional Transmission Additions

ISO-NE operates and plans the transmission grid for most of the six New England states, as shown in Fig. 3. ISO-NE is interconnected with the New York ISO, the Hydro Québec system, and to New Brunswick in Canada. New England imports substantial amounts of electricity, especially from Canada.

Over the period 2003 through October 2018, ISO-NE placed $10.7 billion of new power transmission assets into service, with the majority of this since 2008 and concentrated in a few large projects (ISO New England 2018). ISO-NE calculate the

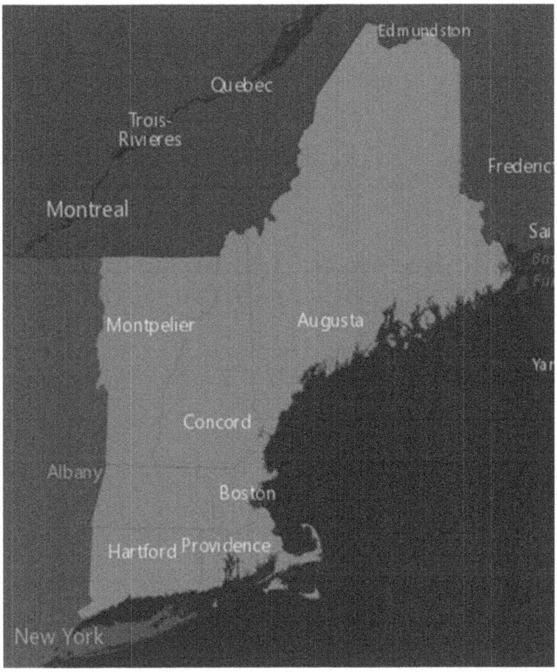

Fig. 3 ISO New England Region. *Source* SNL

sum of congestion, uplift and reliability agreement costs in the region have fallen from approximately $700 million per year to less than $100 million per year in 2017 and 2018 due to transmission investment.

3.2.2 Regional Pipeline Additions

New England is served by three major interstate pipelines that bring gas into the region, which has no gas production and very limited gas storage. There are also LNG import terminals in the region but utilization of these facilities is currently low. As in other regions of the United States, there is no regional pipeline operator—each pipeline is responsible for its own operations under the terms of its FERC tariff (Fig. 4).

Despite successes in other parts of the country, the interstate pipeline industry has not succeeded in building much new capacity in New England, as shown in Fig. 5.

These are relatively small additions to the New England system, where peak day sendout LDC is greater than 4.3 billion cubic feet per day (Bcf/d). By 2020 peak demand for natural gas is expected to near 6 Bcf/d.

Figure 6 shows that this trend has continued for some years, with the recent jump in investment in 2016–17 primarily tied to one pipeline expansion project.

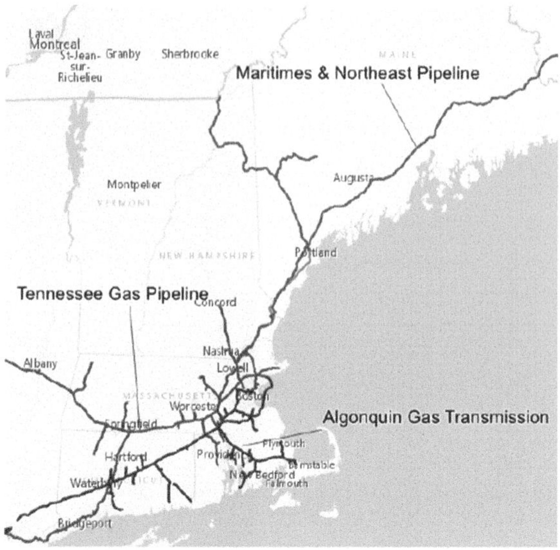

Fig. 4 Major pipelines serving New England. *Source* SNL

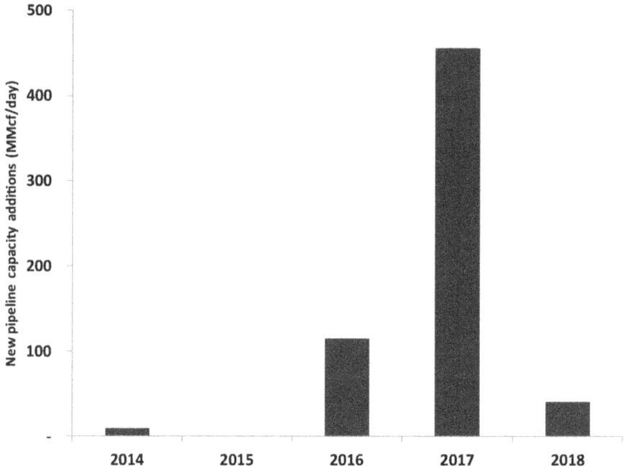

Fig. 5 Pipeline capacity additions in New England (2014–2018). *Source* Authors' analysis of PointLogic data

3.2.3 Price Signals for New Investment

The lack of new pipeline investment is not due to a lack of strong price signals showing the value of new capacity. Pipeline developers and shippers look to gas basis (the difference between locational gas prices in two different markets) as a signal of the potential value of pipeline capacity. Figure 7 shows basis prices from 2010 through 2018 at the Algonquin Citygate market point (the most liquid market point in New England) versus spot prices at Henry Hub in Louisiana, the most commonly referenced gas pricing point in North America. On many days even cheaper gas is

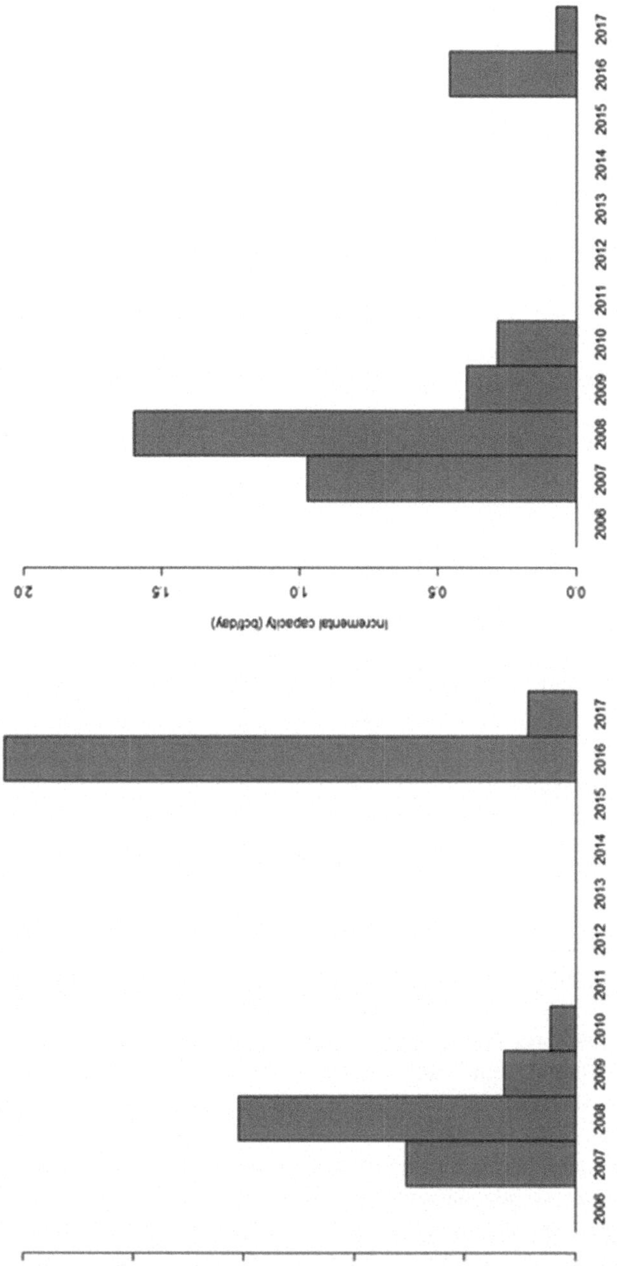

Fig. 6 New England pipeline investment and capacity (2006–2017)

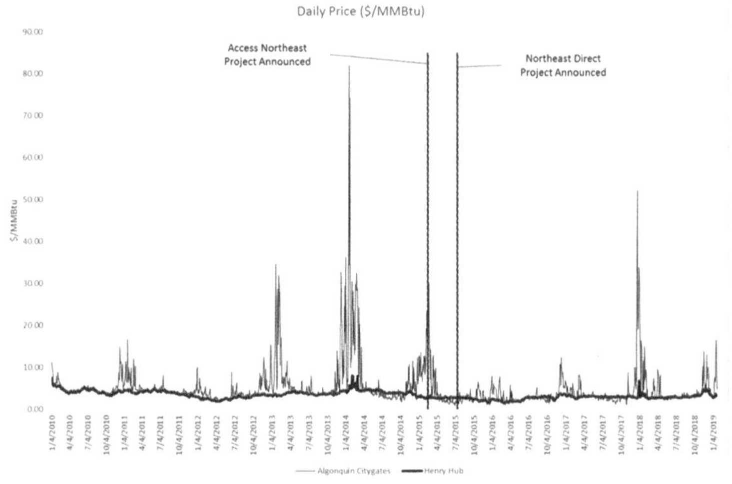

Fig. 7 Algonquin Citygate Price versus Henry Hub Natural Gas Prices. *Source* Data from Bloomberg

available in the Appalachian region (location of the Marcellus and Utica shale plays) which is closer to New England.

A shipper with transmission capacity into New England on an import pipeline would see large profits on days when the Algonquin Citygate price increases sharply. The 2014 "polar vortex" event and other spikes show that these basis spikes occur primarily in periods of very high regional gas demand, usually tied to low temperatures in the region.

Figure 8 presents the basis differential into New England on the Algonquin Pipeline as a function of available pipeline capacity.[2] As the graph shows, prices within the New England region rise materially as available pipeline capacity diminishes.

Gas LDCs typically cover most or all of their core gas demand requirements with firm transportation contracts, and, with regulatory approval, pass these costs through to their core customers. LDC peak gas demand, however, has been growing relatively slowly in the region for years. In contrast, ISO-NE, as the regional electric grid operator, has shown significant concern over the region's dependence on gas-fired generation with potential shortfalls in regional gas deliverability (van Welie 2018). Natural gas is the primary fuel for 45% of the region's generating capacity and sets the locational marginal price (LMP) more than 75% of the time. ISO-NE has stated that ensuring adequate fuel supply for the region's generators is New England's most pressing electric reliability challenge, and that by winter 2024/25 many modeled scenarios showed risk of load shedding due to fuel shortfalls.

[2]The Algonquin Pipeline directly serves approximately 50% of the gas-fired generation in New England and the Algonquin City Gate is the most liquid pricing point in New England.

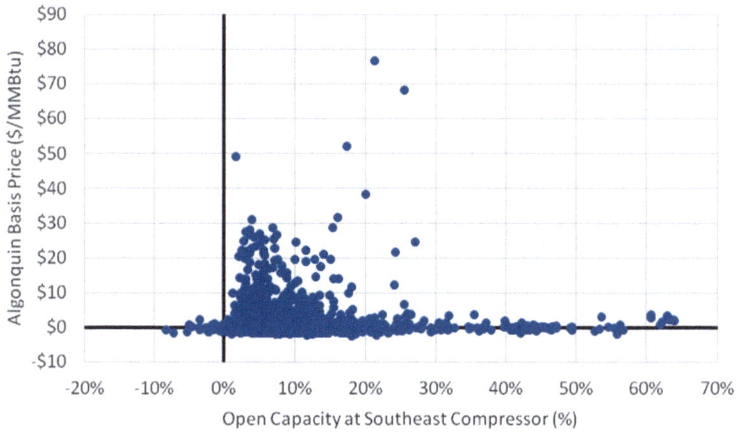

Fig. 8 Algonquin pipeline open capacity and basis differential (December 2012 through January 2019). *Source* Authors' analysis of data from PointLogic and Bloomberg

3.2.4 Contracting Issues Hindered Proposed Pipeline Projects

Several major pipeline projects have been proposed in recent years to bring additional gas into the region. The $3.3 billion Northeast Energy Direct pipeline failed due to a lack of contractual commitments for transportation capacity. The Access Northeast project would have upgraded 125 miles of the Algonquin pipeline which serves most of the region's gas-fired generation. The project developers noted merchant generators in the region had little incentive or ability to sign long-term contracts for pipeline capacity, as, unlike LDCs, there is no pass-through mechanism for these costs (Kruse 2016). A proposed mechanism under which certain pipeline transportation costs could be recovered through utility electric rates was rejected by the Massachusetts Supreme Judicial Court (MSJC 2016).

Increases in peak natural gas demand in New England are driven primarily by the electric generation sector, but in the absence of market incentives, or means to pass-through these costs, generators have been unwilling or unable to sign the long-term pipeline capacity agreements that would allow the pipeline expansion capacity to be built. In conjunction with the difficult environmental and land use issues in the densely populated New England region, the lack of gas pipeline contractual counterparties has led to relatively little new pipeline capacity being built in the region, even with a high "basis" price signal.

3.2.5 Electric Transmission as an Alternative to Pipelines

With its high gas prices, New England also has relatively high electricity prices. In theory, lack of new gas transportation into New England could be, at least partially, ameliorated by new electric transmission facilities from neighboring regions, especially Québec, which has substantial hydropower resources. This has proven quite difficult, however, due to issues associated with siting new large electric transmission facilities.

For example, in March 2017 the Massachusetts state regulator approved a process in which state electricity utilities could procure a large amount of new clean energy under long-term contracts; these contracts could include the costs of building new transmission lines within and into New England (Department of Public Utilities 2017). In January 2018, the "Northern Pass" project was selected. This was to be a 1090 MW high voltage direct current (HVDC) and AC transmission line system designed to move Québec hydropower to Massachusetts, with much of the HVDC line routed through the state of New Hampshire. In March 2018, the Site Evaluation Committee of the State of New Hampshire denied a certificate for Northern Pass to be built through that state (Site Evaluation Committee 2018).

Following the permit denial, the local utilities in Massachusetts terminated the selection of the Northern Pass project and selected an alternate new transmission line project through Maine for delivering hydropower from Québec (Department of Energy Resources 2018). This alternate project has recently received certificate approval by the Maine state utility commission but still faces substantial political and legal opposition (Gheorghiu 2019).

In short, building new large-scale electric transmission into the New England region has proven to be costly, slow and difficult, and hence has not provided an easy alternative to the contractual and other issues associated with new gas pipeline construction.

3.2.6 New England Policy Responses

Given a lack of infrastructure coordination and the growing dependence on natural gas generation in the early 2000s, ISO-NE instituted market changes to support reliability. The initial market mechanisms established capacity payment penalties for generators that were unavailable during "critical" periods. These critical periods were often during the winter when gas market demands were at their highest consuming a large share of the available pipeline capacity to the region. Given the magnitude of the price spikes experienced in New England, ISO-NE determined these measures were insufficient to provide the level of reliability required.

To promote greater grid reliability, in 2013, ISO-NE instituted a new Winter Reliability Program in 2013 in an effort to promote greater grid reliability. This program shifted focus from encouraging gas unit availability to promoting the availability of alternate fuels such as LNG, petroleum, and demand response to manage peak generation demand.

Given ongoing reliability concerns, ISO-NE proposed new market rules to replace the Winter Reliability Program in 2018. The new capacity market rules had two primary components: (i) ISO-NE would integrate demand response resources into its daily energy and reserve economic dispatch on a level comparable to generation resources and (ii) ISO-NE introduced Pay-For-Performance capacity market incentives. These rules essentially shifted payments from under-performing generating resources to over-performing resources. The new incentives were added to the Forward Capacity Market after ISO-NE observed a weak linkage between capacity payments and actual performance by resources during times of system stress (FERC 2018).

Despite these policy initiatives, there is little sign electric power generators in the region are willing or able to sign the types of large, long-term capacity contracts necessary to support new pipeline projects into the region. New pipeline development projects remain speculative and no new major construction projects are on the immediate horizon.

3.3 *Incomplete Intermediate Contract Markets and Investment*

In economic terms, the pipeline investment model has been successful where long-term contract markets are robust and reasonably complete. The largest traditional shippers were regulated gas LDCs who had the ability to pass-through these pipeline transportation costs in their regulated rates, and state regulation often required LDCs to sign such contracts. In this case, pipelines were able to secure the contracts needed to build new transportation capacity to meet growing LDC gas demand.

For pipelines from producing regions (the "producer-push" pipelines), natural gas exploration and production (E&P) companies have strong incentives to secure the pipeline capacity to move their new gas to market, and many E&P companies were capitalized such as to support the credit requirements of long-term pipeline transportation contracts. Forward contracting in the natural gas markets is relatively robust, and E&P's could hedge much of their delivery basis risks through these contracts.

New England provides a case study of how the pipeline investment model is much less effective when marginal demand growth is largely in the merchant power generation sector, where long-term (e.g., 10 years or more) forward contract markets (for energy and capacity) are much less robust.

The electric distribution companies and retailers, who serve electric loads, tend to contract for only a few years at a time. Regional electricity forward contract markets are not highly liquid, and extend out only a few years. The transactions costs for hedging longer-dated forward power contracts are also high.

In these circumstances, it would be difficult or impossible for merchant power generators to hedge the risks of entering into a 10-year or longer gas supply contract or to directly contract with pipelines for such long-term transportation. The strong correlation between regional natural gas prices and ISO-NE LMPs raises the risks to generators in signing such contracts. Thus, it is unsurprising that the merchant generators have not contracted for extensive new pipeline capacity into New England.

4 Conclusions and Policy Implications

The U.S. interstate pipeline industry has been very successful in developing new projects where demand exists *and* shippers are willing to sign the long-term contracts necessary to support large-scale pipeline investment. FERC has used the existence of these contracts, or precedent agreements, to signal need for new capacity (a key requirement under the *Natural Gas Act* in certifying a new project) and pipelines rely on the stable revenues from these long-term contracts to underpin the large sunk cost investments required.

The policy initiatives of ISO-NE discussed in Sect. 3.2.5 have not provided a strong basis for building new regional pipeline capacity to support electric generation. The proposed changes to the regional capacity market, while they may provide some additional incentives to hedge gas exposure among the region's generators, fundamentally do not support the long-term contracting necessary to stimulate new pipeline construction. Over time both the gas and electric industries have recognized the mismatch in investment models. They have made ongoing adjustments, but with marginal success to date, as the changes made have still been based upon, and reside within, their respective investment models.

Given the fundamental mismatch between the regulatory models for investment natural gas and electric transmission planning, policy design to support any large needed gas transmission investment needed to supply electric generation will be difficult. Requiring merchant generators to contract for firm gas transportation would involve large costs which could not likely be recovered in energy market prices, especially since regional LMPs do not reflect fuel supply scarcity (Adamson and Tabors 2013). Adding capacity market qualification requirements would also involve large fixed cost burdens for merchant generators, which were planned and financed without such obligations.

Pipelines would need a strong incentive mechanism to invest in capacity to serve these electric generation markets, given the risks associated with such investment without the usual 10–15 year transportation contracts with creditworthy shippers. The cost of service regulatory regime for interstate natural gas pipelines caps the upside for pipelines on new investments (through the scope for a rate case and lower subsequent authorized returns on equity in the future if the pipeline is deemed to be over-earning on its original capital investment), while exposing them to substantial downside and stranded asset risks if less firm transportation capacity than projected is contracted in the future. Any pipeline investment-based approach to the gas–electric coordination issues raised in this paper will, therefore, require a fundamental shift in FERC policy toward pipeline investment in such an asymmetric risk environment.

References

S. Adamson, Comparing interstate regulation and investment in U.S. Gas and Electric Transmission. Econ. Energy Environ. Policy **7**(1) (2018)

S. Adamson, R. Tabors, Pricing short-term gas availability in power markets, in *Growing Concerns, Possible Solutions: The Interdependency of Natural Gas and Electricity Systems* (MIT Energy Initiative, 2013)

Department of Energy Resources, Commonwealth of Massachusetts, Letter to D.P.U. regarding petitions for approval of proposed long-term contracts for renewable resources pursuant to Section 83D of Chapter 188 of the Acts of 2016, DPU 18–64, 18–65, 18–66, July 23, 2018 (2018)

Department of Public Utilities, Commonwealth of Massachusetts. *Order on Joint Petition*, D.P.U. 17–32, March 27, 2017

Edison Electric Institute, *Transmission Projects at a Glance* (2016)

Energy Information Administration, *U.S. Natural Gas Pipeline Project Database* (2019). Accessed 26 Feb 2019. Available from https://www.eia.gov/naturalgas/data.cfm#pipelines

Federal Energy Regulatory Commission, *Winter 2018–19 Energy Market Assessment*. Item No. AD06-3-000, October 18, 2018 (2018). Available from www.ferc.gov/market-oversight/reports-analyses/mkt-views/2018/10-18-18-A-3.pdf

J. Gheorghiu, Controversial $1B Canada—U.S. Transmission Line gets Maine PUC approval, in *Utility Dive*, April 11, 2019 (2019)

ISO New England, *Energy Policy Update: Winter Reliability Program*. December 22, 2015 (2015)

ISO New England, *Regional System Plan: Transmission Projects and Asset Condition, October 2018 Update* (2018). Available from www.iso-ne.com

R. Kruse, *Algonquin Gas Transmission, LLC*. Presentation at FERC Technical Conference on Capacity Release Tariff Proposal, Docket No. RP16-618-000, May 9, 2016 (2016). Available from https://www.ferc.gov/CalendarFiles/20160509090835-Algonquin%20Presentation.pdf

J. Makholm, *The Political Economy of Pipelines* (University of Chicago Press, 2012)

Massachusetts Supreme Judicial Court, Decision in *Engie Gas & LNG LLC vs. Department of Public Utilities*. SJC-12051, August 17, 2016 (2016)

Site Evaluation Committee, State of New Hampshire, *Decisions and Order Deny Application for Certification of Site and Facility*. Docket 2015–06, March 30, 2018 (2018)

G. van Welie, *ISO New England: State of the Grid 2018* (2018). Available from www.iso-ne.com

The Emergence of Smart and Flexible Distribution Systems

Derek W. Bunn and Jesus Nieto-Martin

1 Introduction

Electricity network operators are entering a period of significant change. The transition to a low carbon economy requires them to accommodate and anticipate the impact that distributed renewable energy resources will have on their transmission system operations. Distribution network operators (DNOs), being responsible for the local, lower voltage networks, are therefore increasingly exploring alternatives to conventional physical asset reinforcements. More flexible alternatives and smart technologies are needed in order to minimise costs, whilst maintaining security of supply, and at the same time facilitate a more unpredictable, decentralised energy mix to be connected to their networks (Pérez-Arriaga et al. 2017). Flexible solutions necessarily require the DNOs to become more active in system operations at the distribution level and thus we see their separation into distribution system operators (DSOs), as organisational entities, to mirror some of the operational activities of the higher voltage transmission system operators (TSOs) (Western Power Distribution 2018b).

The growth of distributed energy resources (DERs), as well as the use of new operational techniques, has motivated a growing interest in market-based solutions for the DSOs to provide energy and flexibility services at local level (Pastor et al. 2018). This represents a new competitive tier in the already complex market arrangements of electricity planning, production and retailing. It opens new opportunities of engagement for producers, consumers, retailers, aggregators and network service operators, (Biegel et al. 2014). Within network operations, in particular, the amount of DERs connected at low voltage levels is creating major disruptions (Eid et al. 2016). For the TSO, DERs increase the sources of ancillary service providers, but also bring greater uncertainty to the extent that DER activities may be less visible and

D. W. Bunn (✉) · J. Nieto-Martin
London Business School, Regents Park, London NW1 4SA, UK
e-mail: dbunn@london.edu

J. Nieto-Martin
e-mail: jmartin@london.edu

M. R. Hesamzadeh et al. (eds.), *Transmission Network Investment in Liberalized Power Markets*, Lecture Notes in Energy 79,
https://doi.org/10.1007/978-3-030-47929-9_17

predictable closer to real-time operations. For DNOs, whilst embedded generators present new stresses on the local infrastructure, they also offer a range of active, flexible options with which to manage their networks. All of which motivates the need for new market arrangements to support that co-ordination, having a direct impact on infrastructure planning and operations.

Some countries have seen large communities of residential, commercial and industrial customers choosing to migrate to off-grid microgrid solutions, using the main grid connection only as a backup service (Villar et al. 2018). California, for example, in 2018, had more than 30 microgrids in operation and contained 40% of all residential PV within the USA. Germany had over 300 MWh of residential storage and could cover 60% of peak demand with its PV capacity. Elsewhere even greater network disruptions have been anticipated: by 2020 all substations in South Australia have been expected to be in reverse flow due the proliferation of DERs, including large-scale batteries.

In Britain, the government, regulator and transmission system operator agreed in 2018 to create a separate electricity system operator (ESO) (Ofgem 2017, 2018), independent of transmission network ownership, partly to encourage more third-party investment in the network assets. In order to achieve its intent, the regulatory framework should provide stakeholders with confidence that the ESO is truly independent, acting in the best interests of all participants and consumers and giving stakeholders a process through which to hold the ESO to account. Offshore connections to windfarms and international interconnectors have brought more private finance into the transmission network and, where appropriate, the regulator encourages construction of new lines to be open to competitive tenders. Integrated private microgrids have been rare in the mainland GB system, but a few offshore examples in the islands have nevertheless evolved as a consequence of their isolation.

In a fully liberalised system, as in GB, keeping the system operator role within the same legal entity as the asset owner would create a conflict of interest not only for asset investment but as more flexibility options emerge. The transmission and distribution asset owners could perhaps exert undue influence over the decision making of the SO and even inhibit the willingness of new entrant providers to invest. Thus, at the lower voltage distribution network level, 14 DSOs have replicated the system operations of the ESO (Capgemini 2018). Whilst the EU has also been pursuing the DSO model to encourage innovation, in many parts of the world the system operations and ownership remain within the same organisation, as indeed do local low voltage distribution and high voltage transmission (EDSO 2018).

Going forward, to facilitate their more active roles, the DSOs are expected to be investing in technologies to give them much greater monitoring of the network. This increased visibility will cover real and reactive power, for both import (demand) and export (generation) connections. As well as ensuring the power flows on the network are monitored with high granularity, their systems will allow the energy distribution patterns to be recorded much more fully. This should help DSOs to forecast flexibility requirements and ensure the network is pro-actively managed in an optimum way.

Apart from the organisational response of creating more actively engaged DSOs, smart techniques offer the means for DSOs to utilise their existing assets more efficiently. Automated load transfers, meshed networks, dynamic asset rating and storage are technological innovations which are changing the operational protocols of distribution networks from being passive infrastructure to becoming an adaptive configuration of real-time resources (Nieto-Martin et al. 2018). In the next section, we review these smart grid innovations in detail and present some lessons on each of them from the trials undertaken by the distribution company, WPD, in Britain. We then return to the more general organisational challenges associated with the emergence of DSOs.

2 Smart Grid Techniques

2.1 Description

The introduction of smart grid techniques can improve operational efficiency, and thereby avoid incremental asset investment in conventional line reinforcements. We consider four ways in which the DSO can pursue these operational efficiency gains:

- **Automated Load Transfer**. When one section of a distribution network is at capacity, another may have spare. Therefore, automated load transfer schemes allow a DSO to move power around to solve constraints.
- **Meshed Networks & Voltage Control**. By slightly manipulating the voltage at which electricity is delivered to customers it has been shown that demand can be increased or decreased. For most of the time, there is scope for a DSO to use voltage control as a form of demand response to provide flexibility.
- **Energy Storage**. The utilisation of storage as a dynamic asset offers a flexible alternative to planning capacity to meet peak demands.
- **Dynamic Asset Rating**. A number of exogenous conditions can affect the physical capacity of network components, and, if forecasted, this can be utilised. For example, under windy and cool conditions, an overhead line can have its rating increased.

We now describe each of the above in more detail and then review some lessons learnt from their implementations.

2.1.1 Automated Load Transfer

Automatic load transfer (ALT) on the distribution network is the process of changing the state of switching devices on the network to shift the location of the normally open points (NOPs),[1] and cause an improvement in the network's performance. Deliberately changing the open point, and consequentially what loads are supplied

[1]Normally, open refers to a switch action in which the current does not flow in its normal state.

from which primary substations, affects the key network parameters of losses, voltage and capacity headroom (Western Power Distribution 2015a).

A large number of circuits in low voltage distribution networks are run in an 'open-ring' configuration. On these circuits, feeders from the same or adjacent primary substations are electrically connected together at the feeder extremity, via a switching device that is normally in the open position. These feeder inter-connection points are the NOPs. All loads on such circuits are ordinarily associated and fed from a specified feeder/primary substation. It is possible to close these normal open points and create an open point elsewhere on the network (maintaining the open-ring nature of the network), and change the feeder/primary substation that a load (or number of loads) are fed from. Figure 1 represents different NOP movement to balance feeder utilisation. Routinely this is done under maintenance or fault circumstances. The positions of NOPs on a mature portion of network have been established for a variety of reasons, including limiting load/number of customers on a single feeder; managing network voltage; and allowing immediate access for switching purposes. In many instances, these NOPs have been in place for lengthy periods of time (years). As such,

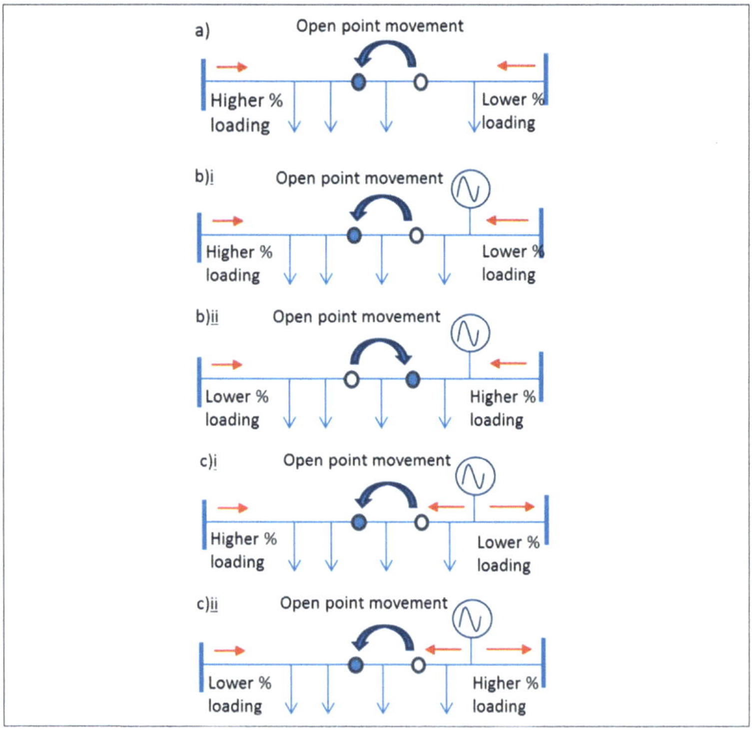

Fig. 1 NOP movement to balance feeder utilisation in the presence of **a** load only, **b** load and low value generation and **c** load and high-value generation. *Source* (Western Power Distribution 2015a)

their position may no longer be optimal with respect to losses, voltage and feeder capacity headroom, particularly, where incremental growth in load on a network (within authorised supply capacities) has occurred.

Automatic load transfer seeks to change the power flows on the network through alternative NOP locations. However, there are other potential benefits that may be gained when considering automatic load transfer as a more flexible operational tool within an electricity distribution network. These benefits include:

- Active management of network feeding arrangements to maximise utilisation of existing capacity.
- Automated load transfer at peak times.
- Voltage regulation.
- A more even load profile of circuits and feeders.
- Management of key performance indices such as customer interruptions (CI) and customer minutes lost (CML).
- Real-time transfer of load or generation across feeders and primary substations.
- A positive impact on climate metrics resulting from reduced losses, due to more even loading, better voltage regulation and reduced reinforcement.

Implementations depend on the network configuration and connected load. Network reconfiguration is a highly complex, non-differentiable, constrained and non-linear (due to the on-off nature of the circuit breakers) mixed integer optimisation problem, due to the high number of switching elements in a distribution network (Narimani et al. 2014). Thus, evaluation of all possible configurations is challenging. Furthermore, with substation loads having stochastic dynamics, the validation of benefits may also be very awkward.

From a theoretical perspective, a network reconfiguration is an optimisation problem that may have different objective functions, such as minimum switching operations, minimum power loss, balanced feeder load balancing or their combination (Tsai and Hsu 2010) to comply with a set of operational constraints such as bus bar voltage limits, line or cable capacity ratings and fault levels. Generally, these methods can be grouped into several categories; classic optimisation techniques, (Botea et al. 2012), (Menders et al. 2013), sensitivities analysis methods, (Jabr et al. 2012), knowledge-based heuristic methods, (Gonzalez et al. 2012, Ferdavani et al. 2013) and genetic algorithms, (Leonardo and Lyra 2009, Oliver and Kipouros 2014). Sensitivities analysis methods and knowledge-based heuristic methods can provide practical results with short computing time but may not be global solutions. Heuristic techniques including ‘sequential switch opening’, (Merlin and Back 1975), (Shimohammadi and Hong 1989) and ‘branch exchange’ (Civanlar et al. 1988, Baran and Wu 1989) deal with a branch at a time. Sequential switch opening is where some of the switches of the network are initially closed, preserving a meshed network, then, to eliminate network loops, the switches are opened sequentially starting with the switch that has the lowest current. The process is repeated until the network reaches a radial structure. Branch exchange methods are different from sequential switching, the method starts from the initial configuration of the network and performs pairs of open/close switching actions to produce new network topologies whilst maintaining the radial nature of the system.

2.1.2 Meshed Networks

Meshing networks (Western Power Distribution 2015b) is the process by which circuit breakers on the network are switched in order to feed loads from a multiple of locations. This approach fundamentally allows the load on each feeder in a meshed circuit to deviate according to variations in the connected load, without the need for pre-existing analysis and changes to switch states. It is evident that closing NOPs exposes more connected customers to supply interruption following a network fault. Therefore, any planned closure of open points for long term operation is routinely accompanied by the installation of along-the-feeder fault sensing and interruption equipment (protection relays and circuit breakers). The installation of along-the-feeder protection devices restores, and potentially improves (i.e. reduces), the probability of customer interruption under fault conditions with meshed operations. Meshing is primarily done to improve the security of supply. However, there are other potential benefits that may be expected when considering a meshed network.

These benefits could include:

- Improved capacity margins.
- Voltage regulation.
- Increased penetration of distributed generation.
- Reduced losses.
- Power quality improvements.

There are however disadvantages to meshing and these include increased fault levels, increased complexity of protection and automation, leading to additional cost.

2.1.3 Energy Storage

At the local level, energy storage (Western Power Distribution 2015c) may offer the following potential benefits:

- Improved capacity margins.
- Increased penetration of distributed generation.
- Deferring network reinforcement by reducing peak loads in branches of the network (above the point of battery connection), where the unmodified peak loads would ordinarily have approached or exceeded effective circuit capacity.
- Power quality and phase balance improvements through active filtering that counters harmonic distortion, and prioritises output to more lightly loaded phases.
- Provision of frequency response and other ancillary services by utilising the stored energy outside times of peak load (primary purpose).

- Improvements in control of voltage at the point of connection.

However, batteries, in particular, have specific operational drawbacks and limitations that include:

- Any reduction in the peak circuit loading is heavily dependent on the prevailing shape and duration of load peaks (e.g. short sharp peaks vs. long relatively flat peaks), the power rating and capacity of the energy storage system and the strategy used to trigger the start of energy output.
- Worsening of network power quality due to the connection of power electronics.
- An operational life that is dependent on the pattern of usage (e.g. repeated high depth of discharge operation).
- Provision of suitable sites.
- Operating noise.
- Overall system efficiency.
- Construction costs.
- Operating costs (maintenance, plus the net cost of electricity for commercial services).

2.1.4 Dynamic Asset Rating

Traditionally overhead lines (Western Power Distribution 2015d), transformers and cables have been assigned capacity ratings intended to ensure operation within safe operating limits, and allow assets to achieve nominal service life. These ratings may be fixed for specific periods of time (e.g. summer and winter ratings), or may relate to a load that has a daily cyclic characteristic (e.g. transformers and cables). However, these ratings essentially do not take the transient environmental conditions, nor the thermal states of the assets, into account. In this respect, the ratings are regarded as 'static'—not responsive to the current thermal or environmental conditions of the asset. These 'static' ratings make assumptions about prevailing environmental conditions (air temperature, wind speed and direction, etc.) and set a limit on electrical current passing through the asset such that safety and service life of the assets are maintained.

Thus, dynamic asset rating (DAR) methods seek to extend the operations of these assets beyond their static limits, through continuous assessment of the asset's actual thermal state (derived from preceding operating circumstances or more extensive metering), and the prevailing environmental factors. Whilst seeking to increase capacity, this approach can also identify periods where the dynamic rating is calculated as less than the static rating, thereby potentially reducing the asset's rating under some circumstances.

The dynamic rating is often referred to as 'ampacity'—the maximum current that can pass through an asset before the temperature limits are reached. The ampacity may be defined as either 'sustained' or 'cyclic' where sustained refers to the asset seeing a steady load, whereas as cyclic refers to the asset seeing an ever-changing load following a set pattern. This technique seeks to properly increase the capacity of assets during peak usage periods to alleviate constraints, whilst maintaining safety and managing impact on asset life. DAR can also constrain use of assets (e.g. generation) when environmental/load conditions are not favourable.

1. Transformers
   For each transformer, the temperature of the insulation (a limiting factor for operation) is governed by the heating effect of current flowing through the windings, and the cooling of the transformer oil. The temperature of the oil (and cooling effect on the insulation) is governed by the ambient air temperature, the heating from the load current and the cooling process according to the cooling arrangement of the transformer (Western Power Distribution 2015e). Sustained load and cyclic load ratings are given by manufacturers, sometimes with different cooling mechanisms, to limit operating insulation temperatures (typically to less than 98 or 110 °C) for a range of ambient temperatures to guarantee that an acceptable service life of at least 20 years can be achieved.
   Primary transformers are typically installed as multiple units, where the loss of one unit from service does not interrupt supply, and many transformers are located outdoors where the ambient temperature (in GB) only infrequently exceeds 30 °C. Therefore, the transformers tend not to be operating close to their temperature limits, resulting in a longer service life span. It is possible to take advantage of these conditions to rate the transformer dynamically based on hotspot temperature, rather than on a static basis. It should be noted that the hotspot temperature exists somewhere around the windings but is difficult to exactly locate. The location and temperature are a function of transformer design and cooling functionality, ambient air temperature, oil temperature and winding losses amongst other parameters. This makes the hotspot temperature difficult to assess with any degree of certainty. Although direct measurement methods do exist, they can only be applied to newly built units, for which the manufacturer can install bespoke, technically advanced and measuring facilities (for instance, sensors with fibre optic cables). Therefore, for existing in-service applications, the temperature may only be estimated. To establish a dynamic asset rating for a transformer, two elements are necessary:

   - A thermal model of the transformer is required to assess prevailing transformer oil and winding temperature given previous load and ambient air temperatures.
   - A process is required that will iteratively increase modelled load current and calculate consequential hotspot temperature (using the thermal model) until the limiting hotspot temperature is reached. The load current that results in this limiting hotspot temperature is the dynamic asset rating, or ampacity of the transformer. This can be either sustained or cyclic.

   The potential benefits that may be expected when considering dynamic asset rating of transformers within an electricity distribution network include:

   - Deferring network reinforcement by allowing more current to pass through the transformer when the weather conditions are favourable to cooling without adversely affecting life.

- Assisting with ratings when highly fluctuating loads are connected (i.e. average rate of loss-of-life of the transformers are still within specified limits even if temporarily the transformer is overloaded compared to nameplate rating).

2. Cables
The static ratings of underground cables (Western Power Distribution 2015f) are based on the rise of temperature of the cable insulation, e.g. 90 °C for cross-linked polyethylene (XLPE) insulation and 65 °C for oil impregnated paper or 75 °C for other paper insulation types. The temperature is limited to avoid insulation breakdown leading to cable failure. The cable temperature increases with the current passing through the cable. This current is limited to a static summer and winter current rating and a cyclic summer and winter rating as defined, for example, in the UK Engineering Recommendation P17, Energy Networks Association 1976. These values are reduced (the cable is de-rated) when the cable is ducted or in close proximity to other cables. The ratings contained within P17 are typically calculated using representative values for soil characteristics, taking the thermal resistivity of soil as a set seasonal value. Although this is fine for a generalised solution that will fit the large majority of cables on the UK distribution network, it does not allow the full realisation of carrying capabilities. The ratings defined within P17 have been used over 30 years by the majority of UK DNOs.
Recommendation P17 consists of three documents relating to the rating of 11 and 33 kV solid paper insulated cables and polymeric cables. The 'distribution rating' is the most common rating basis applied throughout the distribution network (the maximum current that can be carried for five days whilst keeping the insulation below a maximum temperature). In addition, a cable has two static ratings throughout the year, 'summer' and 'winter'. The 'winter' rating takes into account the ability of the cables to carry larger currents, and therefore power flows in winter months due to colder temperatures, and generally wetter ground. This rating is broadly independent of the laying depth of an underground cable, provided the burial depth is at least 600 mm.
Thus, dynamic asset rating of cables seeks to maximise network capacity usage by monitoring soil temperature and moisture. This data would be used to calculate 'real-time' asset capacity, potentially allowing for higher ampacity for limited periods rather than the current 'static rating' current used by distribution network operators. The DAR technique allows the underground cable to be temporarily run above its continuous current rating providing it remains below the critical temperature set out by the manufacturer. A dynamically rated cable would provide the option of running underground cables to incorporate short term increases in load that might defer capital expenditure on network reinforcement. Research into the dynamic capabilities of underground cables undertaken worldwide has led to the development of a number of monitoring techniques and simulation softwares applicable to the transmission and distribution network.

2.2 *Smart Techniques in Practice*

2.2.1 Automated Load Transfer

The objectives of a practical trial (Western Power Distribution 2015a) were to learn from shifting load between high voltage feeders by altering normal open points on two distinct trial areas of a distribution network, based on the prevailing network loads, and thereby explore:

- Potential impacts, both benefits and trade-offs, that could be derived from implementing alternative network configurations (normal open points that are different to the pre-existing set).
- Various types of impact, including feeder load balance; feeder utilisation; circuit losses; circuit voltages.
- Potential to schedule changes to normal open points that deliver material net benefits.

For the comparison, two areas of part of the WPD local network in the UK were considered: one underground section; and one (largely) overhead. It was required to identify suitable substations to install additional remote control prior to being able to carry out an in-depth network study or power system analysis. To aid the selection of suitable sites, a simple power flow study was carried out to identify locations that would allow a variety of different configurations of load and customer numbers by moving open points on the network.

The assessment framework was developed to:

- Model the trial sections of network, to allow the performance of the pre-existing NOPs to be examined.
- Identify alternative NOPs, intended to improve performance.
- Test alternative NOP locations on the network, and through the use of modelling, assess and validate the benefits of the alternative NOPs with respect to the pre-existing NOPs.

These multiple configurations outline the identified NOPs that resulted from analysis of the two separate sections of network (underground and overhead). A preferred configuration was also developed in Fig. 2 of the seven selected NOPs.

The main lessons from these trials were as follows:

- Feeder utilisation
  - Feeder utilisation varies seasonally and over the course of the trial varied by around 10% (Looking at the difference in feeder currents under the nominal configurations over different trial periods). There were also variations due to the addition of more generation.
  - Although it is possible to reduce loading on a feeder, this reduction can be limited by network configuration constraints.

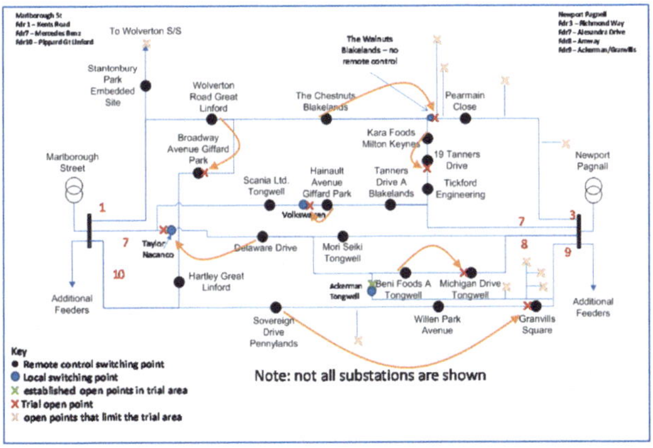

Fig. 2 Stylised network diagram showing operations of normal open points with arrows indicating switching actions. *Source*: Western Power Distribution 2015a

- Losses
  - A revised configuration of NOPs compared to the pre-existing configuration should lead to modest improvements in network losses.
  - Improvements are found to be up to 12% using loss minimisation configurations.
- Minimal node voltage
  - Voltage improvement is possible using automated load transfer under loss minimisation optimisation, but it is more noticeable on a rural overhead line network.
- Customer impact
  - Potentially more customers are at risk of being impacted by a fault if the network is reconfigured to reduce losses or increase capacity headroom.

2.2.2 Meshed Networks

The mesh networks (MN) trial (Western Power Distribution 2015b) consisted of the installation of circuit breakers and associated protection on an urban two feeder mesh fed from the same primary and the installation of monitoring devices. The trial area was near the MK Dons football stadium within the WPD network area comprised of two feeders. The network schematic is shown in Fig. 3.

Results from the trials are presented in Fig. 4. It can be seen that overnight demand from loads nominally connected to the Granby Court feeder (shaded in yellow) is less than the demand from the loads nominally connected to the Dons Fast Food feeder (shaded in green). During this overnight period, the effect of meshing is to raise power through the Granby Court feeder breaker and reduce power through the Dons Fast Food feeder breaker. This can be seen as the difference between the solid and

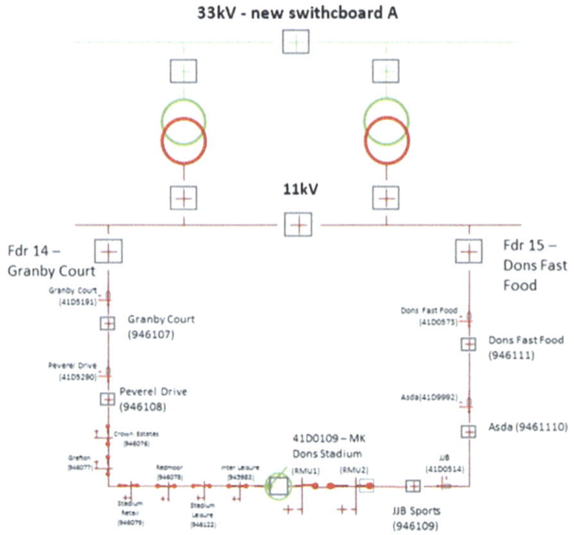

Fig. 3 Meshed network feeders circuit schematic connected at the NOP at MK Dons stadium distribution substation. *Source* Western Power Distribution 2015b

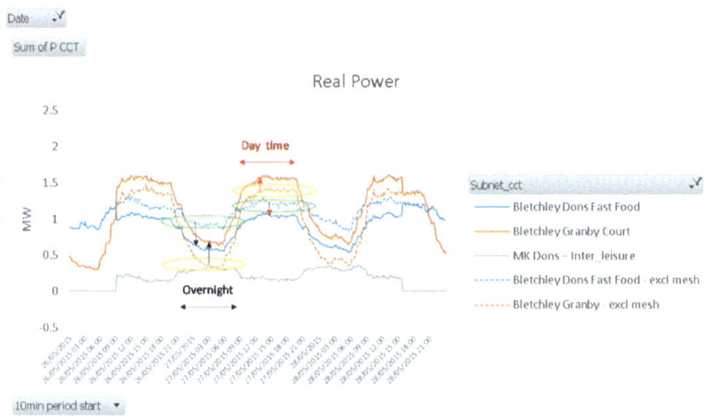

Fig. 4 Measured power throughout mesh trial area highlighting periods of demand feeder loads and nominal loads

the dashed traces, and the black arrows in Fig. 4. The feeder with nominally higher demand (Dons Fast Food) is supported by power transferred from the Granby Court feeder, and the power on each feeder become broadly similar. However, throughout the day, the nominal load on the Granby Court feeder is greater than the demand from the nominal Dons Fast Food loads, yet at the (closed) NOP, power is still transferred from the Granby Court feeder side to the Dons Fast Food side. This creates a greater difference between the loadings of the two feeders and is not obvious: meshing the circuit during these periods increases utilisation on the source that is already supplying more power.

Within the trial, although half-hour periods of improvement in capacity headroom were found (the load transfer was still from Don's to Granby Court, but the loading on Granby Court was lower than Don's resulting in an improvement to the headroom at Don's), these periods only occurred during minima in daily feeder loading. Therefore, no useful improvement in capacity headroom occurred. Whilst the two feeders considered in the trial were the same rating, shifts of load through meshing may cause more significant changes in percentage capacity indications. For this trial network, no distinguishable difference was found between losses in mesh and radial configurations. As expected, fault level rose on the network with mesh configuration; however, this was well within the ratings of the connected switchgear.

2.2.3 Energy Storage

The aim of the trial was to operate a trial 11 kV feeder with battery/inverter units installed at five LV substation locations to provide energy storage (ES) and explore:

- Baseline operation of the LV substations and HV feeder without operation of the energy storage units.
- Peak lopping and trough filling using demand forecasts.
- Voltage response.
- Frequency response.
- Impact on power quality.
- Specific operational circumstances, for example, response to circuit fault/ disturbance.
- Reliability and degradation.

Additional objectives included: investigating optimum charge and discharge windows; available triggers for charging regime; DNO connection requirements for ancillary grid service operation; best placement of storage on the system; changes to battery condition over the course of the operational trials; improvements for equipment specifications.

The installed equipment comprised of five energy storage units (inverter with battery module) connected at existing substations on a single 11 kV feeder from a primary substation. Each site contained: a 50 kW/100 kWh energy storage (sodium nickel) battery, with battery management system; a 100 kVA rectifier/inverter unit; site controller (providing user interface and control functionality); protection connection circuit breaker; and fused connection to the LV distribution network at the adjacent distribution substation. Figure 5 represents an electrical schematic of the ES systems installed.

Figure 6 shows effective peak shaving over the course of 1 week. Thus, it was recommended that if the network has a constraint, the battery could be usefully operated in peak-shaving mode. This also improved battery operating life since the battery only discharges for high-value usage.

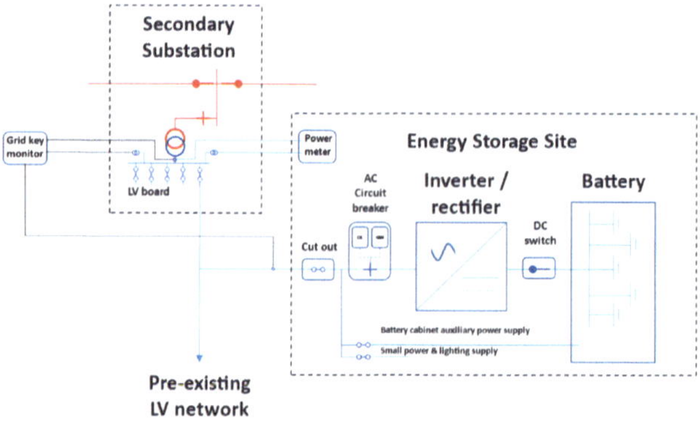

Fig. 5 Electrical schematic of the ES system

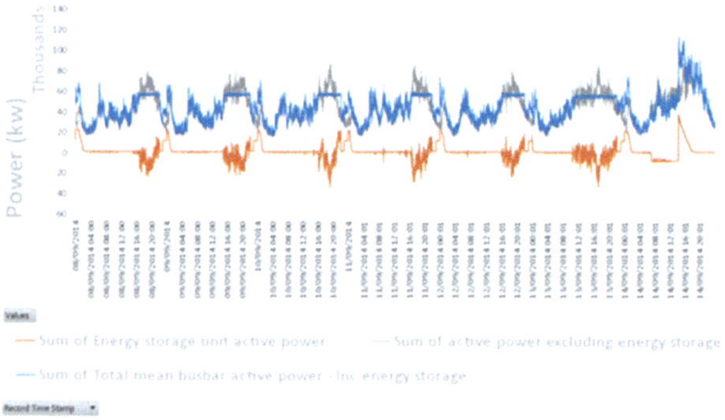

Fig. 6 Peak-shaving operation over a sample week

2.2.4 Dynamic Asset Rating

This trial had a number of key elements:

1. Installation and commissioning of an online relay plus associated input instrumentation.
2. Preparation of an offline thermal model to allow tuning of thermal model parameters.
3. Tuning of thermal model parameters.
4. Assessment of the benefits of instantaneous/of-the-moment dynamic asset rating benefits.
5. Assessment of the dynamic asset rating relay/online or offline methods.
6. Gathering of environmental data.
7. Assessment of forecast benefits.

The main lessons for each asset are summarised as follows:

- Overhead Line Dynamic Asset Rating.
  This trial was implemented on three 11 kV overhead lines coming out from a substation to allow an offline thermal model to be created and validated, and for dynamic asset rating values to be estimated.
  A dedicated dynamic asset rating relay is not essential, other computing devices/systems could perform real-time assessment calculations (e.g. the network management system), if these are required. In addition, offline modelling was particularly important because it allowed extension of the work into forecasting of future ampacity.
  There were significant amounts of time when the dynamic rating was higher than the static rating. Overhead lines are predominately affected by wind speed direction and thus significant variations occur both across seasons and within short time scales (minutes). When coupled with the low thermal capacities of the lines, taking advantage of the benefits this assessment offers is evidently very circumstantial.
  It is clear that dynamic asset rating systems offer potential benefits to distributed generation. A good example would be increasing export from wind farms on a windy day beyond their usual limits. In contrast, there is less scope to enable solar panels to export more as this would be correlated with the thermal radiation reducing the line ratings.
  The trial identified significant average real-time ampacity benefits. However, the ampacity varies over a large range within a very short time frame largely driven by variation in wind speed. This rapid variation in wind speed coupled with the low thermal mass of the asset/short time constants for changes in temperature means that the asset cannot be loaded and relied upon at these identified enhanced average levels for extended periods of time. The potential for this rapid fluctuation in ampacity led the project to investigate the possibility of using forecast weather to estimate forward ampacity. However, cautious conclusions were drawn from this study as it was clear that the potential benefits could not be relied upon over the longer planning horizons as it does not offer a firm alternative to line reinforcements.
- Underground Cable Dynamic Asset Rating. This trial was undertaken on a representative network of 33 and 11 kV underground cables, allowing an offline thermal model to be created and validated, and for cable dynamic asset rating values to be estimated. The dynamic asset rating of cables is dependent on an accurate cable thermal model and a set of soil temperature measurements to allow conductor temperature to be calculated and a dynamic asset rating set that limits this conductor temperature to the specification of 65°.
  Within this technique, the thermal model was validated by comparison with the industry widely accepted CRATER model (Poidevin 2008), and by comparison to measured cable temperatures. Comparisons with CRATER suggested comparable results. Calculated cable temperatures from the thermal model also compared well to measured values over the load ranges experienced during the trials, even with measurement issues (temperature measurement placement and soil resistivity val-

ues). The trial findings suggested that this technique may be able to provide relief to cables hitting thermal limits in some circumstances, especially in winter. It was therefore recommended that a further research should look at the following issues:

- Where the cable is approaching thermal limits.
- The ratings basis is cyclic.
- Further improvements of soil parameter measurement are targeted.
- Full assessment is made of the actual cyclic load shape that the cable is experiencing is conducted.

• Transformer dynamic asset rating. Transformer performance is dependent on thermal models. For any specific transformer, a period of data collection would be required to calculate the necessary model parameters. This complicates the use of dynamic asset rating for a transformer in the planning context: either extensive data collection is required, or conservative generic assumptions of model parameters would have to be used, and the latter may lead to very little benefit.
  However, the period of data collection may not preclude some generalisations. Outdoor transformer ampacity assessments, using predicted weather, resulted in a gain in peak ampacity of up to 10% with a mean of up to 5% for a large proportion of the winter months. This compares to ampacity gains of up to 15% based on measured conditions at the time. Well ventilated indoor transformers show calculated ampacity gains in the winter months of up to 3% for around 70% of the time. Indoor transformers with no ventilation may have no benefits at all. From a planning perspective, the type of housing for the transformers should be considered within any proposed application.
  Planning tools should allow for a wider range of asset attributes to be included, such as housing type and ventilation. The cyclic rating is based on a fixed percentage of the (sustained) name plate rating. Therefore, an increase in the sustained dynamic rating should result in the same percentage increase in cyclic and emergency ratings. However, further research is required. Therefore, the trial indicated that there is potential benefit from the deployment of transformer dynamic asset rating to reassess thermal capacity on a case-by-case basis.
  Such potential could be targeted at existing transformers that are approaching thermal/load limits, involve limited installation of temperature & load monitoring, tuning of transformer specific models and assessment of potential to run at higher than nominal ratings. This approach could include addressing the issue of risk management with respect to transformer life.
  It is unlikely that an asset would be run at full dynamic rating under normal operation. It would be much more likely that this usage of the dynamic rating would occur under a planned outage scenario or emergency operation to carry the load when an adjacent circuit is out.

2.3 *Overall Assessment of Smart Grid Techniques*

Table 1 provides a high-level summary of the above smart grid techniques and their impacts on various network metrics, as described below.

2.3.1 Thermal Limits & Capacity Headroom

- All techniques altered capacity on the network.
- Dynamic asset rating evaluates capacity more accurately than static ratings which may suggest additional or in some cases less capacity. Overhead lines are predominately affected by wind speed/direction and so significant variations occur across seasons and within short time scales (minutes). When this variability of rating is combined with the low thermal capacities of overhead lines, the applicability of this technique is limited to particular circumstances. The dynamic ratings of both cables and transformers are dependent on ambient temperatures, meaning diurnal (for transformers) and seasonal variations are manifest. But the larger thermal capacities of underground cables means short time duration changes in ambient conditions cause less short term variability in asset ampacity.
- Automated load transfer and meshing shift load from one part of a network to another, thereby potentially relieving constraints. Automated load transfer offers a far more transparent mechanism, whilst meshed networks are continually dynamic by their very nature. The extent to which benefits exist is highly dependent on the

Table 1 Impact matrix of smart grid techniques

	DAR-OHL	DAR-Tx	DAR-cables	ALT	Mesh	Energy store
Thermal limits capacity headroom	✓	✓	✓	✓	~	✓
Voltage limits	No impact	No impact	No impact	✓	~	✓
Faut levels	No impact	No impact	No impact	No impact	×	×
PQ	No impact	No impact	No impact	~	~	✓
Enablement of DG	✓	✓	✓	✓	✓	✓
Losses	×	×	×	✓	✓	×
CI/CMLs	No impact	No impact	No impact	~	~	No impact
Grid network services	No impact	No impact	No impact	No impact	No impact	✓

Key: ✓ Positive impact; × negative impact; ~ network dependant may have positive or negative impact

connectivity of any specific network, demand/generation connected to the network, and the extent to which the loads vary.
- Energy storage shifts load over time, reducing load at a capacity-constrained periods, only to increase the load at a less critical points. The specified power and storage energy capacity clearly need to be appropriately matched to the network requirements; and adaptive triggering is required to deal with daily variations to optimise performance. Energy storage may complement dynamic asset rating by providing a mechanism to alter load patterns such that constrained assets might make the best use of available ampacity.

2.3.2 Voltage

- Automatic load transfer demonstrated the largest voltage benefit (4%), on some of the rural circuits that were trialled, but no significant benefit was found on urban circuits.
- Meshed networks considered a small urban trial where there was no significant impact on voltage.
- The trialled energy storage systems achieved little impact.

2.3.3 Fault Levels

- As recognised, introducing storage into a network may increase fault levels, but in this trial, the storage faults were small compared to pre-existing fault levels, and so had negligible impact. Meshed networks will also increase fault level due to the reduced circuit impedance. For the mesh technique trial, this was within the ratings of all circuit equipment.

2.3.4 Power Quality (PQ)

- Mesh trials showed no discernible impact on power quality. Super-position theory and the feeding of harmonic loads via different sources means that harmonics presently fed from one source could be fed from two sources (depending on network impedances), however, it is unlikely that larger scale trials will show any marked appreciable benefits as the majority of loads were within limits defined by standards and as such it would be difficult to differentiate small changes.
- The installed energy storage equipment did not specifically have functionality aimed at improving power quality. At one site, some improvement was noted, however, this was a beneficial coincidence arising from the nature of a local (within standards) power quality disturbance and the inductance/capacitance smoothing network in the storage system.

- More targeted studies of a network that has a known power quality issues could be identified to further examine the potential of automatic load transfer and meshed techniques to beneficially impact this issue.

2.3.5 Enablement of Distributed Generation

- Whilst not explicitly trialled, all of the smart interventions would help the DSO in managing the uncertainties introduced by distributed generation.

2.3.6 Losses

- As discussed in the preceding technique-trial specific section, automatic load transfer and meshing offer some potential, though the magnitude is network specific.
- The trialled storage systems increased losses, and dynamic asset rating will tend to increase losses if higher circuit loads are facilitated.

2.3.7 Customer Interruptions and Minutes Lost

- Automatic load transfer changes NOP positions and consequently affects connected customers per feeder. The trial algorithms:
  - Increased performance by 15% (whilst optimising capacity headroom) on a rural/OHL network.
  - Increased performance by 50% (whilst optimising losses/voltage) on an urban/cable network.
- Meshing networks does not improve customer security as such; the improvement only occurs when additional automatic sectioning occurs beyond that offered by the pre-existing NOP. Due to communication system limitations, the implemented trials did not increase the number of sections, essentially maintaining the pre-existing customer security.

2.3.8 Network Services

- Whilst these trials have demonstrated that frequency response is possible with the storage technique, a marketable service was not fully evaluated. A big question is whether regulatory constraints would permit such ownership by the network owner or system operator, or whether independent new entrants would be sufficiently motivated to provide these assets.

Evidently, smart technologies offer scope for DSOs to manage their distribution networks more actively and efficiently, but their implementations are not always straightforward and the benefits can be mixed. Thus, alongside new operational technologies, DSOs are looking at flexibility services, provided within a market context, again to move away from the conventional passive investment approach of simple reinforcement. Conventional reinforcement will still be needed where the future loads on the network require a significant and predictable capacity improvement. However, for short periods of time or for uncertain conditions, non-asset-based solutions, such as demand side response or other flexibility services, may solve the issues at lower cost. DSOs will therefore be upgrading areas of their business systems to facilitate the dispatch and settlement of flexibility services.

3 Flexibility Services

Flexibility products and initiatives have been used at transmission level since the very beginning of the wholesale market in order to balance the overall electricity system. Traditionally this came from option-like contracts with larger power stations or hydro facilities to increase or decrease their output at short notice (Biegel et al. 2014). As the generation mix has changed, with substantial resources directly connected to the distribution networks, the ESO's flexibility products have become more complex. New providers include smaller renewable and conventional generating units and demand response customers, sometimes brought to market by aggregators. As DNOs transition to DSOs, the DSOs will have their own flexibility and reserve contracts with mostly the same providers. They will also deploy a range of smart grid technical solutions which will automatically control selected customer equipment and reconfigure the network flows (Nieto-Martin et al. 2018).

Flexibility is usually interpreted by the system operators as the ability of a power system to maintain stability in the face of swings in supply or demand. Traditionally, flexibility was provided in power systems almost entirely by controlling the supply side at large power stations. The GB system has seen increasing shares of intermittent renewable generation requiring additional flexibility to maintain system reliability as the variations in supply and demand grew to levels far beyond what was originally conceived. This has led to the introduction of additional flexibility programmes by the ESO (in GB run by National Grid) for short term reserve and fast acting frequency response services. As larger power stations continue to close, electricity generation becomes much more distributed, more flexibility will be needed across the whole system. This flexibility gap will need to be covered by new flexibility options, much of which will be facilitated by a DSO (Pastor et al. 2018).

Locational marginal prices (LMPs) are widely used worldwide in transmission systems to account for geographical variation in costs of electricity primarily due to losses and constraints depending on distances and transmission capacities between consumption loads and generation (ISO New England Inc 2013). LMPs can be defined as the marginal cost of supplying the next unit of demand at a certain location taking into account supply offers, demand bids and network characteristics including losses and limits. The theory underpinning LMPs was outlined in 1988 by (Schweppe et al. 1988), characterised by three components: the energy component, congestion component and loss component. The energy component is the load-weighted average of the system local prices. The congestion component reflects the marginal cost due to binding constraints, e.g. line capacities and reserve requirements. The loss component is the marginal cost of any losses for a specific location. To date, the LMPs have not been widely spread across distribution systems, however, they have potential of solving sequential tightness at local levels promoting the use of alternative connection arrangements, reducing system losses and focussing on operational direct investments in constrained areas where upgrades can be cost-effective or ultimately, the last resource to ensure a reliable supply (Yuan et al. 2016).

Alternative connection agreements have become commonplace for DSOs' distributed generation customers who want to connect to the distribution system, but have been limited by the conventional infrastructure solution. DSOs' innovative solutions allow customers to connect their distributed generation at reduced cost, with quicker time scales, but only if they agree to a higher risk of curtailment when the DSO needs to manage high flows (Laur et al. 2018). There are four variants (Western Power Distribution 2018a):

- Active network management (ANM) is a customer control facility which allows for full dynamic control of the network, generation and demand by the DSO.
- Soft intertrip—Some networks are constrained due to a single upstream asset requiring reinforcement, or a single limit being infringed under certain conditions. This solution has an on-site soft intertrip remote terminal unit which provides two normally open contacts for the customers control system to monitor: Stage 1 and Stage 2. When both sets are open, the connection will be free of constraints. The levels of curtailment corresponding to the operation of the Stage 1 and Stage 2 contacts are defined at the planning stage.
- Timed—This solution is a simple timer-based device that monitors the connection agreement with the customer, which will include some form of curtailment based on time of day. The customers connection agreement will include an operating schedule which will define the times and levels of capacity available to them.
- Export limited—This type of connection enables customers to have their import from or export to the distribution grid capped. This often allows customers to connect renewable generation or storage beyond their meter whilst protecting the distribution network. Measurement and control equipment are used to automatically adjust the customer equipment to ensure they comply with their connection agreement.

These alternative connection solutions will be mostly used over operational time scales on a case-by-case basis. In contrast, customer DER flexibility services may be evaluated in the context of investment decisions to reduce, defer or negate conventional build. Identifying, contracting and operating such solutions are at the centre of the DSO transition.

3.1 DSO Market Models

The increased number and capacity of distributed energy resources connected to the electricity system is leading to an increase in the level of active management of demand and generation seen on the distribution network. This changing system is driving an increase in the interactions between the transmission and distribution networks and there is a growing need for the parties to move away from the current market model. Moving away from traditional roles will allow new markets to be created and accessed by a wider number of participants, helping both existing and new market participants to support network and system operation.

Customer connected flexibility and distribution network smart grid flexibility can help alleviate both transmission and distribution constraints and contribute to releasing additional capacity on both the transmission and distribution networks. There is significant value for both active and passive customers connected to the electricity network in maximising the usage of these flexibility sources where it is effective and economic to do so.

In order to economically achieve this, the greatest number of participants must be able to provide services across a number of market procurements. Achieving this cost effectively, there must be limited conflict between various flexibility providers and network capacity must be sufficient to facilitate the services provided by market participants. Four key principles to achieving this have been discussed in (Western Power Distribution 2018a, Energy Networks Association 2018):

1. Facilitating accessibility to markets. Customers will expect a level playing field access to a wide range of revenue streams and DSOs will have a key role in facilitating fairness. Multiple paths to market could increase participation, but should not lead to conflict or complexity. Customers will expect the complexity to be designed out by the market mechanisms. Ultimately, the efficiency of the route to market will be reflected in the commercial revenues passed through to participants. Distribution network operators will, it is hoped, thereby increase their usage of non-asset solutions, creating new markets for new and existing participants.
2. Increased ESO-DSO co-ordination. Clear co-ordination processes and common methodologies for procurement and dispatch of services will aid efficient local/system wide usage of resources. Principles of access and rights for access will also need to be considered from a whole system viewpoint. Increased information exchange across the transmission and distribution interface will enable conflicts to be managed on an operational timescale. Evolving the existing roles and responsibilities to have a more co-ordinated approach to system resilience, will take advantage of new forms of flexibility on the system.
3. Product/service convergence. Convergence at the design stage of the products and services which utilise flexibility across both transmission and distribution system requirements will reduce the likelihood and impact of any market conflict. Co-ordination across market procurers to define consistent methodologies and principles will help support level playing field access. Providing information to

customers on the prerequisites for service delivery will enable them to assess the suitability of connection types and ensure they can benefit from potential revenue streams. Convergence of services and connection types will aid the simplification of customer offerings and improve the customer experience.
4. Signposting for services. DSOs should publish more information on the availability of capacity across their networks for power delivery. They will also publish information to assess fiability of the network to transmit power and understand the utilisation of assets. Proactive information publishing will provide leading signals on where to connect to maximise system efficiency and charging methodologies will be adaptive to reduce network congestion. This visibility of the existing and future network will help markets deliver the services required. DSOs will further stimulate markets by the signposting of markets for non-asset solutions, opening new revenue streams for participants (Pastor et al. 2018).

We now consider four market designs. To operate the system using the conventional market model where there are no or few conflicts between distribution network constraints and distribution network connected services, it may be feasible to operate and plan the system using the conventional market model (3.1.1). The ESO-led market model (3.1.2) maintains the same system hierarchy and enables all commercial services to be agreed and settled using existing mechanisms. As distribution network constraints increase, curtailment and service conflicts will increase across a number of voltage levels, and the ESO will not be able to optimally manage the dispatch of conflicting services deep within the distribution network. The co-ordinated market model (3.1.3) allows both ESO and DSOs to share a single procurement model but requires complex market design and effective visibility of operations to ensure the model provides efficient outcomes. The DSO will work with the ESO to facilitate shared procurement activities in areas where the level of constraints is beginning to increase. The DSO-led market model (3.1.4) requires a significant shifting of responsibilities for system balancing operations and commercial contract activities, but will result in the most efficient whole system outcome for distribution networks with multiple complex constraints. Where distribution constraints are impacting on service delivery through existing market models, the DSO will work with the ESO and move towards a DSO-led market model.

3.1.1 Conventional Market Model

The conventional market model for procuring services to resolve transmission issues has no direct link with DSO constraint management. This has no effect when solely transmission connected energy resources are utilised, or when the distribution network capacity is assumed to be unconstrained. However, as distribution system operators are increasingly actively managing the network, curtailment due to constraints can cause conflicts and reduce in the effectiveness of services delivered (Fig. 7). As the number and level of constraints increase, the likelihood and consequence of the conflicts will become more apparent.

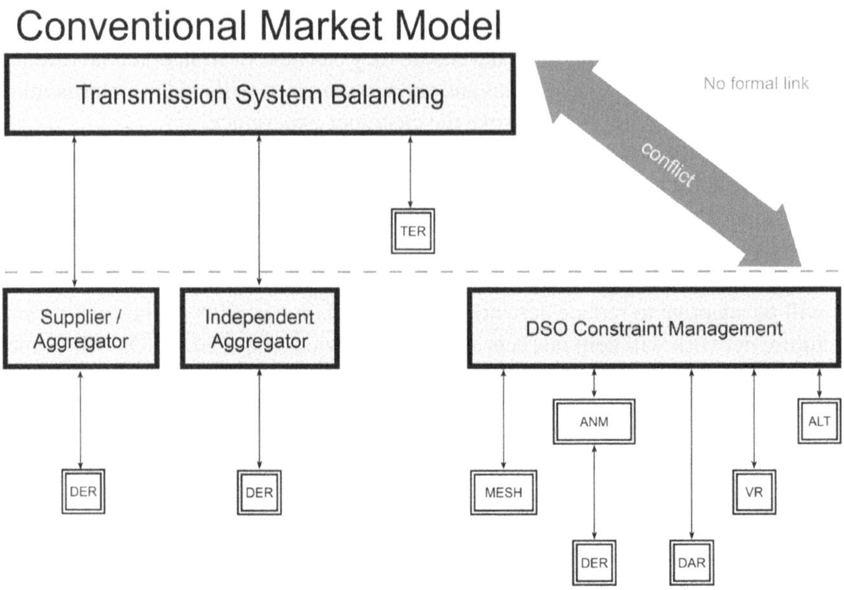

Fig. 7 Schematic representation of the conventional market model

3.1.2 ESO-led Market Model

The ESO-led market model preserves the same arrangements as the current market model, with the ESO directly contracting with distribution connected DERs for services. A visibility link between the ESO and DSO enables the ESO to have oversight of any conflicting actions DSO constraint management may undertake, such as curtailment under active network management systems (Fig. 8). It is also able to call upon distribution network smart grid flexibility through commercial contracts with the DSO. Although this model allows conflicts to be managed and/or mitigated by increasing the visibility between the ESO and DSO, the complexity of this process increases as the number and curtailment of those constraints increases. Supplier and/or aggregator managed DERs within or adjacent to active network management zones will be affected by DSO constraint management and so the ESO will need to inform the DSO of any potential conflicts in order to ascertain the impact. The nested nature of distribution constraints and the interactivity of meshed networks vastly increase the complexity when assessing the effectiveness of DER to deliver services, meaning a full network impedance model with connectivity is required, as well as historical and real-time load flows for real-time and forecast curtailment studies. Any planned network running arrangement alterations the DSO takes or any unplanned outages will potentially undo the contracted service position. Customers in constrained distribution networks may be disadvantaged as the ESO calls on flexibility services in unconstrained areas ahead due to the uncertainty and risk associated with non-delivery of contracted services due to DSO actions. As the smart

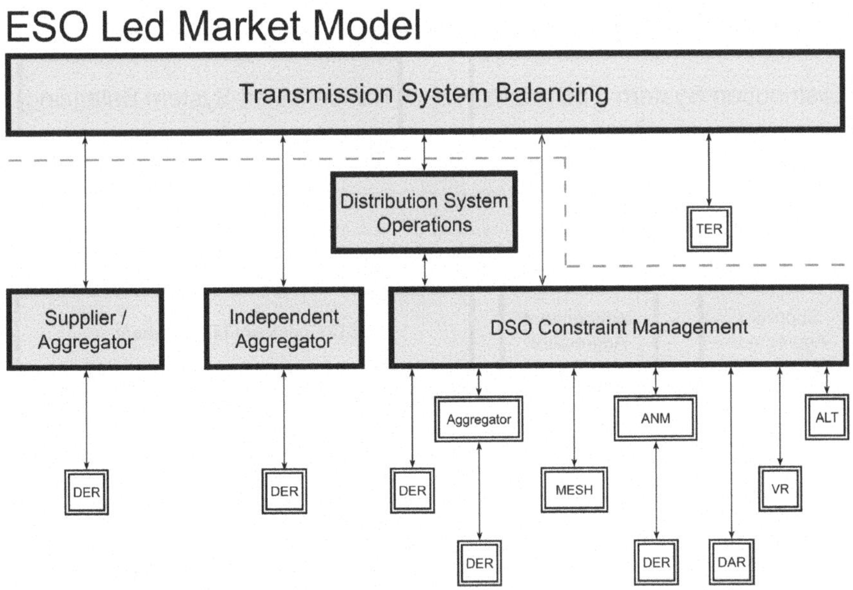

Fig. 8 Schematic representation of the ESO market led model

grid flexibility is embedded within the DSO operations, the ESO may not maximise its usage of the resource, preferring to utilise more expensive contracted flexibility which will be able to commit to operating much further ahead. However, networks with few constraints or smaller numbers of distribution connected energy resources may be simpler to describe and result in an acceptable level of inefficiency due to unplanned or uncoordinated actions.

3.1.3 Multiple Party Co-ordination Model

The multiple party co-ordination model develops a dual procurement approach, which enables both DSO and ESO to directly contract with distribution connected DER through a variety of aggregator and supplier paths. Visibility of contracts placed would be exchanged between the DSO and ESO, enabling conflicts to be understood and managed (Fig. 9). This model allows equal access to flexibility services from a number of market procurers, but requires a sharing of roles and responsibilities, which would need frameworks, principles and methodologies to be developed and agreed prior to operation. Again, the complexity of the visibility platform between DSO and ESO increases as the number and curtailment of the distribution constraints increases. Traditional reinforcement, charging signals and proactive signposting of services can all be used to minimise the likelihood of conflicting actions occurring and help simplify the visibility required between operators. Market mechanisms and incentives that place responsibilities on all parties to ensure visibility of services is

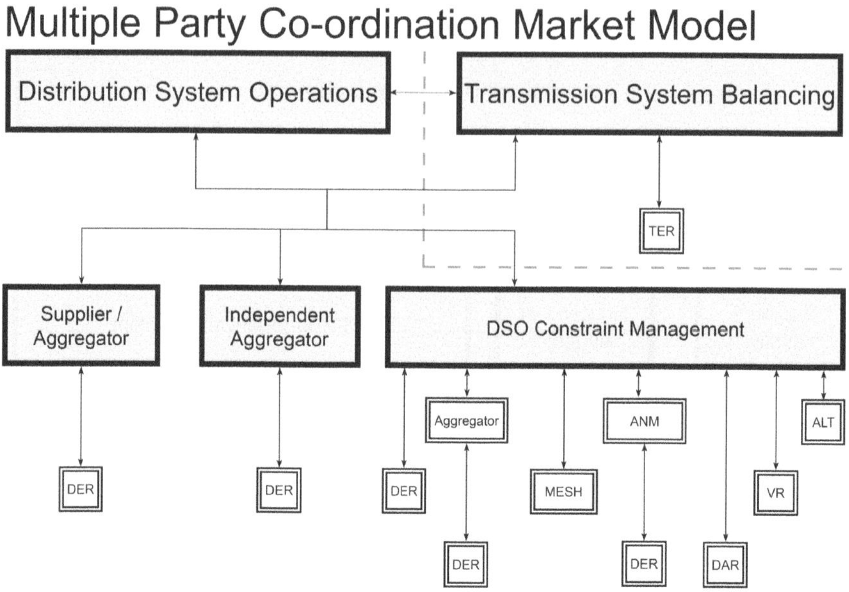

Fig. 9 Schematic representation of the multiple co-ordinated market model

maintained and penalties or balancing mechanisms to ensure the results of unplanned conflicts are equitably shared may help this model run efficiently.

3.1.4 DSO-led Market Model

The DSO-led market model changes the hierarchy of the commercial frameworks and allows the DSO to co-ordinate the prioritisation of flexibility services with respect to the constraints on the distribution network. By operating the model this way, the services offered up to the ESO will inherently be co-ordinated across the transmission-distribution boundary (Fig. 10). The DSO is able to assess not only the effectiveness of services within the distribution network and factor that into the economic cost, but also the impact of running those services on future optionality. By optimising the dispatch of those services based on a holistic impact on the local distribution network, the whole system outcome will be most efficient and cost-effective. The ESO will be able to take balancing actions on a national level within a pool of services competitively procured through a number of DSOs, without adversely impacting localised network constraints. The DSO will be responsible for ensuring the requested response is efficiently delivered through a full range of flexibility services at their disposal. Customers in constrained and unconstrained distribution networks will have their flexibility services called upon equally by the ESO as the DSO would take on the risk associated with non-delivery of contracted services due to DSO actions, which the DSO would be uniquely placed to best assess.

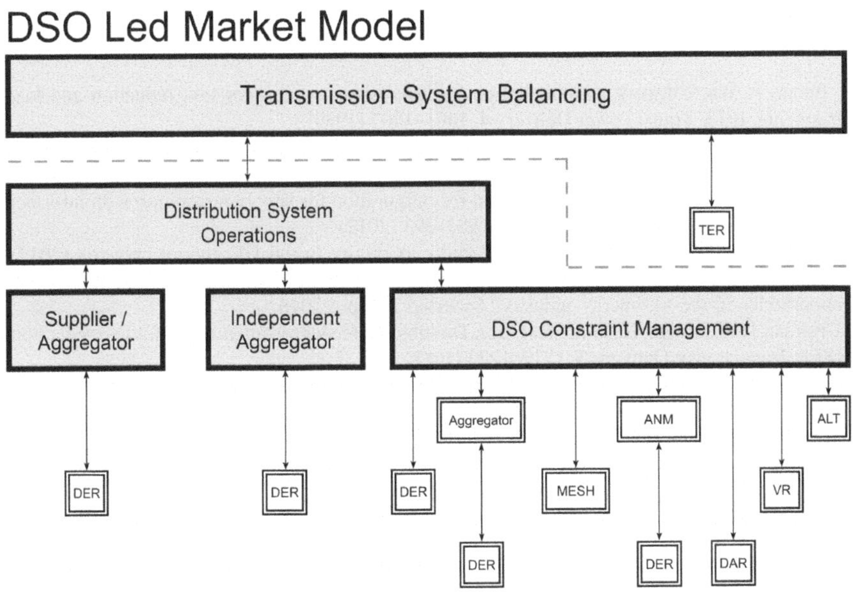

Fig. 10 Schematic representation of the DSO-led market model

4 Summary

We have observed that electricity distribution networks are now more complex to manage, driven by the rapid development of renewable and distributed resources as well as the emerging electrification of transport. This has motivated the emergence of DSOs to take a more active responsibility in managing distribution operations and to contract independently for flexibility services. Smart technologies can provide more efficient asset utilisation and smart markets can facilitate aggregated small-scale participants to sell various flexibility services to the distribution network operators. This chapter has reviewed the approaches that distribution system operators (DSOs) can adopt to enable a greater volume of demand, generation and storage to be connected in a smarter and more active setting. Smart grid technologies can help, but are not a complete solution, whilst the greater commercial role of DSOs in contracting for services places it alongside other stakeholders in the industry, i.e. the transmission system operator, retailers and generators, who may have similar requirements for system and energy balancing. Market co-ordination is therefore an open issue for market regulation and government policy if flexibility services are to be traded efficiently. Access to smart technologies and smart markets will therefore be crucial for DSOs as they actively move from their traditionally passive role in the supply chain.

Acknowledgements Authors would like to thank Western Power Distribution Innovation and the Energy Networks Association for the support and discussion that led to this chapter.

References

M. Baran, F. Wu, Network reconfiguration in distribution systems for loss reduction and load balancing. IEEE Trans. Power Delivery **4**, 1401–1407 (1989)

B. Biegel, L.H. Hansen, J. Stoustrup, P. Andersen, S. Harbo, Value of flexible consumption in the electricity markets. Energy **66**, 354–362 (2014)

A. Botea, J. Rintanen, D. Banerjee, Optimal reconfiguration for supply restoration with informed A* search. IEEE Trans. SmartGrid. **3**(2), 583–593 (2012)

Capgemini, Unlocking the grid: challenge and opportunity in the UK energy industry, (2017). Available at: https://www.capgemini.com/gb-en/resources/unlocking-the-grid-challenge-and-opportunity-in-the-uk-energy-industry. Accessed 24 April 2018

S. Civanlar, J.J. Grainger, H. Yin, S.S.H. Lee, Distribution feeder reconfiguration for loss reduction. IEEE Trans. Power Delivery **3**, 1217–1223 (1988)

EDSO, EDSO presents key recommendations for electricity market design trilogue, (2018). Available at: https://www.edsoforsmartgrids.eu/edso-presents-key-recommendations-for-lectricity-market-design-trilogues. Accessed 17 July 2018

C. Eid, P. Codani, Y. Perez, J. Reneses, R. Hakvoort, Managing electric flexibility from distributed energy resources: a review of incentives for market design. Renew. Sustain. Energy Rev. **64**, 237–247 (2016)

Energy Networks Association, Engineering Recommendation P17 Part 1, in *Current Rating Guide for Distribution Cables* (Energy Networks Association, London) (1976)

Energy Networks Association, Future worlds, (2018). Available at: http://www.energynetworks.org/electricity/futures/open-networks-project/future-worlds/future-worlds-consultation.html. Accessed 18 Aug 2018

A. Ferdavani, A. Asuhaimi, A. Kavousifard et al., Reconfiguration of distribution system through two minimum-current neighbour-chain updating methods. IET Renew. Power Gener. **7**(12), 1492–1497 (2013)

A. Gonzalez, F. Echavarren, L. Rouco et al., A sensitivities compution method for reconfiguration of radial networks. IEEE Trans. Power Syst. **27**(3), 1294–1301 (2012)

ISO New England Inc., ISO New England Manual for Market Operations Manual M-11, Rev. 47. October 2013, Technical Report

R. Jabr, R. Singh, B.C. Pal, Minimum loss network reconfiguration using mixed integer convex programming. IEEE Trans. Power Syst. **27**(2), 1106–1115 (2012)

A. Laur, J. Nieto-Martin, D.W. Bunn, A.V. Pastor, Optimal procurement of flexibility services within electricity distribution networks, in *European Journal of Operational Research*, Special Issue in *Advances in Stochastic Optimization* (2018)

M.O. Leonardo, Q. Lyra, Adaptive hybrid genetic algorithm for technical loss reduction in distribution networks under variable demands. IEEE Trans. Power Syst. **24**(1), 445–453 (2009)

A. Menders, N. Boland, P. Guiney et al., Switch and tap-changer reconfiguration of distribution network using evolutionary algorithms. IEEE Trans. Power Syst. **28**(1), 85–92 (2013)

A. Merlin, H. Back, Search for a minimal-loss operationing spanning tree configuration in an urban power distribution system, in *Proceedings of PSCC 5th Power System Computation Conference* (Cambridge, UK, 1975)

M. Narimani, A. Vahed, R. Azizipanah et al., Enhanced gravitational search algorithm for multi-objective distribution feeder reconfiguration considering reliability, loss and operational cost. IET. Gener. Trans. Distrib. **8**(1), 55–69 (2014)

J. Nieto-Martin, T. Kipouros, M. Savill, J.Woodru, J. Butans, Technoeconomic distribution network planning using smart grid techniques with evolutionary self-healing network states, in *Complexity* (2018)

Ofgem, Upgrading our energy system - smart systems and flexibility plan, (2017). Available at: https://www.ofgem.gov.uk/system/files/docs/2017/07. Accessed 25 July 2018

Ofgem, Electricity SO reform, (2018). Available at: https://www.ofgem.gov.uk//electricity/transmission-networks/electricity-so-reform. Accessed 28 July 2018

J.M. Oliver, T. Kipouros, A.M. Savill, Electrical power grid network optimisation by evolutionary computing. Proc. Comput. Sci. **29**, 1948–1958 (2014)

A.V. Pastor, J. Nieto-Martin, D.W. Bunn, A. Laur, Evaluation of flexibility markets for retailer-DSO-TSO coordination. IEEE Trans. Power Syst. https://doi.org/10.1109/TPWRS(2018), 2880123

A.V. Pastor, J. Nieto-Martin, D.W. Bunn, L. Varga, Design of local services markets for pricing DSO-TSO procurement coordination, in *IEEE Power & Energy Society General Meeting (PESGM)*, pp. 1–5 (2018)

I.J. Pérez-Arriaga, J.D. Jenkins, C. Batlle, A regulatory framework for an evolving electricity sector: highlights of the MIT utility of the future study, in *Economics of Energy & Environmental Policy, International Association for Energy Economics*, vol. 0 (Number 1) (2017)

G.J.L. Poidevin, *Manual for CRATER Current Rating Software for Dynamic Ratings: Version 1b. ST3 Module 3* (2008)

F.C. Schweppe, R.D. Caramanis, R.D. Tabors, R.E. Bohn, in *Spot Pricing of Electricity* (1988)

D. Shimohammadi, H. Hong, Reconfiguration of electric distribution networks for resistive line losses reduction, in *IEEE* (1989)

M. Tsai, F. Hsu, Application of grey correlation analysis in evolutionary programming for distribution system feeder reconfiguration. IEEE Trans. Power Syst. **25**(2), 112–1133 (2010)

J. Villar, R. Bessa, M. Matos, Flexibility products and markets: literature review. Electr. Power Syst. Res. **154**, 329–340 (2018)

Western Power Distribution, Project FALCON. Automated Load Transfer, (2015a). Available at: https://www.westernpowerinnovation.co.uk/Document-library/2015/Project-FALCON-Engineering-ALT.aspx. Accessed 20 October 2015

Western Power Distribution, Project FALCON. Meshed Networks, (2015b). Available at: https://www.westernpowerinnovation.co.uk/Document-library/2015/Project-FALCON-Engineering-Mesh.aspx. Accessed 20 Oct 2015

Western Power Distribution, Project FALCON. Energy Storage, (2015c). Available at: https://www.westernpowerinnovation.co.uk/Document-library/2015/Project-FALCON-Engineering-Trials-Batteries.aspx. Accessed 20 Oct 2015

Western Power Distribution, Project FALCON. Dynamic Asset Rating Overhead Lines, (2015d). Available at: https://www.westernpowerinnovation.co.uk/Document-library/2015/Project-FALCON-Engineering-Trials-DAR-OHL.aspx. Accessed 20 Oct 2015

Western Power Distribution, Project FALCON. Dynamic Asset Rating Distribution Transformers, (2015e). Available at: https://www.westernpowerinnovation.co.uk/Document-library/2015/Project-FALCON-Engineering-Trials-DAR-Distribution.aspx. Accessed: 20 Oct 2015

Western Power Distribution, Project FALCON. Dynamic Asset Rating Cables, (2015f). Available at: https://www.westernpowerinnovation.co.uk/Document-library/2015/Project-FALCON-Engineering-Trials-DAR-Cables.aspx. Accessed 20 Oct 2015

Western Power Distribution, DSO Strategy, (2018a). Available at: https://www.westernpower.co.uk/our-network/strategic-network-investment/dso-strategy. Accessed 19 Sept 2018

Western Power Distribution, Distributed generation EHV constraint maps, (2018b). Available at: https://www.westernpowerinnovation.co.uk/Connections/Generation/Distributed-generation-EHV-constraint-maps.aspx. Accessed 14 Apr 2018

Z. Yuan, M.R. Hesamzadeh, D.R. Biggar, Distribution locational marginal pricing by convexified ACOPF and hierarchical dispatch. IEEE Trans. Smart Grid **9**(4), 3133–3142 (2016)

Practical Experiences with Transmission Investment

Practical Experiences with Transmission Investment in the New Zealand Electricity Market

Lewis Evans

1 Introduction

Transmission shifting electrical energy short and long distances has been an integral part of the electricity industry since the industry's inception, which for New Zealand was in the 1880s. Practical experiences with transmission are influenced by the evolution of the electricity market as a whole. New Zealand now has a wholesale electricity market where all electricity market participants are regulated by codified rules with statutory governance. The codified rules are designed to produce just sufficient institutional coordination to account for electricity's peculiar characteristics. Transmission and distribution networks are regulated as to form of pricing and their level of profit. The performances of other market participants are disciplined by competition.

In reaching this position, New Zealand has passed from geographically isolated innovation in electricity establishment, to national, concentrated political control of a vertically integrated industry, to a decentralized market that is well disposed to respond economically to the present uncertain rapid changes in technology. Transmission investment can only be understood in the context of the institutional settings. In New Zealand, it has been materially affected by these settings.

The following section describes the state of the electricity market in 2018. The third section briefly reviews key features of institutional change, and how they have affected electricity market arrangements. The fourth section examines interactions between the state of transmission and the performance of the markets sitting on the transmission network. The significant effects of regulatory settings on transmission investment are examined in the fifth section. Transmission pricing is evaluated in

---

L. Evans (✉)
Victoria University of Wellington, Wellington, New Zealand
e-mail: lew.evans@vuw.ac.nz

M. R. Hesamzadeh et al. (eds.), *Transmission Network Investment in Liberalized Power Markets*, Lecture Notes in Energy 79,
https://doi.org/10.1007/978-3-030-47929-9_18

the sixth section, and it is followed in the final section with discussion, and some conjecture, about the implications of innovation affecting electricity in 2018.

2 The Market in 2018

2.1 General Structure

The New Zealand electricity market (NZEM) has spot and hedge markets sitting on the transmission platform that is the national transmission network (grid), where the grid injection nodes connect generators, and off-take nodes connect lines-only distribution (lines) networks, that supply smaller commercial and non-commercial entities and households. There are a number of relatively large industrial firms connected directly to the grid. The wholesale market consists of spot and hedge markets taken jointly. Figure 1 schematic description of the grid shows that it is long and thin and has one significant loop; a relatively large proportion of demand is in the northern half of the North Island, a large proportion of supply in the lower South Island and that the transmission networks of the two Islands are linked by a cable supporting high-voltage direct current (HVDC).[1]

Apart from the HVDC, the grid transmits alternating current (AC) electricity via a backbone made up of a network of 220 kV transmission lines in each of the North Island and South Island. These connect to 110, 66 and 50 kV transmission lines that are stepped down for retail companies on lines networks[2] to supply consumers with 240 V at 50 Hz electricity. There are 27 electricity lines-only companies,[3] some industrial entities and generators connected to the national grid at more than 250 grid exit and entry nodes. All market participants are subject to the Electricity Industry Participation Code 2010 (Code).[4]

2.2 Governance

The electricity authority (EA) is the governing body of the entire New Zealand electricity market—of which a subset is the spot market. It manages and enforces the Code and has been in place since 2010. It has the statutory objective

[1]Operation of the HVDC in the early years is described by Peter Taylor, *White Diamonds North: 25 Years' operation of the Cook Strait Cable 1965–1990*, Transpower, 1990, 109p.

[2]In 1999, distribution companies were forced to choose between owning and managing networks or retailing energy. From that year the owners of distribution networks have been termed lines companies: full distribution requires lines and retail company services. The lines-only distribution businesses can own renewable generation and generate within their network up to a capacity of 50 MW and retail on their own network up to 75 GWh.

[3]Electricity Networks Association (http://ena.org.nz/lines-company-map/, accessed 29/4/2018).

[4]The Code is available at https://www.ea.govt.nz.

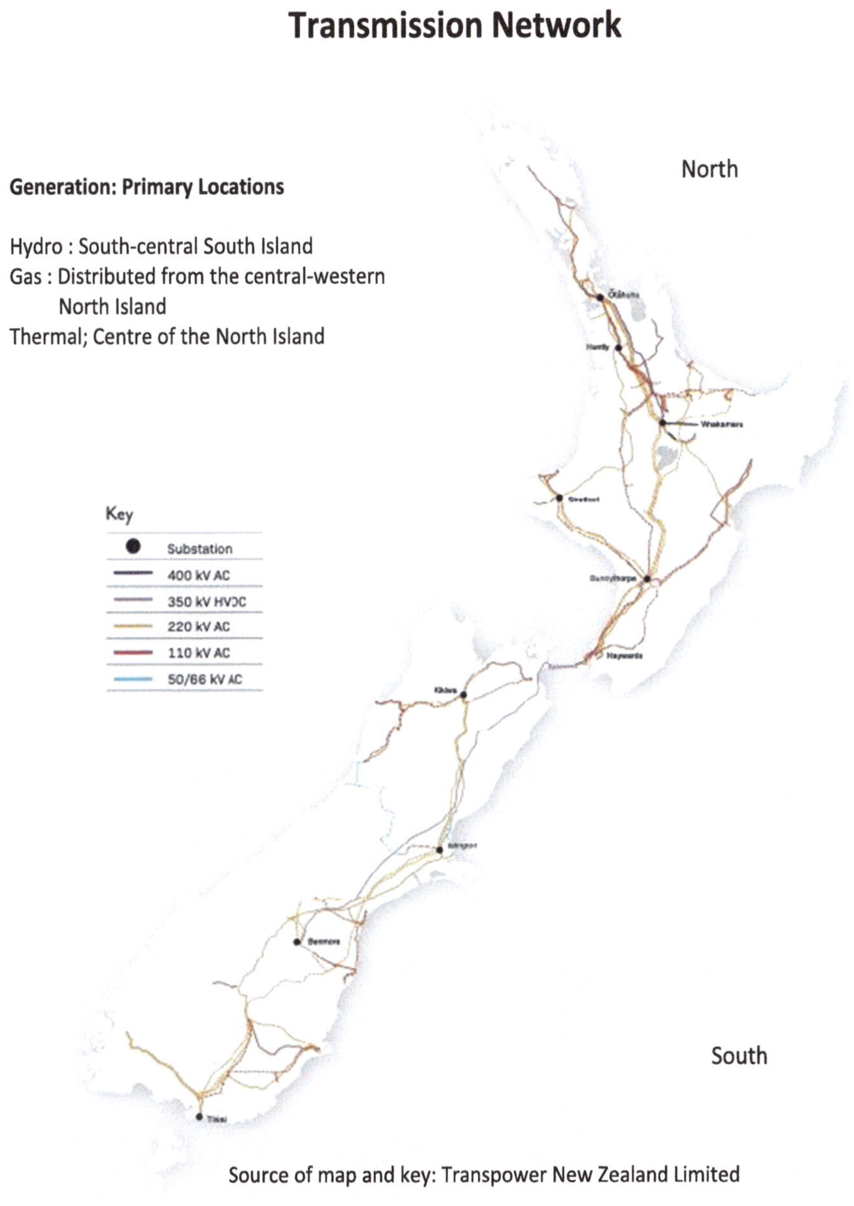

Fig. 1 The New Zealand Electricity Grid

> to promote competition in, reliable supply by, and the efficient operation of, the electricity industry for the long-term benefit of consumers[5];

which is interpreted by the EA as providing the basis for the objective of dynamic efficiency in EA decisions.[6] The EA is an independent Crown Entity[7] with a Board appointed by the Minister of Energy. Independence means that its decisions are not ratified by the government. The prime more-specific functions of the EA are:

- *Governance of industry:*
  - *design and enforce regulation specified in the Code,*
  - *manage any undesirable situation,*
  - *promote competition and an active hedge market.*
- *Monitor and evaluate NZEM service (e.g. NZEM pricing, clearing and settlement) providers, and*
- *Monitor security of supply, and industry, and market performance.*

All elements of the market are subject to consumer and competition law administered by the New Zealand Commerce Commission (CC). Transmission and lines entities are singled out for company-specific price regulation implemented by the CC pursuant to Part IV of the New Zealand Commerce Act (1986).[8] There is a fine line between the responsibilities of the EA and the CC in relation to lines companies and transmission. The EA codifies rules for the structure of transmission and lines company pricing and sets the price structure for grid user-charges. The CC is responsible for control of the level of prices, and approval of grid investment.[9] While it is natural that the level of grid prices be considered jointly with transmission investment, this separation of responsibilities can be a source of tension.

The relationship between the EA and CC regulatory authorities is governed by terms of the relevant Acts, and a 2010 memorandum of understanding that gives effect to this allocation of responsibilities and requires consultation between them.[10] It is facilitated by the almost common statutory objectives of the two authorities; the EA's and CC's objectives both include *the long-term benefit of consumers.*

[5]The Electricity Industry Act 2010, c115. (http://www.legislation.govt.nz/act/public/2010/0116/68.0/DLM2634233.html, accessed 20/7/18).

[6]Dynamic efficiency being the expected present value of social net benefits of the activity or policy being evaluated. It is implemented by cost–benefit analysis. The EA discusses its objective in Interpretation of the Electricity Authority's Statutory Objective, mimeo, the Electricity Authority, 14 February 2011, 21p. (https://www.ea.govt.nz/search/?q=statutory+objective&s=&order=&cf=&ct=&dp=&action_search=Search, accessed 28/5/2018).

[7]Discussion of the Crown Entity form of governance can be found at http://www.ssc.govt.nz/cegmos3, accessed 3/8/2018.

[8]http://www.legislation.govt.nz/act/public/1986/0005/86.0/DLM87623.html/, accessed 20/5/2018.

[9]The transmission regulatory regime is discussed below in the context of investment.

[10]Available at https://www.ea.govt.nz/search/?q=commerce+commission&s=&order=&dp=&files%5Bpdf%5D=pdf&files%5Bdoc%5D=doc&files%5Bxls%5D=xls&files%5Bcsv%5D=csv&files%5Bzip%5D=zip&types%5B2%5D=2&action_doApplyDocs, accessed 2/2/2019.

2.3 *Transmission*

The grid is owned by Transpower New Zealand Limited (Transpower) which is a state-owned enterprise. As such Transpower has a corporate form 100% owned by the New Zealand government.[11] It has the objective to operate as a commercial business and be " as profitable and efficient as comparable businesses that are not owned by the Crown".[12] While it must abide by the Code, Transpower has a *Relationship Charter* with the EA.[13] It reports that while in their codified or legal roles there may be conflict to be resolved, Transpower will deliver long-term benefits to New Zealand consumers by promoting competition in the wholesale market, ensuring reliable supply on a cost–benefit basis and promoting efficient operation of the market. It goes on to report guiding rules for relationship behaviour. From the charter, there is the potential for some conflict with the SOE objective of being a successful business.

Transmission is operated under the Transco model wherein the system, or market, operator (SO) is a separate unit owned by the grid owner, Transpower. The SO is regulated under the Code and is under contract and monitoring with the EA. The SO is responsible for the dispatch of offered generation, securing reserves, and ancillary services such as voltage support. Also, pursuant to requirements of the Code, it regularly reports on current and prospective states of supply and demand on the grid.

Prospects for electricity demand and supply and the future shape and characteristics of the grid are assessed by the Ministry of Commerce from time to time.[14] Relatively small grid investments such as those that improve efficiency of interconnection arrangements may be proposed by Transpower or another party and if agreed implemented under some mutual funding arrangement. Proposals for significant grid enhancement are instigated by Transpower under wide consultation with stakeholders that include spot market participants and other affected parties. The investment process is governed by the Grid Investment Test (GIT) administered by the CC.[15] It is considered further below.

[11]The SOE Act (http://www.legislation.govt.nz/act/public/1986/0124/latest/whole.html#DLM98017, accessed 2/8/2018).

[12]https://treasury.govt.nz/information-and-services/commercial-portfolio-and-advice/commercial-portfolio/types-commercial-crown-entities, accessed 2/12/2018.

[13]https://www.transpower.co.nz/sites/default/files/bulk-upload/documents/Transpower%20ea%20relationship%20charter.pdf, accessed 18/2/2019.

[14]Presently subsumed in the Ministry of Business, Innovation and Employment.

[15]The investment evaluation process has been termed Capex IM (Capital Expenditure Input Methodology) since 2008 when it was formally established in legislation, but this chapter will use the more evocative term GIT.

2.4 Generation and Retail

In 2012, the EA reported that there existed 188 generation stations of all types with operational capacity, and that of these 143 were fully embedded in lines company networks. The other—usually the larger generators—were fully or partially connected to the national grid.[16] There are five relatively large generation companies all of which are vertically integrated with retail to some degree and so are ascribed the term Gentailers.[17]

Hydro-electricity has a large share of New Zealand generation. In consequence, the amounts and proportions of generation fuels used in any year vary with reservoir inflows due to fluctuation in climatic conditions. In 2017, a total of 43,045 GWh was produced of which approximately 60% was hydro, 15% gas-fired, 17% geothermal, and 5% wind: the residual included coal, solar and biomass. The majority of hydro is found in the South-Central South Island, and the majority of wind generation is located in the North Island: geothermal and gas generation are confined to the North Island. For 2017, the composition of demand was approximately 37% industrial, 30% residential and the remainder commercial.[18] The industrial category includes an aluminium smelter which at 12% of demand is New Zealand's largest single consumer of electricity.[19]

The lines companies do not retail electricity, but provide access to retailers and service relatively small generation plants (distributed generation) and networks embedded within their network footprints. They may also engage in other varied activity, such as provision of telecommunications services and, in one distinctive instance, own a vineyard.

In 2014, there were some 48 retail brands marketed to household and commercial entities utilising various competitive strategies.[20] The number of retailers varies across lines networks in response to network-specific demand.[21]

---

[16]GeneratingStationListSep2012.xls (https://www.ea.govt.nz/dmsdocument/13837, accessed 28/5/2018).

[17]The 5 relatively large generators include the public companies Trustpower and Contact Energy and the 3 publicly listed mixed-ownership companies where the government holds a 51 per cent ownership interest: these are Meridian Energy, Genesis Energy and Mighty River Power. The latter 3 were fully state-owned from their inception in 1999 until 2016.

[18]Ministry of Business Innovation and Employment (MBIE), (http://www.mbie.govt.nz/info-services/sectors-industries/energy/energy-data-modelling/statistics/electricity, accessed 28/5/2018).

[19]EA, *Electricity in New Zealand*, January 2018, p. 9, (https://www.ea.govt.nz/about-us/media-and-publications/electricity-nz/, accessed 20/7/18).

[20]Electricity in New Zealand, op. cit. p. 8.

[21]There are some 25 brands to choose from on many lines networks (Electricity in New Zealand op. cit. p. 16). The retail strategies differ widely: for example, households may choose between long-term fixed-price variable volume contracts and spot market pricing.

2.5 Spot and Hedge Markets

Electricity is traded across the grid through the New Zealand electricity market (NZEM) that is a compulsory pool in which trading periods are of half-hour duration. It is an energy-only market, and it has reserves[22] co-optimised and priced jointly with the energy market for each trading period.[23]

The spot market is operated by Transpower's SO, and pricing, clearing and settlement service providers are contracted to other entities: presently the New Zealand stock exchange.

The spot market has been in place since 1 October 1996. Trading periods are ½ hour in duration at some 250 nodes on the grid and final prices are determined ex-post by a scheduling, pricing and dispatch (SPD) model on a uniform-output-price auction basis. Prices are uncapped and derived from voluntary bids and offers.[24] Generators must submit their offers to generate in advance of a two-hour window during which time offers can be altered only if there are unforeseen problems with the performance of generating units. In the initial design of the market, provision was made for a day-ahead market but given the widespread presence of hedges it had little utility and was never commercially used.

The trading-period spot prices are complemented by hedges commonly in the form of contracts-for-differences (CFDs),[25] but there are other forms. Financial Transmission Rights (FTRs) are provided at two North Island and three South Island nodes.[26] Additional risk management instruments are provided by the Australian Stock Exchange.[27] They include futures and options contracts. Risk is also managed through vertical integration of retail and generation. It is difficult to assess the extent to which electricity is traded only on the basis of spot prices: but 15% would be a reasonable assessment.

---

[22]Electrical energy reserves take various forms. For example, they include instantaneous start generation and reserves that achieve an N-1 reliability standard for core elements of the grid. This means that the system is planned to be in a secure state such that if a single contingent event occurs the system continues in a satisfactory operational state: but if a second event occurs, load may have to be shed. (https://www.ea.govt.nz/operations/transmission/grid-reliability-standards/, accessed 2/12/2018).

[23]Transpower, Market 101, Part 3 (https://www.transpower.co.nz/sites/default/files/bulk-upload/documents/Market%20101%20Part%203%20co-optimisation%20of%20energy%20and%20reserves.pdf, accessed 3/3/2019).

[24]Much of load demand used in dispatch has continued to be forecasted by the Market Operator.

[25]A CFD mimics a bilateral contract for some contract quantity q^c over some defined number of trading periods. If p^s is the spot price and p^* the strike price, the CFD price is $p^c = p^s + (p^* - p^s)$, and the value transacted under the contract is $p^c q^c$, or equivalently $p^* q^c$.

[26]The number and locations of FTRs are not fixed but are determined by cost–benefit appraisal by the FTR manager. The FTR balances are funded by the loss and constraint rentals generated by the internode losses on the grid. (https://www.ea.govt.nz/operations/wholesale/hedges/ftr-market/, accessed 2/2/2019).

[27]https://www.asx.com.au/products/energy-derivatives/new-zealand-electricity.htm, accessed 26/2/2019.

3 Evolution

3.1 Central Planning

In 1903, the Government passed the Water Power Act 1903, and from this time, the electricity industry was firmly one of government central planning and implementation, with some delegation of powers and functions to local authorities.

In 1946, the State Hydroelectric Department was established to oversee the development of transmission and generation. There followed in the 1950s and 1960s a period of expanded government investment in hydro-electric schemes and their supporting transmission. In 1958, the first New Zealand geothermal electricity generation plant was established at Wairakei, in the centre of the North Island. Since then, geothermal plants have been established at other locations; and additional geothermal generation possibilities continue to be discovered.

The electricity systems of the North and South Islands were linked in 1965 by the HVDC Inter-Island link between Benmore—the area of the country that holds a large share of hydro-electric generation—and the Haywards substation near Wellington thereby connecting to the North Island national grid network. The HVDC's capacity at establishment was rated at 600 megawatts (MW). In 1965, the prime motivation to link the two markets was to synchronise existing and potential large capacity South Island hydro-generation with rapidly growing North Island demand, particularly that of Auckland. The price of electricity was administratively set, and at a lower level for South Island consumers.

Hydro-electric and geothermal generation had competition with the discovery of commercial amounts of natural gas in 1959 at Kapuni in the province of Taranaki.[28] This led to investment in petrochemicals in that province, and gas pipelines to Wellington and Auckland. The discovery of the offshore Maui field followed in 1969: it was large on an international scale, and some eight times larger than Kapuni.[29] These two fields affected transmission because generation now had fuel and location alternatives. The efficient design and capacity of the national grid did not then depend on the location of rivers and geothermal fuel sources, but involved gas pipe distribution as well as electrical transmission. There have been no commercially viable gas discoveries for supply to the South Island.

The presence of natural gas has substantially benefitted the market operation and choices. The storage and generation of gas-fired plant have complemented hydro in managing the volatility of hydro-electric supplies, and they have affected the location of plant. A prominent example occurred on the formation of the first SOE in 1995: it cancelled a proposed hydro dam in the south of the South Island, at the same time

[28]Roger Gregg and Carl Walrond, Oil and gas—"Early petroleum exploration, 1865–1960", *Te Ara—the Encyclopedia of New Zealand*, (https://teara.govt.nz/en/oil-and-gas/page-4, accessed 28/5/2018).

[29]Roger Gregg and Carl Walrond, "Oil and gas—The Māui gas field", *Te Ara—the Encyclopedia of New Zealand*, (https://teara.govt.nz/en/oil-and-gas/page-5, accessed 28/5/2018).

establishing a combined cycle gas plant in Auckland: effectively bypassing the full grid.[30]

The 1940s–1980s saw the development of large hydropower schemes by a government valuing the expansion of the hydropower industry for its services as well as for employment in construction. There are now some 75 hydro-electric power stations with generation capacity of 0.5 MW or more.[31] The 5 major hydro-electricity schemes utilise lake Manapouri, and the Clutha and Waitaki rivers in the South Island, and Tongariro and Waikato rivers in the North Island.

In the early 1980s, the planning, implementation and operation of large generation schemes and the national grid remained the purview of government. Evaluation process of transmission and generation investment entailed the state of demand for electricity being assessed by a power planning committee consisting of heads of government departments. These included the government department for generation and transmission (EDME), Treasury, the Statistics Department, among others. This committee would reach a view on future demand for electricity and assess it against supply; and recommend or not additional generation and transmission investment. Approved projects would be built by the Ministry of Works and Development[32] and transferred to EDME for ongoing operation and supply. The system was centrally planned and administered.

Decisions about electricity projects and prices were political. There were imbalances over time in supply and demand where shortages were managed by—commonly residential—brownouts, and surpluses were allocated to new projects by political strategy. The political nature of decisions and the concomitant centralised structure led to poor electricity industry performance that was documented in reports[33] of the mid-1980s; providing impetus for change.[34]

---

[30]Another example, is that coincident with the expansion of the grid into Auckland in 2010–14, discussed below, the aged Otahuhu plant has been decommissioned and peaking gas plant established near the gas well-head in Taranaki.

[31]Wikipedia (https://en.wikipedia.org/wiki/List_of_power_stations_in_New_Zealand, 28/5/2018) edited table of "Generation update", mimeo, EA, 2013, 7p. (https://web.archive.org/web/20140416192143, http://www.ea.govt.nz/dmsdocument/11455, accessed 28/5/2018).

[32]The Ministry of Works was present from 1943 to 1973 when it became the Ministry of Works and Development, it had its genesis in 1870, as the Department of Public Works. It was the infrastructure construction arm of government. By the early 1980s, the Ministry of Works and Development and its antecedents had been responsible for the construction of all major electricity works, as well as assets for other industries.

[33]See for example, the Treasury Report, *Review of Electricity Planning and Electricity Generation Costs*, (https://www.dropbox.com/s/rn9ny1blkpdad92/McLachlan_Treasury_Mar_1984.pdf, accessed 2/2/2019), and Grant, David "Aluminium smelter, Tiwai Point, tera.govt.nz. TeAra—*The Encyclopedia of New Zealand*, (https://teara.govt.nz/en/photograph/21183/tiwai-point-aluminium-smelter, accessed 6/6/2018); and Reilly, op. cit. p. 213., that describes the political process of consideration of aluminium smelters and associated inefficient transmission investment.

[34]Evans, Lewis T. and Richard B. Meade, *Alternating Currents or Counter Revolution: contemporary electricity reform in New Zealand*, Victoria University Press, 2005, 356p, Chap. 5.

3.2 Towards Markets

In 1984, central planning was abruptly replaced by the philosophy of decentralised decision-making. Impetus for change was the result of the parlous state of the economy in 1983, and earlier isolated attempts to improve long-standing economic performance.[35] The reform in 1984–91 took the general form of unilateral liberalisation of markets.[36] Concomitantly, competition policy also changed.[37]

Commercial law applies to all commercial activity and the reformed Commerce Act 1986 of that year prohibited conduct that involved unilateral market power, cooperation and arrangements that limited competition.[38] Administered by the CC, it had the purpose to promote competition in markets for the long-term benefit of consumers within New Zealand, and it contained powers for certain actions in relation to services provided by electricity transmission and lines, gas pipeline, and airport businesses. These powers aside, there was no provision for separate industry-specific regulation.[39] Instead firm and industry actions were monitored by Ministries with a threat of regulatory action should competition prove insufficient for desired market performance: the approach was termed light-handed regulation.

In the same year, the State Sector Act 1986 enabled the conversion of government departments, producing goods and services, to SOEs. It was applied to the government department holder and manager of generation and transmission, EDME, when it was transformed into the Electricity Corporation of New Zealand Ltd (ECNZ).[40]

Change came rapidly.[41] By 1998

- Transmission was an SOE (Transpower);
- Lines companies were corporate entities and excluded retail functions;

---

[35]Singleton, John, "An Economic History of New Zealand in the Nineteenth and Twentieth Centuries" EH.Net, *Economic History Association*, (https://eh.net/encyclopedia/an-economic-history-of-new-zealand-in-the-nineteenth-and-twentieth-centuries/, accessed 4/6/2018).

[36]Evans, Lewis, Arthur Grimes, Bryce Wilkinson and David Teece, "Economic Reform in New Zealand 1984–94: The Pursuit of Efficiency", *Journal of Economic Literature*, vol. XXXIV, 1856–1902, Dec. 1996.

[37]The heavily regulated state of New Zealand commerce and the low weight placed on competition in policy prior to 1984 are explained by White, Douglas QC, "Facilitating and Regulating Commerce", *Victoria University of Wellington Law Review*, 33, 2002, 821–839.

[38]Much of the conduct considered by the CC in 1986–98 concerned mergers. As discussed below, for electricity the CC was required to assess the electricity spot market in 1996 for the likelihood it would "substantially lessen competition". The market had been established in 1996 as a multilateral contract and required authorisation by the CC to operate.

[39]There was some firm-specific regulation as when the SOE Telecom New Zealand Limited was sold in 1991 the terms of sale included government retention of a share that required the price of fixed-line service to households to rise no faster than the rate of inflation.

[40]Corporatisation as ECNZ is examined in detail by Spicer, Barry, Michael E. Bradbury, Darian J. Kerkin and Paul Rouse, *The Power to Manage: restructuring the New Zealand Electricity Department as a state-owned enterprise: the Electricorp experience*, Oxford University Press, 1991, 188p.

[41]The nature and sequence of electricity restructuring is described in "Chronology of New

- ECNZ had been split into transmission, 3 generation SOEs and one publicly owned generation company;
- Generation was permitted to own retail[42];
- There existed a spot market (NZEM);
- An information disclosure regime was in place, particularly for deemed natural monopolies. Information disclosure was the essential regulatory component of the approach—termed light-handed regulation—it replaced pre-1984 central planning and held sway until the early 2000s.

3.3 *Transmission Ownership*

The creation of Transpower from ECNZ as a stand-alone SOE took place in 1994. At the time, there was a proposal that Transpower be owned by a club whose members were grid-connected entities; in particular lines and generation companies. The idea was that such a consumer/cooperative form of ownership would internalise market power issues arising from the grid's natural monopoly characteristics; and that in consequence transmission-specific price and investment regulation would not be required.

In principle, the benefits of the club proposal can be inferred from the efficiency of the transmission regulatory scheme, H-R-G-V,[43] in which a monopoly Transco is free to choose its operation and investment, but is constrained to charge a two-part tariff. The first part is a fixed-fee aggregated over all transmission users that cannot exceed the sum of the fixed fee of the previous period plus the incremental consumers' surplus of that and the current period. The second part is the variable fee which is the merchandising surplus of the current period: being the fees charged for internodal transmission less costs. If the fixed-fee constraint is binding in all previous periods, the scheme reduces to the Transco receiving the total surplus from transmission less the amount of consumer surplus at the date the H-R-G-V scheme was started. Even if the fixed-fee constraint has not been binding, the Transco gets the incremental consumer and merchandising surpluses of each period. The benefit of the club proposal was as for the H-R-G-V scheme. It was the internalisation to the transmission owner of the costs and benefits of information, investment costs

Zealand Electricity Reform", Energy Markets Policy, Energy and Resources Branch, MBIE, 2015, (http://www.mbie.govt.nz/info-services/sectors-industries/energy/electricity-market/electricity-industry/chronology-of-new-zealand-electricity-reform/?searchterm=CHRONOLOGY%20OF%20NEW%20ZEALAND%20ELECTRICITY%2A, accessed 7/6/2018).

[42]The generators rapidly acquired retail businesses and became called Gentailers. In separating lines from energy, very few distribution firms retained the retail business; most retained the lines network and sold the more competitive retail businesses. There were by the year 2000 five vertically integrated Gentailers; three of which were SOEs and two held by investors.

[43]The H-R-G-V scheme is analysed in detail by Vogelsang in chapter "A Simple Merchant-Regulatory Incentive Mechanism Applied to Electricity Transmission Pricing and Investment: The Case of H-R-G-V" of this volume.

and outcomes, thereby producing dynamically efficient decision-making. However, regulation is not necessary for the club model.

The costs of the club proposal included operational and investment inefficiencies of the club given its heterogeneous ownership.[44] Heterogeneity would arise in the club because of the mix of owner characteristics; consisting of grid injectors and extractors, and the different locations of users. The counterfactual was the SOE model as there was no firm-specific regulation in 1994, and no possibility of investor-owned transmission at that time. In any event, the distribution companies were not interested in acquiring a stake in the grid and the club proposal was not implemented.[45] Transpower became an SOE.

3.4 Governance Established

The first market governance institution was that of the spot market (NZEM). At its inception, the NZEM was a joint venture with a multilateral contract basis for participation and governance.[46] Entry was open to any party that met the NZEM rules. With membership came, the ability to participate in governance. The NZEM did not own physical assets and it contracted-in some seven service providers including the market administrator.[47] Particularly because it was a contract market as opposed to a statutory body approved by government, the NZEM had to seek certification by the CC that it did not, in the words of the Commerce Act 1986; substantially lessen competition. It was granted authorisation in 1996.

The market initially provided for bilateral contracts, that were closed by the use of the spot market as a load following generator; but this was abandoned when a compulsory pool was introduced.

Issues arose relating to the governance of the electricity market as a whole and the governance of the spot market which was in essence self-governance by market participants: questions were arising about the spot market's ability to take decisions relating to change. Also, there were other market functions to be coordinated by a

---

[44]Conditions that enhance the possibilities of efficient club, or cooperative, structures are discussed in Hansmann, H. (1996). *The Ownership of Enterprise*. The Belknap Press of Harvard University Press: Cambridge, Massachusetts.

[45]Evans and Meade, op. cit., p. 141.

[46]The 1990s contractual structure of NZEM is set out in Arnold, Terence and Lewis Evans, "Governance in the New Zealand electricity market: a law and economics perspective on enforcing obligations in a market based upon a multilateral contract", *The Antitrust Bulletin*, Fall 2001, pp. 611–633.

[47]The other six were: the grid operator, the scheduler, the dispatcher, the reconciliation, pricing and clearing managers.

governance structure.[48] In addition, as explained below, grid and lines pricing and transmission investment decision-making were being challenged by the light-handed regulatory environment.

In 2001, the government required the Commerce Commission (CC), to control price or revenue of Transpower and lines companies. And it required the industry to form an electricity governance board that would oversee NZEM (the spot market), MACQs (management of quality standards) and MARIA (metering including customer switching). If the industry failed to agree a comprehensive governance structure the government would establish a statutory governance board. The industry did fail: and the Government established the Electricity Commission (EC) in 2004. It had the particular features of some direct political input to decisions, and the responsibility of approving Transpower investment proposals. In 2010, it was replaced by the existing EA which does not have these features: the CC governs the grid investment approval process.

4 Transmission and Competition

4.1 One Market

The design of the market and subsequent grid investment decisions were guided by the approach of creating one market, or competitive neutrality, wherein the general "basis" spot and hedge price levels were the same at all major grid entry and exit nodes. Given the interisland connection, this would enable competition in generation particularly, and in demand and thereby enhance a workably competitive[49] electricity market.

The generation sector has been well-approximated by an oligopoly of 5 Gentailer companies with fringe competition in distributed generation.

The Gentailers can be viewed as playing a game repeated every half-hour into the indefinite future. However, coordinated action is limited by features of the New Zealand electricity industry that engender uncertainty and different, hard to predict cost structures of each Gentailer. The diverse plant portfolios of the generators—thermal and hydro particularly—and the following features intrinsic to the New Zealand market render coordinated action by generators less likely. These features include:

---

[48]These included the management of quality standards for the grid—Transpower had initiated a review of setting grid standards and establishing a self-governing process that proposed a Multilateral Agreement on Common Quality Standards (MACQS)—and for metering including facilitating customer switching.

[49]Workable competition is discussed by Markham, Jesse W. (1950), "An Alternative Approach to The Concept of Workable Competition", *American Economic Review*, 40(3), 349–361; which contends that it must enable the dynamic efficiency factors of investment and innovation.

Table 1 Weekly average inflows in Gigawatt hours (GWh). Inflows to Major Hydro-Electricity Producing Reservoirs

Correlations	Taupo	Benmore	Wanaka	Te Anau
Taupo		0.11	0.09	0.06
Benmore			0.79	0.62
Wanaka				0.48
Te Anau				
Mean	176,642	76,583	125,298	164,514
Coeff.-variation	0.39	1.10	0.76	1.24

Table 2 Short-run Characteristics of Inflows (GWh)

Half-Life	Taupo	Benmore	Wanaka	Te Anau
Weekly	1.27	1.5	3.83	1.05
Daily	10.15	3.92	34.5	2.54

Notes
1 Calculated from weekly data supplied by the EA for the period 30/6/1931-30/6/2010
2. Short-run properties assume inflows (y) follow the stochastic process $dy_t = \eta(\mu - y_t)dt + \sigma\sqrt{y_t}d\zeta_t$ where ζ_t is standard normal. The unconditional means and variance are μ and $\frac{\mu}{2\eta}\sigma^2$; with half-life $\frac{-\ln 0.5}{\eta}$

- Generation dominated by hydro which share of total generation (50% dry year–70% wet year) is volatile; as is the gas-fired share of generation as it makes up the 20% variation in hydro-generation.
- Limited water storage capacity: in aggregate less than 10% of demand, and just 4 weeks of peak winter demand.
- Seasonal mismatch: demand is relatively high in winter when inflows are low being locked up as snow and ice.
- Inflows to the major hydro reservoirs that are very different and weakly correlated, as are illustrated by Tables 1 and 2.

These factors mean the cost structures of the competing generators vary over time across plant type, location and weather conditions and are very uncertain. In the repeated game among generators, these characteristics promote competition in generation at the expense of coordinated action.[50] They facilitate workable competition across the market if the grid has adequate capacity.

[50]There has been no evidence of government ownership of three Gentailers inducing affiliated actions. These firms are required to act as independent businesses with specific distinct boards and on the evidence have competed vigorously in the product market and in some cases through court processes.

The extent of one market was assessed for the period 1996–2005 by Evans et al. (2008)[51] using principal component analysis applied to prices of the daily 48-trading periods at 7 widely geographically positioned nodes of the grid. The first principal component was viewed as representing the basis level of market price and the second component deviations from basis brought about by market separation. The seven nodes were chosen on the grounds that they covered core parts of the grid that would have a relatively high volume of trade and throughput: they included nodes at each end of the HVDC, Auckland and nodes in the centre of the North Island. There can be expected to be price separation for nodes further away from those with core transmission traffic; reflecting congestion costs associated with capacity limitations of non-core grid lines.

It found that both components explained some 98% of variation in prices; and that almost all of the explained variation was due to the first, or basis price, component: although in some particularly off-peak periods, there was some small separation of prices between certain nodes. For the bulk of transmission over the period to 2005 there had been one market.

Adequate capacity of the HVDC Inter-Island link is important for confidence by generation and retail to take hedge positions in both islands and consequently for competition in retail and the existence of one market. Congestion constraints at the HVDC induce price separation between the two islands. In addition, congestion affects spot prices by constraining the locations of contracted reserves—that are co-optimised in pricing with the spot energy market. The existence of the interisland connector the HVDC improves competition to supply reserves, but its capacity is also a potential contingent event to be managed by the SO.

In 2009 one market was at risk due to the limited capacity and risk of non-performance of the HVDC link. It reflected different growth in demand at different locations, but also that the HVDC had been running at restricted capacity since 2007, because of the age—some 40 years—of critical operational equipment. It reflected the limited investment of Transpower discussed below in the context of regulatory arrangements.

The 2009 Government-convened Electricity Technical Advisory Group (ETAG) acknowledged the state of the HVDC, the absence of FTRs at that time, and the consequently limited presence of competing generation and thereby retail in some, particularly South Island, retail markets.[52] It made the following recommendations:

- The state-owned primarily South Island generator Meridian Energy (Meridian) swap a 1000 GWh per year South Island hedge with the state-owned solely North Island generator, Mighty River Power (MRP); and
- Meridian swap a 450 GWh South Island hedge with the North Island state-owned generator Genesis Energy (Genesis).

[51] Evans, Lewis, Graeme Guthrie and Steen Videbeck, "Assessing the Integration of Electricity Markets using Principal Component Analysis: Network and Market Structure Effects", *Contemporary Economic Policy*, Vol. 26(1), January 2008, pp. 145–161.

[52] In addition, the 5th—ranked by generation—Gentailer Trustpower had relatively small South Island plants.

The swaps were made in 2010 for a period of 15 years. The hedges had the effect of bypassing the HVDC for the swap participants. The South Island hedge, for example, provided the North Island Gentailer MRP with an assured supply of generation in the South Island at a price independent of volatile movements in the spot price. The swap effectively enhanced the variety of generation sources promoting additional competition in local retail markets in both islands: thereby facilitating workable competition.

4.2 Transitory Events and the Code

The presence of one market does not rule out temporary market power episodes and certain of these are directly linked to the management of the grid. A good example arose in 2011 when the Gentailer Genesis found itself in a net-pivotal position—that is, it was the marginal generator and could produce more energy than that required to supply its own customers in a situation of no competitive supply. The situation arose because of transmission restrictions brought about by Transpower carrying out transmission line maintenance. Because NZEM has no spot-price cap, and because the maintenance period was planned and notified some 4 months in advance of the maintenance work, the case sparked concern about market power opportunities under the Code.

On Saturday 26 March 2011, the Grid had 2 of 6 220 kV, and 3 of 5 110 kV lines out of service restricting supplies from the South Island and rendering Genesis, net-pivotal north of Hamilton. Its bids engendered a rise in the spot price of electricity for Hamilton and regions north; from NZ$367/MWh to between $21,000/MWh and $23,000/MWh for 7 hours.[53] The effect was exacerbated by the failure of generation plant of another Gentailer, in the critical trading periods. It raised electricity spot prices very substantially in the whole of the North Island and produced complaints from many demand-side market participants. The EA responded by declaring an undesirable trading situation (UTS) and setting final prices for the affected periods at NZ$3,000/MWh being the price that its modelling indicated was the marginal cost of new entrant generation under normal bidding at the relevant location.[54] The EA UTS declaration and decisions taken under it were upheld in an appeal to the High Court by supply-side companies.

The transmission restrictions were perfectly predictable for all market participants as they had been posted by Transpower well in advance. Indeed, Genesis's potential

[53] Reported in The High Court of New Zealand Wellington Registry Decision CV-2011-485-1371 [2012] NZHC 238. There was also an upward adjustment due to a spring washer effect arising because of a transmission constraint within a loop that has links of different impedance. To avoid overloading, the lower capacity link may require dispatch of additional—higher priced—generation around the lower capacity link, resulting in price separation either side of the constrained link.

[54] EA summary of decisions on actions to correct 26 March UTS.pdf (https://www.ea.govt.nz/code-and-compliance/uts/undesirable-trading-situations-decisions/uts-26-march-2011/final-undesirable-trading-situation-decision-and-proposed-actions/, accessed 1/9/2018).

bidding behaviour was signalled to the market two days earlier when it bid unusually high prices; however, on this day without the network constraint, competition meant that the high-priced tranches were not dispatched and so did not affect spot prices. The event raised issues about predictable grid events and market participant behaviour under the Code.

Drawing an analogy with contract theory, the presence of a UTS in the Code enables the "completion of an incomplete contract". The UTS provides that the EA as governor of the Code holds decision rights[55] about matters that arise because of the incompleteness of the Code. In implementing these rights, the EA is constrained by its legal objective. The presence of the UTS in the Code therefore enables minimising restrictions on market participants imposed through the Code, while catering for the myriad potential unanticipated, or anticipated but unpredictable-in-time events that are—in some cases deliberately—not provided for in the Code. The UTS provision has been activated only once in the 22 years of the market, although there have been occasions in which its use has been contemplated.

The EA consulted with market participants on the terms of the Code for predictable constrained transmission situations following resolution of the event. Suggestions included increased situation-specificity of the Code: for example that prices in transmission-maintenance-constrained areas be set for the constrained period as though at some specification of "normal prices". Changes were made to the Code. They indicated broad behaviours of generators that would and would not be deemed undesirable; but they did not include situation-specific rules. Instead, the Code maintains there be no spot-price cap and that the UTS provision be a threat to behaviour that is beyond the specific rules of the Code and deemed adverse to the performance of the market in furthering the EA's legal objective of enhancing the long-term benefit of consumers.

4.3 Transmission and Wholesale Market Performance

The spot and hedge markets jointly with the capacity and operation of the grid determine the performance of the wholesale market. They reflect entry and exit of generation and load from the market as well as the performance of the spot market. Given the NZEM's dependence on hydro-generation, spot prices measure the opportunity cost of a unit of delayed generation; or equivalently the expected value of the marginal unit of stored water.[56] They are volatile for the reasons given above. The fluctuations are to a very considerable extent hedged out rendering the hedge

[55]Hart reviews issues in allocating decision rights in incomplete contracts in private and public-private settings. Hart Oliver, "Incomplete Contracts and Control", *American Economic Review*, 107(7) 2017, 1731–1752.

[56]Described in Evans Lewis and Graeme Guthrie, "How Options Provided by Storage Affect Electricity Prices", *Southern Economic Journal*, 75(3), 2009, 681–702, and Evans, Lewis, Graeme Guthrie and Andrea Lu, "The Role of Storage in a Competitive Electricity Market and the Effects of Climate Change", *Energy Economics*, 36, 2013, 405–418.

price the base-price for retail and other market demand, and the spot price the price of imbalanced generation and retail. As mentioned above, some 85% of electricity traded through the spot market is hedged.[57]

The hedge price of electrical energy in an efficient stable or growing market should approximate the cost of electricity delivered by the most efficient generation plant next to be installed. The process that produces this result was explained by the Gentailer Trustpower in evidence in a case it brought against the Commissioner of Inland Revenue[58] in 2013. At that time, its demand for energy to meet retail commitments was considerably in excess of the amount it generated.

Trustpower had established a portfolio of sites consented for construction of new plant—typically hydro or wind—and treated these as options to build. It would first seek hedges for future supply, and if these were priced higher than the cost of new installed generation it would exercise an option to build. Periodic investigations, such as that of the 2009 electricity task-force[59], have shown that the hedge price has approximated the electricity-unit cost of new plant; and that there has been adequate investment in new, and renewal of old, generation plant in NZEM. This is a measure of the efficiency of the combined spot and hedge—or wholesale—market.

Spot market modelling studies have their place in studying issues of market design, although they are less useful in credible measurement of market performance. They attempt to infer *ex ante* decisions in an *ex post* setting and have challenges in modelling the dynamic system taking account of expectations and constraints in forward-looking sequential offer and bid decisions. In addition, hedge positions are important for spot market performance as well as for the wider wholesale market.[60] Hedges affect both bidding and offering in the spot market, and by their purposeful design alter the incidence of volatile revenue and cost allocations among spot market participants.

The spot market study by Wolak (2009)[61] conducted for the CC found no evidence for affiliated actions: its other results were confounded by a failure to recognise that the spot prices measured the value of stored water.[62] There have been a number of operational research studies that seek to mimic the spot market structure by modelling

[57]The spot market plays a vital role in scarcity-pricing imbalances in drought periods, particularly.

[58]Trustpower appealed the Commissioner's ruling on whether certain expenditure on creating an option to build was tax-deductible. It failed in its appeals to the High Court, the Appeal Court and finally the Supreme Court (accessed at: SC 74/2015 [2016] NZSC 91).

[59]A 2018 version is available at https://www.mbie.govt.nz/info-services/sectors-industries/energy/electricity-price-review/consultation/first-report.pdf, accessed 29/9/2018.

[60]Whether the hedges are by contract or vertical integration.

[61]*An assessment of the performance of the New Zealand wholesale electricity market*, https://www.researchgate.net/publication/228378779_An_assessment_of_the_performance_of_the_New_Zealand_wholesale_electricity_market, accessed 7/3/2019.

[62]Evans, Lewis, Seamus Hogan and Peter Jackson, "A Critique of Wolak's evaluation of the NZ electricity market: introduction and overview", *New Zealand Economic Papers*, 46(1), 2012, 1–11. Evans, Lewis and Graeme Guthrie, "An examination of Frank Wolak's model of market power and its application to the New Zealand electricity market, *New Zealand Economic Papers*, 46(1), 2012, 25–235.

and infer spot market performance by assessing spot market outcomes against the modelled counterfactual. They have the difficulties already described.[63]

5 Transmission Investment and Regulation

5.1 Light-Handed Regulation

Investment by Transpower has been significantly affected by the regulatory regimes in place, and these changed significantly as the New Zealand economy advanced from the economic reform period. With the exception of investment approval—discussed below—application of the regimes was similar as between Transpower and the lines companies; because of their common natural monopoly characteristics.

Although, from 2001 Transpower had been assigned to CC monitoring, light-handed regulation continued to apply for a period and when price regulation was introduced there was experimentation in the form of price setting. It induced regulatory risk and adversely affected investment.

Light-handed regulation was initially implemented by the Ministry of Commerce as it then was, and afterwards under the CC's jurisdiction. It consisted of assessing company performance using data on measures of efficiency and financial performance that firms were required to report; with the threat of (heavy) firm-specific regulation should there be non-performance.

The financial performance indicators placed heavy emphasis on the so-called Accounting Rate of Profit, being the financial rate of return earned by Transpower and lines networks on the optimised deprival value (ODV) of their networks. The ODV measurement used the optimised depreciated replacement cost (ODRC) which was an estimate of the current replacement value of network assets based only on an assumed necessary configuration of system assets to meet demand at of the date of measurement: with a depreciation adjustment based on existing asset ages and lives. In addition, the economic value (EV) was calculated as the present value of the income expected to be earned on the network assets. The ODV was then the lesser of the ODRC and EV.

The use of ODRC as a capital base for natural monopoly assessment and regulation had problems that included the number and nature of assumptions required to construct it, and additional shortcomings for its use in price regulation.[64] If there were economies of scale in construction, as is common in greenfield and maintenance projects in capital intensive network industries, the optimisation process of ODRC would result in a regulatory measure of capital stock that is less than that of the firm. Where there is uncertainty in demand, irreversible investment and scale economies

[63]For modelling analysis without hedges see, for example, Philpott, Andy and Ziming Guam, *Models for estimating the performance of electricity markets with hydro-electric reservoir storage*, mimeo, Electric Power Optimisation Centre, 2013, 46p.

[64]Evans and Meade, op. cit.

of investment, the firm's sequential investment decision-making over time trades-off capital cost reductions from scale, against the likelihood that demand will ultimately support the investment. Scale economies suggest investing in lumpy amounts that produce expected excess capacity in the short, but not the long term.[65] Whereas uncertainty in demand and irreversibility suggest smaller more frequent investment lumps that take less risk of shortfalls in demand, investment scale economies imply larger lumps of investment.[66]

A regulator computing the optimised replacement cost does so knowing what demand is at the time of computation: and thus with information the firm did not have when building its network in a sequence of forward-looking decisions. Optimising the cost of service with respect to a given known demand must produce a lower-cost capital stock than a sequence of investment decisions where demand is uncertain. Evans and Guthrie (2006)[67] show that, economies of scale in investment imply an extremely high allowable rate of return if the firm is to contemplate investment under replacement cost-incentive regulation. While ODRC does have depreciated assets rather than full network replacement; least cost asset configurations at regulatory dates will generally reduce the capital stock over that which a dynamically efficient firm will have made. Unadjusted ODRC regulated price or revenue paths will be lower than that required to sustain dynamically efficient investment.

The issue is important, for ODRC-based regulation was used in information disclosure relating to light-handed regulation; and as late as 2004, as the basis of CC regulation of Transpower.

The ODRC prices were never actually implemented[68]; but in repeatedly, formally seeking to apply them the CC generated a great deal of uncertainty. The CC regulatory proposals for these prices engendered disputes and financially infeasible price paths. They precipitated consequent ad hoc arrangements settled between firms and the CC.

An alternative pricing process was made law in the Commerce Act Amendment 2008. It remains in place in 2018. It required the CC to develop what was termed

[65] In transmission, capacity excesses and shortfalls are reflected in the spot price.

[66] A simple representation of grid expansion into Auckland proposed in 2003 illustrates the issue: should Transpower invest in a 400 kV line initially, or 220 kV initially and an additional 220 kV later depending upon what is learned about the growth in demand. If there are no economies of scale in investment, there is no cost saving in the two-step process, but if there are economies of scale the single-step expansion and some excess capacity for a period—perhaps forever if uncertain demand does not appear—may be the economic investment. (Glenn Boyle, Graeme Guthrie, and Richard Meade, Real Options and Investment: the New Zealand Grid Investment Test, *NZ Institute for the Study of Competition and Regulation*, mimeo, 2003, 17p).

[67] Evans, Lewis and Graeme Guthrie, "Incentive Regulation When Costs are Sunk", *Journal of Regulatory Economics*, 29(3), 2006, 239–264.

[68] In 2006, the lines network company Unison Networks Limited negotiated an administrative control arrangement with CC in response to a CC regulation proposal based upon ODRC (https://comcom.govt.nz/__data/assets/pdf_file/0028/88381/Unison-Decision-Not-to-Declare-Control-11-May-2007.pdf, accessed 1/9/2018).

"input methodologies" that were legal rules, requirements and procedures for calculating inputs to setting price-quality paths. They included such matters as the WACC, leverage and the asset base. They are applicable to Transpower.[69]

The input methodologies specify a form of building block regulation in which the measured asset base is annualised using a regulatory rate of return; both calculated under input methodology specification. It provides for the recovery of investment expenditure and some incentive for efficiency gains and meeting performance-quality thresholds over the regulatory period. The quality-price path is reset 5 times yearly. A formal process for Transpower investment appraisal and approval called the Grid Investment Test (GIT) became an input methodology. The GIT and Transpower investment are discussed below.

The revised Act permitted appeal to the High Court as merit review for a limited period on the structure of "input methodologies". Previously, all CC firm-specific regulatory decisions had been subject to judicial review only. There has been more certainty in regulated prices for Transpower and companies since the regime was introduced in 2008.

5.2 *Transpower Investment and Light-Handed Regulation*

Transpower was solely responsible for transmission investment from the inception of the market in 1995/6 until it became subject to the industry-specific Electricity Commission's (EC) governance in 2004.[70] During 1995/6–2004, Transpower was in a position similar to merchant transmission in that it could reach agreements for investment and concomitant charges; but the commercial environment was weak in that the light-handed regulation information disclosure regime provided Transpower with no certainty of recompense for investment initiated by it. It was a period of regulatory risk that coincided with the Transpower Chief Executive's view that distributed generation was about to emerge that would (partially) strand investment in the grid.[71]

Transpower investment was low. Annual investment averaged less than NZ$100 m[72] per annum in the period 1995/96–2004/5[73]. It rose to something over

[69] All lines companies would be subject to information disclosure, and those that were not consumer-owned would be subject to the default/customized, price-quality regulation.

[70] Established in 2004, it was the first industry-wide governing body of the electricity market. In relation to transmission, its purposes were: develop grid reliability standards and a Grid Investment Test, assess Transpower's proposed grid upgrade plan, including against alternatives to specific investments, provide binding approval or otherwise, of Transpower investment proposals (approval of the EC was necessary and sufficient for Transpower to recover investment cost), and approve a transmission pricing policy and develop a benchmark pricing agreement.

[71] Reilly op. cit. pp. 198–199.

[72] All financial data are in nominal terms unless otherwise stated.

[73] Improving Electricity Market Performance Vol. 1 (http://www.mbie.govt.nz/info-services/sectors-industries/energy/previous-reviews-consultations/review-of-the-electricity-market-2009/documents-image-library/Improving%20Electricity%20Market%20Performance%20-%20Volume%20One.pdf, accessed 1/8/2018).

NZ$500 m in 2009/10. In 2009, Transpower proposed a programme of significantly greater investment for the following 10 years.

In the 1995/6–2004 period, maintenance and minor capacity enhancement investments were made that sought to just maintain or enhance the performance of the existing grid. There were concerns about security of supply to Auckland and further north especially following a line failure in 1998 that precipitated a 5-week long blackout of the Auckland CBD.[74] Where constraints were causing a security of supply issue Transpower would seek a means to recover cost of essential expenditure. If there was no security of supply issue Transpower would propose putting in place investments sought and paid for by market participants.[75]

There was some low-cost expansion of grid capacity. One source of capacity increase resulted from the use of aerial laser technology in 2000/01 to map precisely the location of lines relative to vegetation and the land.[76] The information it provided enabled Transpower to identify where it may raise the capacity at low cost by increasing line clearances. Other, individually relatively minor investments—for example, establishment of new substations—were made in conjunction with grid customers. There was a background of some growth in electricity production and consumption over the period,[77] but economising on major grid investment was aided by the location of additional generation plant in the central and upper North Island where the bulk of electricity demand lay. However, by 2001 an active policy of renewal and expansion was justifiable. Reflective of Transpower's management of the grid during 1994–2001, transmission charges had fallen by 30% in real terms over the period.[78]

Under light-handed regulation, the Grid owner did not have the protection of a regulatory compact. Customers could, and did, refuse to pay charges based on posted prices for transmission services; and the only recourse Transpower had was application to the courts. Transpower took disputed non-payment with two gentailers to the High Court in 2000, arguing that the services were those of an essential facility[79], and therefore, Transpower should be able to post prices for them. The High Court did not uphold the application but, in its judgement, said that there was a matter for government to address. Although a negotiated outcome was found,[80] it was short-term and the disputes continued to pose revenue uncertainty for Transpower that was only resolved when statutory firm-specific regulation was put in place in

---

[74]The Blackout was due to multiple line failures managed by the Distribution company, *Final Report Ministerial Enquiry*, 1999 (http://www.med.govt.nz/templates/Page____12136.aspx, accessed 1/9/2018).

[75]Transpower Annual Report 2000/01, p. 17.

[76]Reilly op. cit. p. 214.

[77]The rate of growth in electricity generation averaged 1.8% per annum during 1994–2004 (calculated from MBIE electricity generation data (https://www.mbie.govt.nz/info-services/sectors-industries/energy/energy-data-modelling/statistics/electricity, accessed 1/9/2018).

[78]Colin Maiden, chairman of Transpower, Transpower New Zealand Limited Annual Report, 2003/04, p. 5.

[79]Essential Facility is not an established concept in New Zealand law.

[80]Reilly op. cit. p. 212

2004. From that time, Transpower's posted prices were part of the CC's approval of Transpower pricing and so were recoverable, as were costs of investment approved by the EC, and after 2010, by the CC.

5.3 Firm-Specific Regulation and the Grid Investment Test

Industry-specific regulation did not provide an investment framework immediately it was introduced in 2004. Indeed, it introduced a turbulent period for Transpower which had identified a portfolio of grid projects that it considered justified investment. It included, as most in need of upgrade transmission into (i) Auckland and the upper North Island, (ii) Christchurch and the upper South Island and iii) the HVDC Inter-Island link. It formally placed before the regulator—the EC in 2005—a proposal for a 400 kV transmission line to expand transmission capacity into Auckland. It had been through the processes required to meet the environmental tests under the Resource Management Act 1991 (RMA).[81] In principle, the RMA processes balance the social costs and benefits of effects of investment on the environment broadly defined. The institutional process for assessment includes consultation with affected parties and, for large projects, consideration by adjudicative bodies, in part to resolve the conflict between local and national interests. The resource consent-to-build process is time-consuming and costly. In the case of new grid investment, the local environmental effects of the new above-ground grid were a local cost—for example, to landowners—to be balanced against national benefits.

The EC rejected Transpower's proposal on the grounds it would fail a Grid Investment Test (GIT) in that there were alternative projects that "maximise expected net market benefits when compared to the proposal".[82] In making its decision, the EC applied a GIT that it had developed for the Code of the time. The GIT was applied by the EC when developing grid reliability and other grid investments and reviewing transmission alternatives, and by Transpower when preparing a grid investment report on proposed investments for inclusion in a proposal grid upgrade plan.

The GIT is a methodology for scoping the market opportunities that are relevant to a cost–benefit analysis of a transmission proposal and for the approach to analysing the proposal in the context of the statement of market opportunities. It sets out inputs to the analysis such as the discount rate and the value of lost load, and the types of costs and benefits that were to be considered. It distinguished between reliability and economic benefits. Reliability investments were held to satisfy the GIT

[81]"The RMA is New Zealand's main piece of legislation that sets out how we should manage our environment. The RMA is based on the principle of sustainable management which involves considering effects of activities on the environment now and in the future when making resource management decisions" Ministry for the Environment (http://www.mfe.govt.nz/rma/introduction-rma, accessed 1/9/2018).

[82]Electricity Commission Draft Decision on Transpower's 400 kV grid Investment Proposal, April 2006. (https://www.ea.govt.nz/about-us/what-we-do/our-history/archive/operations-archive/grid-investment-archive/git/, accessed 2/8/2018).

if the investment maximised benefit less costs, relative to alternatives. The proposed investment must, in addition, have net market benefits that are positive. In each case the assessments to be based on the associated market development scenarios and probabilities of occurrence. The GIT also requires a consultative process of affected parties within and without the electricity market.

On the EC's rejection under its newly formed GIT, Transpower almost immediately submitted an alternative proposal that was accepted by the EC in 2007; the 2-year delay largely resulting from the EC and Transpower-wide stakeholder consultation that was required under the electricity governance rules.

The statutory assignment of grid investment approval to the electricity regulator of the time, the EC, and price control to the competition regulator (CC) soon produced conflict.[83] In 2005, Transpower announced plans to raise prices by 13% per annum for 4 years in order to fund grid investment. This conflicted with the price thresholds set down by the CC and the CC announced its intention to take financial control of Transpower. The result was a negotiated administrative settlement that enabled price increases. Significant investment approvals followed: by January 2010 the EC had approved 15 grid upgrade projects totalling NZ$2.4 Billion.[84] However, the conflict between the EC and CC portended regulatory change. Investment approval and pricing require coordination and so should rest with one governing body. Given the CC's legal responsibility for Transpower price control, the CC was the more natural location for the governance of the investment evaluation process; the GIT.

Transpower's investment proposal notified in 2005 included significant upgrades of the HVDC inter-island link. The value of the HVDC lay in the economic efficiency gains of transfer of electricity from lower to higher value use, and competition in amount and diversity of electricity for energy trading and reserves.[85] At the time of the investment proposal electricity mostly flowed north, but there were also times of significant southward flows, particularly in drought years that limited South Island generation. The HVDC investment proposal called for the retirement of Pole[86] 1, the installation of Pole 3 and upgrading of the 20-year-old Pole 2. The resulting two-way capacity of the link would then be 1200 MW. Reflecting the previous investment strategies already described, Pole 1 was 39 years old, and any new or used replacement parts were hard to find: as mentioned above, Pole 1 ran at restricted capacity from 2007. The approved cost of the upgraded project was $672 m.[87] The allocation of

[83]Reilly op. cit. p. 212.

[84]The Electricity Commission, Approved Transmission Projects 14-Jan-2010 (https://www.ea.govt.nz/about-us/what-we-do/our-history/archive/operations-archive/grid-investment-archive/list-of-transmission-projects/, accessed 3/9/2018).

[85]Interestingly, the proposal put forward by Transpower was accompanied by a net benefit calculation based upon reliability only and not capacity expansion, although competition benefits were noted (*HVDC Inter-Island Link Upgrade Project Investment Proposal* Executive Summary, Transpower, 2005).

[86]The Pole is the equipment that converts DC/AC current at either end of the HVDC cable.

[87]The investment is discussed at file:///Users/evansle/Desktop/New%20HVDC%20Pole%203%20Commissioned%20%7C%20Transpower.webarchive accessed 5/3/2019

HVDC costs has been determined by the Transmission Pricing Methodology (TPM) discussed below.

In 2010, the EC regulator was re-structured by removing the requirement of Ministerial approval of Code changes, discretion to make any generation decisions.[88] The replacement regulator was the present EA. It's particular mandates on establishment included the development of the hedge market, concomitantly financial transmission rights, and retail competition.[89] Application of the GIT and the level of grid revenue has been the responsibility of the CC, but the EA retains responsibility for determining the form of transmission pricing.

6 Transmission Pricing

6.1 Introduction

The transmission pricing methodology (TPM) sets the form of charges that market participants pay for the use of the grid. It affects the distribution of payments among them but the average quantum of grid charges is fixed by the price-level constraints imposed by the regulation settings of the CC's grid regulation. The prices for electrical energy transported across the grid are the nodal prices determined in the spot market. They reflect relative congestion across nodes on the grid and so signal potential investments by market participants and the grid owner that may be economic. Under the EA's legal objective, it is the dynamic efficiency of the totality of the spot, hedge, and grid markets that is relevant for the assessment of the TPM. Taking the spot market as workably competitive and so relatively efficient allows focus on the grid pricing structure (TPM) effect on investment and operation of the grid, although an issue remains as to whether or not TPM settings should augment locational price signals of the spot market.

The TPM was never settled satisfactorily by the earlier EC regulator, although it approved a form of TPM in 2007. It continues to be the subject of debate and study by the EA and market participants to the present day. In the decentralised market transmission, prices are important to dynamic efficiency: in particular, for signalling efficient investment in, and off, the grid; for its users to minimise inefficient use of the platform that is the grid; and for the stability of the regulatory compact that is represented by the EA and the Code. The stability is enhanced by a Code that is transparently based upon the EA's legal objective and thus likely to be time

[88]The 2009 Government appointed electricity taskforce (ETAG) concluded that the ability of the industry regulator (EC) to control generation—even if to run at emergency situations only—created uncertainty for generator investors because of inability of the regulator to commit to defined "emergency" situations.

[89]The EA has reported on the state of retail competition at https://www.ea.govt.nz/monitoring/retail-market-snapshot/, accessed 15/9/2018.

consistent and provide surety of institutional arrangements for irreversible long-term electricity investments.[90]

6.2 Grid Charges

The TPM[91] produces Transpower charges under the three categories: grid connection, interconnection and HVDC assets.[92] Although the HVDC is an interconnection asset connecting two sets of grid interconnection assets, its separate charging categorisation has been in place since 1999 when it was set by Transpower. At that time, there was no applicable industry-specific regulator.

Specifically, TPM elements in 2010–12 were as follows.[93]

1. A connection charge to recover the costs of dedicated AC (connection) assets connecting a distributor, major user and/or generator to the grid. Where there is more than one entity using the connection, the cost is shared according to anytime maximum injection (AMI) or demand (AMD).
2. An interconnection charge that recovers the costs of AC interconnection assets and the proportion of overhead and corporate costs paid by distributors lines companies and grid-connected major users; it is allocated by contribution to regional coincident peak demand (RCPD): that is, a customer's offtake at that connection location during a regional peak demand period.
3. A HVDC charge that recovers the costs of the HVDC Inter-Island link. It was paid by South Island generators based upon their share of peak injections to the grid in the South Island—termed the historical anytime maximum injection (HAMI). The HAMI for a South Island generator means either the average of 12 highest injections at its location during the capacity measurement period of a pricing year, or the average of the highest injections at that location during any of the 4 immediately preceding pricing years, whichever is the highest. Beginning in late 2015, the HAMI calculation is being progressively replaced by the mean 5-year injection to the grid of each South Island generator (SIMI).[94]

---

[90]The EA sets out its view that there exist inefficiencies in the TPM in *Transmission Pricing Methodology: Problem definition relating to interconnection and HVDC assets*, Electricity Authority, Working Paper, 16 September 2014. 123p.

[91]https://www.ea.govt.nz/code-and-compliance/the-code/part-12-transport/schedule-12-4/, accessed 3/9/2018.

[92]Transpower's 1999 rationale for the separate HVDC charge is provided at: https://www.transpower.co.nz/industry/transmission-pricing-methodology-tpm/transpower-hvdc-sunk-cost-recovery-paper---April-1999, accessed 7/2/2019. It is shaped by the regulatory setting of the time which under light-handed regulation was not industry-specific. It was dependent on competition and contract law for an enforceable arrangement.

[93]Electricity Authority, *Transmission Pricing Methodology: issues and proposal*, consultation paper, 10 October 2012.

[94]At the same time, the Code was modified changing the definitions of RCMP—and so Kvar charges—for each of the voltage constrained northern North Island and northern South Island.

Overlaying these charges was the TPM prudent discount policy in which charges are discounted to avoid uneconomic bypass of the grid. These have been rare.[95] The charges were ameliorated by Transpower payment of loss and constraint rentals[96]—arising as inframarginal rent generated by spot market prices in constrained sections of the grid—to grid customers apportioned in proportion to transmission charges. In addition, HVDC loss and constraint rentals were paid to the South Island generators.[97] These rentals are collected by the NZEM and transferred to Transpower at the time of trading-period settlement.

For the year 2010/11, the three charges totalling grid revenue were: the connection charge (NZ$126 m), interconnection charge (NZ$413 m) and the HVDC (NZ$85 m). The programme of prospective grid investment forecast to 2017/18 little change in connection charges, but that HVDC charges would double and interconnection charges increase by 76% reflecting the planned grid enhancement. The relative importance in the TPM of the HVDC is indicated by the Transpower's 2018 regulated-asset valuations of: HVDC, $0.6b and connection and interconnection assets of $4.0b.[98]

6.3 *Evaluation of Transmission Pricing*

The continuing TPM debate has concerned three key elements: the form of the HVDC charge, the form of the interconnection charge and the extent to which TPM should augment spot market price network-congestion signals. The process is led by the EA with much consultation with market participants, including Transpower. The EA judges alternatives according to its statutory objective of dynamic efficiency which it seeks to implement by cost–benefit analysis. Review and consultation have continued since 2010, and just recently, in June 2018, the EA announced it would provide a further formal proposal for consultation.[99]

The first report on TPM under the auspices of the EA was by the working group TPAG[100]; a body consisting largely of market participants. It set down the following specific efficiency considerations to judge TPM performance, in concept at least,

[95] Transpower reports 4 of these discounts, 2 of which are entitled notional embedding (https://www.transpower.co.nz/industry/revenue-and-pricing/revenue-and-pricing, accessed 2/9/18). The form of prudent discounts is under consideration in the present TPM review.

[96] The rentals paid are net of FTR balance charges.

[97] Of course given demand, these rentals are inversely related to the capacity of the HVDC. Investment in the HVDC discussed above increased the capacity significantly and led to an order of a doubling of HVDC charges.

[98] https://www.transpower.co.nz/sites/default/files/uncontrolled_docs/Securing%20our%20Energy%20Future%20RCP3%20Proposal.pdf, accessed 10/3/2019.

[99] https://www.ea.govt.nz/development/work-programme/pricing-cost-allocation/transmission-pricing-review/development/next-steps-june2018/, accessed 2/9/18.

[100] Transmission Pricing Advisory Group (TPAG), Transmission Pricing Discussion paper, 7/6/2011, 161p.

they are largely self-explanatory[101]: beneficiary pays; locational price signalling; unintended efficiency impacts that, for example, affected demand-side management; competitive neutrality; implementation and operational costs; and good regulatory practice.

Drawing on a model that included choice about plant location by grid-connected parties, TPAG concluded that system-wide pricing methodologies designed to enhance spot market prices in signalling relative transmission costs across locations—for example, tilted postage stamp pricing as between the North and South Islands—yielded sufficiently small benefits that they were outweighed by uncertain costs. These costs included: consequences from unmodeled issues, adjustment costs and the time consistency of locational grid charges.[102] However, grid charges and recommendations for modifications of them are affected by congestion[103] and in conjunction with spot prices incentivise the benefit of congestion mitigation of various sorts at any point on the grid. Also, the EA's current proposal for beneficiary pays interconnection asset charges, discussed below, does induce area-specific charges for the cost of interconnection assets for those areas where the benefits mainly lie.

TPAG also considered issues of static reactive power in regions where transmission capability was limited by voltage stability limits[104] and the balance of deep and shallow connection in defining Connection Assets. The deep connection charging regime includes assets at connection points (shallow connection assets) and assets required by individual connecting parties. Costs are shared according to peak demand. Alternatives considered by TPAG included Flow Tracing wherein subject to a deemed cut-off boundary separating connection and interconnection assets the energy flow determined the charge, and b) the existing charging arrangement but with beneficiaries determined under the GIT process paying for the capacity they require. TPAG identified efficiency gains in moving to either of these alternative approaches, mainly because of the engagement they would produce by incentivising the connecting party to be concerned about grid asset investments. It made no firm recommendation.

The form and function of the HVDC are those of an interconnection asset. Given grid-asset-locational pricing is not efficient, the purpose of the HVDC is to link the networks of both islands to facilitate competitive neutrality for the country as a whole. The HVDC charge assignment to South Island generators does not do this.

The inefficiencies of the HVDC arrangement identified by TPAG were: disincentive for South Island relative to North Island generation investment; competition effects wherein because the charges are a fixed cost to generators investment in the generation of a level of MW would be more economic for an incumbent generator

---

[101]TPAG. s.4.3.2

[102]TPAG Appendix C. The model assessed the minimum system cost of particular configurations of load and generation and variations in these attributed to response to locational grid charges. TPAG suggested that the result was affected by the cost of transmission being relatively low as compared to generation and the fixity of location for hydro and geothermal generation.

[103]For example, the RCPD- and HAMI-based charges reflect peak usage and therefore grid capacity.

[104]TPAG. s.8.2

than a smaller entrant[105]; and investment and dispatch inefficiencies arising from the HAMI-based charges affecting offer strategies and providing a disincentive to invest in peaking capacity.[106] A quantitative cost–benefit evaluation of the HVDC charges by TPAG that subsumed competitive neutrality for the country as a whole indicated a net benefit to changes to treating the HVDC as an interconnection asset folding HVDC charges into the totality of interconnection charges.

Another point made by TPAG and the EA was that the HVDC charges violated the beneficiaries pays principle. The flows on the link are generally south to north but increasing demand in the South Island means that in years of drought in the south significant reverse flows occur.[107] Even without north and south directional flows there are demand as well as supply beneficiaries in linking the two geographically distinct markets with the HVDC.

For interconnection assets, the EA advocates a charge based upon beneficiary pays and a postage stamp charge on load for the remainder. By definition, interconnection assets have externalities in their use that render an absence of property rights over elements of them. Indeed, a key argument for interconnection assets to be held by only one entity is in order that these externalities[108] are internalised in decision-making. In consequence, it can be anticipated that the ability to assign specific benefits and therefore benefit-specific costs to market participant users of the grid will be limited, and that another—postage-stamp type—charge will be important.

Once in place, transmission assets are largely sunk, and hence, in principle, efficient pricing post-investment calls for fixed charges that cover investment (on an annuity basis) and maintenance costs, and thereby minimally affect spot market decisions. However, as a practical matter "fixed" charges need be allocated based upon some characteristic—e.g. size of load/off-take—and hence, there will be some effects of grid charges on grid usage and market entry and exit.

---

[105]The system also provided an incentive, depending upon nature of the plant and structure of networks, for small generators to embed in lines network footprints and thereby mitigate grid charges.

[106]Scientia Consulting reports that replacement of HAMI with SIMI pricing is leading to previously with-held generation being offered and higher net benefits from the same configuration of plant. Thus, the switch to SIMI pricing may reduce, but not eliminate, the benefit of treating the HVDC as an interconnection asset. (https://www.transpower.co.nz/sites/default/files/publications/resources/Scientia_Market_Impact_HAMI_SIMI.pdf, accessed 3/3/2019).

[107]NERA p. 57 reports an increasing flow southwards during 2004–2008 culminating in equal North-South flows. (New Zealand Transmission Pricing Project: a report to the New Zealand Electricity Steering Group, NERA economic consulting, 2009, 142p.

[108]These externalities reflect Kirchhoff's laws (stated at pp. 375–6 of Stoft, Steven, *Power System Economics*, IEEE Press, Wiley-Interscience,2002, 468 pp.) in the flow characteristics of AC electricity.

6.4 Beneficiary Pays

The EA proposes to measure the amount and incidence of the benefit of an investment and assign interconnection charges in proportion to benefit. It plans to do this by discovering benefits using the linear-programme temporary-equilibrium SPD model used to calculate final prices. Run separately with and without the assets of the investment, the SPD-calculated benefit to a market participant—specifically all grid-connected parties—is the change in willingness to pay for the network due to the presence of the asset, where this is positive.

The EA's initial proposal[109] had beneficiary calculations being made every trading period. The bids and offers of the actual trading-period run would be taken and applied in a second SPD run where the asset was removed from the total of grid assets available in the initial run. This provided a measure of benefit for all connected parties; who would then be charged a trading-period fee equal to its share of the fixed (per trading period) revenue requirement of the asset-cost, where its share is its share of total positive benefit. The accuracy of measurement plainly depends upon energy spot market bids and offers not being affected by prospective beneficiary-pay charges and by the extent to which these offers and bids would be different in the counterfactual of the absent asset.

The ultimate incidence of the beneficiary pays charge would depend on the connected party and whether that party could or would pass through the charge to that party's customers: as for existing grid charges. If generation and directly connected load are in competitive markets the incidence will lie with them, and they can be expected to respond economically to changed grid charges. But for lines companies the extent of pass-through will be affected by applicable price regulation rules, and the response to change that in turn will reflect the extent to which their customers are diffuse. Because the grid interconnection assets to which beneficiary-pays charges would be applied affect different geographic areas of the grid differently, these charges could be expected to produce grid locational charges. In an area materially benefitting from a grid investment, concomitant charges will be higher than they would be if investment was funded under a postage stamp. The charges would vary with the variations in network flows: including North and South flows on the interisland link.

Market participant benefit estimation might be thought of as an approach to measuring the welfare increment of the H-R-G-V regulatory scheme.[110] For this scheme, the increase in consumers' surplus from the previous year relaxes the pricing constraint on the regulated Transco. The EA's proposal was to assess consumer surplus from particular grid investments using the SPD tool and then reduce it by per-period investment charges for those that benefitted. Measurement of consumer surplus increment is required in both schemes; but SPD is most unlikely to reveal consumer surplus change with a useful degree of accuracy; even in short-run time

[109]EA presentation on the SPD method: Transmission Pricing Methodology Modelling workshop, December 2012.

[110]Vogelsang, op. cit.

frames for which it is designed. It may signal significant conjestion and have helpful information for small changes located in the commonly observed transaction price/quantity range but plainly it does not reveal demand and welfare for the full market. Indeed, for New Zealand there is limited demand revelation in the spot market and the SO plans for dispatch use demand forecasts rather than revealed bids.[111]

The EA argues that the beneficiary-pays approach for interconnection assets produces dynamic efficiency benefits in that the charges are user charges and as such will heighten market participant interest and participation—and hence information revelation—in the GIT process that Transpower and the CC use for new investments.[112] The argument may be particularly relevant given that Transpower is assured recovery of cost of investment approved by the CC. In this setting, Transpower may well have a lower investment hurdle rate[113] for investment proposals than other market participants.

Other efficiency benefits relative to a pure postage stamp, for example, are not obvious. They will be affected by the way the (fixed) beneficiary charges of willingness to pay are actually translated into market participant characteristics, for example, load size and the particulars of any usage response.

6.5 *The TPM Proposal*

The EA has chosen in 2018 to abandon the assessment of investment beneficiary-pays charges by trading period. The decision followed extensive study of and consultation about the trading-period approach and the volatile results it produced. Instead, it proposes to require Transpower at the time of investment evaluation to use practical methods, including runs of SPD, to measure the incremental benefit of any investment for grid-connected parties on a forward-looking basis. Transpower is then to set and allocate fixed charges—called AoB charge—for the life of the asset based upon the assessed net private benefit. There are numerous details of implementation to be resolved: they include circumstances of maintenance of the charge in real terms, and how tightly the charge is tied to the original market participant beneficiaries.[114]

In sum, the EA second issues paper reports a positive cost-benefit outcome for treating the HVDC as an interconnection asset and:

- a connection charge similar to the status quo;

---

[111]Dispatchable demand of relatively large load is offered into the spot market and managed under the market rules.

[112]There is a connection between the evaluation of investment proposals as these too commonly have SPD evaluation runs as input.

[113]Although transmission investment proposals have to pass the CC-hurdle rate for approval.

[114]EA, Transmission Pricing Methodology: second issues paper: (https://www.ea.govt.nz/development/work-programme/pricing-cost-allocation/transmission-pricing-review/consultations/#c15999, accessed 9/9/2018).

- a beneficiary charge for interconnection assets set as an ex ante determined fixed charge with a range of implementation issues to be left to Transpower, loss and constraint rentals to be paid to parties with grid charges; and
- a residual charge for interconnection assets: applied to load only, similar to the status quo form of interconnection charge.

The EA estimates that the direct effect of its proposed changes relative to the status quo will entail a reallocation of final demand prices across grid-connected parties that have a standard deviation of 2%.[115]

The proposed TPM has much complexity and devolves much of the responsibility for implementation to Transpower. The economically significant development of the proposal is the treatment of the HVDC as just another interconnection asset.

7 Transmission in 2018 and Beyond

Transmission and the TPM review processes now have to contend with changing opportunities brought about by rapidly innovating technology and the growth of renewable generation. There is a development of more energy-efficient devices—for example, LED lighting—digital information and control technologies, economically viable batteries (mobile and fixed), modes of generation and storage, and even a potentially different balance of AC and DC electricity in the offing. These changes and expectations about them are affecting decisions throughout the market. In conjunction with enhanced distributed generation, including solar and wind power, they are affecting the long-accepted institutional structure for electricity production and delivery. At one level, the changes are facilitating further decentralisation—for example, micro-grids—that bypass some established networks; but the economic position of the new technologies is not yet clear. For example, the extent of scale economies in battery technology, and the extent to which digital communications lower the cost of centralised as well as decentralised control are unsettled matters.

The decentralised energy-only New Zealand electricity market structure is in a good position to allow innovation and allow an orderly transition to a new market structure whatever that may be. The EA has instigated a review and study of lines network pricing this year aimed at pricing closer to the costs of the services provided and concomitantly admitting economic uptake of new technology.[116] This year it examined the Code in response to a proposal for a retail company to connect a scalable battery to the grid at the point of connection of a small thermal generator in Auckland City. Battery storage of energy directly with the grid is a first for New Zealand, and the existing Code permitted it. The Auckland lines company uses batteries in its network and supplies them and solar generation to households under the present Code.[117]

[115]Data were taken from Table 5 of Appendix F of the EA's second issues paper op. cit.

[116]https://www.ea.govt.nz/development/work-programme/pricing-cost-allocation/distribution-pricing-review/, accessed 27/9/2018.

[117]https://www.vector.co.nz, accessed 10/9/2018

Change may be having some effect on New Zealand electricity consumption already, for whereas real GDP grew by 24% between 2007 and 2017 electricity generation increased by just 1%.[118]

Perhaps presently the most immediate area of electricity innovation is being felt by lines and retail companies as they manage developments in distributed generation and networks that are decentralised to the level of businesses and households. There is competition with transmission and lines-based generation that may, or may not rely on the services of transmission networks. The composition of the average household electricity bill in 2017 was 10.5% transmission, 27% lines, 13% retail, 32% generation and 13% value-added tax (GST)[119] which illustrates that decentralised consumers can most lower costs with an alternative system by management of—or in the extreme not taking—grid and lines network services.

The grid owner's management strategy must recognise the potential disruption posed by the myriad of electricity-related developments. Real options literature tells us that increased uncertainty generally increases the social benefit of waiting. Such is the uncertainty attending transmission that large investments that will be sunk and which may have long planning, and build, periods and investment-economies of scale are likely to be less efficient than smaller just-in-time flexible investments made as technology and transmission demand structure evolves. Certain future investment will be of a different nature from the past—it will include batteries for example.

The SO's real-time management of transmission will change as offers to generate assume different characteristics. Both because of relative cost declines in wind and solar generation and regulatory policy designed to limit CO_2 emissions, there will be an expansion of renewable generation at the expense of non-renewable thermal generation. With the exception of hydro, presently renewable generation offers to the market are made at their operational marginal cost of zero. This means they have priority in dispatch, which is not of itself an issue for the market.[120] However, renewable generation such as wind and solar, that have no storage and depend intimately on the weather for their fuel, offer relatively uncontrollable generation which the SO manages at a cost to the market. Such generation is presently only some 5% of total generation in New Zealand; but should it expand the concomitant costs will grow. However, now these costs are being internalised to the generator by bundling generation plant and battery storage,[121] rendering consequent offer strategies based on relatively predictable supplies: just as for hydro.

[118]The GDP data are obtained from https://www.rbnz.govt.nz/-/media/ReserveBank/Files/Statistics/tables/m5/hm5.xls and https://www.mbie.govt.nz/building-and-energy/energy-and-natural-resources/energy-statistics-and-modelling/energy-statistics/electricity-statistics/, accessed 3/3/2019.

[119]EA (https://www.ea.govt.nz/about-us/media-and-publications/market-commentary/events/telling-new-zealands-electricity-story/, accessed, 10/9/2018).

[120]The energy-only market determination of capacity through the existing form of spot and hedge markets without alteration of the Code would be efficient for an energy-only market of solely renewable generation characterised by zero operational marginal cost. (Lewis Evans, "The electricity spot market: is it future proof"?, *The Electricity Journal*, 30, 2017, 25–29).

[121]McCulloch et al. (2019) report that these bundles are now inexpensive even relative to gas plant (Robert McCullough, Eric Shierman, Michael Weisdorf and Louis Bengston, "The End of Big Iron:

These innovations will likely affect spot market demand as well as supply and change the task of the SO. When many market participants have storage and control, their bids and offers will be based upon their individual strategic view of the future prices and demand and supply situations. Some decision-making will be automated, and there will be decentralised reactions to events that the SO previously has had to manage. Consequently, forecasting trading-period demand and management of contingent events will likely be more challenging. The demand for, and nature of, SO services as well as the demand for transmission investment will be significantly affected by the new technology.

how Wind and Solar Became Cheaper than Hydro, Coal and Nuclear", *Public Utilities Fortnightly*, March 2019, pp. 38–48).

Transmission Network Investment Across National Borders: The Liberalized Nordic Electricity Market

Lars Persson and Thomas P. Tangerås

1 Introduction

The world's first liberalized multinational electricity market was created in 1996 when Norway and Sweden opened a power exchange for trading wholesale electricity between the two countries. Finland joined in 1998, and Denmark in 2000. The *Nord Pool* power exchange later expanded to incorporate also Estonia, Latvia and Lithuania, and was coupled with the North European power market in 2014.

The backbone of the integrated market is the high-voltage transmission network that enables electricity to flow from power plants in one country to consumers in another. Figure 1 shows a map of the network infrastructure in Northern Europe. In an integrated market, removal of network bottlenecks affects energy flows and prices across the entire market and therefore has implications also for surrounding counties. Welfare-improving network investment requires accounting for these indirect effects of capacity expansion. This chapter analyzes international infrastructure investment in the context of the Nordic market.

We give in Sect. 2 a brief account of the historical background for liberalization of the Nordic countries. Main arguments in favor of deregulating the wholesale electricity market were to improve short-run incentives to produce electricity efficiently and create informative price signals to govern long-run investment decisions. Investments in hydro and thermal capacity have been limited the last 25 years. Possible explanations can be excess capacity and subsidies to renewable investment that have

L. Persson · T. P. Tangerås (✉)
Research Institute of Industrial Economics, Grevgatan 34, P.O. Box 55665, SE 102 15 Stockholm, Sweden
e-mail: thomas.tangeras@ifn.se

L. Persson
e-mail: lars.persson@ifn.se

M. R. Hesamzadeh et al. (eds.), *Transmission Network Investment in Liberalized Power Markets*, Lecture Notes in Energy 79,
https://doi.org/10.1007/978-3-030-47929-9_19

Fig. 1 Transmission network map for Northern Europe 2018. *Source* www.entsoe.eu/data/map/

pushed down electricity prices. Most of the capacity expansion has been in renewable electricity. The picture is different for transmission network capacity, which nearly doubled the first ten years after liberalization. Congestion rent earned on interconnections has contributed to the profitability of network investment.

An important benefit of market integration in the Nordic market has been to take advantage of regional differences in the generation mix that generate gains from trade. Connecting a diverse portfolio of generation assets through a transmission network reduces the risk of supply shortages and reduces the cost of maintaining supply security. Because of geographical concentration of asset ownership, the Nordic electricity market has been vulnerable to the exercise of market power. Market power can be mitigated by network investment and market integration. Market integration reduces greenhouse gas emissions by stabilizing production and thereby reducing the need for fossil fuel generation units to handle local demand peaks. These and other main economic arguments for market integration are reviewed in Sect. 3.

We discuss in Sect. 4 transmission network planning in the Nordic market. A main driver of current investment is the transition from a hydro-/nuclear-based system to one with large shares of intermittent renewable electricity, which requires network reinforcement. Ambitions to export excess production and increase hydropower access to foreign markets drive integration with the rest of Europe. When calculating benefits from network expansion, network owners emphasize gains from trade, security of supply and network losses. Gains from trade are measured on the basis of local production and consumption imbalances and the frequency with which prices differ across regions. Security of supply is measured by reserve margins—production and import capacity relative to peak demand. Such gains from network expansion are in different units of measurement and therefore not directly comparable with one another, with other potentially relevant economic effects of market integration or with project costs. It is essential to apply a unified framework in which it is possible to aggregate all consequences of an expansion in network capacity, to be able to assess the full welfare economic consequences of investment.

Section 5 analyzes countries' incentives to undertake welfare-increasing transmission network investment in a multinational electricity market. Investment incentives generally are distorted because of third-party country effects and because countries planning a project not always have an incentive to cooperate with third-party interests. Inefficiencies can persist even if countries manage to cooperate. Investments often are interrelated so that the profitability of one project depends on the (non) completion of other projects. In that case, procedural differences in how projects are decided can play a fundamental role in the outcome. Also, monopoly power can render network investment decisions inefficient. We discuss two approaches for increasing efficiency and ensuring countries' willingness to cooperate. Under a decentralized approach, projects originate in pairwise negotiated outcomes between the investing parties. Third-party countries can propose changes, but project modifications are voluntary. Under a centralized approach, projects are developed at the central level. Distributing surplus by the extent to which countries contribute to value creation improves incentives to participate and represents an equitable distribution of gains from market integration.

Section 6 concludes the chapter with a discussion of the integrated Nordic market in relation to the political ambition of the European Union (EU) to create a well-functioning internal electricity market.

2 The Liberalized Nordic Electricity Market

2.1 History of Liberalization

Liberalization of the Nordic electricity market began with the Norwegian Energy Act of 1990 that laid the foundation for a deregulated wholesale electricity market in Norway.[1] The decision to restructure the electricity market was based on an increasing discontent in Norway with the economic inefficiency of the domestic electricity system. In particular, there was no link between marginal costs of expanding production and network capacity and marginal benefits of doing so under the regulated system. Instead, capacity had been allowed to increase in an effort to supply energy-intensive manufacturing industries with (for them) inexpensive electricity. These firms were on long-term supply contracts with generation companies under regulated prices that fell short of covering long-run marginal production costs. Households and the service industry made up for some of the difference by paying as much as four times the price of electricity paid by energy-intensive industries. To achieve the desired investments, much of the capacity expansion was undertaken by the state-owned company *Statkraft*. There were also direct subsidies to those firms (mainly municipal) that were not state-owned.

In the regulated Norwegian market, producers had few outside options once they had fulfilled their supply obligations under the long-term contracts. For instance, producers were prevented from exporting their electricity because they did not have complete access to the high-voltage transmission grid. Most of the Norwegian production is hydropower and therefore subject to random variation. When producers had nowhere to sell excess power in wet years, they ran water past the turbines. Toward the end of the 1980s, annual spillage amounted to around 5% of total production. Some excess electricity was exported to Denmark and Sweden, but at prices much below those paid by Norwegian consumers. It was obvious to many that there was scope for improvement in an electricity system that regularly threw away a resource with zero short-term marginal production cost.

Main objectives of the electricity market reform of 1990 were to[2]:

- Establish a platform for trading wholesale electricity in the short-term market—a power exchange—supplemented by financial markets and capacity adjustment mechanisms.
- Achieve complete and non-discriminatory access to the transmission network.
- Vertically separate the state-owned incumbent into:
  - A generation and retail unit: Statkraft.
  - A transmission network owner and system operator (TSO): *Statnett*.

---

[1] The historical account of Norwegian liberalization is based on Bye and Johnsen (1991), Bye and Hope (2005, 2007) and Bredesen (2016).

[2] Restructuring of the Norwegian electricity market did not involve privatization, unlike in the UK. See, for instance, Armstrong et al. (1994) for an overview of regulatory reform in the UK.

- Impose regulation on network companies designed to increase economic efficiency.

The transition to a liberalized market was facilitated by the fact that Norwegian producers had already garnered experience with market-based trading platforms for short-term power before the reform took place. Already in 1972, a coalition of producers had developed *Samkjøringen* as a tool for reallocating electricity among themselves. Samkjøringen was a power exchange that collected bids and offers and cleared them by way of an equilibrium price. Hence, producers had already seen the benefits of market pricing of electricity. This precursor to the current spot market was insufficient because it covered only 10% of annual production. The new power exchange, *Statnett Marked*, encompassed the entire geographical market in Norway. It was organized as a subsidiary of the Norwegian TSO, Statnett.

In Sweden, there was a consensus view that short-term gains of liberalization were small because the electricity market already operated in a cost-efficient manner.[3] The concern was more with long-term efficiency: Sweden also seemed to suffer from having overinvested in production capacity. The ambition was for deregulation to deliver better price signals that would translate into more efficient investment decisions further down the road. In the beginning of the 1990s, Sweden had taken similar structural steps as Norway. For instance, vertical separation between generation and retail (*Vattenfall*) on the one hand and a TSO (*Svenska Kraftnät*, SvK) on the other had been accomplished by 1992. Based on the Statnett Marked power exchange, Sweden and Norway formed a jointly owned power exchange, *Nord Pool*, for trading wholesale electricity within and between the two countries. The world's first multinational wholesale electricity market started operation in 1996.

Finland followed suit and joined Nord Pool two years later, in 1998, and then Denmark in 2000. A main motivation for Finland to join Nord Pool was to increase efficiency and competitiveness of the energy sector.[4] An interesting difference between Denmark, Finland and most other countries that deregulated was a division of transmission network ownership prior to liberalization. In Finland, *Imatran Voima* (now *Fortum*) and *Pohjolan Voima* both owned substantial generation and transmission assets. As part of the restructuring of the industry, the two firms separated transmission from generation to create one single and jointly owned TSO, *Fingrid*. For those historical reasons, Fingrid has always been partially privately owned. In 2011, Fortum and Pohjolan Voima sold their shares in Fingrid to comply with EU regulations concerning ownership unbundling. The majority of Fingrid now is state-owned, with a minority share held by private companies without ownership shares in the electricity sector. Denmark originally had two transmission networks without direct physical connection. The western network covered the Jutland Peninsula and was integrated with Germany. The eastern network supplied Zealand and was integrated with Sweden. The western and eastern networks were owned and operated by two companies *Eltra* and *Elkraft System*, both of which were vertically separated

[3]This historical account of Swedish liberalization is taken from Högselius and Kaijser (2007).

[4]See Pienau and Hämäläinen (2000) for an account of Finnish electricity market deregulation.

from generation. The two merged in 2005 to create one single TSO, *Energinet.dk*, owned by the Danish state. The two grids then became interconnected in 2010.

Estonia was incorporated into Nord Pool in 2010, Lithuania in 2012 and Latvia the following year. Nord Pool was then coupled with the other Northern European power markets in 2014.

2.2 The Nordic Power Exchange: Nord Pool

The cornerstone of the Nordic wholesale electricity market is the power exchange, *Nord Pool*. The most important trading platform on the power exchange is the day-ahead market, *Elspot*. Elspot traded 394 terawatt-hours (TWh) electricity in 2017, which amounted to 94% of total production of the Nord Pool member countries.[5]

Elspot currently spans the Nordic countries Denmark, Finland, Norway, Sweden and the three Baltic states Estonia, Latvia and Lithuania.[6] It is also coupled with the Northern European power market. Elspot is divided into 15 price areas, five of which are in Norway, four are in Sweden, and two are in Denmark. The other countries comprise one price area each. The number of price areas has changed over time. Sweden, for instance, was one single price area until 2011. The Norwegian price areas have changed several times.

Every day before noon, the transmission network owners (TSOs) submit to Nord Pool for each of the 24 h of the following day the trading capacities on the transmission lines that connect the different price areas within Nord Pool. The export and import capacities of the transmission lines from surrounding countries directly connected to Nord Pool are similarly reported. These countries are Germany, the Netherlands, Poland and Russia. Electricity producers submit offers to Nord Pool for each of the 24 h and for every price area. Similarly, electricity retailers and large industrial consumers submit bids of how much electricity they are willing to purchase during the different hours in the different price areas. Producers are only allowed to participate in the local markets (price areas) where they have physical production capacity. The same is true for consumers: Retailers and industrial consumers can only participate in markets where they have physical consumption capacity.[7]

Transmission capacity is bid inelastically into Elspot; i.e., bids are price independent. All other market participants can submit up to 62 offers/bids for every hour and price area, specifying how much electricity they are willing to sell/purchase at different prices in each area. The price cap is 3000 Euros per megawatt-hour (EUR/MWh). Firms can bid negative prices, but not below −500 EUR/MWh. After

[5]Trading data are from the Nord Pool Annual Report 2017, which can be accessed at www.nordpoolgroup.com. Production data are from www.nordpoolgroup.com/Market-data1/#/nordic/table.

[6]The transmission grid of the fifth Nordic country, Iceland, is physically disconnected from all other countries' transmission networks. Iceland operates its own market.

[7]Virtual (convergence) bidding (Jha and Wolak 2015) therefore is currently not allowed on Nord Pool.

gate closure of Elspot, Nord Pool combines the producer quantity/price offers by linear interpolation to generate an hourly supply curve for each of the 15 price areas. Demand curves for every hour and price area are constructed on the basis of the price/quantity bids of retailers and industrial consumers. Nord Pool then adds all area supply (demand) curves to generate an hourly *system* supply (demand) curve for the Nordic market. The intersection of those two curves establishes the *system price*. Nord Pool applies the system price to the area curves to calculate supply and demand volumes in each price area. Nord Pool then checks to see whether the resulting flows between the price areas and the import/export capacities lie within the network capacities that were supplied by the TSOs. If so, then the system price is the equilibrium price for the Elspot market that hour. Often, network capacity is insufficient to handle trade flows. Such bottlenecks occur, for instance, during peak hours when an increase in consumption creates excess demand in densely populated areas. Nord Pool then uses the area supply and demand curves to clear each price area separately, subject to the binding area network constraints. Hence, the Elspot equilibrium is characterized by up to 15 hourly area prices depending on the severity of network constraints. Equilibrium prices are higher in import-constrained than export-constrained areas.[8]

Retailers and industrial consumers (producers) pay (receive) the hourly area price for all electricity they purchase (sell) within the price area for delivery that specific hour. If there are no bottlenecks in the system, so that the system price also is the equilibrium price, then all payments and revenues balance out net of trade with countries outside Elspot. When bottlenecks occur, then consumers on Elspot pay more for the electricity they purchase than what the producers receive in compensation. The difference is the total *congestion rent* that is generated on Nord Pool that hour. This rent is distributed across the transmission network owners on Nord Pool depending on where in the system the congestion has occurred, the traded volumes and other factors, such as ownership.

The day-ahead market can be cleared as much as 36 h before actual delivery, and a lot can happen that may cause market participants to want to rebalance their portfolios relative to the day-ahead allocations. To allow for such redispatch, Nord Pool operates also an intraday market, *Elbas*. This market opens two hours after gate closure of the day-ahead market and closes one hour prior to physical delivery. Elbas features continuous trading and therefore essentially is a pay-as-bid market where the same product is traded at multiple prices over the course of the trading period as new market information arrives. Within the hour of delivery, the national TSOs in each of Nord Pool's member countries take over and clear actual production and consumption by way of different balancing markets.

[8]Nord Pool's clearing procedure implies that the supply and demand functions generally have well-defined and nonzero point elasticities, unlike in electricity markets that feature supply and demand step functions.

2.3 Generation Capacity in the Liberalized Nordic Electricity Market

Arguments in favor of liberalization were based on the notion that generous regulation had generated excessive investment in generation and network capacity. Figure 2 shows how installed generation capacity has evolved from the start of liberalization in 1991 until 2015. The figure displays the annual capacities in megawatts (MW) aggregated over Denmark, Finland, Norway and Sweden, for the most important energy sources.

The sum of nuclear, hydro and other thermal capacity (coal-, gas-, bio-fueled condensing power and combined heat and power plants) has remained relatively constant. Total growth over the 25 sample years was approximately 9%, almost all of which occurred after year 2000. Indeed, there has been investment in hydropower and thermal power, but most of it has gone into replacing decommissioned capacity. For instance, the two 600 MW Barsebäck nuclear reactors in southern Sweden represented nearly 10% of total installed nuclear capacity in 1991. Both were closed by governmental decree, reactor 1 in 1999 and reactor 2 in 2005. By 2015, nuclear capacity had recovered and was slightly above the 1991 level. Two-thirds of the 7 400 MW increase in non-intermittent capacity was hydropower. The picture is consistent with a story in which generation companies entered the era of deregulation with excess hydro and thermal capacity. Market prices were below the long-run cost of capital for a long period of time. Producers had to wait more than ten years for demand to catch up, and capacity expansion become profitable.

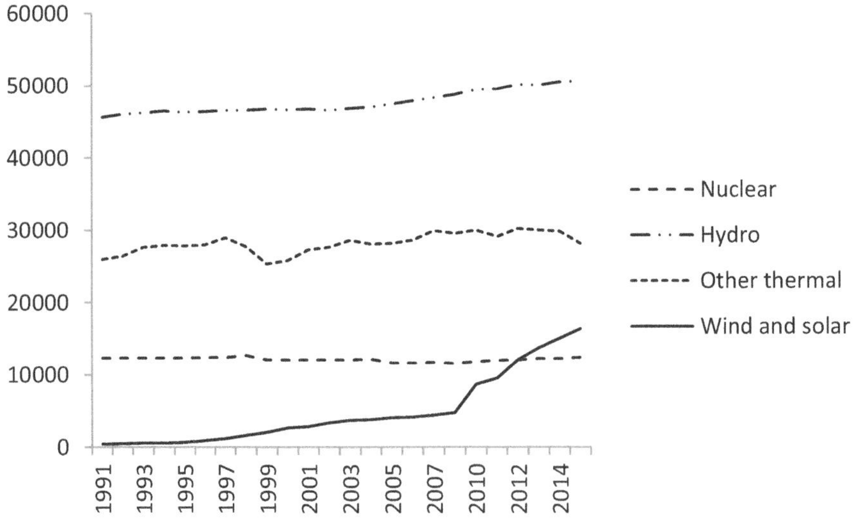

Fig. 2 Installed generation capacity (MW) in Nordic countries 1991–2015. *Source* Nordel Annual Statistics 1991–2008. Data for 2010–2015 are at https://www.entsoe.eu/data/data-portal/. Year 2009 is missing

Evaluating the underlying reasons why capacity has changed so little is complicated by the fact that liberalization has had effects besides exposing firms to market prices. In particular, vertical separation of generation and transmission asset ownership and the incorporation of incumbent firms created market participants on Nord Pool with market power and incentives to use it. Those incumbent firms own the majority of Nordic generation capacity even today (NordREG 2014): Vattenfall (Sweden) has 19%, Statkraft (Norway) 14%, Fortum (Finland) 12% and Ørsted (Denmark) 6%. With the purchase of Sydkraft in 2001, E.ON of Germany has been the only new large player to enter the Nordic market since deregulation, with 7% of total generation capacity. Market power is accentuated by a geographical concentration of asset ownership. Vattenfall, for instance, owns 37% of Swedish generation capacity. Joint ownership, in particular of Swedish nuclear power, creates additional, collective market power. Producers with market power have an incentive to withhold output to increase prices. Recent evidence based on the bidding behavior on Nord Pool suggests that firms indeed behave in such a way as to increase prices (Lundin 2020; Lundin and Tangerås 2020; Tangerås and Mauritzen 2018). Firms that exercise short-term market power in general have distorted long-run incentives even to invest. Hence, liberalization has shifted the investment paradigm from one of incentives to overinvest to one of incentives to underinvest. Without further analysis, there is no way of telling whether the lack of investment has been an efficient response to price signals or an attempt by firms to drive the price of electricity up above long-run marginal cost.

The most striking feature of Fig. 2 is the growth in renewable generation capacity that has occurred mainly after the turn of the millennium. Denmark was a front-runner in the development of wind power, with production starting already in the late 1970s. A major change occurred when Sweden in 2003 became the second country in the Nordic market to launch an ambitious support system for renewable electricity. Average annual growth in solar and wind capacity has been 17.5% since 1991. In 2012, solar and wind power overtook nuclear power in terms of capacity.

Under the Swedish tradable green certificate system, producers earn one certificate per MWh certified renewable electricity they supply. Certificates are sold to retail companies mandated to cover a share of final consumption by renewable electricity. This additional source of revenue to what producers earn on selling the electricity in the wholesale market, stimulates renewable investment. The certificate system was designed to be technology-neutral in the sense of targeting the most cost-efficient production instead of specific technologies. In the beginning of the period, firms invested both in bio-fueled combined heat and power plants and in wind power. But after 2010, wind power has been the dominating source of new renewable capacity brought online; see Fig. 3.

To develop a better understanding of the gains from liberalizing the Nordic market, and the value of transmission investment, it is important to pay attention to not only the long-run trends in energy production, but also important geographical differences across the Nordic region in how electricity is produced.

Table 1 shows hydropower, nuclear, nonnuclear thermal and solar/wind power production and consumption measured in TWh in the four Nordic countries in

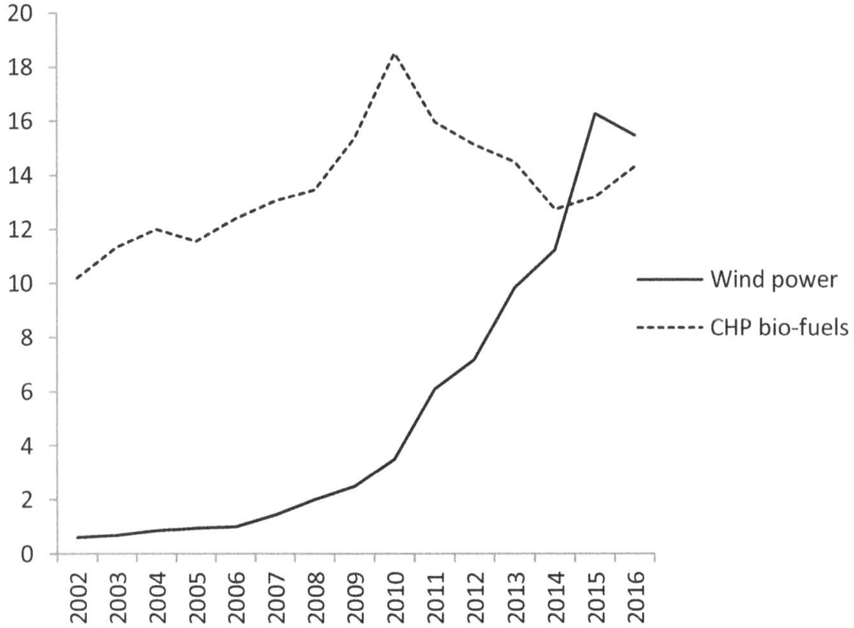

Fig. 3 Annual renewable electricity production (TWh) in Sweden 2002–2016. *Source* Statistics Sweden (www.scb.se)

Table 1 Electricity production and consumption (TWh) in the Nordics 2015

	Denmark	Finland	Norway	Sweden	Total
Hydro	<0.1	16.6	139.0	74.0	229.6
Nuclear	0.0	22.3	0.0	54.4	76.7
Thermal fossil fuel	10.5	13.4	3.5	3.8	31.2
Thermal biofuel	2.3	10.7	0.0	9.8	22.8
Solar/wind	14.7	2.3	2.5	16.6	36.1
Total	27.5	65.3	145.0	158.6	396.4
Consumption	32.4	82.5	128.3	135.9	379.1

Source ENTSO-E (www.entsoe.eu/data/data-portal/) and ENTSO-E (2015)

2015. Hydropower located mainly in Norway and northern Sweden is the dominating energy source and accounted for nearly 60% of Nordic production. Nuclear power in Finland and southern Sweden is the second most important source with 20% of production in 2015. Thermal fossil and biofuel condensing and combined heat and power plants in Denmark, Finland and southern Sweden accounted for 14% of production. Wind power located predominantly in Denmark and Sweden has

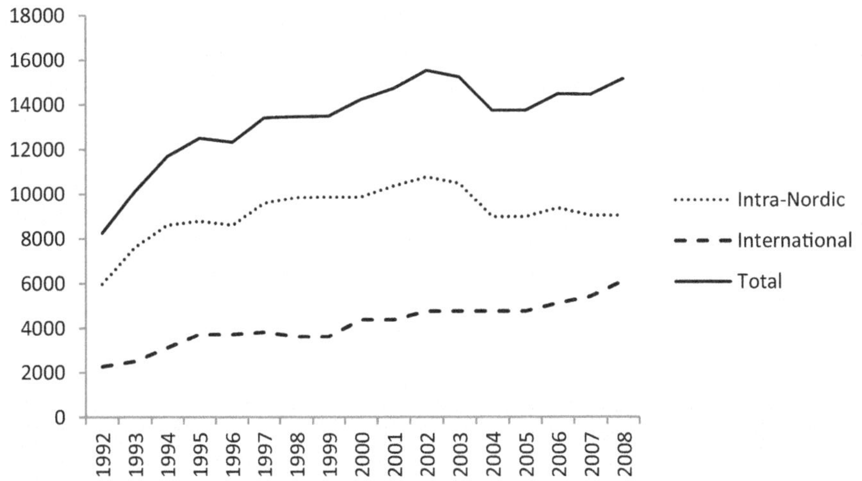

Fig. 4 Installed transmission network capacity (MW) in Nordic countries 1992–2008. *Source* Nordel Annual Reports (entsoe.eu/). Because of missing data, the 1992 international capacity is from 1991. We use the maximum of import and export capacity for individual lines, where the two differ

grown to become a more important source of electricity production from the early 2000s and onward.[9]

The trade flows of electricity were as follows (ENTSO-E 2015): Denmark was a net importer from Norway and Sweden and a net exporter to Germany. Finland was a net importer from Sweden and Russia and a net exporter to Estonia. Norway was a net exporter to Denmark, Sweden and the Netherlands. Sweden was a net exporter to Denmark, Finland, Germany and Poland and a net importer from Norway. Overall, the Nordic countries were net exporters of electricity in 2015.

2.4 Transmission Capacity in the Liberalized Nordic Electricity Market

Whereas generation capacity, with the exception of wind power, has increased marginally since deregulation, the picture is completely different for transmission network capacity.

Figure 4 shows how network capacity (MW) evolved between 1992 and 2008.[10] The dotted line depicts the intra-Nordic capacity, i.e., cross-border transmission capacity between the four countries Denmark, Finland, Norway and Sweden. The dip

[9]Notice the difference in capacity utilization between solar/wind and nuclear power.

[10]Data from 2009 and onward are currently unavailable from ENTSO-E.

that occurred 2002–04 was temporary and due to replacement of a 300 kV transmission line between Norway and Sweden by a 420 kV line that required disconnection of the initial transmission line. Intra-Nordic transmission capacity increased by 80% the first ten years after liberalization. The dashed line shows transmission capacity between the four Nordic countries and surrounding ones. International network capacity more than doubled between 1992 and 2008. Sweden was responsible for most of this expansion by interconnecting with Germany and Poland. Denmark almost doubled interconnection capacity with Germany, and Finland increased the import capacity from Russia. Still, the Nordic countries remain more connected with each other than with surrounding countries, measured in terms of transmission capacity. The solid line in Fig. 4 is the total cross-border capacity of the Nordic market during the years 1992–2008.

Transmission capacity nearly doubled the ten years following liberalization. There could have been underinvestment in cross-border relative to domestic transmission prior to regulation that was subsequently corrected. But it is important to bear in mind that the TSOs were regulated before and after liberalization, so investment was driven by regulatory incentives even in the new market. So if capital returns to regulation were consistently high, then there was no reason for TSOs to reduce investment after liberalization. Furthermore, deregulation of the wholesale market generated an additional source of revenue, congestion rents, that the TSOs could use to expand existing capacity and interconnect with new countries. We return to the issue of network congestion rent in Sect. 3.1.

3 Economics of Market Integration

A well-functioning multinational electricity market relies fundamentally on the transmission network having the capacity to transport electricity reliably and cost efficiently from power plants to consumers across the full geographical footprint of the market. This section contains an overall description of the economic benefits associated with market integration, the correct estimation of which should underlie well-informed decisions to expand capacity.

3.1 *Gains from Electricity Trade*

Figure 5 illustrates the direct gains from trade in the Nordic wholesale electricity market associated with an increase in cross-border transmission capacity between two countries.

The x-axis shows quantities in MWhs during one hour on the day-ahead market Elspot, and the y-axis shows Elspot prices in EUR/MWh for that hour. The export (import) country is identified by superscript $E(I)$. The market-clearing price in the export country is given by p_0^E in the default situation when there is no transmission

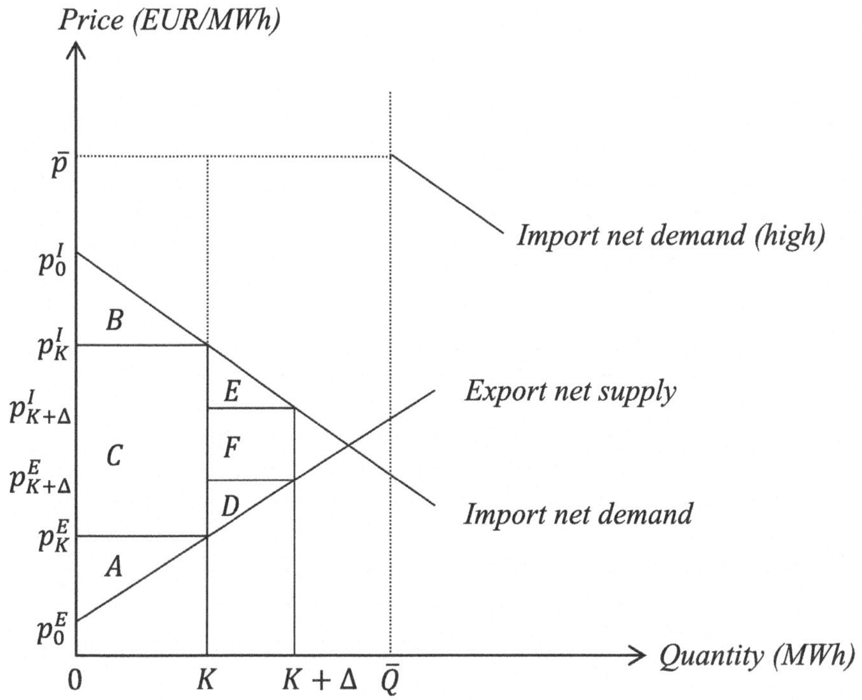

Fig. 5 Gains from electricity trade

capacity between the two countries. For prices above p_0^E, the export country has excess supply of electricity. This domestic imbalance is increasing in the price and is illustrated in the figure by the *export net supply* curve. The market-clearing price in the import country is equal to p_0^I if there is no cross-country transmission capacity. The import country has domestic excess demand of electricity for prices below p_0^I, illustrated in the figure by the *import net demand* curve.

For the sake of exposition, let the two countries be Denmark and Norway. Recall from Table 1 that Denmark relies heavily on wind power for domestic electricity supply. A reason why Denmark would benefit from market integration is because trade allows Denmark to import hydropower from Norway in situations with little wind in Denmark. Suppose now Denmark and Norway build a transmission line with capacity K between the two countries. Norwegian producers would require a price equal to p_K^E to be willing to cover both domestic demand and export K to Denmark. Inexpensive hydropower from Norway causes the price in Denmark to drop down to p_K^I, where domestic supply and imports are just sufficient to cover domestic Danish demand. Norwegian producers benefit both from the increase in the wholesale price of electricity in Norway from p_0^E to p_K^E and from a net increase in output because of exports to Denmark. Conversely, the price increase and reduction in the domestic use of electricity hurt Norwegian consumers. The sum of the two is positive because Norway is a net exporter to Denmark. The increase in total consumer and producer surplus in Norway is measured by the triangle A in Fig. 5. Danish consumers benefit from the drop $p_0^I - p_K^I$ in the domestic wholesale price and from any resulting increase in the domestic use of electricity. Danish producers lose because of the reduction in the domestic price and because some of the domestic output is replaced by imports from Norway. The sum of consumer and producer is positive because Denmark is a net importer from Norway. The increase in total surplus in Denmark is measured by the triangle B in Fig. 5. The network owner buys the K MWh electricity at price p_K^E in Norway and sells at price p_K^I in Denmark. The rectangle $C = (p_K^I - p_K^E)K$ measures the *congestion rent* of the investment. The total gains from trade this particular hour equal $A + B + C$.

There are other gains from trade than importing electricity from Norway to resolve a domestic supply shortage in Denmark. In some situations, there can be so much wind in Denmark as to create a domestic supply surplus. The cross-border connection then allows Denmark to export cheap electricity, thus enabling Norway to save hydropower for future consumption. Such trade reversal is captured in Fig. 5 simply by a reinterpretation of superscripts, so that E now refers to Denmark and I to Norway.

Rather than increasing export possibilities from one country to the other in a uniform manner, an important benefit of market integration has been to take advantage of the geographically diverse generation mix in the Nordic market; see Table 1, which makes it easier to correct local short-term imbalances between demand and supply through cross-border trade. For instance, market integration between Norway and Denmark essentially allows Norwegian hydropower to act as a battery for Danish wind power. Offsetting local fluctuations through trade reduces the cost of equating supply and demand because the total production capacity in the market can be smaller.

There are also redistribution effects compared to when market integration essentially serves to increase trade in one direction. The benefits of market integration are better aligned between consumers and producers and across countries when trade flows go in both directions. In principle, consumers and producers can all benefit from improved market integration. The annual gain from trade over the cross-border interconnection between Denmark and Norway is the sum of all hourly trade surplus increases. Building an interconnection between the two countries is profitable strictly on trade terms if the total increase in trade surplus in a representative year covers the annual variable and fixed cost of the investment.

Consider now an additional investment that increases capacity even further to $K + \Delta$. The price difference in the wholesale price of electricity then falls to $p_{K+\Delta}^I - p_{K+\Delta}^E$ between the two countries. Improved market integration redistributes some of the congestion rent on the initial capacity K to producers and consumers in Denmark and Norway. Net surplus in the exporting country increases by the additional triangle D and by the triangle E in the importing country. The change in congestion revenue equals the rectangle $F = (p_{K+\Delta}^I - p_{K+\Delta}^E)\Delta$. The increase in total trade surplus per unit of incremental capacity equals:

$$\frac{D+E+F}{\Delta} = p^I_{K+\Delta} - p^E_{K+\Delta} + \frac{D+E}{\Delta}$$
$$= p^I_K - p^E_K - \frac{p^I_K - p^I_{K+\Delta} + p^E_{K+\Delta} - p^E_K}{2}$$

The price differences $p^I_{K+\Delta} - p^I_{K+\Delta}$ and $p^E_{K+\Delta} - p^E_{K+\Delta}$ become negligible for Δ sufficiently small. It follows that the hourly trade gain from expanding cross-border capacity marginally above K is approximately equal to the price difference $p^I_K - p^E_K$ between the two countries that hour. Average annual price differences thus measure the trade benefits of marginal investments in transmission capacity between local markets.

3.2 Security of Supply

There are economic consequences of transmission network investment that cannot be captured by comparing incremental increases in spot market trade surplus with investment cost. Return to Fig. 5, and let Norway be the export and Denmark the import country. Suppose Denmark has very unfavorable wind conditions and that *import net demand (high)* represents the excess demand curve for electricity in Denmark. Looking at the figure, A plus the congestion revenue $(\bar{p} - p^E_K)K$ appears to measure the total surplus increase associated with Denmark importing K MWh electricity from Norway. But in Fig. 5, $\bar{p}$ is the maximal price in the Danish market. It is a regulated or implied price, not one that equates imports and domestic supply with domestic demand. Denmark has excess domestic demand $\bar{Q} - K$, even after all available domestic generation capacity and all transmission capacity have been bid into the spot market. In situations where the market fails to clear, the TSO activates capacity reserves to cover the difference $\bar{Q} - K$.[11] In such cases, one cannot estimate the value of transmission capacity on the basis of supply and demand curves in the spot market. Instead, the incremental benefit is the reduction in the TSO's cost of maintaining security of supply.

Figure 6 replicates the top part of Fig. 5. Net demand $\bar{Q}$ is the sum of household and business demand minus the sum of wind power output and thermal capacity available to the spot market. *Marginal thermal cost* in Fig. 6 shows the variable production cost of the generation units in the domestic TSOs portfolio of capacity reserves. This reserve consists of generation capacity owned by the TSO itself plus generation capacity procured from the market. The latter are high-cost units that producers cannot profitably bid into the market given the price cap $\bar{p}$.

[11] In the terminology of Nordel (2008), a "market failure" characterizes a situation in which the day-ahead wholesale market fails to clear. A "system failure" occurs when there is insufficient physical capacity to cover demand at the delivery hour without curtailment of consumption.

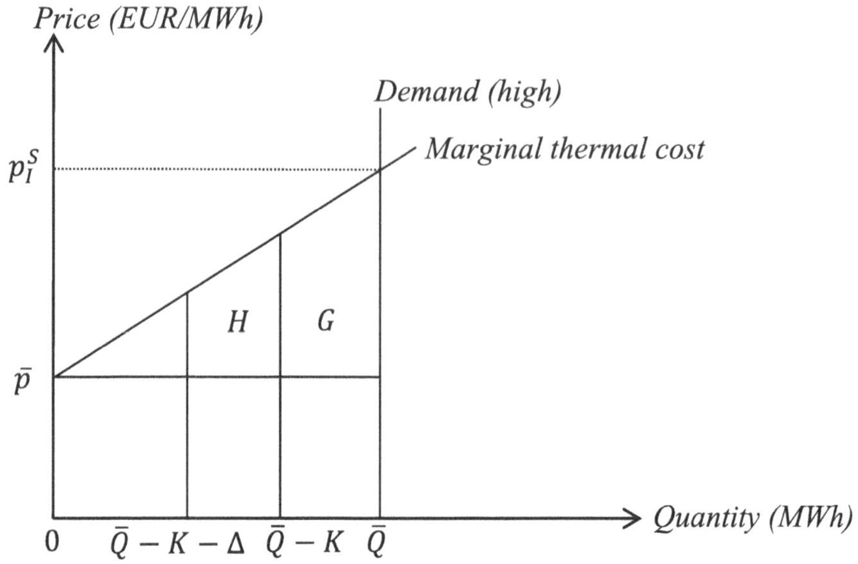

Fig. 6 Gains from supply security

Market integration increases the security of supply in two ways. First, the likelihood of a resource-constrained situation is lower because the total supply of electricity available at the price cap is higher. Second, when a resource-constrained period does arise, the amount of domestic capacity reserves the TSO needs to activate is smaller when transmission capacity is larger. In Fig. 6, the TSO only has to activate $\bar{Q} - K$ instead of $\bar{Q}$ when cross-border transmission capacity is K instead of zero. The cost saving equals the trapezoid G, which comes in addition to the surplus in the day-ahead market. The trapezoid H measures the incremental cost saving when additional Δ MW are added to cross-border transmission capacity. In the long run, the improved security of supply associated with market integration implies that the TSO can reduce its capacity reserve.

3.3 Market Power

The analysis in Sect. 3.1 of the trade gains from market integration rests on the assumption that the wholesale markets for electricity in the two countries are competitive. Ownership of generation capacity is concentrated in the hands of a few producers in the Nordic electricity market; see Sect. 2.3. Hence, they have the possibility to withhold production and thereby drive up the wholesale price of electricity. Producers' exercise of market power causes two main problems for assessing the benefits of transmission investment.

Under imperfect competition, observed export net supply and import net demand curves are misrepresentations of marginal social values and social costs of consuming and producing electricity. Market power in the import country drives p_0^I up above the competitive level and renders *import net demand* in Fig. 5 an upward-biased version of the marginal social value of imports. Taking the supply and demand curves at face value would then exaggerate the benefit of network investment. Market power in the export country renders *export net supply* in Fig. 5 an upward-biased version of the marginal social cost of export by driving the domestic price p_0^E of electricity up above the competitive level. In this case, there would be underinvestment based on the supply and demand curves. With market power in both countries, the two distortions would offset one another with ambiguous net effect.

Market integration affects market performance under imperfect competition by affecting the intensity with which producers compete against one another across the two countries. An increase in competitive pressure would materialize as a downward shift in *import net demand* and an upward shift in *export net supply* in Fig. 5. On the basis of marginal changes in trade surplus, network investment could be more or less beneficial under imperfect competition compared to the case of perfect competition. However, intensified competition would improve domestic resource allocation in both countries and thus yield an added benefit to network expansion. Under plausible circumstances, therefore, network investment is more beneficial when producers exercise market power than would otherwise be the case. However, one has to have detailed information about the extent to which firms exercise market power in the wholesale market to be able to quantify the effect on market performance of transmission network investment.

3.4 Investment in Generation Capacity

Figure 5 illustrates short-run effects of transmission network investments. In the long run, the price changes in the electricity market affect generators' incentives to invest in capacity and industries' incentives to invest in more energy-intensive production facilities. The decrease in the price from p_0^I to p_K^I in the import country renders it less profitable to bid generation capacity into the spot market and more profitable to expand industry production. The opposite holds in the export country. Consequently, the long-run import net demand curve lies above *import net demand* in Fig. 5, whereas the long-run export net demand curve is below *export net supply*. Investment decisions based on the observed demand and supply curves therefore underestimate the long-run trade gains of expanding transmission network capacity. The magnitude depends on the long-run elasticities of demand and supply. The interdependence of transmission and generation capacity and coordination of such investment is analyzed numerically, for instance, in Tohidi et al. (2017a, b).

3.5 Environmental Effects

The burning of fossil fuels for electricity and heat is the world's largest source of greenhouse gas emissions. Improved market integration has the potential to cut emissions by reducing the need for generation capacity reserves to handle local demand peaks. Reserve units often are gas-fired power plants because they must be flexible and available at short notice. But emission reductions can also occur if market integration evens out fluctuations in production.

Figure 7 illustrates how emission of greenhouse gases (GHGs, mainly CO_2) on the *y*-axis fluctuates with electricity production from coal- and gas-fired power plants on the *x*-axis. Emissions measured in tons per MWh increase with electricity production because conversion of fossil fuels to power is less efficient at high levels of output. Assume that there is no cross-border transmission capacity between Denmark and Norway. If wind conditions are poor, then the generation companies in Denmark must dispatch a large amount Q_h^I of thermal power to satisfy demand, which releases GHG_h^I tons of greenhouse gases into the atmosphere. But if there is an abundance of wind, then only very little thermal power Q_l^I has to be used in order to meet demand. Danish electricity production is nearly fossil free in this case, with emissions dropping all the way down to GHG_l^I. The average thermal output is equal to Q_a^I, and the average emissions are GHG_a^I.

A cross-border transmission line between Denmark and Norway provides hydropower that can be used so as to even out fluctuations in wind power output. As an indirect consequence, thermal output also stabilizes across periods. Assume for the sake of presentation that thermal production is the same in both periods and equal to Q_a^I when transmission capacity is equal to *K*. Although thermal production is the same before and after transmission investment, it would be incorrect to

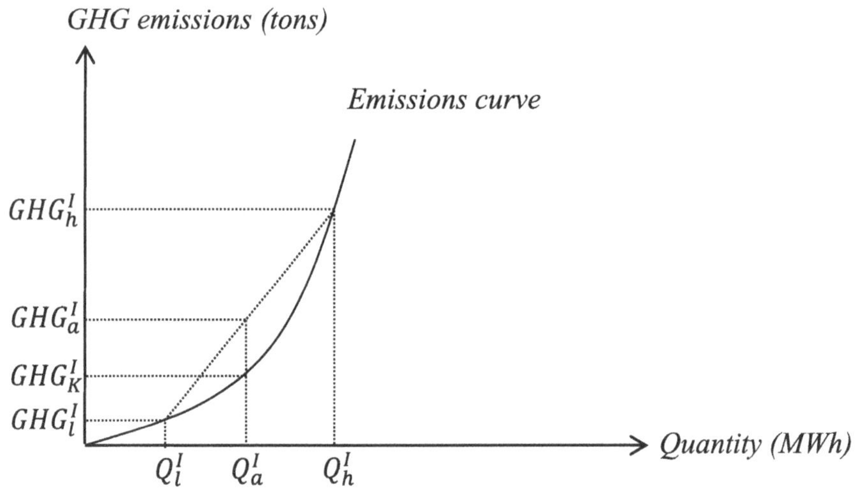

Fig. 7 Greenhouse gas emission and market integration

conclude that market integration had zero effect on GHG emissions. The mere fact that thermal production is stabilized drives average emissions down to GHG_K^I. To estimate the marginal environmental benefit of transmission investment requires a price on reducing greenhouse gas emissions. Such a price exists since the introduction of the EU-ETS trading platform for greenhouse gas emissions.

Whereas transmission network investment can have global environmental benefits in terms of reducing emissions by coal- and gas-fired power plants, there are local negative externalities, for instance, because of visual degradation. Efficient investment requires that local inhabitants be compensated for such damages, but they are in general difficult to quantify.

3.6 *Network Losses*

Transmission of electricity across long distances leads to electrical power losses. Additional transmission capacity affects the way in which electricity flows through the entire network and therefore the total network losses. The effect of transmission on electricity transportation costs should be included in the analysis to get a picture of the full social benefit of network investment.

4 Transmission Network Planning in the Nordic Market

The Nordic TSOs apply cost-benefit analysis to identify viable transmission network investment projects within the Nordic market and to surrounding countries. A review of methods and results is found, for instance, in the Nordic Grid Master Plan 2008.[12] The Nordic Grid Development Plan 2017 describes ongoing and planned future investments in the Nordic transmission network.[13]

A current main driver of network investment is the fundamental transformation in the Nordic generation mix that is expected to occur over the decades to come. Increased intermittent electricity production foremost in Denmark and Sweden requires network reinforcement in particular in the north–south direction in Norway and Sweden to remove domestic bottlenecks and thereby unlock hydro production from the northern part of the market. System planners envision the increase in renewable production to yield a net production surplus in the Nordic market to be exported abroad via new or improved cross-border connections to Germany from Denmark, Norway and Sweden, to Great Britain from Denmark and Norway, to the Netherlands from Denmark and to Lithuania from Sweden. Nuclear power in Finland and

[12]www.entsoe.eu/fileadmin/user_upload/_library/publications/nordic/planning/080300_entsoe_nordic_NordicGridMasterPlan2008.pdf.

[13]The plan can be found, for instance, at www.svk.se/siteassets/om-oss/rapporter/2017/nordic-grid-development-plan-2017-eng.pdf.

Sweden faces a large-scale phaseout that will be partially replenished by new nuclear power in Finland at old and new locations. Increased network capacity is planned in Finland and Sweden to maintain reliability subsequent to the restructuring of Nordic nuclear power. A second driver of network investment is the need to upgrade and replace a large part of the Nordic transmission grid that was built in the 1950s and 1960s. Finally, the demand for electricity is expected to grow substantially in the northernmost part of Norway with the development of Barents Sea oil fields and the electrification of oil production, which will require network capacity expansion.

Actual cost-benefit calculations begin with specifications of particularly interesting scenarios that differ from one another concerning assumptions about economic growth, fuel prices, and energy and climate policy. Based on these scenarios, projections are derived for consumer demand, generation capacity and the fuel mix in electricity generation for the following 15–20 years.

Cost-benefit analysis has emphasized market integration, security of supply and energy losses. Benefits of market integration are derived on the basis of the effects of network reinforcement on the system *energy balance* and on the *market situation*. The energy balance at a specific location is measured by the volume difference between annual production and consumption. The market situation is measured by the frequency with which prices between local markets differ from one another during the year. Security of supply is defined in terms of the *loss of load probability*, which is required not to exceed 0.1% on an annual basis. In practice, security of supply is evaluated by first simulating an extreme peak demand situation—an event that occurs at most once every ten years. By comparing peak demand with installed generation, transmission and import capacity, one can then see if there is sufficient power in the system to satisfy local demand. The *reserve margin* is the difference between generation and import capacity and peak demand. The system satisfies security of supply if reserve margins are positive at all key locations. A reason why security of supply can be jeopardized is because of equipment failure, related to either the production or transmission of electricity. The Nordic market applies the $N - 1$ criterion, meaning that the system should maintain full functionality even if one (near) essential unit of equipment breaks down.

In scenario analysis, it is relatively straightforward to integrate the $N - 1$ reliability criterion and to calculate how new interconnections contribute to the energy balance and security of supply. But the benefits from market integration are measured in volumetric terms and/or frequencies that are not directly comparable with each other or with other important factors such as investment cost. TSOs use proprietary simulation tools to attach monetary values to the benefits of network investment. However, it is unclear how volume changes are converted and whether consumer and producer surplus effects are accounted for. Analyses account for changes in generation capacity, but treat generation investment as exogenous to network investment. Potentially relevant factors, such as market power or environmental effects, are typically ignored. Network losses typically are included in the simulation studies.

An alternative and more transparent way to estimate gains from market integration would be to use actual supply and demand curves from Nord Pool and the various short-term balancing markets. One could then recalculate market equilibria on the basis of proposed network reinforcements and thus estimate gains from trade. One way to quantify in monetary terms how a specific network project contributes to supply security would be to estimate the least-cost combination of generation and transmission capacity investment required to achieve the same reserve margin and compare this number with the investment cost of the project.

5 Incentives to Undertake Cross-Border Transmission Projects

Consider a stylized Nordic electricity market consisting of the four countries Denmark, Finland, Norway and Sweden. Assume that there are aggregate welfare gains of improving market integration by increasing cross-border transmission network capacity. In an integrated market, removal of network bottlenecks affects energy flows and prices across the entire market and therefore has implications also for surrounding counties. Welfare-maximizing network investment requires accounting also for those indirect effects of capacity expansion.

We will use elements of coalition theory to gain better understanding of incentives to, and problems associated with, carrying out international infrastructure projects in a multinational electricity market. Because of state sovereignty, no country can build a transmission line and unilaterally connect it to a foreign network. By the nature of such projects, implementation requires cooperation, specifically that countries jointly decide which projects to undertake, the technical properties of these projects and how to share investment proceeds and costs. Coalition theory deals with such cooperative decisions. The ability of involved parties to negotiate and write legally enforceable contracts renders a cooperative framework particularly useful.[14] We will follow the approach in Horn and Persson (2001) that proposes a method to systematically compare different coalition configurations and to generate predictions about coalition formation.[15]

A *coalition structure* in the Nordic market is a partitioning of the four countries into different coalitions that cooperate on cross-border network projects. A coalition structure comprises one, several or zero investment coalitions. With every coalition structure is associated a total surplus of each coalition in the structure. Each coalition consists of at least two countries, and not all countries need to be part of a coalition. For instance, the coalition structure $\{DN, S, F\}$ is the one where Denmark and Norway cooperate to build a transmission line between the two countries, but Sweden and Finland cooperate with no one.

In a four-country electricity market, there are 15 potential coalition structures, not all of which are likely to occur. For instance, the coalition structure $\{DF, N, S\}$

[14]See overviews of cooperative theory in Greenberg (1994), and Osborne and Rubinstein (1994).

[15]Domestic transmission investment may also have third-party effects, but the network owner can unilaterally decide on such investments. Tohidi and Hesamzadeh (2014) discuss such domestic transmission investment in a network with multiple network owners.

would involve building a transmission line at the bottom of the Baltic Sea between Denmark and Finland; see Fig. 1. This is probably a very expensive project. A better alternative could be to include Sweden, form the coalition $\{DFS, N\}$ and improve market integration through joint reinforcements of the Danish–Swedish and Finnish–Swedish connections. The structure $\{DF, N, S\}$ is *unstable* if the joint surplus of Denmark, Finland and Sweden is higher under $\{DFS, N\}$ than what they could attain under $\{DF, N, S\}$. Coalitions can be unstable also in the other direction. Consider a coalition $\{DFN, S\}$ between Denmark, Finland and Norway that involves an interconnection between northern Finland and northern Norway; see Fig. 1, additional to one between Denmark and Norway. The value of a Finnish connection is limited by domestic north–south congestion in Norway that would have to be resolved. A better idea for Denmark and Norway could be to form a sub-coalition and leave Finland out of the picture by not building any line to Finland. The structure $\{DFN, S\}$ is unstable if $\{DN, F, S\}$ renders Denmark and Norway a higher joint surplus than what they could attain under $\{DFN, S\}$.

More formally, a particular coalition structure is *dominated*, and therefore will not occur in equilibrium, if at least one coalition in the structure is unstable either to inclusion of one or more countries in that coalition or to formation of sub-coalitions. A coalition is *stable* if there are no inclusions of outside countries and no formation of sub-coalitions that would increase the joint surplus of the directly involved parties. A particular coalition structure is *undominated* if all coalitions in the structure are stable. The set of undominated coalition structures defines the *core*. In the application to transmission network investment, the core identifies the different collections of transmission projects that can be implemented in equilibrium. If the core consists of only one single structure, then this coalition structure is the unique prediction of the theory. The core can consist also of multiple structures, in which case coalition theory can sustain multiple network configurations as equilibria. Finally, the core can be empty, in which case standard coalition theory has no predictions.

Gains from trade and increased security of supply probably are the most important economic effects of improving market integration of the ones categorized in Sect. 3 of this chapter. At least, those are the effects TSOs usually emphasize in their cost-benefit analyses. One might expect two TSOs negotiating over whether to build an interconnection to internalize the bilateral economic effects of the investment, provided the underlying cost-benefit analysis has been thorough and provided it lies in the two TSOs best interest to maximize the joint surplus of the two countries.

To fix ideas, suppose Denmark and Norway discuss building a new cross-border transmission line with capacity x between southern Norway and western Denmark to facilitate the use of Norwegian hydropower as a battery for Danish wind power. Let $V_i(x)$ be the total surplus in country $i = D, N$ if the line is built, net of any congestion rent and under the assumption that the two countries each pay half of the investment cost. Assume that the cost-benefit analysis has shown the investment to be jointly welfare improving for the two countries: $V_D(x) + V_N(x) > v_D^0 + v_N^0$, where v_i^0 is the surplus of country i under the status quo. Even so, it might be that one of the parties benefits a lot whereas the other loses from the investment. In coalition theory, such imbalances are overcome by reallocating congestion rent or investment cost between the two parties, or by way of compensation payments. Hence, the status quo structure $\{D, F, N, S\}$ is dominated by $\{DN, F, S\}$ in this example.

Let T_i be the (potentially negative) compensation payment received by $i = D, N$. Let the two countries negotiate the size x of the project and compensation payments (T_D, T_N) to maximize the following *Nash product*

$$\left(V_D(x) + T_D - v_D^0\right)^{\alpha_D} \left(V_N(x) + T_N - v_N^0\right)^{\alpha_N}$$

subject to each country i's participation constraint, $V_i(x) + T_i \geq v_i^0$, and subject to budget balance $T_D + T_N = 0$. In the above Nash product, $V_i(x) + T_i - v_i^0$ measures the net benefit to country $i = D, N$ of the transmission project relative to the outside option of no project. The parameter α_i measures the bargaining strength of i in the negotiation. Nash (1950) provided the foundation for this approach to bargaining by showing that the outcome of a two-person bargaining game satisfying certain efficiency and independence axioms precisely maximizes the Nash product. In particular, the negotiated solution is constrained efficient because the capacity $\tilde{x}$ of the transmission line maximizes the joint surplus of Denmark and Norway:

$$V_D'(\tilde{x}) + V_N'(\tilde{x}) = 0$$

5.1 Sources of Investment Distortions

Consider third-party effects in the Denmark–Norway example. Assume that Denmark is a net importer of electricity both from Norway and Sweden; see Sect. 2.3. Increased market integration with Norway then causes electricity prices in Denmark to fall, which in turn reduces congestion rent earned on interconnections between Denmark and Sweden. They also lead to a reduction in wholesale prices in Sweden whenever transmission constraints do not bind. Price reductions benefit consumers, but hurt producers in Sweden. Trade surplus falls because of the country's position as a net exporter of electricity to Denmark. Hence, Sweden is likely to be worse off by the Danish–Norwegian investment, i.e., $\tilde{v}_S < v_S^0$, where $\tilde{v}_S = V_S(\tilde{x})$ is the total surplus in Sweden if Denmark and Norway build a cross-border transmission line with capacity $\tilde{x}$. If Sweden loses from all marginal capacity expansion, i.e., $V_S'(x) < 0$ for all $x > 0$, then the transmission project will be excessive from the aggregate viewpoint of the three countries.

In the above example, Sweden would lose from what it perceives to be excessive investment in cross-border connections between Denmark and Norway. Assume instead that Denmark is a net exporter to Norway, but keep everything else as before. Increased market integration then drives up the wholesale prices of electricity in Denmark to the benefit of Sweden. If $V_S'(x) > 0$, then the coalition structure $\{DN, F, S\}$ would yield too little investment from the joint viewpoint of Denmark, Norway and Sweden. In fact, a lack of coordination could imply that welfare-improving investments might not occur at all. This can happen if investment is jointly unprofitable for Denmark and Norway, $V_D(x) + V_N(x) < v_D^0 + v_N^0$ for all $x > 0$, but collectively welfare improving in the sense that there exists an $x' > 0$ such that

$$\sum_{i \in \{D,F,N,S\}} \left(V_i(x') - v_i^0\right) > 0.$$

The problem here is that the *pivotal* countries—the ones that determine whether investment is surplus-increasing or not—are third-party countries. They are not involved in project planning and assessment unless by invitation of Denmark and Norway.

We have so far assumed that there is only one cross-border investment project. But a grid extension plan contains multiple projects. Because of the size of cross-border projects and their international effects, the incremental value of one specific project may increase or decrease depending on the extent to which other network reinforcements are undertaken. For instance, the expansion in variable renewable energy sources such as solar and wind power at the European continent has increased the value of integrating Norway with Northern Europe to use Norwegian hydropower as balancing power against intermittent energy sources. One possibility is to expand Danish–Norwegian–German network capacity. Increasing cross-border capacity between Denmark and Norway is more profitable if the Danish–German line is built and vice versa. An example where network investments are substitutes instead of complements is market integration with Great Britain. The value of the proposed 1400 MW *Viking Link* connecting the Danish and British transmission systems is smaller if the *NorthConnect* project between Great Britain and Norway is completed and vice versa because they serve a similar purpose. Coordination problems associated with interrelated investment decisions may give rise to inefficiencies that are not always solved by forming infrastructure coalitions. Consider the following example.

Let there be four transmission projects: Number one is a Norway–Sweden connection, number two is between Denmark and Sweden, number three is a Denmark–Germany connection, and number four links Sweden and Germany through a cross-border interconnection. Let there be parallel coalitions: The Denmark–Germany and Norway–Sweden connections are built under structure $A = \{DG, NS\}$. Under structure $B = \{DN, GS\}$, the Denmark–Norway and Germany–Sweden connections are built. Assume the following surplus relations

$$v_{DG}^A = v_{NS}^A = v^A, v_{DN}^B = v_{GS}^B = v^B > v^A,$$

where index G refers to Germany and v_{ij}^z is the joint surplus of countries i and j under structure z. Structure B is better from an aggregate surplus perspective than A.

Assume that for coalition structures $C = \{D, G, NS\}$ and $E = \{DG, N, S\}$, the following surplus relations hold:

$$v_{NS}^E > v^A > \max\left\{v_N^0 + v_S^0, v_{DG}^C\right\}, v_{DG}^E < v_D^0 + v_G^0, v_{NS}^C < v_N^0 + v_S^0.$$

Assume that decisions are taken sequentially. Let Norway and Sweden first decide whether to build a connection, and then Denmark and Germany decide. We solve the game by backward induction. It is optimal to build the *DG* line in the second stage if the *NS* line is built in the first stage because $v^A > v^C_{DG}$. But if the *NS* line is not built in the first stage, then it is not optimal to build *DG* in the second stage by $v^E_{DG} < v^0_D + v^0_G$. In the first stage, Norway and Sweden anticipate that *DG* is built if and only if they themselves build the *NS* connection. Coalition structure *A* then is the equilibrium by $v^A > v^0_N + v^0_S$. Now reverse the sequence of decisions. In this case, Norway and Sweden will *not* build the *NS* connection either way, by $v^A < v^E_{NS}$ and $v^C_{NS} < v^0_N + v^0_S$. Upon realizing that Norway will not build the *NS* line, Denmark will not build any line either by $v^E_{DG} < v^0_D + v^0_G$. The status quo is the equilibrium in this second case. The problem here is that Norway and Sweden are the countries that benefit the most from market integration *A* compared to the status quo. And because of complementarities in investment decisions, they can induce Denmark and Germany to follow. For investment to come about, it is therefore essential that Norway and Sweden move first. The general insight is that reversing the order of investment decisions can fundamentally alter the equilibrium network structure in the market.

To continue the example, assume that Norway and Denmark first decide whether to build a connection and then Sweden and Germany. Under the assumptions

$$v^B > \max\{v^0_D + v^0_N;\, v^F_{GS}\},\, v^H_{GS} < v^0_G + v^0_S$$

for coalition structures $F = \{DN, G, S\}$ and $H = \{D, N, GS\}$, it is easy to verify that coalition structure *B* is the equilibrium of this particular game. Hence, differences in the order in which different countries negotiate infrastructure projects can also affect the equilibrium outcome.[16]

The sequence of electricity market liberalization in the Nordic countries probably had an effect on the order in which transmission network investment decisions were taken. Under a different liberalization sequence, the network structure in the Nordic market perhaps would had been different than today. There is no way of knowing, and we cannot say whether the current structure is more or less efficient than what another counterfactual structure network would have been. Either way, that discussion is esoteric because liberalization is irreversible. But the above analysis *does* point to network investments being sensitive to the decision-making process, and seemingly innocuous procedural differences can be important. Joint network planning in the Nordic electricity market therefore has the potential to increase efficiency beyond what the countries could achieve by bilateral negotiations.

[16]Nilssen and Sørgard (1998) consider merger formation games along these lines.

5.2 The Value of Cooperation

Assume that the *grand coalition* $\{DFNS\}$, in which all Nordic countries cooperate on network investment, always can replicate the project portfolio of any other coalition structure and potentially do better. Coalition theory then predicts the grand coalition to be the only equilibrium candidate because all other coalition structures are dominated (Horn and Persson 2001). However, this result is uninformative regarding how to achieve the desired level of cooperation under voluntary participation.

Under joint decision making, all parties to a negotiation must agree to build the transmission line. The status quo prevails if at least one country vetoes the project. In other words, all participation constraints must be met for the project to go through. In the bilateral negotiation, Norway and Denmark can achieve participation through compensation payments (T_D, T_N). In a joint negotiation between Denmark, Norway and Sweden, the three countries similarly negotiate x and compensation payments (T_D, T_N, T_S) to maximize the Nash product:

$$\left(V_D(x) + T_D - v_D^0\right)^{\alpha_D}\left(V_N(x) + T_N - v_N^0\right)^{\alpha_N}\left(V_S(x) + T_S - v_S^0\right)^{\alpha_S}$$

subject now also to Sweden's participation constraint $V_S(x) + T_S \geq v_S^0$ and budget balance $T_D + T_N + T_S = 0$. The outcome of this negotiation is the efficient capacity x^* that accounts also for the marginal effect on Sweden of increasing transmission capacity:

$$V_D^{'}(x^*) + V_N^{'}(x^*) + V_S^{'}(x^*) = 0.$$

But why would Denmark and Norway invite Sweden into the negotiation and award a third (or fourth) country veto right over the project? In our benchmark example, the negotiated transmission line has a smaller capacity than the two countries would jointly prefer: $x^* < \tilde{x}$. What is more, Denmark and Norway would have to *pay* Sweden not to veto even the modified outcome because Sweden would be better off without any additional transmission line between Denmark and Norway: $V_S(x^*) < v_S^0$ implies $T_S^* > 0$ by Sweden's participation constraint. Coalition formation would not seem to work here because Denmark and Norway would rather leave Sweden outside the discussion.

The question is whether we can find rules for Nordic cooperation that would be acceptable to all parties and lead to an efficient outcome, in this case a cross-border transmission line with capacity x^*? Instead of vetoing the entire project, assume that each country can only veto the outcome of that specific negotiation. In particular, Denmark and Norway are free to negotiate any outcome between the two of them if joint negotiations break down. Denmark and Norway would then choose $\tilde{x}$. The default outcome, or threat point, of the three-party negotiation then changes from the status quo to $\left(\tilde{x}, \tilde{T}_D, \tilde{T}_N\right)$. In a joint negotiation between Denmark, Norway and Sweden, the three countries would now negotiate x and compensation payments (T_D, T_N, T_S) to maximize:

$$(V_D(x) + T_D - \tilde{v}_D)^{\alpha_D}(V_N(x) + T_N - \tilde{v}_N)^{\alpha_N}(V_S(x) + T_S - \tilde{v}_S)^{\alpha_S}$$

with participation constraints $V_i(x) + T_i \geq V_i(\tilde{x}) + \tilde{T}_D = \tilde{v}_i$, $i = D, N$, and $V_S(x) + T_S \geq \tilde{v}_S$. Even this negotiation implements the efficient investment x^* because negotiations maximize the joint surplus of the involved parties. But compensation payments differ. In the negotiation, Denmark and Norway are willing to implement x^* if and only if Sweden pays enough compensation to give the two countries a higher surplus than $\tilde{v}_D$ and $\tilde{v}_N$. Sweden would be willing to compensate both countries those required amounts to reduce capacity in the project from $\tilde{x}$ to x^*.[17]

Under the modified rule for negotiations among countries, Denmark and Norway *do* have an incentive to invite Sweden into the negotiation because of Sweden's willingness to pay to avoid the unfavorable outcome $\tilde{x}$. One way to implement such a rule would be to allow in a first stage all countries in the Nordic market to agree on the desired cross-border investments in pairwise negotiations between the directly involved parties. When these projects are on the table, then third parties can request renegotiation of each project, with veto right for all negotiating parties of the renegotiated solution.[18]

So far, we have kept Finland out of the equation. There could be good reasons for doing so in our example. Assume that transmission constraints between Denmark and Sweden are always binding, both before and after the expansion to x^* of cross-border capacity between Denmark and Norway. The only consequence for Sweden of this investment is a change in congestion rent on cross-border trade with Denmark. Transmission bottlenecks effectively isolate Finland and Sweden from the Danish and Norwegian market. Hence, electricity prices, production, consumption and trade in and between Finland and Sweden remain unchanged. Changes in the cross-border transmission capacity between Denmark and Norway have no economic consequences for Finland in this case.

Things are different in the more plausible case where transmission constraints are not always binding between Denmark and Sweden. Sweden is a net exporter of electricity to Finland; see Sect. 2.3. Assume for the sake of the argument that transmission capacity between Sweden and Finland is sufficiently high that the two countries always are fully integrated. Improved market integration between Denmark and Norway reduces wholesale prices in the net importing country Denmark, and by way of market integration prices in Sweden and Finland. Sweden is hurt by this price decrease, in its capacity of being a net exporter of electricity to both Denmark and Finland. Although having no direct network connections with either Denmark or

[17]To see that such compensation is feasible, let $T_i = \tilde{v}_i - V_i(x^*)$, $i = D, N$. Sweden's net surplus then equals $V_S(x^*) + T_S - \tilde{v}_S = V_S(x^*) - T_D - T_N - \tilde{v}_S = V_D(x^*) + V_N(x^*) + V_S(x^*) - V_D(\tilde{x}) - V_N(\tilde{x}) - V_S(\tilde{x}) > 0$.

[18]Could not Denmark and Norway game such rules by initially proposing a very large project just to extract a large compensation payment from Sweden? No, because a large project would not pose a credible threat point. If Sweden issued its veto right in the joint negotiations, then it would be in Denmark and Norway's joint best interest to renegotiate their initial bilateral agreement to $(\tilde{x}, \tilde{T}_D, \tilde{T}_N)$. Hence, $(\tilde{x}, \tilde{T}_D, \tilde{T}_N)$ is the only credible threat point, and all parties should rationally foresee it.

Norway, Finland nevertheless experiences a positive net benefit from the investment because the country is a net importer of electricity, and market integration drives down the price of electricity in Finland. Even Finland should be allowed to participate in the planning of new transmission capacity between Denmark and Norway to ensure an efficient total outcome of the negotiations.

The possibility that distant third-party countries are affected by the investment is not a reason to change the sequential planning structure where countries first bilaterally negotiate cross-border transmission investment and then renegotiate. Consider a joint negotiation of the cross-border line between Denmark and Norway across all four Nordic countries. Assume that the bilateral outcome $\left(\tilde{x}, \tilde{T}_D, \tilde{T}_N\right)$ is the default outcome. The Nash bargaining solution is the combination of x and compensation payments (T_D, T_F, T_N, T_S) that maximizes

$$
\begin{aligned}
&(V_D(x) + T_D - \tilde{v}_D)^{\alpha_D}(V_F(x) + T_F - \tilde{v}_F)^{\alpha_F}(V_N(x) + T_N - \tilde{v}_N)^{\alpha_N} \\
&\quad (V_S(x) + T_S - \tilde{v}_S)^{\alpha_S}
\end{aligned}
$$

subject also to Finland's participation constraint, $V_F(x) + T_F \geq V_F(\tilde{x}) = \tilde{v}_F$, and budget balance $T_D + T_F + T_N + T_S = 0$. The multilaterally negotiated cross-border transmission capacity x^{fb} between Denmark and Norway maximizes joint surplus:

$$
V_D'\left(x^{fb}\right) + V_F'\left(x^{fb}\right) + V_N'\left(x^{fb}\right) + V_S'\left(x^{fb}\right) = 0.
$$

Finland benefits from capacity expansion, $V_F'(x) > 0$, and Sweden loses from it, $V_S'(x) < 0$, so we cannot say on the basis of our current assumptions whether transmission capacity increases or decreases under renegotiation relative to the bilateral benchmark: $x^{fb} \gtrless \tilde{x}$. Consequently, it is ambiguous whether Finland or Sweden will be net contributors of capacity payments. However, it is easy to verify that there exists a vector of compensation payments $\left(T_D^{fb}, T_F^{fb}, T_N^{fb}, T_S^{fb}\right)$ that implements the efficient investment x^{fb}, and that transfers are budget-balanced and render all Nordic countries weakly better off than they would have been under bilateral negotiations alone. Hence, Nordic cooperation is individually rational under the sequential procedure.

Improved market integration with continental Europe implies that investment in the Nordic network can have ramifications for a larger number of countries further south. Even the broader Nordic perspective then runs the risk of generating distorted investment incentives. As we saw above, all parties affected by an expansion of transmission capacity should be invited to participate in the negotiations for the outcome to implement a jointly efficient solution. Negotiations between multiple interested parties are manageable when there are only a few of them, but become increasingly complicated with the inclusion of more and more parties around the table. One way to proceed would be to have transmission planning at multiple levels. First, individual countries negotiate bilateral projects. These projects are then potentially renegotiated at regional level. For instance, the Nordic market could be one such region. Then, the remaining projects are lifted to the European level with final possibility for renegotiation.

5.3 The Value of Centralization

The grand coalition, which involves cooperation of all affected parties, maximizes total welfare under plausible assumptions. Section 5.2 proposed a sequential procedure that gives coalition incentives to include additional members and then establish the grand coalition as the only possible equilibrium structure. However, the observation that there exists no other equilibrium coalition structure does not imply that the grand coalition itself is stable. Member countries that do not get enough out of the coalition can have incentive to block investment projects by forming sub-coalitions. There are at least two approaches to solving this problem. The first is a bottom-up approach in which countries negotiate projects in a bilateral manner and use third-party compensation payments to internalize the full effect of their decisions. The second is a top-down approach in which one tries to implement full cooperation by distributing the *coalition worth*, i.e., the value created by the coalition, across members in such a way as to maintain stability of the grand coalition.

To analyze third-party compensation, assume that negotiations are bilateral between Denmark and Norway. Let third-party countries $j = F, S$ receive net compensation $T_j = V_j(x) - v_j^0$. Denmark and Norway negotiate x and (T_D, T_N) to maximize

$$\left(V_D(x) + T_D - v_D^0\right)^{\alpha_D} \left(V_N(x) + T_N - v_N^0\right)^{\alpha_N}$$

subject to $V_i(x) + T_i \geq v_i^0$, $i = D, N$, and budget balance $T_D + T_F + T_N + T_S = 0$. This solution internalizes all marginal third-party effects through the compensation mechanism and therefore implements the jointly efficient capacity x^{fb}.

If Denmark is a net importer of electricity from both Norway and Sweden, and Sweden is a net exporter of electricity to Finland, then Denmark, Norway and Finland would voluntarily contribute to building the interconnection, but Sweden would not. Instead, the three countries would have to compensate Sweden essentially for transiting electricity to Finland. Moreover, the marginal cost of compensation would cause Denmark and Norway to build relatively less transmission capacity. Hence, it would not be individually rational for the three countries to enter into an arrangement of third-party compensation with Sweden. As a consequence, the implemented project would be inefficient from an aggregate viewpoint.

A scope for centralization arises on the basis of the challenges associated with establishing a voluntary third-party compensation system. European Parliament (2009) contains a mechanism for compensating costs associated with cross-border flows of electricity. Compensation is to be paid to transiting countries by the countries in which the flows originate and terminate. In the example above, Denmark and Finland should compensate Sweden for transiting electricity. According to the mechanism, such costs shall be estimated on the basis of long-run average incremental costs. Benefits that a network incurs as a result of hosting cross-border flows shall be taken into account. In sum, third-party countries shall be compensated for their net cost of market integration. If country i's net cost is calculated as $T_i = V_i(x) - v_i^0$, then the investment will be efficient at the aggregate level under bilateral negotiation.

It is relatively straightforward to derive optimal compensation mechanisms for third-party countries when there is only one potential investment. Things are more complicated if there are multiple and interrelated projects. Then, third-party effects depend on all project decisions. We have also shown that the order in which projects are decided can affect the equilibrium allocation and therefore third-party surplus. In such complex settings, a better approach than bilateral bargaining might be to impose in a centralized manner the solution that maximizes total welfare and then construct compensation payments that distribute coalition wealth in such a way as to achieve stability of the grand coalition—the top-down approach.

For a set of compensation payments to be able to sustain the fully cooperative outcome, a reasonable starting point would be to require that member countries receive a share that is positively related to the value added they contribute to the coalition. The purpose of the Shapley value is to achieve such equitable distribution of coalition surplus.

To proceed, we let $H = \{1, \ldots, h\}$ be a coalition of h countries (for instance, the Nordic). There are many ways in which this coalition can form, and the contribution of each individual country $i \in H$ generally depends on the order in which it joins the coalition and which other projects have already been decided. We therefore need to calculate the expected contribution of country i. Let coalition H form sequentially by adding one country at a time. Assume that all sequences are equally likely so that each sequence occurs with probability $1/h!$. The likelihood that country i joins an arbitrary coalition $S \subset H$ of size s equals $s!(h-s-1)!/h!$ because there are $s!$ possible ways to reach coalition S before player i joins S and there are $(h-1-s)!$ possible ways to continue thereafter. Let $v(S)$ be the total surplus the members of coalition S can achieve by cooperating given that the countries in H that are not members of S do not cooperate with anyone. The incremental value of adding country i to coalition S then equals $v(S \cup i) - v(S)$. Adding i to coalition S is valuable, for instance because i and country $j \in S$ can decide on a cross-border interconnection once i joins the coalition, but not sooner. Sum over all possible coalitions S that do not contain country i, to get the expected contribution

$$\varphi_i(v, H) = \sum_{S \in H \setminus i} \frac{s!(h-s-1)!}{h!} [v(S \cup i) - v(S)]$$

of country i to coalition H. This is the *Shapley value* of country i. The Shapley value has a number of desirable properties. It distributes all the surplus of the coalition. Transfers depend on individual contributions to wealth creation, but not on identity. In particular, countries that do not contribute to creating wealth do not receive any of the surplus.

To visualize the Shapley value, consider a coalition between Denmark, Norway and Germany to remove bottlenecks associated with exporting Norwegian hydropower to Germany through Denmark; see Fig. 1. Countries cannot decide unilaterally to build interconnections to other countries. Therefore, $v(i) = 0$ for all three countries $i = D, G, N$. Assume that it would be too expensive for Germany and Norway to bypass Denmark, so that $v(GN) = 0$. However, each country would unilaterally benefit from increasing market integration with Denmark, with the Danish–Norwegian connection creating a larger joint surplus to the coalition than the Danish–German connection: $v(DN) = 12 > v(DG) = 6$. The coalition of all three countries increases joint surplus even further: $v(H) = 18 > v(DN)$.

Applying the Shapley formula yields the following distribution of surplus in the coalition:

$$\text{Denmark: } \varphi_D = \frac{1}{6}[v(DG) + v(DN) + 2v(H)] = 9$$

$$\text{Germany: } \varphi_G = \frac{1}{6}[v(DG) + 2(v(H) - v(DN))] = 3$$

$$\text{Norway: } \varphi_N = \frac{1}{6}[v(DN) + 2(v(H) - v(DG))] = 6.$$

Norway gets a larger share of total surplus than Germany because the value of connecting Denmark with Norway is larger than the value of connecting Denmark with Germany: $\varphi_N - \varphi_G = \frac{1}{2}[v(DN) - v(DG)] = 3$. Denmark obtains the largest share of surplus because of its pivotal position as a transit country that can veto all projects: $\varphi_D - \varphi_N = \frac{1}{2}v(DG) = 3$.

The Shapley value distributes most of the coalition worth to the coalition members that contribute the most to the coalition. Such a division is equitable, but is it enough to guarantee stability? In the above example, Denmark has no incentive to deviate from the grand coalition because cooperation then would completely collapse. For the same reason, Germany and Norway have no joint incentive to deviate. Germany and Norway have no unilateral incentive to deviate if the stand-alone value to both countries is small, i.e., $v_G^{DN} < 3 = \varphi_G$ and $v_N^{DG} < 6 = \varphi_N$. It is also easy to verify that Denmark and Germany have no joint incentive to deviate if Denmark and Norway have no joint incentive to deviate. The joint net value

$$\varphi_D + \varphi_N - v(DN) = \frac{1}{6}[4v(H) - 4v(DN) - v(DG)] = 3$$

to Denmark and Norway of staying in the coalition with Germany, relative to deviating and forming their own coalition, is positive by our assumption that value of building the two interconnections is sufficiently large. Hence, the grand coalition is stable in our example even if the surplus distribution is uneven. Kristiansen et al. (2018) simulate offshore transmission projects in a model of the Northern European electricity market and calculate associated Shapley values. They show that the core is non-empty in their model.

But Shapley value surplus division can be unstable and seemingly unable to sustain the grand coalition. This occurs in the above example if the value of building both lines is small. For instance, $v(H) = 13$ implies $\varphi_D = 7\frac{1}{3}$, $\varphi_G = 1\frac{1}{3}$ and $\varphi_N = 4\frac{1}{3}$.

Denmark still earns most of the surplus, followed by Norway and Germany. But now it is jointly better for Norway and Denmark to deviate from the grand coalition and leave Germany out: $v(DN) - \varphi_D - \varphi_N = \frac{1}{3}$. This does not exclude the possibility that there are *other* divisions of surplus that sustain the grand coalition. But Horn and Persson (2001) look one step ahead and require of proposed sub-coalitions that even they must be stable to pose as credible deviations. In particular, remaining members have an incentive to persuade deviators to return to the grand coalition. They have the means to do so if the grand coalition maximizes total surplus. If so, there is no credible deviation from the grand coalition.

The grand coalition is efficient under the plausible assumption that it can replicate investment portfolios for every coalition structure that is more fractioned. Based on the above arguments, there are also arguments as for why there are reasons to believe that an investment coalition based on full cooperation of all countries has the potential to be stable if the grand coalition maximizes total surplus. In that case, the cooperative game with Shapley value surplus division would obtain exactly the same total level of surplus that would have been possible in a model in which one single entity chose investments to maximize joint surplus. But observe that Shapley value surplus division could underestimate the value of some coalition members' strategic positions. One way forward to solve the predicament could be to start negotiations with calculated Shapley values as benchmarks for surplus division and then use a negotiation process to adjust compensation payments. This could be a faster process that leads to a more efficient and equitable outcome than sequential negotiations.

5.4 Country Versus TSO Incentives

The finding in Sect. 5.2 that negotiated outcomes maximize the total welfare of the parties to the negotiations rests on the assumption that national decision makers maximize domestic welfare. Infrastructure investment decisions are taken by the TSOs, either unilaterally (for the domestic network) or in cooperation with other TSOs (for international connections). A government believing that the TSO would always maximize domestic welfare could leave the TSO alone to act completely on its own devices. In reality, TSO operations are almost always under some form of government supervision and regulation. Regulation is typically imposed to curb the monopoly power of the TSO, which it could otherwise exploit to its own purposes. If there is a discrepancy between regulatory objectives and TSO incentives, one cannot presume that the unilateral or negotiated TSO investment decisions maximize either domestic or aggregate welfare with or without side payments between TSOs. How to implement efficient transmission network investment then boils down to devising and enforcing well-designed regulatory policies for the integrated electricity market.

In the Nordic market, TSO regulation has been national in scope. With too narrow a focus on domestic effects, national regulatory agencies run the risk of ignoring externalities abroad when devising regulatory policy. Tangerås (2012) considers transmission governance in a multinational energy market.

Centralized versus decentralized regulation involves a trade-off between the benefits of internalizing cross-border externalities of market integration and coordinating network investment, on the one hand, and the risk of distorted centralized regulation, on the other. The latter occurs if, for instance, a country with little to gain from market integration exercises a dominating influence over the common regulatory policy with the result that the regulatory policy provides insufficient incentives for TSOs to improve network performance. A well-functioning common regulatory agency requires balancing political powers across member countries to prevent any of them from tilting the regulatory policy too far in the own direction.

Whether to collect network ownership and operations in the hands of one single TSO or to maintain multiple TSOs (recall the examples of Denmark and Finland) depends on network topology. In a radial network, the value of expanding capacity in one part of the network goes up if capacity is higher in other parts of the network. Such network complementarity increases the benefit of coordinating network investment, which speaks in favor of having one single TSO. Under a meshed network structure, the value of expanding capacity in one part of the network is lower if capacity is higher elsewhere. Under network substitutability, the benefit of coordinating network investment is smaller. Maintaining network ownership in the hands of a multiple TSOs instead increases efficiency by limiting monopoly power.

The general insight is that no single governance structure outperforms all others. The optimal governance structure depends on political factors, network structure and economic factors, for instance how the gains from energy market integration vary across countries.

5.5 Merchant Transmission Investment

The analysis has relied on an assumption that transmission investment projects are planned and executed by regulated TSOs. Indeed, most of the transmission infrastructure in the Nordic market and to neighboring countries is owned by the Nordic and neighboring TSOs. In Sweden, for instance, the current legislation until recently required SvK (the Swedish TSO) to hold a majority stake in all new cross-border transmission lines connected to the Swedish network. Only reinforcements of the Swedish network deemed optimal by SvK could then be built.

Monopoly of network ownership and coordination facilitates system operation, but has costs associated with monopoly power. A TSO can underestimate the social value of investment if the underlying cost-benefit analysis is incorrect, or has insufficient incentives to invest because of regulatory policy or its market power. In a situation of underinvestment, merchant investment can potentially increase market integration beyond what could be achieved under regulated monopoly. Return to Fig. 5, and suppose a merchant investor contemplates the cross-border line between Denmark and Norway. The merchant only cares about congestion revenue C and is likely to understate the full social value of the projects by neglecting A and B and uncompensated security of supply benefits. In this simplified setting, any project pursued by the merchant necessarily is socially optimal. Under plausible circumstances, therefore, merchant transmission investment can increase efficiency.[19]

The Nordic countries have legal rules that limit who can own cross-border interconnections. Still, some of the transmission capacity that has been installed since 1992 is controlled by commercial players. The 600 MW *Baltic Cable* between Lübeck in Germany and Trelleborg in Sweden is owned and operated by the Norwegian generation company Statkraft.[20] The 600 MW *SwePol Link* between Karlshamn in Sweden and the Bruskowo Wielkie power plant in Poland became operational in 2000. SwePol Link is jointly owned by SvK and the Polish TSO, *PSE-Operator*. The cable is operated by the Swedish generation company Vattenfall. *NorthConnect* is a 1400 MW cable between Norway and Scotland, with planned construction start in 2020. It is jointly owned by Vattenfall and the three Norwegian generation companies *Agder Energi*, *E-CO* and *Lyse Produksjon*. Another example is the 1400 MW interconnector project *NorGer* between southern Norway and northern Germany. This started as a private undertaking, but now has mixed ownership with the Norwegian TSO Statnett controlling 50% and Agder Energi, Lyse Produksjon and the Swiss energy trading company *EGL* sharing the other half.

In a compromise with the political opposition to ensure majority in the Parliament for the EU Third Energy Package, the Norwegian government decided in 2018 to disallow private ownership of international interconnectors. Consolidation of monopoly power is unfortunate on the basis of the potential efficiency gains of allowing private investment and the demonstrated private interest in developing such projects in the Nordic market.

5.6 Network Investment When Transfers Are Restricted

We have based the analysis on the assumption that compensation payments are sufficiently flexible to sustain investment levels that maximize joint welfare of the negotiating parties as equilibrium outcomes. Assume now that transmission investment has domestic consequences that cannot be internalized by way of transfer compensation payments. We frame the analysis around the question of how network investment contributes to security of electricity supply.[21]

There are two main ways how countries can protect consumers against supply shortages in the spot market. One is to reduce the risk of shortages by improving market integration. The other is for the TSO to maintain backup capacity—a strategic reserve. Network expansion is costly. Strategic reserves, by being priced outside the spot market, distort prices by driving a wedge between the long-run marginal utility

[19]Merchant transmission is analyzed in detail by Joskow and Tirole (2005).

[20]Merchant ownership was not in conflict with Swedish law in this particular case because the connection began operating in 1994.

[21]The analysis in this section builds on Tangerås (2018).

of consumption and the marginal cost of capacity. Cost-efficient security of supply is achieved at the optimal balance between market integration and capacity reserves.

Whereas network investment decisions are taken jointly by the TSOs, capacity reserves are decided at the national level. In a multinational electricity market, an increase in the capacity reserve at home has a positive externality abroad insofar as the domestic capacity is available as backup in other countries. But price distortions at home associated with a larger capacity reserve are exported abroad in an integrated market, which represents a negative externality. For any degree of market integration, a decentralized capacity decision can imply downward- or upward-distorted capacity reserves depending on whether the net foreign externality is positive or negative. Network investment and capacity reserves are strategic complements (substitutes) if the net externality is positive (negative). Market integration will therefore be insufficient. To see why, suppose strategic reserves and network capacity are at their first-best efficient levels. If the foreign net externality is positive (negative), then countries in equilibrium reduce (increase) the strategic reserve relative to the first-best. Network investment falls by complementarity (substitutability).

The lack of coordination of capacity reserves leads to downward distortions in market integration even if network investment decisions are taken to maximize the total surplus of the two countries. The problem is exacerbated if network investment decisions are uncoordinated. Absent coordination of capacity reserves, one way to increase the efficiency related to supply security, is to impose regulations that induce network owners to attach a stronger weight to the value of market integration relative to the cost of network expansion and thus overinvest all else equal. A requirement that TSOs should use most of their congestion rent to reinforce the network is an example of such a regulation.

6 Discussion

A main objective of EU energy policy is to develop a well-functioning internal market for electricity (Directive 2009/72/EC). Norway and Sweden took the first steps toward creating an internal Nordic market for electricity already in 1996 when they established a power exchange for trading electricity between the two countries. This market soon expanded to encompass Finland, Denmark and later the Baltic countries.

The Nordic countries realized the value of cooperation and coordinated system operation and development, in particular transmission network investment, in a joint organization, *Nordel*.[22] Transmission network management and system operation are decentralized to the transmission system operators (TSOs), but cooperation extends well beyond non-committed statements to improve system performance. For instance, the regulatory brief of the Swedish TSO, SvK, explicitly requires that SvK

[22]Nordel was dissolved in 2009 when the Nordic countries joined *ENTSO-E*, the European organization for cooperation of transmission system operators for electricity.

cooperates with the other Nordic TSOs to develop network projects that increase total economic welfare in the Nordic countries. Furthermore, project plans shall specify the welfare implication for each country.

Mutual economic benefits and an equal distribution of gains from electricity trade across countries probably have contributed to the strong commitment to market integration and robustness of the Nordic electricity market. For instance, the interconnected network has enabled Finnish and Swedish nuclear power to supply the entire Nordic market with base-load generation, which has released Norwegian and Swedish hydropower capacity for smoothing out fluctuations in net demand. A system with large amounts of hydropower stabilizes prices and reduces the need for thermal backup capacity to ensure security of supply. In particular, access to hydropower from its neighbors has allowed Denmark to develop wind power to an extent that might not otherwise have been economically feasible.

The Nordic electricity market evolved in a decentralized manner with countries volunteering to become members. Instead, the creation of a European integrated electricity market has been much more a centralized project. In other parts of Europe, the benefits of market integration have been less obvious, or the gains more asymmetrically distributed, which perhaps has contributed to a lack of devotion to the plan for an internal electricity market. We have discussed approaches to cooperation that can reinforce countries' incentives to contribute to market integration and increase efficiency. Under a bottom-up approach, cross-border projects are developed via bilateral negotiations between the directly involved parties, in a first step. Neighboring countries are then invited to propose changes in a second step, and both incumbent countries have veto rights over any modifications. This sequential procedure ensures countries' incentives to cooperate because they can always get at least as much through cooperation as from standing alone. In addition, the possibility to renegotiate gives room for efficient project alterations. Projects can be lifted to the aggregate European level for final renegotiation in a third step. Under the top-down approach, cost-benefit analysis is applied at the centralized European level to identify the portfolio of projects that is jointly optimal from a total welfare economic viewpoint. Total surplus is divided among countries on the basis of their individual contributions to value creation. While being equitable in this manner, Shapley value surplus division could still underestimate the benefit of member countries to defect. Compensation payments may therefore have to be adjusted to meet countries' participation constraints.

Examples from the Nordic market show that merchant transmission investment has been a viable alternative to TSO investment in cases where the latter has been reluctant to invest. For instance, merchant investors have initiated a number of cross-border infrastructure projects in Norway in periods where private ownership of interconnectors has been legal in the country. Merchant investors, by focusing on network profit alone, are likely to underestimate the aggregate economic benefits of infrastructure investment under a host of circumstances. In such cases, merchant investment is welfare improving, if allowed. From that perspective, it is regrettable that many countries severely limit the scope for merchant investment, for instance through legal barriers of entry that require TSOs to be majority owners of all cross-border interconnectors. In a situation of underinvestment and lack of cooperation on infrastructure projects, improving the business climate for merchant network investment can be a valuable complement to more centralized investment policy.

Acknowledgements We thank Alexandra Allard for research assistance and the Swedish Energy Agency for financial support.

References

M. Armstrong, S. Cowan, J. Vickers, *Regulatory Reform: Economic Analysis and British Experience* (MIT Press, Cambridge, MA, 1994)

H.-A. Bredesen, *The Nord Pool Market Model*. Nord Pool Consulting (2016). http://www.asean-aemi.org/wp-content/uploads/2016/03/AEMI-Forum-November-2015-Bredesen-Feb2016.pdf

T. Bye, E. Hope, Deregulation of electricity markets: the Norwegian experience. Econ. Polit. Weekly **40**, 5269–5278 (2005)

T. Bye, E. Hope, Deregulering av elmarkedet—Norske erfaringer. Økonomisk Forum **1**, 17–25 (2007)

T. Bye, T.A. Johnsen, Effektivisering av kraftmarkedet.´Statistisk Sentralbyrå, Rapport 91/13 (1991)

ENTSO-E, *Statistical Factsheet* (2015). www.entsoe.eu/publications/statistics-and-data/#statistical-factsheet

European Parliament, Regulation No 714/2009 on conditions for access to the network for cross-border exchanges in electricity. Official Journal of the European Union, L **211**, 15–35 (2009)

J. Greenberg, Coalition structures, in *Handbook of Game Theory*, vol. 2, ed. by R.J. Aumann, S. Hart (Elsevier, Amsterdam (The Netherlands), 1994), pp. 1305–1337

H. Horn, L. Persson, Endogenous mergers in concentrated markets. Int. J. Ind. Organ. **19**, 1213–1244 (2001)

P. Högselius, A. Kaijser, *När folkhemselen blev internationell* (SNS Förlag, Stockholm, 2007)

P. Joskow, J. Tirole, Merchant transmission investment. J. Ind. Econ. LIII 233–264 (2005)

A. Jha, F.A. Wolak, *Testing for Market Efficiency with Transaction Costs: An Application to Financial Trading in Wholesale Electricity Markets* (Stanford University, Manuscript, 2015)

M. Kristiansen, F.D. Muñoz, S. Oren, M. Korpås, A mechanism for allocating benefits and costs from transmission interconnections under cooperation: a case study of the North Sea offshore grid. Energy J. **39**, 209–234 (2018)

E. Lundin, Market power and joint ownership: Evidence from nuclear power plants in Sweden. J. Ind. Econ., forthcoming (2020)

E. Lundin, T. Tangerås, Cournot competition in wholesale electricity markets: The Nordic power exchange, Nord Pool. Int. J. Ind. Organ. **68**, 1–20 (2020)

J.F. Nash, The bargaining problem. Econometrica **18**, 155–162 (1950)

T. Nilssen, L. Sørgard, Sequential horizontal mergers. Eur. Econ. Rev. **42**, 1683–1702 (1998)

Nordel, *Nordic Grid Master Plan 2008* (2008). www.entsoe.eu/fileadmin/user_upload/_library/publications/nordic/planning/080300_entsoe_nordic_NordicGridMasterPlan2008.pdf

NordREG, Nordic Market Report 2014. Report 4/2014 (2014)

M.J. Osborne, A. Rubinstein, A Course in Game Theory (MIT Press, Cambridge, MA, 1994)

P.-O. Pineau, R.P. Hämäläinen, A perspective on the restructuring of the Finnish electricity market. Energy Policy **28**, 181–192 (2000)

T. Tangerås, Optimal transmission regulation of an integrated energy market. Energy Economics **34**, 1644–1655 (2012)

T. Tangerås, Equilibrium supply security in a multinational electricity market with renewable generation. Energy Economics **72**, 416–435 (2018)
T. Tangerås, J. Mauritzen, Real-time versus day-ahead market power in a hydro-based electricity market. J. Ind. Econ., **LXVI**, 904–941 (2018)
Y. Tohidi, M.R. Hesamzadeh, Multi-regional transmission planning as a non-cooperative decision-making. IEEE Trans. Power Syst. **29**, 2662–2671 (2014)
Y. Tohidi, M.R. Hesamzadeh, F. Regairaz, Sequential coordination of transmission expansion planning with strategic generation investments. IEEE Trans. Power Syst. **32**, 2521–2534 (2017a)
Y. Tohidi, L. Olmos, M. Rivier, M.R. Hesamzadeh, Coordination of generation and transmission development through generation transmission charges—a game theoretical approach. IEEE Trans. Power Syst. **32**, 1103–1114 (2017b)

The manufacturer's authorised representative in the EU is Springer Nature Customer Service Centre GmbH, Europaplatz 3, 69115 Heidelberg, Germany. If you have any concerns regarding our products, please contact ProductSafety@springernature.com

Printed and bound by CPI Group (UK) Ltd, Croydon, CR0 4YY
30/09/2026
02225689-0018